Ecology of Wildlife Diseases in the Neotropics

Gerardo Acosta-Jamett • Andrea Chaves
Editors

Ecology of Wildlife Diseases in the Neotropics

Springer

Editors
Gerardo Acosta-Jamett
Facultad de Ciencias Veterinarias
Austral University of Chile
Valdivia, Chile

Andrea Chaves
Escuela de Biología
University of Costa Rica
San Jose, Costa Rica

ISBN 978-3-031-50533-1   ISBN 978-3-031-50531-7 (eBook)
https://doi.org/10.1007/978-3-031-50531-7

This Springer imprint is published by the registered company Springer Nature Switzerland AG
The registered company address is: Gewerbestrasse 11, 6330 Cham, Switzerland

Paper in this product is recyclable.

Contents

About the Authors

Gerardo Acosta-Jamett is Full Professor and member of the Center for Surveillance and Evolution of Infectious Diseases (CSEID) at the Facultad de Ciencias Veterinarias, Universidad Austral de Chile. He is a veterinarian, has an MSc in Ecology, and has a PhD in Veterinary Epidemiology. His research focuses on disease ecology at the wildlife interface with emphasis on the epidemiology of zoonotic diseases, spatial analyses, and the conservation of endangered species.

Mario Alvarado-Rybak is an academic researcher and member of the Núcleo de Ciencias Aplicadas in Veterinary and Agronomic Sciences at the Universidad de las Américas, Chile. He is a veterinarian and has a PhD in Conservation Medicine. He is a member of the IUCN/SSC Amphibian Specialist Group for Chile. His research focuses on the ecology of wildlife diseases in aquatic ecosystems and biodiversity conservation.

Paulina Álvarez-Mendizabal is a Wildlife veterinarian with interests in disease ecology and conservation of birds.

Randall Arguedas is Professor at the Universidad Técnica Nacional, Costa Rica and Director of the Exotic and Wildlife Division of VetLab Laboratory, Costa Rica. His area of interest includes exotic animal and wildlife clinical pathology, especially reptile clinical pathology.

Milena Arguello-Sáenz is a PhD student at the National Autonomous University of México, Faculty of Veterinary Medicine. Her research focuses on the ecology of zoonotic vector-borne diseases at the human-domestic-wildlife interface and its implications for public health and wildlife conservation.

Rafael Avila-Flores is full-time Professor at the Universidad Juárez Autónoma de Tabasco, Mexico. His research focuses on the behavioral and population performance of bats in human-modified landscapes, with emphasis on vampire bats in agroecosystems and insectivorous bats in cities.

Juan Ricardo Sánchez Ayala, DVM, is a PhD student in the School of Veterinary Medicine at Universidad Nacional Autónoma de México (UNAM). His research focuses on the association between ecological and anthropogenic variables and the presence of infectious agents in non-human primates in anthropized and conserved environments to generate information that can be taken into account for the conservation of these species.

Claudio Azat Director of the Sustainability Research Center and PhD in Conservation Medicine at Universidad Andres Bello in Chile. Co-Chair of the IUCN/SSC Amphibian Specialist Group for Chile. Expert member of the Species Classification Committee of the Chilean Ministry of the Environment. His research focuses on wildlife health and biodiversity conservation.

Margarida F. Barber is a Research Assistant at the Center for Research and Advanced Studies of the National Polytechnic Institute (CINVESTAV), Merida, Mexico. She has a master of science degree in Natural Resource Management. Her research focuses on prey–predator interactions, camera traps, and mammal occupancy modeling.

Pablo Beldomenico is an Argentinean disease ecologist, Director of the Disease Ecology Laboratory (Laboratorio de Ecología de Enfermedades – LEcEn) at the Instituto de Ciencias Veterinarias del Litoral (UNL-CONICET), Esperanza, Argentina, Principal Investigator of the Argentine Council for Research and Technology, and Professor at the Universidad Nacional del Litoral.

Fabián Beltrán-Saavedra is a Researcher at the Wildlife Conservation Society specializing in Disease Ecology, One Health, Biological Conservation, and Parasitology. He possesses extensive experience working with wild species, including vicuñas, small cats, flamingos, and micromammals, in the highlands, forests, and savannahs of Bolivia. Additionally, he as an invited Professor in postgraduate courses related to ecology and evolution of parasite-host interactions and conservation and management of the health of wildlife in their natural habitats.

Laís Kelly Amâncio Ribeiro Berenguer is a PhD candidate in Animal Bioscience at the Federal Rural University of Pernambuco, Brazil. She is a veterinarian and has a master's in Animal Bioscience. Her research focuses on the diagnosis of leishmaniasis and gastrointestinal parasites in canines and felines, with a focus on the One Health approach.

Sergio Bermúdez C is an Entomologist at the Medical Entomology Department, Gorgas Memorial Institute for Health Research, Panama. His research focuses on the ecology, distribution, and taxonomy of ticks and tick-borne diseases.

Sofía Bernal-Valle is a PhD candidate at Santa Cruz State University, Bahia, Brazil. She is a veterinarian, and her research focuses on wildlife pathogens and diseases that pose a threat to conservation, with a particular focus on the eco-epidemiology of arboviruses and parasite ecology using the One Health approach.

Júlio César Bicca-Marques is Professor of Conservation Science and Ecology at the School of Health and Life Sciences of the Pontifical Catholic University of Rio Grande do Sul (PUCRS), Brazil. His research focuses on the ecology, behavior, cognition, and conservation of nonhuman primates, including primate-parasite interactions, the effect of disease outbreaks on population persistence in fragmented landscapes, and self-medication. He is a member of the IUCN Primate Specialist Group and the 2021 Distinguished Primatologist of the American Society of Primatologists.

Maria Mercedes Cafrune is a Graduate and Professor in Biological Sciences. She is a Specialist in communicable tropical diseases and a Researcher at the National Institute of Agricultural Technology of Argentina. She works on parasitic diseases of small ruminants (sheep and goats) and South American camelids from the Argentine Northwest.

Francisco Chacón DVM, is a PhD Teacher Support at the University of Chile, Faculty of Veterinary and Animal Sciences. He has worked on zoonotic diseases, with an emphasis on Chagas disease and behavioral changes in insect vectors responsible for the transmission of *T. cruzi*.

Andrea Chaves is Professor at the Biology School, University of Costa Rica. Her focus is to understand the general variables favoring the presence of infectious diseases in diverse communities across multiple spatial scales affected by anthropogenic activities. These consider the ecological interactions of parasites, humans, domestic, and wild animals and the intrinsic risk of zoonotic transmission. She has developed her research in several taxa (especially in nonhuman primates) and with a wide spectrum of infectious agents (viruses, bacteria, parasites), mainly from Latin American wild environments.

Oscar M. Chaves is Professor and Researcher at the Universidad de Costa Rica, Costa Rica. He is a biologist and has a PhD in the ecology of tropical mammals. His research focuses on the behavioral ecology and flexibility of free-ranging primates in human-modified landscapes, primate conservation biology, and plant-animal interactions.

Galaxia Cortés-Hinojosa is Assistant Professor in Veterinary Virology at the Pontificia Universidad Católica de Chile. She is a veterinarian and has a master's in marine sciences and a PhD in veterinary sciences with a research focus on wildlife virology. Additionally, she is a certified aquatic veterinarian and diplomate of the American College of Veterinary Microbiologists. Her ongoing research includes the discovery of wildlife viruses in remote locations with a focus on the effect of anthropogenic impacts on viral diversity, the effect of environmental variables on the viral diversity and immunogenetics of aquatic animals, and viral diagnosis.

Marco Antônio Barreto de Almeida is a Biologist, Doctor in Zoology, Epidemiologist, and Graduate of the Field Epidemiology Training Program of the Ministry of Health of Brazil and Centers for Disease Control and Prevention (CDC). He has an MSc in Ecology. He works as a National Consultant in the Coordination of Surveillance, Preparedness and Response to Emergencies and Disasters at the Pan American Health Organization/World Health Organization – PAHO/WHO, Representation of PAHO/WHO in Brazil.

Jaqueline Bianque de Oliveira is Full Professor of Parasitology at the Federal Rural University of Pernambuco, Brazil. She is a veterinarian and has a PhD in Veterinary Science. Her research applies and integrates multidisciplinary knowledge, such as Parasitic Ecology, Trophic Ecology, Ecotoxicology, Neglected Tropical Diseases, Zoonoses, and Conservation Medicine, in the context of the One Health approach.

Dênisson da Silva e Souza Master in Tropical Animal Science at the Federal Rural University of Pernambuco, Brazil. He is a veterinarian specializing in Clinical Medicine and Management of Wildlife and his research is focused on Conservation Medicine and One Health.

Vanessa Campelo de Souza holds a master's in Ecology at the Federal Rural University of Pernambuco, Brazil. Her research focuses on the influence of anthropization on bird communities and their interaction with parasites.

Rodolfo Dirzo is Associate Dean at Doerr School of Sustainability, Bing Prof in Environmental Science, Professor of Earth System Science and Senior Fellow at the Woods Institute for the Environment. His scientific work examines the study of species interactions in tropical ecosystems from California, Latin America, and other tropical areas of the world. His recent research highlights the decline of animal life ("defaunation") and how this affects ecosystem processes/services (e.g., disease regulation).

A. A. Faccini-Martínez is a physician at Servicio de Infectología, Hospital Militar Central Bogotá, Colombia, and at Servicios y Asesorías en Infectología – SAI (Bogotá, Colombia). Graduated Medical Doctor, Master in Science and PhD in Infectious Diseases. Postdoctoral fellowship as International Visiting Scientist at Pathology Department, University of Texas Medical Branch (Galveston, TX, USA). Clinical, research and scientific publication skills in the fields of infectious diseases, tropical medicine, zoonotic and vector borne diseases, and travel medicine.

Gabriela Felix-Nascimento is Associate Professor of Animal Morphophysiology at the Federal University of San Francisco Valley, Brazil. She is a veterinarian and has a PhD in Animal Bioscience. Her research applies and integrates research techniques from across different disciplines, such as animal morphophysiology, ecotoxicology, parasite ecology, conservation medicine, and One Health.

Hebe Ferreyra is a wildlife veterinarian with a master's degree in Wildlife Management, currently serving as Professor of Wildlife Health at the National University of Villa Maria, Cordoba, Argentina. Her interests lie in the fields of wildlife disease, conservation, and disease management.

Nuri Flores-Pérez is a veterinarian from the National Autonomous University of Mexico (UNAM). Now she is pursuing her master's in Preventive Veterinary Medicine at UC Davis. Her research interests include studying infectious diseases and interactions at the human, domestic animal, and wildlife interface, along with ecosystem conservation.

Gustavo A. Gutiérrez-Espeleta is an Actual President of the University of Costa Rica (2021–2024) and Professor at the Biology School, University of Costa Rica. His research focuses on the conservation genetics of wildlife vertebrates, especially in neotropical primates and wild felids.

C. Guzmán-Cornejo She has a PhD on Biological Sciences and is a full time Professor at the Faculty of Sciences, Universidad Nacional Autónoma de México. Her research has focused on the study of taxonomy and diversity of ticks, on tick-borne diseases, and on the study on mites associated with mammals.

Felipe A. Hernández is Assistant Professor at the Facultad de Ciencias Veterinarias of the Universidad Austral de Chile. He is a veterinarian and has a PhD in Interdisciplinary Ecology at the University of Florida, USA. His research focuses on the ecology of invasive mammals and their role as reservoirs/hosts of infectious agents affecting domestic/wild animal and human health.

Carlos N. Ibarra-Cerdeña is Professor and Academic Coordinator at the Human Ecology Department of The Research Center for Advanced Studies (Cinvestav). His research is broadly on the relationships between human modification of tropical landscapes and disease ecology with an emphasis on the One Health approach and Chagas disease.

Martha P. Ibarra-López is an Academic Technician at the University of Guadalajara and a PhD candidate in Center for Research and Advanced Studies of the National Polytechnic Institute (CINVESTAV). Her research focuses on evaluating the effects of deforestation and selective defaunation on mammal populations as regulators in disease transmission in tropical deciduous forests.

Andrés M. López-Pérez is a Researcher and Professor in the Conservation and Biology of Vertebrates Department at Institute of Ecology A.C. (INECOL) in Mexico. As a veterinarian and disease ecologist, his research is focused on conservation biology and infectious diseases. His research experience includes molecular, ecological, and entomological factors surrounding diseases that impact humans, domestic animals, and wildlife.

Ian MacGregor-Fors is Professor of Urban Biodiversity and Ecosystems at the University of Helsinki. His primary research interest lies in studying the responses of wildlife, primarily birds and plants, in human-dominated landscapes, with a significant emphasis on urban areas.

Gisela Marcoppido is a veterinarian and Professor at the Faculty of Veterinary Medicine (FCV), University of Buenos Aires (UBA), Argentina. Additionally, she is a Researcher at the Institute of Pathobiology, National Institute of Agricultural Technology (INTA), National Scientific and Technical Research Council, Buenos Aires, Argentina.

Karen Uxue Martínez-Pérez is a master's student in Biological Sciences at the National Autonomous University of Mexico (UNAM). Her research focuses on the taxonomy and ecology of helminth parasites in wild mammals.

Thiago F. Martins is a Researcher at the Pasteur Institute of the São Paulo State Department of Health, Brazil, and Curator of the National Tick Collection "Danilo Gonçalves Saraiva" of the Department of Preventive Veterinary Medicine and Animal Health, Faculty of Veterinary Medicine, University of São Paulo, Brazil. He holds a PhD in Experimental Epidemiology Applied to Zoonoses. His research focuses on the taxonomy, biology, ecology, and systematics of ticks with medical and veterinary importance in wildlife from South America.

Pablo Moreno is a Researcher of the Argentine Council for Research and Technology, and Professor at the Facultad de Ciencias Veterinarias y Ambientales of the Maza University, Mendoza, Argentina. His research focuses on the dynamics of health and disease in arid land mammals, with an emphasis on the eco-epidemiology of parasites and the dynamics of stress that impact mammal populations.

Sebastián Muñoz-Leal is Assistant Professor of Zoology at the Faculty of Veterinary Sciences, University of Concepción, Chile. He is Curator of the Chilean Collection of Ticks "Daniel González-Acuña." He conducts research on ticks and tick-borne diseases in South America, with a particular focus on the taxonomy of the Argasidae family of ticks.

Constanza Napolitano is Associate Professor in the Department of Biological Sciences and Biodiversity at Universidad de Los Lagos in Osorno, Chile. Her research focuses on conservation genetics, the impacts of human landscape alteration on wild carnivore populations, and pathogen transmission at the domestic-wildlife interface.

Eduardo E. Palomo-Arjona is a Research Assistant at the Center for Research and Advanced Studies of the National Polytechnic Institute (CINVESTAV), Merida, Mexico. He has a Master of Science degree in Human Ecology. His research focuses on the impact of land-use change on the transmission of parasites in wildlife and the risk to human health.

María Virginia Rago is a Principal Professional at the National Council for Scientific and Technical Research in Argentina. She is a veterinarian with expertise in biodiversity conservation from the University of Buenos Aires. Her focus is in the health of wild animals concerning biodiversity conservation adhering to the principles of One Health.

Oscar Rico-Chávez is full-time Professor at the Laboratory of Disease Ecology and One Health, Facultad de Medicina Veterinaria y Zootecnia, UNAM, Mexico. His research focuses on studying the factors that explain the relationship between host diversity and pathogen diversity.

César R. Rodríguez-Luna is a Postdoctoral Researcher at the Center for Research and Advanced Studies of the National Polytechnic Institute (CINVESTAV) in México. His research focuses on wildlife ecology, with a specific focus on spatio-temporal interactions among species and community ecology.

Octavio Rojas-Soto is a Researcher at the Institute of Ecology, A.C. – CONACYT in Mexico. He is interested in the evolutionary biology of vertebrates and their relationship with geography and environment at different temporal and spatial scales.

André V. Rubio is Assistant Professor at the University of Chile, Faculty of Veterinary and Animal Sciences. His main research focuses on the study of the ecology of wild reservoir hosts and their pathogens associated with the impact of land use changes in South America.

Diego Santiago-Alarcon Assistant Professor at the Department of Integrative Biology of the University of South Florida. He is a biologist with a PhD and M.Sc. in Ecology, Evolution and Systematics from the University of Missouri-St. Louis, USA. His research focuses on the ecology and evolution of host-parasite interactions across environmental gradients and at different spatiotemporal scales.

Mauricio Seguel is Assistant Professor at the Ontario Veterinary College, University of Guelph. His research seeks to understand the environmental and host factors that drive immune phenotypes in wildlife systems. He has led for more than 10 years a team of scientists performing field wildlife studies in Patagonia, Chile, describing several new species of infectious agents and collecting an important archive of biological samples from several animal species.

Brenda Solórzano-García is Associated Professor at the Laboratory of Parasitology and Conservation Medicine, Universidad Nacional Autónoma de Mexico, ENES-Mérida. Her research focuses on the evolutionary ecology of host and parasite associations, with a specific focus on terrestrial mammals. She is also interested in monitoring wildlife health through non-invasive methods and assessing the effects of climate change and habitat modification on ecosystem health in accordance with the One Health approach.

Gerardo Suzán is Professor at the Facultad de Medicina Veterinaria y Zootecnia, Universidad Nacional Autonoma de Mexico. His line of research has focused on Disease Ecology and One Health, and his studies have been developed in different countries, including the USA, Costa Rica, Cuba, Panama, and Mexico. He is Mexico's representative to the Intergovernmental Platform on Biodiversity and Ecosystem Services (IPBES) on Biodiversity and Pandemics and is Co-director of the International Joint Laboratory on Biodiversity, Deforestation and Emerging Diseases (IRD/UNAM). He is a member of the steering committee of the PREZODE (Preventing Zoonotic Disease Emergence) Program representing Latin America.

E. Tarragona She is a Research Assistant at the Instituto de Investigación de la Cadena Lactea and Instituto Nacional de Tecnología Agropecuaria (INTA-CONICET). Her research focuses on ticks and tick-borne diseases occurring in Argentina.

Richard Thomas is a PhD candidate in Veterinary Sciences at the University of Concepción, Chile. He is a biologist from the University of Magdalena, Colombia. His research focuses on the systematics of parasites and vector-borne agents of public and veterinary health concern.

María José Tolsa is a Postdoctoral Researcher at the Institut de recherche pour le développement of France (IRD) and the International Joint Laboratory ELDORADO, IRD/UNAM, México. She is a biologist in Biological Sciences. Her research interest focuses on the link between species diversity and disease risk using flaviviruses, their vectors and hosts as a study model, using macroecological approaches.

Juan Carlos Troiano is Professor at the School of Veterinary Medicine, University of Buenos Aires, Argentina. She Graduated as a veterinarian at the Universidad Nacional del Litoral. Her research interest lies in exotic animals and clinical pathology, particularly hematology in reptiles.

Marcela Uhart is a wildlife veterinarian at the One Health Institute, University of California, Davis. Her focus is on free-ranging wildlife health research and conservation in Latin America, addressing risks from infectious diseases and toxins under a One Health approach.

María del Carmen Villalobos-Segura is a PhD student in the School of Veterinary Medicine at Universidad Nacional Autónoma de México (UNAM). Her research focuses on the influence host traits and landscape attributes have on pathogen transmission. She is also interested in the compilation and use of databases to study host–pathogen interactions through bioinformatic analyses.

E. A. Zieman Is an Assistant Professor of Biological Sciences at Eastern Illinois University. He conducts research on vector-borne pathogens in wildlife and domestic animals and the epizootiology of these pathogens.

Chapter 1
An Introduction to the Ecology of Wildlife Diseases in the Neotropics

Andrea Chaves and Gerardo Acosta-Jamett

Covering just 7% of the continental surface, tropical forests contain more than half of the Earth's species. The neotropical region, spanning from Mexico to Argentina and Chile, including the Antilles, comprises a wide diversity of habitats with extensive rainforests, high-altitude grasslands, seasonal dry forests, and diverse mountain systems (Antonelli et al. 2018). However, the accelerated and poorly planned growth of socioeconomic development activities in countries with largely unstable economies and policies has caused the neotropical landscape to face socioeconomic growth solutions at the expense of habitat destruction, fragmentation, and loss of wildlife. These include the construction of highways, hydroelectric dams, intensive logging, mining operations, extensive livestock farming, monocultures, indiscriminate use of pesticides, and contamination of rivers and other tributaries (Ceballos et al. 2009). On a local scale, neotropical wildlife populations have also been directly and indirectly affected by the unplanned development of Latin American countries. Common examples include illegal trafficking, illegal hunting, interactions between wild animals, humans, and domestic animals, and the transmission of infectious diseases, including zoonotic diseases.

In Latin America, the climatic and environmental conditions of neotropical forests, with an enormous diversity of microclimates, provide ideal environments in which innumerable infectious agents thrive and interact with a great variety of competent hosts, reservoirs, and vectors. In addition to these natural conditions, changes in land use (e.g., agriculture and poorly planned urbanism) and an increase in human

A. Chaves
Escuela de Biología, Universidad de Costa Rica, San José, Costa Rica

G. Acosta-Jamett (✉)
Instituto de Medicina Preventiva Veterinaria y Center for Disease Surveillance and Evolution of Infectious Diseases, Facultad de Ciencias Veterinarias, Universidad Austral de Chile, Valdivia, Chile
e-mail: gerardo.acosta@uach.cl

G. Acosta-Jamett, A. Chaves (eds.), *Ecology of Wildlife Diseases in the Neotropics*, https://doi.org/10.1007/978-3-031-50531-7_1

presence in natural, marine, and terrestrial environments (e.g., tourism) have favored the interaction between humans and their domestic animals with wildlife. Therefore, it is necessary to know and understand the dynamics of infectious diseases considering natural landscapes and anthropogenic disturbance. This is urgent from many perspectives, which can be summarized as follows: (i) pathogens can exacerbate the decline of endangered wildlife species (Nunn and Altizer 2006), and (ii) 62% of zoonotic diseases come from wildlife that come into direct or indirect contact with humans due to anthropogenic disturbance (Childs et al. 2019).

With evidence that anthropogenic disturbances may be the cause of biodiversity disasters and the reason why there has been an increase in disease emergence, high biodiversity has been hypothesized to protect against and reduce the spread of infectious diseases (Ostfeld and Keesing 2000; Zohdy et al. 2019). Reduced intraspecific encounter and regulation of generalist susceptible hosts may be the ways by which high diversity reduces pathogen transmission. The spatial distribution of hosts has been recognized as an important aspect of disease transmission (Escobar 2020; Escobar et al. 2015; Hess et al. 2002). Many species are aggregated in subpopulations, which can influence how pathogens spread within and between populations and will have important consequences for how they are maintained in such populations (Hess 1996). The changes in the environment and loss of biodiversity do not affect all species in the same way. It appears that generalist species are unaffected and in some cases even favored by the anthropization of the landscape. However, specialist species are affected and threaten to a the risk of extinction. For that reason, when we speak of wild species and their functional in disease transmission, it cannot be generalized either. There are groups such as synanthropic species (e.g., rodents: *Mus musculus*) that are recognized for a high capacity to maintain infectious agents. However, another specialist species (e.g., New World primates: *Alouatta palliata, amphibians: Incilius periglenes*) could be recognized as a sentinel species because they are mortal affected by a specific disease (e.g., *Alouatta palliata* with yellow fever) which acts as a warning announcing disease emergence. Or their populations have become extinct due to the presence of lethal pathogens imported to the neotropics (e.g., *Incilius periglenes* and chytridiomycosis).

Native biodiversity (species, community or metacommunity) is the result of biogeographic processes (e.g., speciation, diversification) that occur over evolutionary time (Ostfeld and Keesing 2012). In contrast, dynamic changes in biodiversity are dominated by anthropogenic processes, which act locally and rapidly and consist of biodiversity losses rather than gains. The relevance of species diversity (the number and relative abundance of different species within an ecological community) and their performance in ecosystem functions has traditionally been assessed in terms of the number of species that make up a community (Schwartz et al. 2000). It is currently questioned whether changes in ecosystem functions are due to the loss of any species or functional groups with specific roles. Greater diversity in functional roles leads to greater resource use efficiency, increases community resilience, and interferes with the ability of competitively superior species to become numerically dominant (Tilman et al. 1996).

Species diversity with specific roles is also important in the ecology of infectious diseases, particularly those involving more than one host species (Ostfeld and Keesing 2000). It has been shown that high biodiversity often reduces transmission rates of infectious agents and decreases disease risk (Keesing and Ostfeld 2015). This is explained because (1) most infectious agents are generalist multi-hosts; (2) these hosts differ in their susceptibility and transmission potential; and (3) generally, hosts most likely to acquire and transmit infections (reservoirs) are both abundant and generalist species that are resistant to anthropogenic disturbances (Keesing and Ostfeld 2015; Johnson et al. 2015). For infectious agents that are specialist hosts and that can infect, replicate, and be transmitted in a single host species, increases in diversity also tend to decrease the density of suitable hosts, decreasing their transmission (Ostfeld and Keesing 2012). Many infectious agents considered specialists in addition may temporarily invade other hosts, proliferating within several species of a community, which may differ strongly with respect to susceptibility, transmission, and maintenance of pathogens in distinct landscapes (Young et al. 2017). Currently, the number of introduced species is often numerically greater than native biodiversity and is closely associated with emerging diseases, affecting disease pressure in aspects such as contact rate and strain circulation (Hatcher et al. 2012). Consequently, species living in ecological communities with high diversity tend to dilute the effect of reservoir species and reduce disease risk (dilution effect) (Ostfeld and Keesing 2000). Host communities characterized by high species richness are likely to contain a high proportion of hosts that are inefficient in transmitting the infectious agent (Schmidt and Ostfeld 2001). Therefore, the higher the abundance of weakly competent reservoir species, the greater the dilution effect and the lower the probability of disease transmission.

Although ecological factors are important variables that could influence disease transmission in fragmented habitats, epidemiological theories are also important to understand how pathogens can be transmitted within and between species. Central to epidemiology understanding are the factors that cause and facilitate the transmission and maintenance of infectious agents in host populations. The interaction between infectious agents, the host and the environment can lead to the occurrence of infectious diseases (Thrusfield 2005).

Infectious agents can be divided into two groups: micro- and macroparasites (May and Anderson 1979). The former multiply inside the host and include viruses, bacteria, and protozoa. The latter, in contrast, do not multiply within the host but multiply by producing infective stages that are shed by the host to infect new hosts, including helminths and arthropods. To explain microparasites transmission and maintenance, epidemiological theory relies on the notion of threshold and fadeout theories (Anderson and May 1991; Swinton et al. 2002). The former refers to the host population size or host density, high enough to ensure a basic reproductive number (i.e., number of secondary cases caused by the first infectious individual in a susceptible population), R_0, ≥ 1, in which an infection can successfully invade a population (Anderson 1991; Anderson and May 1979, 1991; May and Anderson 1979).

Fadeout (extinction) theory refers to what happens after an epidemic and whether a microparasite has been able to affect all susceptible individuals, and in that case, the infection will tend to extinction (Grenfell et al. 2002; Swinton et al. 2002). According to these theories, pathogens that are maintained in a population through a high birth rate and never go locally to extinction are termed endemic. On the other hand, infections that only sporadically affect a population are termed epidemics, which, after affecting all susceptible hosts, will tend to local extinction, as the supply of susceptibles from births is not enough to maintain the chain of transmission (Grenfell and Harwood 1997). Overall, pathogens that show prolonged infectiousness and/or that can be maintained in the environment are more likely to exhibit an endemic prevalence (Anderson and May 1979). In contrast, those that are directly transmitted have a short infectious period and cause high host mortality (Anderson 1991; Anderson and May 1979, 1991; May and Anderson 1979). The ability of some pathogens such as canine parvovirus to persist in the environment for prolonged periods could help their maintenance in areas with a smaller population size than that needed, for example, for a directly transmitted pathogen such as CDV, which requires close contact in time or space with a sick animal shedding virus.

Theoretical studies and empirical data suggest that highly infectious pathogens should only persist in populations larger than a threshold or critical community size (CCS) (Bartlett 1957, 1960; Black 1966; Lloyd-Smith et al. 2005; Swinton et al. 2002), where the pathogen is maintained by an input of susceptible by birth and/or immigration (Dobson and Grenfell 1995; Swinton et al. 2002). On the other hand, in populations below the CCS, such pathogens cannot persist because of the low probability of contact between infectious and susceptible hosts (Begon et al. 2003; Tompkins et al. 2002). However, even those populations that are under a CCS (non-maintenance populations), if they are epidemiologically connected with other non-maintenance or maintenance populations (e.g., through immigration), can be part of a larger complex meta-population that can be part of a reservoir in which the pathogen can persist in the long term (Haydon et al. 2002; Viana et al. 2014).

Many authors have recognized that the epidemiology and metapopulation theory have addressed the same issues (Hanski and Gilpin 1997; Harrison 1991; May and Novak 1994; Nee 1994) since populations of hosts can be homologated to habitat patches and the transmission from infected populations to uninfected ones can have counterparts in the colonization of patches by migration of individuals from large to small patches. Thus, a metapopulation is made up of subpopulations or "patches" connected by immigration or dispersal (Hanski 1998; Hanski and Gilpin 1997; Levins 1969). In the case of the "city-village" model, this has been recognized to be very similar to a mainland-island metapopulation, since subpopulations can correspond to host populations, mainlands to cities, and islands to villages, in which the infection is maintained in the "city-village" complex through emigration of infected hosts from a large patch (city) that is above the CCS (maintenance population) and harbors a high density of susceptibles that are infected and can migrate to non-maintenance populations (town or rural areas), maintaining the infection within the metapopulation at a broader spatial scale (Anderson and May 1991; Grenfell and Harwood 1997). This similitude has allowed the use of the tools developed in metapopulation theory to model and predict the spread of diseases (Grenfell and Harwood 1997; Grenfell and Bolker 1998). Thus, if a disease fades out in a small subpopulation within the metapopulation, individuals from other populations can recolonize and maintain the infection (Grenfell and Harwood 1997; Keeling 1997; Keeling et al. 2004). In addition, if the immigration of infected animals into rural areas is a function of the distance to the source population (i.e., cities), we should expect more infected individuals when they are closer to the source population (Grenfell et al. 2001; Keeling et al. 2004).

Anderson and May (1991) included the spatial distribution of the persistence and transmission of diseases in a heterogeneous landscape of varying population sizes when they developed the "cities and villages" model. This model was later confirmed to be a good predictive model for measles by empirical studies in the United Kingdom (Grenfell et al. 2001; Grenfell and Bolker 1998) and the United States (Cliff et al. 1992, 1993), at a broader scale and for pertussis in the United Kingdom (Broutin et al. 2004b; Rohani et al. 1998, 1999, 2000) and at a finer scale in Senegal (Broutin et al. 2004a). These studies highlighted the importance of migration between large (i.e., cities) and small (i.e., towns or rural areas) populations in the maintenance of infection, showing that infection is transmitted following a size hierarchy from large cities to small villages and finally to rural areas, having an endemic state in large populations and an epidemic state with more fade-outs in small ones.

In our region, the effects of environmental degradation, biodiversity loss, and the interactions between infectious agents and native biodiversity have enormous information gaps, so it is urgent to increase research in this field. Only until all the components involved in the problem are understood it is possible to think about intervention. It is therefore necessary to first compile an integrated picture from the perspective of different areas of knowledge (Schneider et al. 2009) and then generate integrated initiatives to control the emergence of infectious diseases (Fig. 1.1). This book seeks to highlight the primary ecological, environmental, social, and economic variables associated with the risk of maintenance, transmission, and dissemination of emerging, re-emerging, and neglected infectious diseases, in which neotropical vertebrates are involved. It allows the compilation of the knowledge that we have up to this date for the neotropical region as well as the current needs of knowledge improvement. This book covers topics including amphibians, birds, reptiles, marine, and terrestrial mammals, rodents, bats, neotropical primates, carnivores, and camelids from terrestrial, aquatic, and marine environments that inhabit a vast area in the neotropical region.

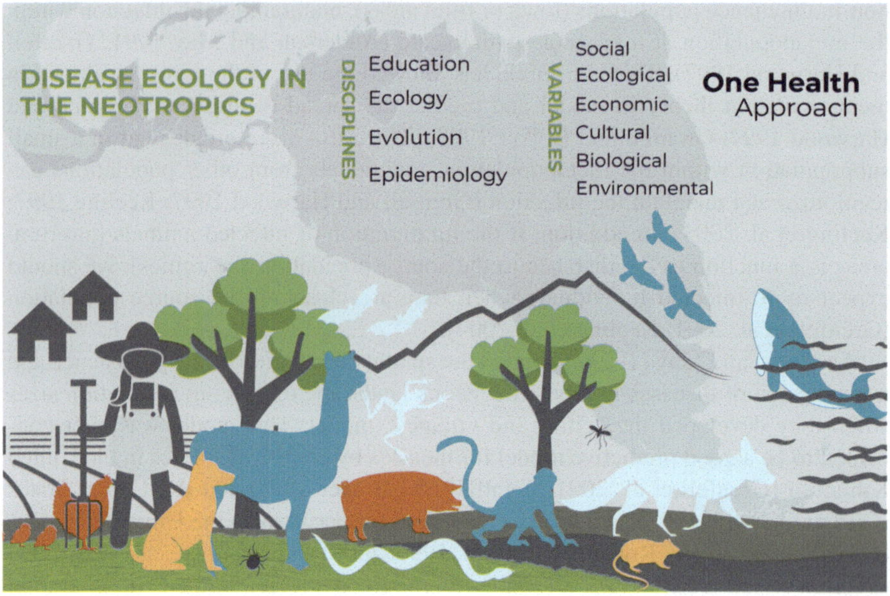

Fig. 1.1 The development of disease ecology in the neotropics requires the conjunction of several disciplines with the objective of understanding all variables involved, while taking into account a One Health approach for a biodiverse environment

References

Anderson RM (1991) Populations and infectious diseases: ecology or epidemiology? J Anim Ecol 60:1–50

Anderson RM, May RM (1979) Population biology of infectious diseases. Nature 280:361–367

Anderson RM, May RM (1991) Infectious diseases of humans: dynamics and control. Oxford University Press, Oxford

Antonelli A, Ariza M, Albert J, Andermann T, Azevedo J, Bacon C, Faurby S, Guedes T, Hoorn C, Lohmann LG, Matos-Maraví P, Ritter CD, Sanmartín I, Silvestro D, Tejedor M, Ter Steege H, Tuomisto H, Werneck FP, Zizka A, Edwards SV (2018) Conceptual and empirical advances in Neotropical biodiversity research. PeerJ 4(6):e5644. https://doi.org/10.7717/peerj.5644

Bartlett MS (1957) Measles periodicity and community size. J R Stat Soc Ser A Gen 120:48–70

Bartlett MS (1960) The critical community size for measles in the United States. J R Stat Soc Ser A Gen 123:37–44

Begon M, Hazel SM, Telfer S, Bown K, Carslake D, Cavanagh R, Chantrey J, Jones T, Bennett M (2003) Rodents, cowpox virus and islands: densities, numbers and thresholds. J Anim Ecol 72:343–355

Black FL (1966) Measles endemicity in insular populations: critical community size and its evolutionary implication. J Theor Biol 11:207–211

Broutin H, Elguero E, Simondon F, Guegan JF (2004a) Spatial dynamics of pertussis in a small region of Senegal. Proceedings of the Royal Society of London Series B-Biological Sciences 271:2091–2098

Broutin H, Simondon F, Guegan JF (2004b) Whooping cough metapopulation dynamics in tropical conditions: disease persistence and impact of vaccination. Proceedings of the Royal Society of London Series B-Biological Sciences 271:S302–S305

Ceballos G, Vale MM, Bonacic C, Calvo-Alvarado J, List R, Bynum N, Medellín RA, Simonetti JA, Rodríguez JP (2009) Conservation challenges for the Austral and Neotropical America section. Conserv Biol 23:811–817. https://doi.org/10.1111/j.1523-1739.2009.01286.x
Childs ML, Nova N, Colvin J, Mordecai EA (2019) Mosquito and primate ecology predict human risk of yellow fever virus spillover in Brazil. Philos Trans R Soc B 374:20180335. https://doi.org/10.1098/rstb.2018.0335
Cliff AD, Haggett P, Smallman-Raynor M (1993) Measles: an historical geography of a major human viral disease from global expansion to local retreat, 1840–1990 (Oxford, Blackwell), p. xx + 462
Cliff AD, Haggett P, Stroup DF (1992) The geographic structure of measles epidemics in the northeastern United States Am J Epidemiol 136:592–602
Dobson AP, Grenfell BT (1995) Introduction. In: Grenfell BT, Dobson AP (eds) Ecology of infectious diseases in natural populations. Cambridge University Press, Cambridge, pp 1–19
Escobar LE (2020) Ecological niche modeling: an introduction for veterinarians and epidemiologists. Front Vet Sci 7:519059
Escobar LE, Peterson AT, Papes M, Favi M, Yung V, Restif O, Qiao HJ, Medina-Vogel G (2015) Ecological approaches in veterinary epidemiology: mapping the risk of bat-borne rabies using vegetation indices and night-time light satellite imagery. Vet Res 46:92
Grenfell B, Harwood J (1997) (Meta)population dynamics of infectious diseases. Trends Ecol Evol 12:395–399
Grenfell BT, Bjornstad ON, Finkenstadt BF (2002) Dynamics of measles epidemics: scaling noise, determinism, and predictability with the TSIR model. Ecol Monogr 72:185–202
Grenfell BT, Bjornstad ON, Kappey J (2001) Travelling waves and spatial hierarchies in measles epidemics. Nature 414:716–723
Grenfell BT, Bolker BM (1998) Cities and villages: infection hierarchies in a measles metapopulation. Ecology Letters 1:63–70
Hatcher MJ, Dick JTA, Dunn AM (2012) Disease emergence and invasions. Funct Ecol 26:1275–1287
Haydon DT, Cleaveland S, Taylor LH, Laurenson MK (2002) Identifying reservoirs of infection: a conceptual and practical challenge. Emerg Infect Dis 8:1468–1473
Hess G (1996) Disease in metapopulation models: implications for conservation. Ecology 77:1617–1632
Hess GR, Randolph SE, Arnebergh P, Chemini C, Furlanello C, Harwood J, Roberts MG, Swinton J (2002) Spatial aspects of disease dynamics. In: Hudson RJ, Rizzoli AP, Grenfell BT, Heesterbeek H, Dobson AP (eds) The ecology of wildlife diseases. Oxford University Press, Oxford, pp 102–118
Hanski I (1998) Metapopulation dynamics. Nature 396:41–49
Hanski IA, Gilpin ME (1997) Metapopulation biology: ecology, genetics, and evolution Academic Press, San Diego, CA
Harrison S (1991) Local extinction in a metapopulation context: an empirical evaluation. Biological Journal of the Linnean Society 42:3–16
Johnson PRJ, Ostfeld RS, Keesing F (2015) Frontiers in research of biodiversity and disease. Ecol Lett 18:1119–1133
Keeling MJ (1997) Modelling the persistence of measles. Trends Microbiol 5:513–518
Keesing F, Ostfeld R (2015) Is biodiversity good for your health? Science 349:235
Keeling MJ, BjØrnstad ON, Grenfell BT (2004) Metapopulation dynamics of infectious diseases. In: Hanski IA, Gaggiotti OE (Eds), Ecology, Genetics and Evolution of Metapopulations. Elsevier, Burlington, pp. 415–446
Lloyd-Smith JO, Cross PC, Briggs CJ, Daugherty M, Getz WM, Latto J, Sanchez MS, Smith AB, Swei A (2005) Should we expect population thresholds for wildlife disease? Trends Ecol Evol 20:511–519
Levins R (1969) Some demographic and genetic consequences of environmental heterogeneity for biological control. Bull Entomol Soc Am 15:237–240

May RM, Anderson RM (1979) Population biology of infectious-diseases. Nature 280:455–461
May RM, Novak M (1994) Superinfection, metapopulation dynamics, and the evolution of diversity. J Theor Biol 170:95–114
Nunn CL, Altizer S (2006) Infectious diseases in primates: behavior, ecology, and evolution. Oxford University Press, Oxford
Nee S (1994) How populations persist. Nature 367:123–124
Ostfeld RS, Keesing F (2000) Biodiversity and disease risk: the case of Lyme disease. Conserv Biol 14:722–728
Ostfeld RS, Keesing F (2012) Effects of host diversity on infectious diseases. Annu Rev Ecol Evol Syst 43(1):157–182
Rohani P, Earn DJ, Finkenstadt B, Grenfell BT (1998) Population dynamic interference among childhood diseases. P Roy Soc B-Biol Sci 265:2033–2041
Rohani P, Earn DJ, Grenfell BT (1999) Opposite patterns of synchrony in sympatric disease metapopulations. Science 286:968–971
Rohani P, Earn DJ, Grenfell BT (2000) Impact of immunisation on pertussis transmission in England and Wales. Lancet 355:285–286
Schmidt KA, Ostfeld RS (2001) Biodiversity and the dilution effect in disease ecology. Ecology 82(3):609–619
Schneider MC, Romijn PC, Uieda W, Tamayo H, Fernandes- da Silva D, Belotto A, Barbosa-da Silva J, Leanes LF (2009) Rabies transmitted by vampire bats to humans: an emerging zoonotic disease in Latin America? PanAmerican journal of public health 25(3):260–269
Schwartz M, Brigham C, Hoeksema J et al (2000) Linking biodiversity to ecosystem function: implications for conservation ecology. Oecologia 122:297–305. https://doi.org/10.1007/s004420050035
Swinton J, Woolhouse MEJ, Begon ME, Dobson AP, Ferroglio E, Grenfell BT, Guberti V, Hails RS, Heesterbeek JAP, Lavazza A, Roberts MG, White PJ, Wilson K (2002) Microparasite transmission and persistence. In: Hudson RJ, Rizzoli AP, Grenfell BT, Heesterbeek H, Dobson AP (eds) The ecology of wildlife diseases. Oxford University Press, Oxford, pp 83–101
Thrusfield M (2005) Veterinary epidemiology, 3rd edn. Blackwell Scientific Publications Ltd., Oxford
Tilman D, Wedin D, Knops J (1996) Productivity and sustainability influenced by biodiversity in grassland ecosystems. Nature 379:718–720
Tompkins DM, Dobson AP, Arneberg P, Begon ME, Cattadori IM, Greenman JV, Heesterbeek JAP, Hudson PJ, Newborn D, Pugliese A, Rizzoli AP, Rosa R, Rosso F, Wilson K (2002) Parasites and host population dynamics. In: Hudson RJ, Rizzoli AP, Grenfell BT, Heesterbeek H, Dobson AP (eds) The ecology of wildlife diseases. Oxford University Press, Oxford, pp 45–62
Viana M, Mancy R, Biek R, Cleaveland S, Cross PC, Lloyd-Smith JO, Haydon DT (2014) Assembling evidence for identifying reservoirs of infection. Trends Ecol Evol 29:270–279
Young HS, Parker IM, Gilbert GS, Guerra AS, Nunn CL (2017) Introduced species, disease ecology, and biodiversity – diseases relationships. Trends Ecol Evol 32:41–54. https://doi.org/10.1016/j.tree.2016.09.008
Zohdy S, Schwartz TS, Oaks JR (2019) The coevolution effect as a driver of spillover. Trends Parasitol 35(6):399–408. https://doi.org/10.1016/j.pt.2019.03.010

Chapter 2
The Development of Disease Ecology as a Science in Latin America and the Caribbean

Milena Arguello-Saenz, Francisco Chacón, Andrea Chaves, André V. Rubio, and Gerardo Suzan

Milena Arguello-Saenz and Francisco Chacón contributed equally with all other contributors.

2.1 Introduction

The Latin American and Caribbean regions harbor rich biological diversity, where approximately 60% of terrestrial life on the planet is found (Unep 2016). As in many other parts of the world, biodiversity in Latin America and the Caribbean (LAC) presents multiple threats, including infectious diseases that affect wildlife conservation, such as chytridiomycosis, distemper canine virus, and sarcoptic mange (Martinez-Gutierrez and Ruiz-Saenz 2016; De León et al. 2019; Escobar et al. 2022). The LAC region also presents several geographical areas considered hotspots of zoonotic risk, mainly in tropical areas of southern Mexico, Bolivia, and Brazil, among others (Allen et al. 2017; García-Peña et al. 2021). In this context, research on disease ecology has become crucial for understanding host–pathogen

---

Supplementary Information The online version contains supplementary material available at https://doi.org/10.1007/978-3-031-50531-7_2.

M. Arguello-Saenz · G. Suzan (✉)
Departamento de Etología, Fauna Silvestre y Animales de Laboratorio, Facultad de Medicina Veterinaria y Zootecnia, Universidad Nacional Autónoma de México, Ciudad de México, Mexico

F. Chacón · A. V. Rubio
Departamento de Ciencias Biológicas Animales, Facultad de Ciencias Veterinarias y Pecuarias, Universidad de Chile, Santiago, Chile

A. Chaves
Escuela de Biología, Universidad de Costa Rica, San José, Costa Rica

G. Acosta-Jamett, A. Chaves (eds.), *Ecology of Wildlife Diseases in the Neotropics*, https://doi.org/10.1007/978-3-031-50531-7_2

interactions in dynamic environments within the region, which can be useful information for preventing and controlling disease outbreaks.

The LAC region encompasses numerous countries with diverse environmental and socioeconomic contexts. Therefore, research related to disease ecology may have different approaches and high contrasts in productivity and collaboration among countries. In this introductory chapter, we performed a literature review in the three languages (Spanish, English, and Portuguese) in LAC and a bibliometric analysis to identify how the field of disease ecology has been developed in LAC, recognizing major fields, hosts, and infectious diseases studied, as well as identifying important collaborations and networks among countries and institutions that have significantly contributed to this discipline.

2.2 Literature Review

2.2.1 Data Retrieval

We used a systematic bibliographic search method, following The Preferred Reporting Items for Systematic Reviews and Meta-Analyses (PRISMA) (Page et al. 2021).

The initial search was carried out on PubMed using the keywords "disease ecology," "ecología de enfermedades," and "ecologia de doenças" in combination with "Latin America."

This search obtained 1440 results, which were imported into a graphical interface for the R package litsearchr (Grames et al. 2019), to extract the keywords and generate a co-occurrence network.

The PRISMA protocol allows the search for articles to be extended using complementary methods to the search in databases, and it allows searching the references of review articles to identify original sources that have not been included in the original search. Therefore, we conducted an additional review based on articles obtained from the Web of Science.

This helped us to conduct a second search equation, which included publications in three languages: Spanish, English, and Portuguese (Box 2.1), obtaining 4139 results when searching the Web of Science databases.

Box 2.1: List of Keywords Used in the Final Search. These Keywords Were Generated with the R Package "litsearchr" (Grames et al. 2019)
Search equation:

TS = (("disease ecology" OR "ecología de enfermedades" OR "ecologia de doenças" OR "wildlife diseases" OR "enfermedades de vida silvestre" OR "doenças da vida selvagem" OR "neglected diseases" OR "enfermedades

(continued)

Box 2.1 (continued)

desatendidas" OR "doenças negligenciadas" OR "zoonotic diseases" OR "enfermedades zoonóticas" OR "doenças zoonóticas" OR "emergent diseases" OR "enfermedades emergentes" OR "doenças emergentes" OR "re-emerging diseases" OR "enfermedades reemergentes" OR "doenças re-emergentes" OR "infectious disease*" OR "enfermedades infecciosas" OR "pathogen" OR "patogeno" OR bacteria OR virus OR protozoa OR helmint*) AND (transm* OR ocurrenc* OR incidenc* OR prevalenc* OR risk* OR riesgo OR infec*) AND ("central* america*" OR "Centroamérica" OR "América Central" OR "south* america*" OR "Sudamérica" OR "América del Sur" OR "latin* america*" OR "América Latina" OR neotropic* OR Caribbean OR Caribe OR Mesoamérica OR Tropics OR Argentina OR Bolivia OR Brazil OR Brasil OR Chile OR Colombia OR "Costa Rica" OR Cuba OR Ecuador OR "El Salvador" OR Guatemala OR Honduras OR Mexico OR Nicaragua OR Panama OR Paraguay OR Peru OR Puerto Rico OR "Dominican Republic" OR Uruguay OR Venezuela OR Haiti OR Belize OR Belice OR "Puerto Rico"))

2.2.2 Standardization and Cleaning of the Data

These results were imported into Sysrev, a web platform for data curation and systematic review (Bozada Jr. et al. 2021), defining inclusion and exclusion criteria (Box 2.2) to be able to filter and select each article to finally extract and collect the necessary data to develop this chapter.

2.2.3 Bibliometric Analysis

A total of 4139 articles were obtained as a result of the search in Web of Science. When reviewing the articles, a large majority of clinical and pharmacological index articles were observed. We recognize that our search strategy may miss articles from LAC, but we believe that the results present an overall picture of temporal, spatial, and thematic trends of publications from LAC.

The concordance of the selection of articles was 99.6% between two reviewers, where the final inclusion or exclusion decision was determined by a third reviewer, obtaining a total of 14 conflicting articles. In total, 620 documents met our inclusion criteria, including 619 articles and 1 book chapter. The main data of the set of documents included in the analysis are found in Table S1.

Box 2.2: Criteria for Inclusion and Exclusion of Scientific Articles

(a) *Inclusion criteria*

(i) It was developed directly in or with samples/data from Latin America and the Caribbean.
(ii) It was published in the last 30 years (1992–2022).
(iii) It is an original article, research article, short communication, or case report.
(iv) It is written in English, Spanish, or Portuguese.
(v) The research covers some fields of knowledge in disease ecology (mechanisms, patterns, and effects of host–pathogen interactions, links between ecological interactions and disease evolution, effect of changes in ecosystems on prevalence, incidence, spatial distribution, etc.).
(vi) Specify authors, partner institutions and countries where the research was conducted.

(b) *Exclusion criteria*

(i) The research was developed outside of Latin America or the Caribbean.
(ii) The research was published before 1992.
(iii) The investigation corresponds to a review, local or regional.
(iv) The research is strictly laboratory, unrelated to the dynamics of the disease (cultures, comparison of diagnostic methods, drugs, etc.).

We used R 4.2.2 software (R Core Team, Vienna, Austria) and the Bibliometrix 4.1.1 package (Aria and Cuccurullo 2017) for bibliometric data analysis.

2.3 Results and Discussion

2.3.1 Temporal Distribution of Scientific Publications Related to Disease Ecology in Latin America and the Caribbean

Regarding scientific production in disease ecology and associated topics, we can observe a considerable increase in publications from 2013 to 2022. Initially, from 1992 to 2012, the maximum number of publications per year was 25 articles in 2012; however, after this period, it began to increase. From 2013 to 2022, the minimum number of publications per year was 24 (2013), with a maximum of 72 articles (2020), reaching a total of 521 articles in this period, which is equivalent to 84% of the total publications selected (Fig. 2.1). This result is congruent with the increase

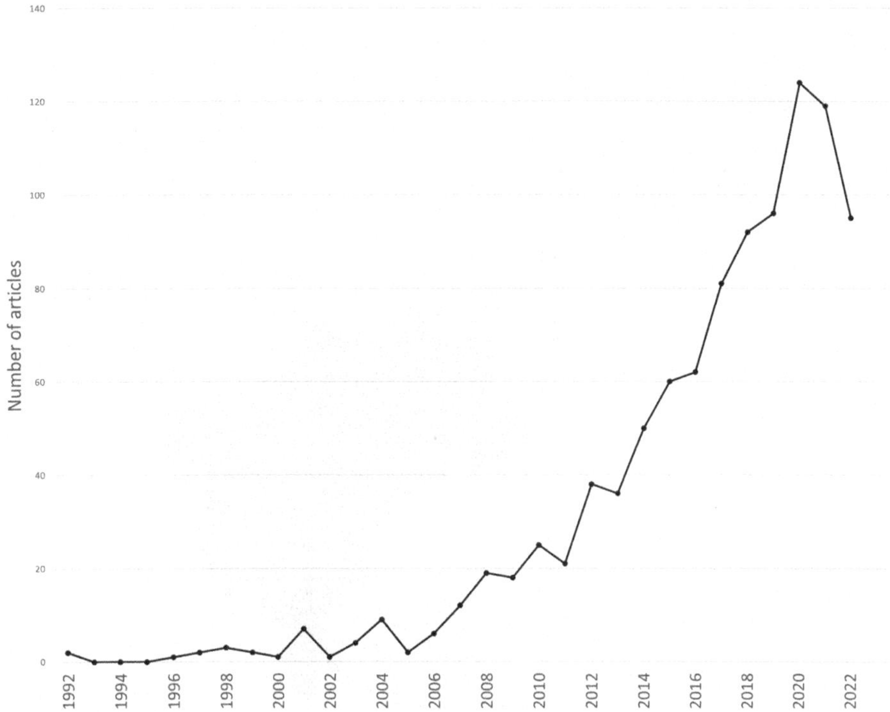

Fig. 2.1 Annual production of scientific publications in Disease Ecology (1992–2022)

in disease ecology worldwide (Brandell et al. 2021; Koprivnikar and Johnson 2016). Furthermore, the accelerated growth of research in the last 12 years (Fig. 2.1) seems to be a reflection of the growth of scientific publication in LAC, which has been increasing over the last 15 years in most countries of the region (Lewis et al. 2021). This may be associated with both an improvement in graduate education and an increase in research investment in LAC (Lewis et al. 2021).

2.3.2 Scientific Publication Production and Collaboration by Region/Countries

Brazil is the country with the highest production of publications in the field of Disease Ecology in LAC (317), followed by México (96) and Argentina (91) (Fig. 2.2). The development of scientific research conducted by LAC in the area of disease ecology seems to reflect the general behavior of scientific research in the region. In general terms, scientific production shows a concentration of knowledge disseminated by a few countries. Therefore, if we group the countries that have published the most scientific papers in a period of 30 years, Brazil, Argentina, and

Fig. 2.2 Scientific production per Latin American and Caribbean countries

Mexico stand out, followed by Chile and Colombia, which are separated from Uruguay, Costa Rica, Panama, and Peru with far fewer publications, to finish with the other LAC countries in a last group that lags further behind. This reflects a gap in research capacity that can be associated with the unstable social, economic, and political realities of many LAC countries that may have irrefutable implications for scientific development (Ciocca and Delgado 2017).

One way to measure collaboration between the authors of each country is by measuring the degree of local production or in conjunction with other countries,

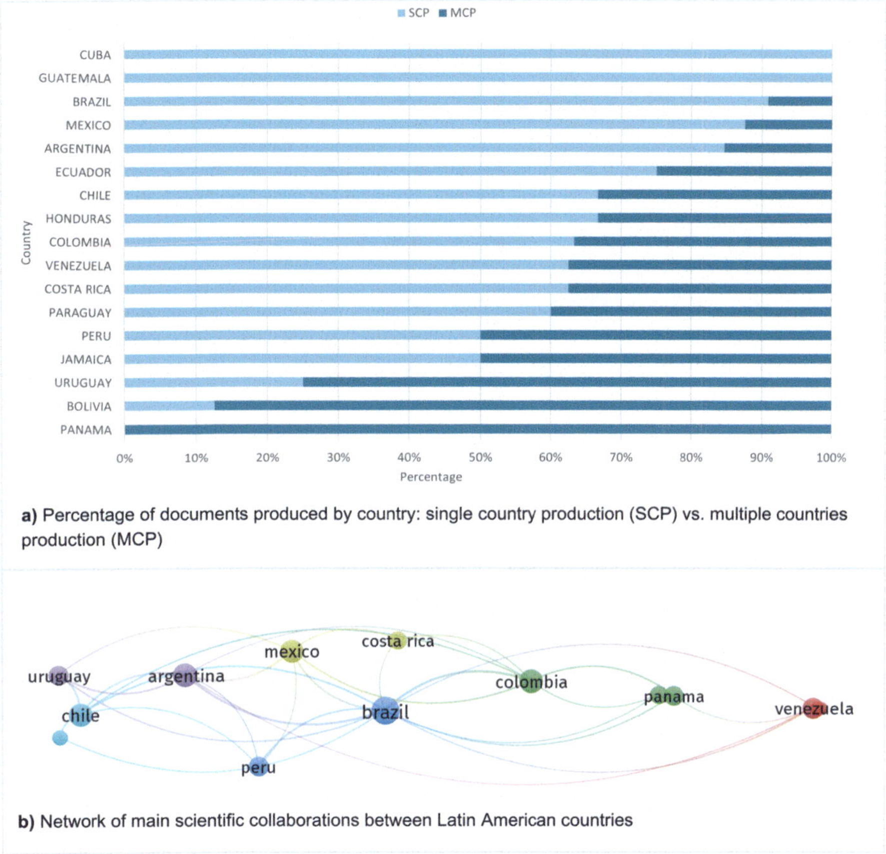

Fig. 2.3 Scientific collaboration between Latin American and Caribbean countries

sharing authorship, and generating networks between the institutions in charge of creating knowledge in each country. We observe that Brazil ranks as the central source in the production of documents on the ecology of diseases, together with the production of the rest of the region's countries (Fig. 2.3b); however, the dominant percentage in most of the countries of the region is local production, especially in Brazil, where the generation of publications involving authors from various countries is less than 9%, being the lowest percentage in the region. Other countries appear with 0% Cuba and Venezuela. On the other hand, Uruguay, Ecuador, and Costa Rica have over 20% production compared with other countries (Fig. 2.3a). Bolivia and Panama show the most publications in collaboration with other countries.

These differences may reflect the local amount of funding for research and the necessity of collaboration to achieve the research goals. Additionally, the political relationships between countries may affect these percentages; for example, Cuba has 0% international collaboration in this search, at least with other Latin American countries.

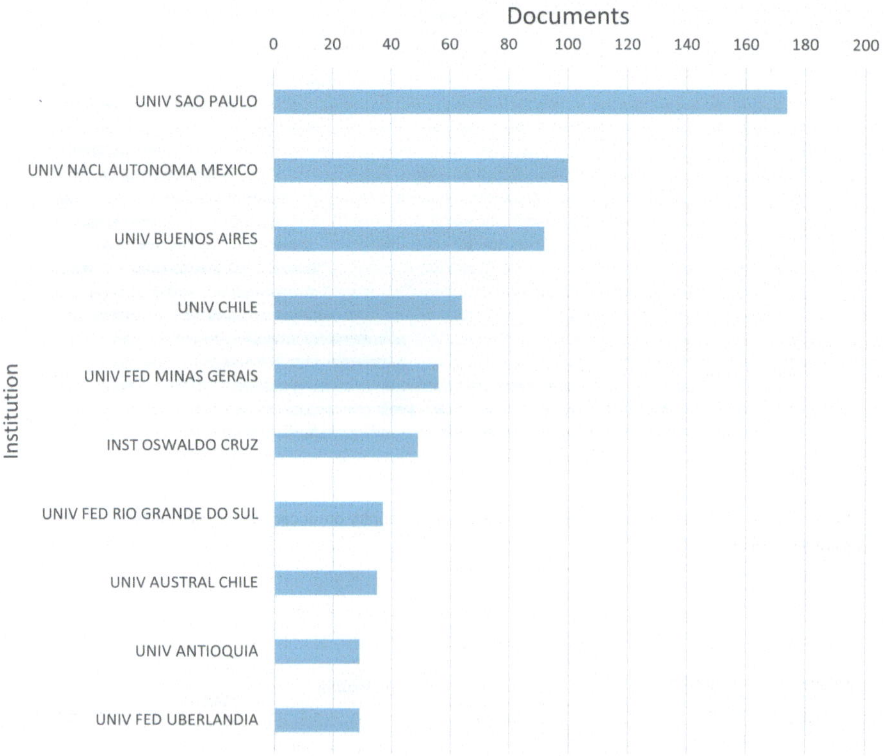

Fig. 2.4 Distribution of relevant institutions (documents >28) that contribute to the disease ecology in Latin America and the Caribbean

2.3.3 *Research by Affiliation*

Ten institutions were selected with the largest number of publications, with more than 29 publications each. As a result, an important domain of the University of Sao Paulo (Brazil), followed by the Universidad de Buenos Aires (Argentina) and the Universidad Nacional Autónoma de México (Mexico), can be observed. It should be noted that these 10 institutions – five Brazilian institutions, one Mexican, one Argentinean, two Chilean, and one Colombian – belong to only five countries: Brazil, Argentina, Mexico, Chile, and Colombia (Fig. 2.4).

According to UNESCO data from 2018 to 2020, the countries that invest the most in science, technology, and innovation in Latin America in net contributions are Brazil (1.28% of GDP) and Mexico (0.30% of GDP), contributing 54% and 26%, respectively, of the total investment in the region, followed by Argentina (0.63% of GDP) and Chile (0.34% of GDP). In LAC, the average GDP allocated to research is less than 0.70%, a percentage that is below the world average, which was

1.7% in 2017, where the funding base to conduct research in LAC is through public governmental funds (Lewis et al. 2021; UNESCO 2021).

Some of the institutions that finance scientific research in these countries are as follows:

Brazil: The Ministry of Science, Technology, and Innovation (MCTI), the National Council for Scientific and Technological Development (CNPq), the Foundation for Research Support of the State of São Paulo (FAPESP) and the National Bank for Economic and Social Development (BNDES).

Mexico: The National Council of Humanities, Science and Technology (CONAHCYT), the Sector Fund for Research in Education (SEPEDUC), the Sec-tor Fund for Research in Health and Social Security (SALUD) and the Sector Fund for Energy Research (SENER-CONAHCYT).

Argentina: The Ministry of Science, Technology and Productive Innovation (MINCYT), the National Council for Scientific and Technical Research (CONICET), the National Agency for Scientific and Technological Promotion (ANPCyT) and the Fund for Scientific and Technological Research (FONCyT).

Chile: The Ministry of Science, Technology, Knowledge, and Innovation (MINCIENCIA), the National Agency for Research and Development of Chile (ANID), the National Fund for Scientific and Technological Development (FONDECYT) and the Corporation for the Promotion of Production (CORFO).

Since 2007, the University of Sao Paulo has remained the institution with the highest number of publications in LAC. Similar to the production by country, institutions such as Universidad de Sao Paulo (Brazil), Universidad de Buenos Aires, Universidad de Chile, and Universidad Nacional Autónoma de México are the ones that generate most of the publications (Fig. 2.4) and are the ones that generate the strongest contact networks (the thicker the lines, the higher the collaboration relationship; Fig. 2.5), showing once again that the generation of publications is at the local level above all, leaving international collaboration on a secondary level and little visible.

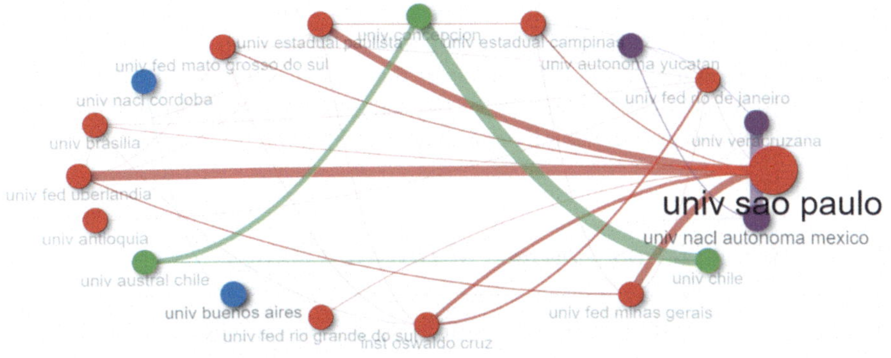

Fig. 2.5 Mapping of institutional collaboration and social networks

In our region, the establishment of networks aimed at exchanging information between institutions and countries is essential since it allows scientists from countries with scarce resources and little public investment to have more dynamic participation, considering that private-market support alone is insufficient (Lewis et al. 2021; UNESCO 2021). Undoubtedly, research in this area in Latin America and the Caribbean is increasing, but international collaborations and their reflection in the journals with the greatest impact still do not reach all the countries of the region equally (Atencio-Paulino et al. 2020). In general terms, in Latin America, international cooperation occurs mostly with the United States and some European countries (mainly Spain) over LAC countries (Lewis et al. 2021). In contrast, collaboration in Europe occurs mainly between European countries, as well as with other leading research countries such as the United States (Oregioni and López 2013). Europe has more institutional and financial support than our region, which faces greater challenges in accessing financial, human and technological resources to promote research in disease ecology and its transfer to society (Torres-Pascual et al. 2021; Sancho et al. 2006; Oregioni and López 2013). See vector graphics instead (see Fig. 2.1).

2.3.4 Average Citations Per Year

Regarding the number of citations per year, the most cited article was in 2003 (Cupolillo et al. 2003), which had an average of 6.5 citations per year (Table S2), indicating the value of this article, marking a breaking point since its publication, although the average number of citations significantly falls the following year, from 2003, it remains relatively constant at values close to 1.5 (Fig. 2.6). This article provides epidemiological information on the relationship between the pathogen (*Leishmania* (*Viannia*) *braziliensis*), hosts, and geographic regions in Brazil (the country with the largest number of publications in LAC; Fig. 2.2), serving as a primary source of information in the field of disease ecology. The other most cited articles are focused on a variety of pathogens/parasites (e.g., yellow fever, *Trypanosoma* spp.), vectors (e.g., mosquitoes, flies, ticks), and hosts (mainly mammals). Vector-borne diseases are the most prominent research topic, which is consistent with their high prevalence and impact in LAC (Apt Baruch 2013; WHO 2023), reflecting the search by researchers to cover the need in the region (PAHO 2023). Additionally, the focus of this study goes beyond traditional study, seeking to relate these issues to problems in natural environments (Table S2).

2.3.5 Most Relevant Journals in Disease Ecology

Researchers from LAC have published in a variety of journals, with 15 journals being the most common for publications in the area (Table 2.1). The journal with the most articles is *Ticks and Tick-borne Diseases*, with 28 articles published (Table 2.1).

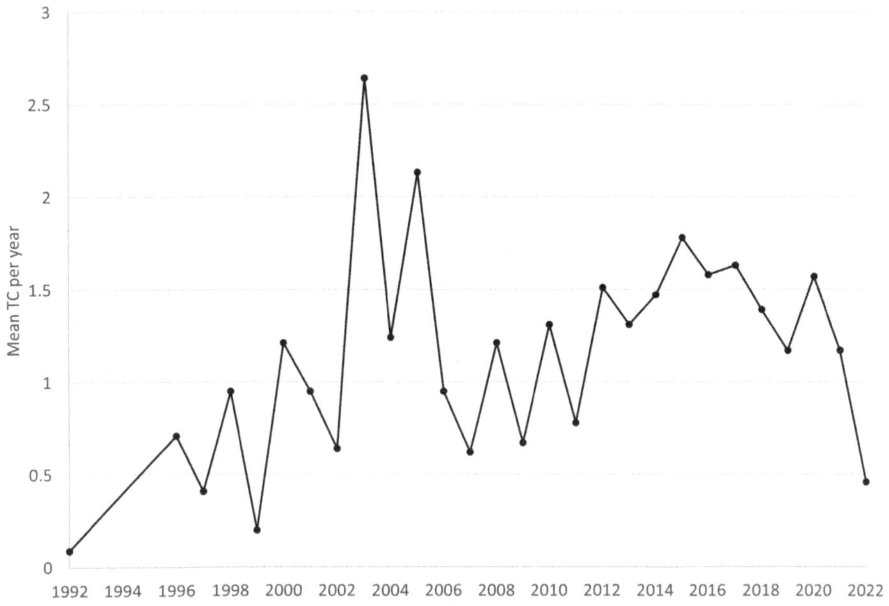

Fig. 2.6 Average citations per year

Table 2.1 Fifteen most relevant journals in disease ecology

Sources	Articles	Impact factor	Open access cost (USD)	Publisher country
Ticks and Tick-borne Diseases	28	3.82	2550	Germany
International Journal of Infectious Diseases	26	12.07	2500	Netherlands
Pesquisa Veterinaria Brasileira	25	0.66	480	Brazil
Acta Tropica	24	3.22	3040[a]	Netherlands
Revista Chilena de Infectologia	23	0.64	No cost	Chile
Journal of Medical Entomology	22	2.43	3963[a]	United States
Revista Brasileira de Parasitologia Veterinaria	22	1.41	300	Brazil
Brazilian Journal of Microbiology	20	2.47	300[a]	Brazil
Revista da Sociedade Brasileira de Medicina Tropical	20	1.58	No cost	Brazil
Revista do Instituto de Medicina Tropical de Sao Paulo	19	1.84	400[a]	Brazil
Plos Neglected Tropical Diseases	18	4.78	2495	United States
Infection, Genetics and Evolution	17	4.39	2250	Netherlands
Transboundary and Emerging Diseases	17	5.00	2500	United Kingdom
Memorias do Instituto Oswaldo Cruz	16	2.74	No cost	Brazil
Parasitology Research	16	2.38	2910[a]	Germany

[a]Hybrid Open Access journal

The journals on this list have a highly variable impact factor (IF), where the lowest was 0.636, which corresponds to the Revista Chilena De Infectologia, of LAC origin and focused on medical sciences and infectious diseases. Its counterpart is a journal with an IF of 12.07 (2021/2022), which corresponds to the *International Journal of Infectious Diseases*, a journal of European origin and, as in the previous case, focuses on medical sciences, clinical, and infectious disease reports.

The cost of publishing in Open Access is a very present issue and is, at least, controversial due to the high costs that researchers must assume, but it is also very varied, from journals such as Memorias Do Instituto Oswaldo Cruz, which provides Open Access at no cost to authors, and other journals such as *Journal of Medical Entomology* with an approximate cost of 4000 USD to publish in Open Access. This may explain why six of the journals with the highest number of publications are LAC journals (mostly Brazilian) with an impact factor that competes with high impact journals from the first world, but with much more accessible publication costs. However, it is not necessarily a matter of cost; it can also reflect a local focus of the investigations, especially in Brazil.

2.3.6 The Most Cited Pathogens, Vectors, and Hosts in Disease Ecology

Pathogens are usually a central topic in disease ecology publications; for that reason, we first analyze the occurrence of keywords coming from the author and/or generated by the journal (Fig. 2.7). In this case, we see that the most frequent words are generic, such as "infection," "diversity," "transmission," "prevalence," or "disease," relegating topics such as "ecology" or "epidemiology" to a second place and words that provide information about the agents in the third place or host, such as "virus," "vector," "acari," or "diptera."

Inquiring into the subject categories addressed by each journal, we observed that the most important subjects are "infectious diseases," followed by "parasitology"

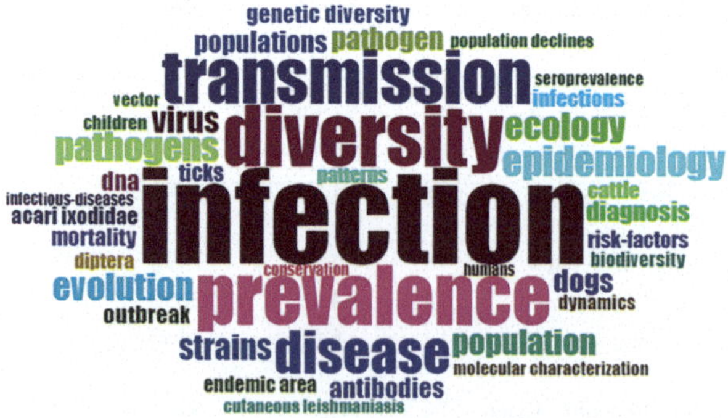

Fig. 2.7 World cloud generated from keywords of the articles

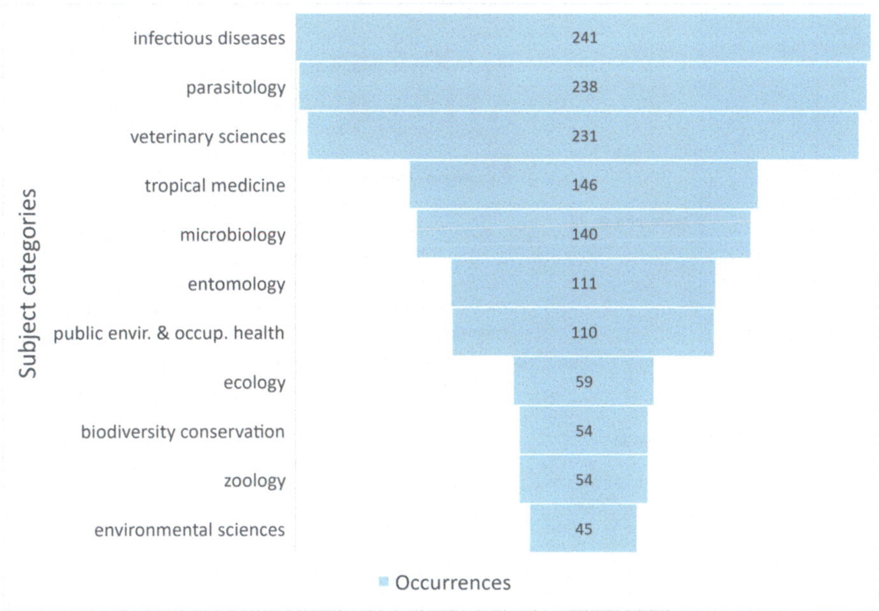

Fig. 2.8 Occurrences of subject categories in the articles

and "veterinary sciences" (Fig. 2.8), leaving "ecology" and "biodiversity conservation" in eighth and ninth place, respectively.

2.3.7 Pathogens

Exploring the content of each article, we were able to extract terms that helped us to know which are the most frequently mentioned pathogens at the genus or species level. We observed that *Rickettsia* sp., *Leishmania* sp., *Leptospira* sp., dengue virus, *Batrachochytrium dendrobatidis,* and *Trypanosoma cruzi* were the species with the highest number of mentions (Fig. 2.9). Bacterial species, viruses, fungi and protozoan parasites.

Regarding the temporal distribution of the study of parasites, some patterns can be related to the increase in cases or outbreaks of these pathogens. For example, in the genus *Rickettsia,* which is composed of species of the group of spotted fevers, there has been a significant increase in publications since 2015, reaching its maximum in 2016. This may be due to an increase in cases of spotted fevers in the United States, from 2802 in 2011 to 4470 in 2012 (CDC 2023), which may have influenced a greater interest in studies across Latin America. Another example is yellow fever studies; the largest outbreak of yellow fever in America for the last 50 years was in Brazil between 2016 and 2018 (PAHO 2022), which may reflect the increase of at least a doubling of publications on this topic in 2020.

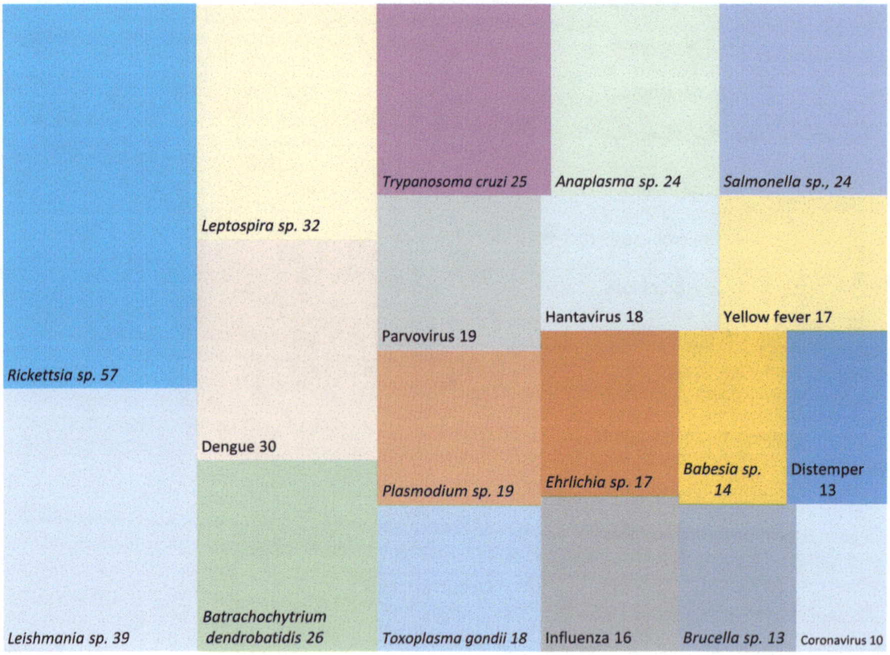

Fig. 2.9 Most cited pathogen species in disease ecology studies in LAC

When we searched for the origin of host samples in research in disease ecology, we found that domestic and wild animals were the most frequent, followed distantly by synanthropic and captive animals (Fig. 2.10). It should be noted that this classification is given by the classification of each author in their publications, classifying animals in captivity mainly only those that were in some type of protection or confinement by humans.

Working with the wild-domestic interface in disease ecology can help to understand the dynamics of disease transmission between wildlife, domestic animals, and humans, preventing and controlling outbreaks that can affect human health, the environment and wildlife conservation (Barroso et al. 2021; Phukon 2015).

The sampling and monitoring of wildlife for research imply the use of a series of technical resources for its development, for example, sampling design, transport, food, capture equipment, handling, identification, measurement, marking, anesthetics, microchips, reagents, and laboratory analysis (Santos and Gaiotto 2020; Marta et al. 2019). These resources may imply high costs for researchers and their work teams whose experimental models involve the use of wild animals, which may limit their development, prioritizing funding to other lines of research, considering the scarcity of funding in LAC (Cepal 2021).

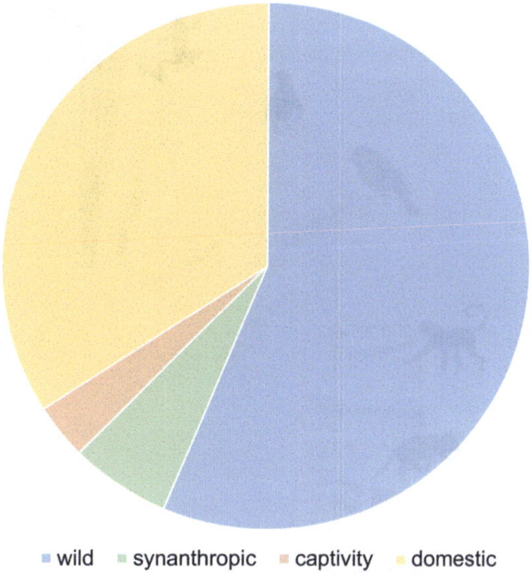

Fig. 2.10 Origin of host samples among degree of captivity/domestication in disease ecology studies in LAC

Our results show that mammals are the hosts with the highest representation and interest for research (Fig. 2.11a), where rodents occupy the second majority, including synanthropic species (Ecke et al. 2022), closely followed by carnivorous domestic animals (dogs and cats) and cattle. However, the first majority were mammals of the order Chiroptera. This interest is probably due to their proximity to some human populations and their importance as reservoirs of re-emerging zoonotic diseases, including viral, bacterial, and parasitic diseases (Baker et al. 2022; Spernovasilis et al. 2022). They are followed by orders of wild species, such as wild carnivores and nonhuman primates.

After mammals, Aves are the second most represented group as hosts due to their potential as a reservoir of zoonotic diseases, such as avian influenza, where the most common route by which the virus enters a territory is through migratory wild birds, which can infect different species of birds and occasionally humans (WOAH 2023).

Amphibians are the third most referenced class of animals as hosts, mainly due to studies on chytridiomycosis, a disease caused by chytrid fungi of the genus *Batrachochytrium*, which is linked to significant declines in the number of amphibian species in LAC and the rest of the world, having a profound negative impact on amphibian biodiversity in the region (Olson et al. 2013).

Reptiles also have a prominent place as a research focus, and reptiles are reservoirs of a wide range of pathogens, including viruses, bacteria, protozoa, helminths, pentastomids, and parasitic arthropods, representing a public health problem due to their zoonotic risk. Some of the zoonotic diseases associated with reptiles are salmonellosis, spotted fever (*Rickettsia rickettsii*), and parasitic diseases such as sparganosis or pentastomiasis (Mendoza-Roldan et al. 2020).

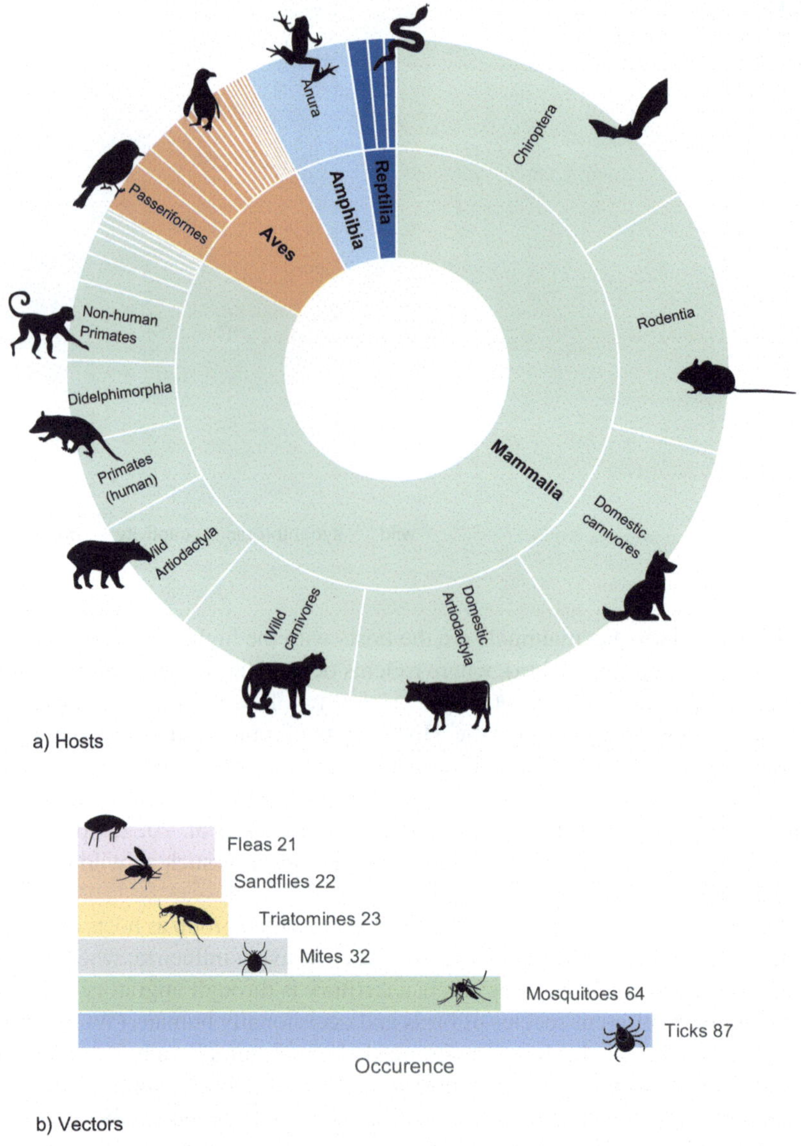

Fig. 2.11 Frequency of hosts and vectors in disease ecology studies in LAC. (**a**) host animals with the highest frequency in the articles. (**b**) Vectors with the highest number of occurrences in the articles

In vectors, the most studied groups are ticks and mosquitoes, which are responsible for the transmission of bacterial diseases such as rickettsiosis (ticks) and high-impact viral diseases in the region, such as dengue, Zika, yellow fever, or malaria (*Aedes*, *Anopheles* mosquitoes) (PAHO 2023). This was followed by mites, triatomines, sandflies, and fleas, with a lower number of occurrences but with important development by researchers (Fig. 2.11b).

Vector-borne diseases are a very present and important problem in LAC, forming part of the list of neglected tropical diseases, boosted by the scarcity of resources in the region, affecting the poorest human populations (PAHO 2023; WHO 2023). These diseases are more frequent in tropical countries, where environmental and socioeconomic conditions favor the presence and reproduction of vectors and animal reservoirs (Monsalve et al. 2009). Deforestation, biodiversity loss, and human migration also influence the distribution and risk of these diseases (Altagracia-Martínez et al. 2012; Gutiérrez and Herbosa 2008).

The prevention and control of these diseases require a multifactorial and comprehensive approach, strengthening research on the ecological, social, and economic determinants that influence the transmission of these diseases (Ault and Nicholls 2010). Therefore, as previously mentioned, the large number of works on vectors that we see in our data shows the interest on the part of researchers to propose solutions to endemic problems in our region.

Our results may have revealed the limitations of the PRISMA method since there are authors/articles of our knowledge that did not appear during the search in the databases (WoS and PubMed). These results may also reflect the need to include better word combinations in the publications (especially in the titles and abstracts), which reflect that they are part of disease ecology topics, achieving better exposure in the databases, to stand out from other areas that also include "ecology," "diseases," and "conservation" in their keywords.

2.4 Conclusions and Recommendations

We conducted a bibliometric analysis to identify how the field of disease ecology has been developed in LAC, recognizing major fields, hosts, and infectious diseases studied, as well as identifying important collaborations and networks among countries and institutions that have significantly contributed to this discipline.

It is evident that Latin America and the Caribbean have the appropriate conditions and knowledge for the development of local and regional research in disease ecology and related topics, favoring research on needs exclusive to the Neotropics. This will promote knowledge on emerging or neglected infectious agents of relevance for the Neotropical region. In this way, environmental, ecological, and social conditions that are unique and have a particular influence on the cycles of these infectious agents favoring the emergence and maintenance of pathogens will be considered.

Despite the evident scientific competence, research in Latin America and the Caribbean seeking to respond to local needs requires support, financing, and local policies that encourage research. Increasing capacities favor long-term research, regional research, interdisciplinary research, and intersectoral research. This ensures the creation and implementation of tools for the control, prevention, and monitoring of emerging and re-emerging neotropical diseases, and in turn, this research generates a change in communities and public policies.

Training and interchanges between countries with a recognized trajectory in disease ecology and/or related branches (such as Brazil, Argentina, and Mexico) with those countries with less research in these areas can facilitate the development of knowledge by reducing costs for the improvement in capacities, particularly the training of specialized professionals, while favoring the creation of work networks and the development of regional knowledge.

Although there is still much to learn, the socioeconomic conditions as well as the rich neotropical biodiversity make Latin America and the Caribbean hotspots for further research to answer ecological, epidemiological, and biological questions that contribute to scientific knowledge of worldwide relevance.

It is a great opportunity to develop collaborative strategies and reinforce aspects of common objectives. The collaboration would facilitate the identification of competent reservoirs and vectors in LAC, standardize methods, make contact networks and be more efficient in diagnosis and prevention and propose incidences in public policies to work with wildlife. In the same way, it allows us to prioritize pathogens since many are understudied and cause unattended diseases, so we understand how the infection and disease emergency systems work and not just limit ourselves to identifying pathogens.

References

Allen T, Murray KA, Zambrana-Torrelio C, Morse SS, Rondinini C, Di Marco M et al (2017) Global hotspots and correlates of emerging zoonotic diseases. Nat Commun 8(1):1124. https://doi.org/10.1038/s41467-017-00923-8

Altagracia-Martínez M, Kravzov-Jinich J, Moreno-Bonett C, López-Naranjo F, Martínez-Núñez JM (2012) Las enfermedades "olvidadas" de América Latina y el Caribe: un problema de salud pública global. Revista mexicana de ciencias farmacéuticas 43(1):33–41

Apt Baruch W (2013) Parasitología humana. McGraw-Hill Interamericana Editores, México

Aria M, Cuccurullo C (2017) Bibliometrix: an R-tool for comprehensive science mapping analysis. J Informet 11(4):959–975. https://doi.org/10.1016/j.joi.2017.08.007

Atencio-Paulino J, Condor-Elizarbe I, Paucar-Huaman W (2020) Desarrollo de la colaboración en publicaciones sobre salud pública en Latinoamérica y el Caribe. Gac Sanit 34:214–215

Ault SK, Nicholls RS (2010) El abordaje integral de las enfermedades tropicales desatendidas en América Latina y el Caribe: un imperativo ético para alcanzar la justicia y la equidad social. Biomedica 30(2):159–163

Baker RE, Mahmud AS, Miller IF, Rajeev M, Rasambainarivo F, Rice BL et al (2022) Infectious disease in an era of global change. Nat Rev Microbiol 20(4):193–205. https://doi.org/10.1038/s41579-021-00639-z

Barroso P, Acevedo P, Vicente J (2021) The importance of long-term studies on wildlife diseases and their interfaces with humans and domestic animals: a review. Transbound Emerg Dis 68(4):1895–1909. https://doi.org/10.1111/tbed.13916

Bozada T Jr, Borden J, Workman J, Del Cid M, Malinowski J, Luechtefeld T (2021) Sysrev: a FAIR platform for data curation and systematic evidence review. Front Artif Intell 4:685298. https://doi.org/10.3389/frai.2021.685298

Brandell EE, Becker DJ, Sampson L, Forbes KM (2021) Demography, education, and research trends in the interdisciplinary field of disease ecology. Ecol Evol 11(24):17581–17592. https://doi.org/10.1002/ece3.8466

CDC (2023) About the division of vector-borne diseases. Retrieved from https://www.cdc.gov/ncezid/dvbd/about.html

Cepal NU (2021) Financiamiento para el desarrollo en la era de la pandemia de COVID-19 y después

Ciocca DR, Delgado G (2017) The reality of scientific research in Latin America; an insider's perspective. Cell Stress Chaperones 22(6):847–852. https://doi.org/10.1007/s12192-017-0815-8

Cupolillo E, Brahim LR, Toaldo CB, de Oliveira-Neto MP, de Brito MEF, Falqueto A et al (2003) Genetic polymorphism and molecular epidemiology of Leishmania (Viannia) braziliensis from different hosts and geographic areas in Brazil. J Clin Microbiol 41(7):3126–3132

De León ME, Zumbado-Ulate H, García-Rodríguez A, Alvarado G, Sulaeman H, Bolaños F, Vredenburg VT (2019) Batrachochytrium dendrobatidis infection in amphibians predates first known epizootic in Costa Rica. PLoS One 14(12):e0208969. https://doi.org/10.1371/journal.pone.0208969

Ecke F, Han BA, Hörnfeldt B, Khalil H, Magnusson M, Singh NJ, Ostfeld RS (2022) Population fluctuations and synanthropy explain transmission risk in rodent-borne zoonoses. Nat Commun 13(1):7532. https://doi.org/10.1038/s41467-022-35273-7

Escobar LE, Carver S, Cross PC, Rossi L, Almberg ES, Yabsley MJ et al (2022) Sarcoptic mange: an emerging panzootic in wildlife. Transbound Emerg Dis 69(3):927–942. https://doi.org/10.1111/tbed.14082

García-Peña GE, Rubio AV, Mendoza H, Fernández M, Milholland MT, Aguirre AA et al (2021) Land-use change and rodent-borne diseases: hazards on the shared socioeconomic pathways. Philos Trans R Soc B Biol Sci 376(1837):20200362. https://doi.org/10.1098/rstb.2020.0362

Grames EM, Stillman AN, Tingley MW, Elphick CS (2019) An automated approach to identifying search terms for systematic reviews using keyword co-occurrence networks. Methods Ecol Evol 10(10):1645–1654. https://doi.org/10.1111/2041-210X.13268

Gutiérrez MJT, Herbosa RO (2008) Enfermedades tropicales transmitidas por vectores: medidas preventivas y profilaxis. Offarm: farmacia y sociedad 27(6):78–87

Koprivnikar J, Johnson PTJ (2016) The rise of disease ecology and its implications for parasitology – a review. J Parasitol 102(4):397–409. https://doi.org/10.1645/15-942

Lewis J, Schneegans S, Straza T (2021) UNESCO science report: the race against time for smarter development. UNESCO Publishing

Marta S, Lacasella F, Romano A, Ficetola GF (2019) Cost-effective spatial sampling designs for field surveys of species distribution. Biodivers Conserv 28(11):2891–2908. https://doi.org/10.1007/s10531-019-01803-x

Martinez-Gutierrez M, Ruiz-Saenz J (2016) Diversity of susceptible hosts in canine distemper virus infection: a systematic review and data synthesis. BMC Vet Res 12:78. https://doi.org/10.1186/s12917-016-0702-z

Mendoza-Roldan JA, Modry D, Otranto D (2020) Zoonotic parasites of reptiles: a crawling threat. Trends Parasitol 36(8):677–687. https://doi.org/10.1016/j.pt.2020.04.014

Monsalve S, Mattar S, Gonzalez M (2009) Zoonosis transmitidas por animales silvestres y su impacto en las enfermedades emergentes y reemergentes. Revista MVZ Córdoba 14(2):1762–1773

Olson DH, Aanensen DM, Ronnenberg KL, Powell CI, Walker SF, Bielby J et al (2013) Mapping the global emergence of Batrachochytrium dendrobatidis, the amphibian chytrid fungus. PLoS One 8(2):e56802. https://doi.org/10.1371/journal.pone.0056802

Oregioni MS, López MP (2013) Cooperación Internacional en Ciencia y Tecnología: La voz de los investigadores. Revista iberoamericana de ciencia tecnología y sociedad 8(22):57–73

Page MJ, McKenzie JE, Bossuyt PM, Boutron I, Hoffmann TC, Mulrow CD et al (2021) The PRISMA 2020 statement: an updated guideline for reporting systematic reviews. BMJ 372:n71. https://doi.org/10.1136/bmj.n71

PAHO (2022) Yellow fever, the returning epidemic. Retrieved from https://www.paho.org/en/stories/yellow-fever-returning-epidemic
PAHO (2023, 2023/03/03/) Vectors: integrated management and public health entomology. Retrieved from https://www.paho.org/en/topics/vectors-integrated-management-and-public-health-entomology
Phukon D (2015) Wildlife – livestock interface: role of disease and its conservation implication
Sancho R, Morillo F, De Filippo D, Gómez I, Fernández MT (2006) Indicadores de colaboración científica intercentros en los países de América Latina. Interciencia 31(4):284–292
Santos AS, Gaiotto FA (2020) Knowledge status and sampling strategies to maximize cost–benefit ratio of studies in landscape genomics of wild plants. Sci Rep 10(1):3706. https://doi.org/10.1038/s41598-020-60788-8
Spernovasilis N, Tsiodras S, Poulakou G (2022) Emerging and re-emerging infectious diseases: Humankind's companions and competitors. In: Microorganisms, vol 10, Switzerland
Torres-Pascual C, Sánchez-Pérez HJ, Ávila-Castells P (2021) Geographical distribution and international collaboration of Latin American and Caribbean scientific publications on tuberculosis in Pubmed. Rev Peru Med Exp Salud Publica 38(1):49–57. https://doi.org/10.17843/rpmesp.2021.381.5726
Unep. (2016) Protected planet report 2016: how protected areas contribute to achieving global targets for biodiversity. In: World conservation monitoring. Centre Cambridge
UNESCO (2021) Latin America. Retrieved from https://www.unesco.org/reports/science/2021/en/latin-america
WHO (2023) Vector-borne diseases. Retrieved from https://www.who.int/news-room/fact-sheets/detail/vector-borne-diseases
WOAH (2023) Avian Influenza. Retrieved from https://www.woah.org/en/disease/avian-influenza/

Chapter 3
Emerging Infectious Diseases and Their Impacts on South American Amphibians

Claudio Azat and Mario Alvarado-Rybak

3.1 Diversity and Conservation of South American Amphibians

With over one-third of amphibian species, South America is the greatest amphibian-rich region in the world (Jenkins et al. 2013). It is home to 2972 anurans, 42 caudates, and 103 caecilians (Frost 2023). Due to the high biodiversity in the Amazonian basin and the Atlantic Forest, only Brazil has 1252 spp. (Azevedo-Ramos and Galatii 2002; da Silva et al. 2014; Frost 2023). Other megadiverse amphibian countries in the region include Colombia (902 spp.), Peru (691), and Ecuador (684). In contrast, countries such as Chile, although having a smaller diversity (61), have high endemicity (72%, Soto-Azat et al. 2015). At the global level, amphibians are undergoing an extinction crisis with no precedent (Collins and Crump 2009). With 40.7% of species threatened with extinction, amphibians are the most threatened vertebrate taxa on Earth (IUCN 2023). Moreover, if data-deficient species are included, an estimated 50% of the world's amphibian species are threatened with extinction (González-Del-Pliego et al. 2019). Latin America is one of the regions showing the greatest loss of amphibian biodiversity, with most declines occurring in high-elevation aquatic environments (Young et al. 2001; Lips et al. 2006). Of the assessed South American species in the IUCN Red List, 34.2% are in a threatened category among extinct (2 spp.), critically endangered (247), endangered (400), and vulnerable (252). In addition, 20.1% of species (530) are considered data deficient (IUCN

C. Azat (✉)
Sustainability Research Center & PhD Program in Conservation Medicine, Life Sciences Faculty, Universidad Andres Bello, Santiago, Chile
e-mail: claudio.azat@unab.cl

M. Alvarado-Rybak
Núcleo de Ciencias Aplicadas en Ciencias Veterinarias y Agronómicas, Facultad de Medicina Veterinaria y Agronomía, Universidad de las Américas, Santiago, Chile

G. Acosta-Jamett, A. Chaves (eds.), *Ecology of Wildlife Diseases in the Neotropics*, https://doi.org/10.1007/978-3-031-50531-7_3

2023). Among the main threats identified for amphibians in South America are agriculture, deforestation, urbanization, and invasive species (IUCN 2023). Other emerging threats include climate change and infectious diseases (Foden et al. 2013; Uribe-Rivera et al. 2017). Low connectivity among populations and low abundance are major intrinsic conservation problems associated with amphibian declines and local extinctions in the southern cone of South America (Kakoliris et al. 2022). Increasing in situ conservation in South America is a major goal to protect habitats, but captive-rearing programs for species subject to imminent extinction have also been encouraged (Young et al. 2001; Harding et al. 2006). Habitat management to restore habitats, recover populations, and eliminate threats such as invasive species is urgently needed to reverse the loss of amphibians in this neotropical region (Kakoliris et al. 2022).

3.2 Emerging Infectious Diseases and Population Declines in Amphibians

3.2.1 Amphibian Chytridiomycosis

Caused by the non-hyphal zoosporic chytrid fungus *Batrachochytrium dendrobatidis* (*Bd*), this is a highly pathogenic, virulent, and readily transmissible and panzootic fungus with low specificity among amphibians (Berger et al. 1998; Longcore et al. 1999; Gower et al. 2013). Five distinct lineages have been recognized so far: *Bd*GPL, *Bd*CAPE, *Bd*ASIA-1, *Bd*ASIA-2/*Bd*BRAZIL, and *Bd*-Asia 3 (O'Hanlon et al. 2018; Byrne et al. 2019; Azat 2021). A second species, *B. salamandrivorans* (*Bsal*), appears to be restricted to Asia, where it is endemic, and Europe, where it has been recently introduced (10 years) and has high impacts on local salamanders (Martel et al. 2014). The life cycle of *Bd* consists of aquatic, motile, uniflagellated zoospores that invade keratinized tissues of amphibians (outer layers of the skin in adults and mouthparts in tadpoles), where they develop into a stationary, intracellular zoosporangium (Fig. 3.1). Each zoosporangium produces large numbers of zoospores (4–150), which are subsequently released via discharge tubes into the environment (Longcore et al. 1999). At 22 °C in vitro, the life cycle is completed in 5–7 days (Berger et al. 2005). In susceptible individuals, *Bd* can colonize large areas of the skin of adult amphibians, causing damage to epidermal cells, electrolyte depletion, and osmotic imbalance, leading to death (Voyles et al. 2009). Surviving infected tadpoles can act as reservoirs of infection (Berger et al. 1998; Schloegel et al. 2006). The pathogen *Bd* can be viable for up to 12 weeks in sterile water (Johnson and Speare 2003) and survive freezing conditions, possibly only when infecting its amphibian host (Ouellet et al. 2005). Such persistence in the environment and via aclinical infections lowers the threshold host density for the disease and allows it to cause the observed catastrophic declines and local and global extinctions (Daszak et al. 1999). The persistence of *Bd* in the environment and its

Fig. 3.1 Life cycle of *Batrachochytrium dendrobatidis*. At 22 °C, the life cycle in vitro takes 4 or 5 days to complete

subsequent impacts on amphibian populations can be attributed to the introduction of the pathogen into naïve host populations, the existence of a free-living infective stage, and the presence of reservoirs (James et al. 2015; Berger et al. 2016; Soto-Azat et al. 2016).

3.2.2 Amphibian Ranaviruses

Ranaviruses are large, icosahedral, double-stranded DNA viruses that belong to the family Iridoviridae (Williams et al. 2005). Ranaviruses can infect and cause disease in amphibians, fish, and reptiles (Hyatt et al. 2000; Johnson et al. 2008). Since the first isolation of a *Ranavirus* from the northern leopard frog (*Lithobates pipiens*) in 1965 (Granoff et al. 1965), an increasing number of infections caused by ranaviruses have been described in ectothermic vertebrates (Hyatt et al. 2000; Johnson et al. 2008). Many *Ranavirus* species infecting amphibians around the world have been identified, including *Bohle iridovirus* (Cullen and Owens 2002), *Ambystoma tigrinum virus* (Bollinger et al. 1999; Jancovich et al. 2003), *Rana esculenta virus* (Fijan et al. 1991; Stöhr et al. 2013), *Rana grylio virus* (Zhang et al. 2001), *tiger frog virus* (He et al. 2002), *common midwife toad virus* (Mavian et al. 2012), and *Chinese giant salamander virus* (Zhou et al. 2013). In contrast to the relatively limited geographic distribution of these ranaviruses, *frog virus 3* (FV3) and several FV3-like viruses are globally widespread (Lesbarrères et al. 2011). The ecology of ranaviruses likely involves a complex interaction of reservoir species, transmission routes, environmental persistence, stressors, and host immunity (Gray et al. 2009). The pathogen infects multiple amphibian hosts, including tadpoles and adults, and may persist in aquatic and terrestrial environments through amphibian, fish, and reptile reservoirs (Hyatt et al. 2000; Lesbarrères et al. 2011). Tadpoles seem to be more susceptible to disease than adults (Green et al. 2002). *Ranavirus* virions can have long environmental persistence; for instance, outside a host, they may be viable for up to 182 days in unsterile pond water maintained at 4 °C (Nazir et al. 2012). Transmission occurs by indirect and direct routes and includes exposure to contaminated water or moist soil, direct contact with infected individuals, and ingestion of infected tissue during predation, cannibalism, or necrophagy (Jancovich et al. 1997; Brunner et al. 2007). The occurrence of recent *Ranavirus*-associated amphibian population die-offs may be an interaction of suppressed and naïve host immunity, anthropogenic stressors, and novel strain introductions (Gray et al. 2009).

3.2.3 Ecoimmunology of **Bd** *and* **Ranavirus**

Infected amphibian species vary in their susceptibility to *Bd* and ranaviruses from asymptomatic carriers to lethal disease (Hyatt et al. 2000; Fisher et al. 2009; Gray et al. 2009; Hoverman et al. 2010). Previous exposure to *Bd* or *Ranavirus* may determine the outcome of subsequent infections, possibly due to the development of an effective pathogen-specific acquired immune response (Carey et al. 1999; Haislip et al. 2011; Rollins-Smith et al. 2011). Frogs with previous *Bd* exposure can survive secondary infections better than can immunologically naïve frogs (Richmond et al. 2009), and amphibian species inhabiting permanent wetlands with larvae that have long development times are less likely to experience morbidity from ranaviruses due to acquired immunity from repeated exposure (Gray et al. 2009). Immune function in amphibians is closely dependent on environmental temperature (Murphy et al. 2011). Low temperature has been associated with lower survival in *Bd*-exposed frogs under laboratory conditions (Andre et al. 2008; Bustamante et al. 2010; Murphy et al. 2011), and chytridiomycosis die-offs have often been associated with high elevation, low temperature, and the winter season (Berger et al. 2004; McDonald et al. 2005; Drew et al. 2006; Kriger et al. 2007; Savage et al. 2011). In addition, a higher prevalence of *Ranavirus* infection in some amphibian populations in the United States has been associated with cold temperatures (Rojas et al. 2005; Gray et al. 2007), and a higher mortality has been reported in experimentally *Ranavirus*-infected red-eared sliders (*Trachemys scripta elegans*) maintained at lower temperatures (Allender et al. 2013). Anthropogenic stressors such as wetland degradation due to cattle farming have been identified as a contributing factor in increased amphibian mortality caused by ranaviruses (Jancovich et al. 1997; Gray et al. 2007). Additionally, analyses at the landscape level have found higher *Bd* occurrence near urban centers and human-modified environments (Bacigalupe et al. 2017; Alvarado-Rybak et al. 2021b).

3.2.4 *Emerging Infectious Diseases and Amphibian Population Declines*

The chytrid fungus *Bd* has been increasingly recognized as an important driver behind alarming amphibian population declines and extinctions across the world (Berger et al. 1998; Lips 2016; Bielby et al. 2008). In fact, *Bd* is recognized as the single pathogen responsible for the greatest loss of biodiversity known to science (Scheele et al. 2019), and its role in the decline and extinction of numerous amphibian species has led to a paradigm shift toward wildlife diseases as a conservation issue (Berger et al. 2016). Thus far, *Bd* has been implicated in the decline of >500 amphibian species and the extinction of 100 (Scheele et al. 2019). While many species are declining, some to the point of extinction, others persist with the enzootic presence of *Bd* (Bielby et al. 2008; Berger et al. 2016). The relationship between *Bd* and amphibians is complex, with wide variability in population declines exhibited by different amphibian host species in response to *Bd*, for which a phylogenetic signal has been demonstrated (i.e., phylogenetically related amphibians show similar patterns of chytridiomycosis susceptibility; Bielby et al. 2008; Baláž et al. 2014; Bacigalupe et al. 2017). Less studied, lethal outbreaks caused by *Ranavirus* have been reported in many parts of the world in both wild and captive amphibians (Cunningham et al. 1996; Jancovich et al. 1997; Green et al. 2002; Greer et al. 2005; Fox et al. 2006; Muths et al. 2006; Une et al. 2009; Balseiro et al. 2010; Geng et al. 2011; Kik et al. 2011; Stöhr et al. 2013). Additionally, long-term amphibian population declines have been confirmed in the common frog (*Rana temporaria*) in the United Kingdom (Teacher et al. 2010), and severe multispecies host declines due to CMTV have been described in Spain (Price et al. 2014).

3.3 Bd in South America

3.3.1 *Distribution*

The fungus *Bd* is widespread in South America, with infection being reported in 522 amphibian species (including anurans, salamanders, and caecilians) from all countries except Paraguay (James et al. 2015; Azat et al. 2022). Based on 21,648 individual PCR analyses, Bd prevalence in South America has been estimated to be 23.2%; however, traditional studies do not account for individual pseudoreplication and imperfect pathogen detection, leading to an underestimation of infection prevalence (Sentenac et al. 2023). Taxa exhibiting higher *Bd* prevalence were in most cases associated with aquatic environments, including Telmatobiidae, Ranidae, Calyptocephalellidae, Hylodidae, and Pipidae (Azat et al. 2022). Areas with the greatest occurrence of *Bd* are associated with environments at high elevations in the Andes (Central high Andes, Puna, and Altiplano) and Brazil's Atlantic Forest (Catenazzi et al. 2011; James et al. 2015; Burrowes and De la Riva 2017; Carvalho et al. 2017; Alvarado-Rybak et al. 2021b; Azat et al. 2022). Two invasive amphibian species have been described to act as reservoirs for both *Bd* and *Ranavirus* in South America (Soto-Azat et al. 2016; Ruggeri et al. 2019). The North American bullfrog (*Lithobates catesbeianus*), with known established wild populations in Venezuela, Colombia, Ecuador, Peru, Brazil, Uruguay, and Argentina (IUCN 2023), and *X. laevis are* widespread in Chile (Mora et al. 2019). Both species have shown evidence of high prevalence of both pathogens, while not undergoing the impacts of disease (Robert et al. 2007; Schloegel et al. 2010b; Peñafiel-Ricaurte et al. 2023). Furthermore, *L. catesbeianus* is intensively farmed in Brazil, Uruguay, and Argentina (Garner et al. 2006; Schloegel et al. 2010b), and exports of wild *X. laevis* from Chile occur (Azat, unpublished data). These species can act as pathogen spreaders locally as they invade new areas (Soto-Azat et al. 2016; Borzée et al. 2017) or at the intercontinental level associated with international trade (Garner et al. 2006; Schloegel et al. 2012; Carvalho et al. 2017; Valenzuela-Sánchez et al. 2018). A map showing the widespread occurrence of *Bd* and surveyed areas in South America is shown in Fig. 3.2.

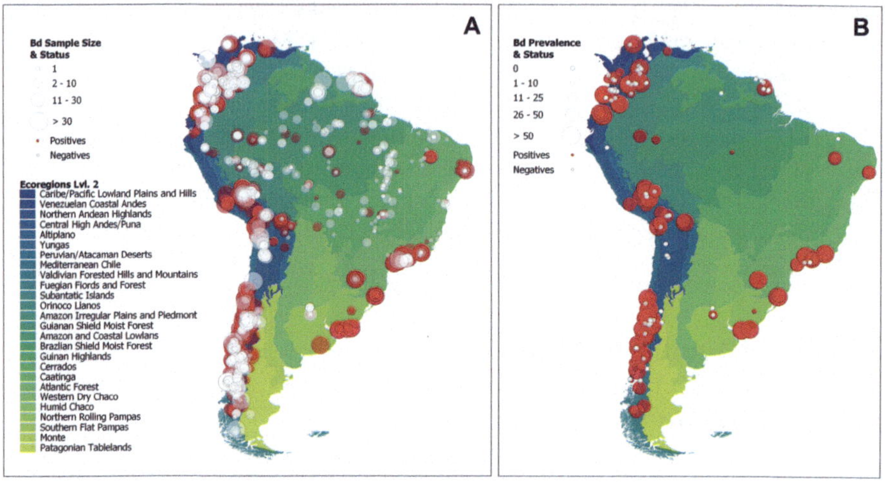

Fig. 3.2 Distribution of *Batrachochytrium dendrobatidis* (*Bd*) in South America. (**a**) Presence and absence of *Bd* per surveyed site. (**b**) Prevalence of *Bd* (considering a minimum sample size of ≥10 individuals per site). The size of the circles is proportional to the number of individuals sampled. Red and white circles represent *Bd*-positive and Bd-negative sites, respectively. The presented data are based on 21,648 *Bd*-specific PCR analyses

3.3.2 Genetic Diversity

Isolates of *Bd* have been genetically characterized in South America, with the hypervirulent Global Panzootic Lineage (*Bd*GPL) being the most widespread genotype (Schloegel et al. 2012; Flechas et al. 2013; Rosenblum et al. 2013; James et al. 2015; Jenkinson et al. 2016; Valenzuela-Sánchez et al. 2018; Burrowes et al. 2020). Based on whole-genome sequencing, O'Hanlon et al. (2018) revealed a recent intercontinental expansion of *Bd*GPL from East Asia within the past century. The recent introduction of *Bd*GPL into South America (Lips et al. 2008; Valenzuela-Sánchez et al. 2018) coincides with the onset of amphibian enigmatic declines in this region beginning in the late 1970s (Lips et al. 2008; Soto-Azat et al. 2013a; Carvalho et al. 2017; Valenzuela-Sánchez et al. 2018). Additionally, the endemic genotype *Bd*ASIA-2/*Bd*BRAZIL has been described to be restricted to the Atlantic Forest of Brazil (Rosenblum et al. 2013; Rodriguez et al. 2014). Of concern, *Bd*ASIA-2/*Bd*BRAZIL has been shown to hybridize with *Bd*GPL-originating hypervirulent hybrid genotypes (Jenkinson et al. 2016; Greenspan et al. 2018; O'Hanlon et al. 2018).

3.3.3 Associated Population Declines

South America is the region suffering the greatest impacts from chytridiomycosis (Bielby et al. 2008; Lips et al. 2008; Rödder et al. 2009; Scheele et al. 2019). Several examples exist for this emerging disease and severe population declines and extinctions in the region, notably the cases of Harlequin toads (*Atelopus* spp.; Ron et al. 2003; La Marca et al. 2005; Lampo et al. 2006), Andean water frogs (*Telmatobius* spp.; Barrionuevo and Ponssa 2008; Burrowes and De la Riva 2017) and Darwin's frogs (*Rhinoderma* spp.; Soto-Azat et al. 2013b; Valenzuela-Sánchez et al. 2017, 2022). Recently, Alvarado-Rybak et al. (2021a) described a lethal chytridiomycosis outbreak in captive Chilean giant frogs (*Calyptocephalella gayi*). This is of concern, as this species is vulnerable and has declined rapidly in recent decades. The pathogen *Bd* has also been described in *Telmatobius* (63 spp.), but only a few studies have linked chytridiomycosis with population-level impacts (Burrowes and De la Riva 2017). Chytridiomycosis due to *Bd* infection has been associated with mortality in *T. pisanoi* and *T. atacamensis* from northern Argentina (Barrionuevo and Mangione 2006), *T. marmoratus* from Peru (Seimon et al. 2007; Catenazzi et al. 2011), and the disappearance of two species of *Telmatobius* from the Upper Manu National Park in southeastern Peru (Catenazzi et al. 2011). In addition, Burrowes and De la Riva (2017) found an association between *Bd* infection and severe population declines since the 1990s, particularly in *Telmatobius* spp. Although *Bd* occurrence is generally low in terrestrial amphibians, high *Bd* prevalence has been detected in Hylidae and Hemiphractidae (Azat et al. 2022), with chytridiomycosis being implicated in population declines of the horned marsupial frog (*Gastrotheca cornuta*) and of three species of *Hyloscirtus* spp. treefrogs (Lips et al. 2006). Terrestrial Darwin's frogs (*Rhinoderma*, two spp.) and harlequin frogs (*Atelopus*, 99 spp.) have been severely impacted by chytridiomycosis (Lampo et al. 2017; Valenzuela-Sánchez et al. 2017). *Rhinoderma darwinii* has direct development in a situation that prevents these species from frequent contact with aquatic habitats and, hence, potentially encounters high amounts of *Bd* (Azat et al. 2021). In contrast, *Atelopus* has an aquatic breeding style associated with small streams in tropical wet forests and humid paramos in Central and South America (La Marca et al. 2005). With many species disappearing and others having suffered drastic declines across their whole distribution (from Costa Rica to Bolivia), the situation of the genus *Atelopus* is possibly the most dramatic example of biodiversity loss due to an infectious disease (La Marca et al. 2005; Lips et al. 2008; Catenazzi et al. 2011; Scheele et al. 2019; Valencia and Fonte 2021). Of 94 *Atelopus* spp. listed in the IUCN Red List, 3% are currently categorized as extinct, 69% as critically endangered, and 13% as endangered (IUCN 2023). Population crashes of *Atelopus* spp. were first recognized in the 1980s in South America, coincident with the proposed arrival of *Bd* (Lips et al. 2008). Severe chytridiomycosis-associated declines have been described in *A. chiriquensis*, *A. varius*, and *A. zeteki* in Panama and Costa Rica (La Marca et al. 2005; Lips et al. 2006), and the combined effects of chytridiomycosis and climate change have been implicated in the disappearance of *A. ignescens* from Ecuador (Ron et al. 2003) and *A. carbonerensis*, *A. mucubajiensis*, and *A. sorianoi* from Venezuela (Lampo et al. 2006). Believed extinct for 30 years, surviving populations of *A. ignescens*, *A. longirostris,* and *A. mindoensis* have been recently rediscovered, and individuals of the first two species translocated to a captive breeding program in Ecuador (Tapia et al. 2017; Barrio Amorós et al. 2020). Native to southern Chile and Argentina, both species of *Rhinoderma* have undergone severe population declines (Soto-Azat et al. 2013a). Once abundant, *R. rufum* is thought to have become extinct in approximately 1982, while *R. darwinii* has recently disappeared from much of its range (Soto-Azat et al. 2013a; Azat et al. 2021). In both cases, chytridiomycosis due to *Bd* infection has been proposed as a driver of these population declines (Soto-Azat et al. 2013b; Valenzuela-Sánchez et al. 2017, 2022). In addition, studies of wild populations have found significantly reduced survival probabilities of *Bd*-infected vs. *Bd*-uninfected frogs for *A. cruciger* in Venezuela (Lampo et al. 2017) and *R. darwinii* in Chile (Valenzuela-Sánchez et al. 2017). Lampo et al. (2017) estimated that infected *A. cruciger* were, on average, four times more likely to die during the study period (6 years) than uninfected individuals. Similarly, Valenzuela-Sánchez et al. (2017) used empirical evidence to model slow declines in the extinction of infected *R. darwinii* populations in the absence of mass mortality and despite low *Bd* prevalence.

3.4 *Ranavirus* in South America

3.4.1 Distribution

While most reports on *Ranavirus*-associated outbreaks and mass mortalities come from North America and Europe (Gray et al. 2009; Duffus and Cunningham 2010; Price et al. 2014), ranaviruses remain largely understudied in South America (Brunner et al. 2021). The first report of *Ranavirus* in the region was made by Zupanovic et al. (1998a) in cane toads (*Rhinella marina*) in Venezuela by detection of antibodies from serum samples and subsequent *Ranavirus* isolation from *L. catesbeianus* and *Leptodactylus* sp. (Zupanovic et al. 1998b). Later, Fox et al. (2006) described a *Ranavirus* outbreak causing high mortality in the Patagonia frog (*Atelognathus patagonicus*) in Argentina, with the virus being characterized as FV3 with 100% identity. Co-infection by *Bd* and *Ranavirus* has been detected in Peruvian Andes frogs, including live-traded *Telmatobius marmoratus* sold at a market in Cusco, wild *Hipsiboas gladiator*, and several species of *Pristimantis* and *Rhinella manu* from Kosñipata valley (Warne et al. 2016). *Ranavirus* infection was first described in Chile in wild anurans from two populations near the capital city of Santiago, involving seven individuals of the invasive *X. laevis* and one individual of the native *C. gayi* living in sympatry (Soto-Azat et al. 2016). Additional work has confirmed that *Ranavirus* is widespread in central Chile but found at very low prevalence in *X. laevis* (Peñafiel-Ricaurte et al. 2023), and genetic analyses have revealed 100% similarity with FV3 (Soto-Azat et al. 2016; Peñafiel-Ricaurte et al. 2023). Although intensively researched, *Ranavirus* appears to be intimately associated with *X. laevis* invasive distribution in Chile, and no evidence of infection has been found in sympatric fish or additional native amphibians (Peñafiel-Ricaurte et al. 2023). Infection of free-living tadpoles of both native species and invasive *L. catesbeianus* and a case of mass mortality involving wild *L. catesbeianus* and an unidentified fish species attributed to *Ranavirus* have been reported in southern Brazil (Ruggeri et al. 2019). In the highland Andes of Ecuador, *Ranavirus* has been detected at one site in *Pristimantis orestes* and *P. phoxocephalus* (Urgiles et al. 2021). As detected in eight localities and several species, including native *Osornophryne* sp., *Pristimantis bogotensis*, *P. elegans*, *Leptodactylus fragilis*, *L. validus*, and invasive *L. catesbeianus*, *Ranavirus* appears to be widespread across diverse habitats in Colombia (Flechas et al. 2023). Additionally, *Ranavirus* causing disease and mortality in tadpoles of *L. catesbeianus* has been confirmed in frog farms in Brazil and Uruguay (Galli et al. 2006; Mazzoni et al. 2009; Candido et al. 2019), with viruses showing high similarity with FV3 (Mazzoni et al. 2009) but also signs of recombination with other ranaviruses possibly associated with international trade (Candido et al. 2019). Details of the geographic distribution of *Ranavirus* in South America are shown in Fig. 3.3.

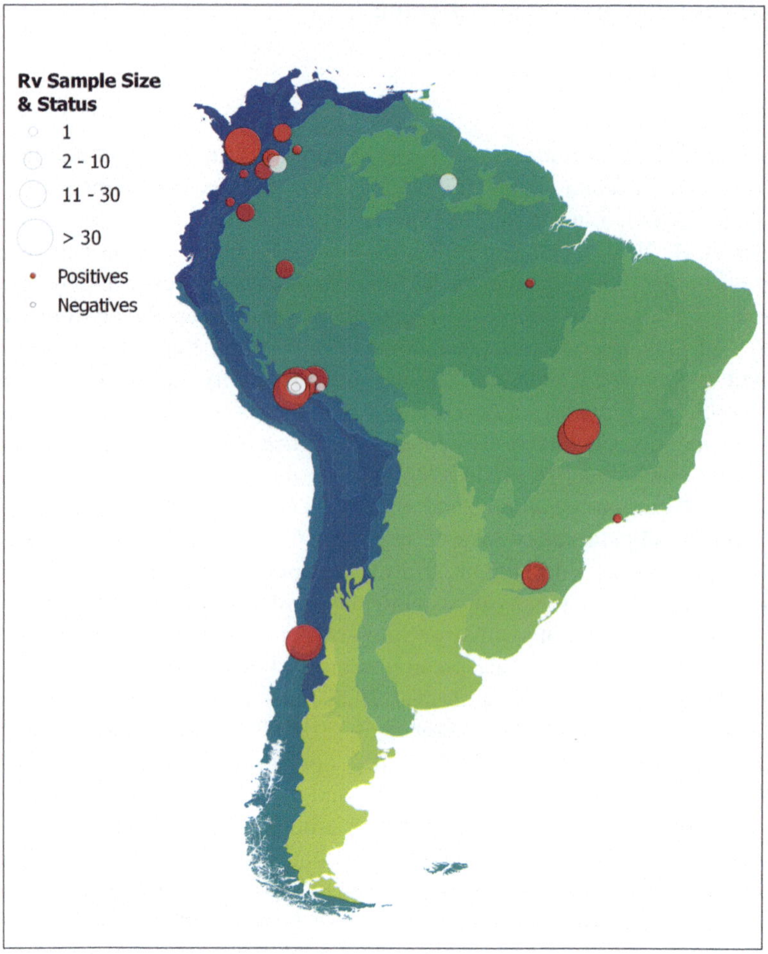

Fig. 3.3 Distribution of *Ranavirus* in South America. Presence and absence of *Ranavirus* per surveyed site. The size of the circles is proportional to the number of individuals sampled. Red and white circles represent *Ranavirus*-positive and *Ranavirus*-negative sites, respectively. The presented data are based on 952 *Ranavirus*-specific PCR analyses

3.4.2 Impacts

Little information on the impacts of *Ranavirus* is available for South America, and no evidence for *Ranavirus* die-offs in native amphibians exists in the region, except for the high mortality and morbidity reported in *A. patagonicus* in northern Patagonia, Argentina (Fox et al. 2006). Most studies developing active surveillance for *Ranavirus* in South America report a low prevalence of 0.1–5% (Urgiles et al. 2021; Flechas et al. 2023; Peñafiel-Ricaurte et al. 2023; but see Warne et al. 2016). In 2011, a dead 2.2 kg female *C. gayi* from a pond near Santiago in Chile was submitted for post-mortem examination. This frog presented lesions consistent with ranaviral disease, i.e., abundant serosanguineous fluid was found subcutaneously and intracelomically, internal organs were edematous (Soto-Azat et al. 2016), and *Ranavirus* was confirmed by qPCR with a high viral load (>3000 viral copies; Peñafiel-Ricaurte et al. 2023). Over half of *Pristimantis* spp. (251 of 497) in South America are threatened with extinction according to the IUCN Red List (IUCN 2023), and it is worrisome that several studies in South America described *Ranavirus* in populations of this genus in Peru, Ecuador, and Colombia (Warne et al. 2016; Urgiles et al. 2021; Flechas et al. 2023). Another group that has been rapidly declining in the region is *Telmatobius* spp., and *Ranavirus* has been described in traded *T. marmoratus* for human consumption in Peru (Warne et al. 2016). Whether *Ranavirus* is a contributing factor to the decline of threatened *C. gayi*, *Telmatobius* spp., and *Pristimantis* spp. is unknown and requires further examination (Urgiles et al. 2021).

3.5 Future Directions

3.5.1 Disease Mitigation

South America has experienced devastating amphibian biodiversity loss due to chytridiomycosis (Scheele et al. 2019), and while *Ranavirus* has increasingly been reported, there is a lack of information on its impacts in the region (Soto-Azat et al. 2016; Flechas et al. 2023). Methods to mitigate the impacts of amphibian emerging infectious diseases are urgently needed to combat the loss of biodiversity (Bosch et al. 2020). Disease prevention by reducing the risk of pathogen introduction is often the best mitigation strategy against *Bd* and *Ranavirus*. Although only a few regions of the world remain *Bd* or *Ranavirus*-free (Bower et al. 2017), informing biosecurity recommendations at the country level is crucial, given the potential for pathogen introduction and strain recombination, which can result in increased virulence (Berger et al. 2016). Chytridiomycosis and ranaviral disease have been listed as notifiable diseases since 2008 by the World Organization of Animal Health (WOAH), but the enforcement of *Bd* and *Ranavirus* control in the amphibian trade has not been implemented by WOAH member states (Schloegel et al. 2010a). Once established, eradication of *Bd* and *Ranavirus* would be an ideal outcome; however, it is unrealistic to think it would be possible to achieve at a large scale (Bosch et al. 2020). Therefore, other mitigation strategies are the most common option available to prevent further species declines. Although mitigation has been better studied in *Bd* than *Ranavirus*, many actions apply to both pathogens, while others are more pathogen specific. Strategies for long-term *Bd* mitigation include management of reservoir and superspreader hosts, reintroduction or reinforcement of susceptible species, *Bd* environmental management, treatment of individuals with antifungals, and augmentation of the host skin microbiome with probiotics (Woodhams et al.

2011; Garner et al. 2016). Additionally, abundant chytridiomycosis-resistant species (including invasive *L. catesbeianus* and *X. laevis*) can be subjected to management for *Bd* and *Ranavirus* mitigation. For example, Scheele et al. (2017) demonstrated that the abundant and least concerning eastern froglet (*Crinia signifera*), which exhibits a high *Bd* prevalence (78.7%) in Australia, acts as a competent pathogen reservoir for the endangered northern corroboree frog (*Pseudoprhyne pengilleyi*). Another strategy involves the identification of low pathogen suitability areas, as they represent pathogen refuges for native amphibians and, for instance, candidate sites for conservation translocations (Puschendorf et al. 2009; Bacigalupe et al. 2019). In the wild, there is only one successful example of *Bd* eradication, which occurred on an island with a single amphibian host (the Mallorcan midwife toad [*Alytes obstetricans*]), based on the use of antifungals and environmental disinfectants (Bosch et al. 2020). However, the use of antifungals in more complex systems has only short-term *Bd*-protective beneficial effects (Hudson et al. 2016). Thus, the approach based on the use of probiotics has gained popularity as an attractive tool for *Bd* control in nature (Bletz et al. 2013; Küng et al. 2014; Rebollar et al. 2016; Niederle et al. 2019). Probiotic augmentation has potential advantages over antifungals, which are difficult to apply in the wild and may have profound effects on the native microbiota of a host or ecosystem (Antwis et al. 2015; Bosch et al. 2020). The use of probiotics for *Bd* mitigation should be context dependent; consequently, isolation and testing of local microbes with *Bd*-inhibition potential should be encouraged widely (Woodhams et al. 2016; Bletz et al. 2013; Kueneman et al. 2016).

3.5.2 Research Needs

While *Ranavirus* research is in the early stages in South America (Fig. 3.3), further research is required to establish the potential impacts of this pathogen on wild populations. Native amphibians living in contact with *Ranavirus* reservoirs, such as the invasive *L. catesbeianus* and *X. laevis*, have been a starting point (Ruggeri et al. 2019; Soto-Azat et al. 2016). Whether *Ranavirus* in the region has a patchy distribution associated with its potential recent introduction or a more continuous presence that has been underestimated is currently unknown. The development of *Ranavirus* niche modeling could help to answer this question, as well as guiding areas that warrant future pathogen surveys (Becker et al. 2016; Bacigalupe et al. 2019), which acquire more relevance in the context of climate change (Price et al. 2019). To our knowledge, *Ranavirus* has not been reported in South American wild or farmed fish or reptiles; thus, it is highly necessary to investigate the multi-host epidemiology of *Ranavirus* in the region, for example, the widespread rainbow trout (*Oncorhynchus mykiss*; Peñafiel-Ricaurte et al. 2023). Genetic characterization of ranaviruses is scarce (Fox et al. 2006; Mazzoni et al. 2009; Soto-Azat et al. 2016; Candido et al. 2019; Peñafiel-Ricaurte et al. 2023) but is highly needed to assess the risk of *Ranavirus* to local populations. In contrast, *Bd* has been more intensively studied in the region (James et al. 2015; Azat et al. 2022). However, some areas remain understudied; thus, future efforts to understand the impacts of *Bd* in native amphibians should focus on areas such as the Amazon basin, Caribbean lowlands, Guianas, Orinoco Llanos, Chaco, and Patagonia (James et al. 2015; Becker et al. 2016; Azat et al. 2022; Fig. 3.2). As limited information is available on the genetic variation of *Bd* and *Ranavirus* in the region and with molecular methods becoming more widely accessible, future efforts should focus on pathogen genotyping (Azat 2021). The introduction or spread of novel *Bd* or *Ranavirus* genotypes, as well as the emergence of hybrids that may have higher pathogenicity or transmissibility, is a matter that should be of high concern (Rosenblum et al. 2013; Jenkinson et al. 2016; Greenspan et al. 2018; Peñafiel-Ricaurte et al. 2023). Genetic studies will allow us to fill gaps in evolutionary history, epidemiology, and impacts of *Bd* and *Ranavirus* on amphibian populations, which is needed to design timely mitigation strategies (Azat 2021). Finally, although *Bsal* appears to be restricted to Asia and Europe, further dispersal with potential impacts on neotropical amphibian diversity is possible (García-Rodríguez et al. 2022). Strict biosecurity measures and active pathogen surveillance remain the most recommended tools to prevent further amphibian biodiversity loss in this megadiverse region.

References

Allender MC, Mitchell MA, Torres T, Seskoska J, Driskell EA (2013) Pathogenicity of frog virus 3-like virus in red-eared slider turtles (*Trachemys scripta elegans*) at two environmental temperatures. J Comp Pathol 149(2–3):356–367. https://doi.org/10.1016/j.jcpa.2013.01.007

Alvarado-Rybak M, Acuña P, Peñafiel-Ricaurte A, Sewell TR, O'Hanlon SJ, Fisher MC, Valenzuela-Sánchez A, Cunningham AA, Azat C (2021a) Chytridiomycosis outbreak in a Chilean giant frog (*Calyptocephalella gayi*) captive breeding program: genomic characterization and pathological findings. Front Vet Sci 8:733357. https://doi.org/10.3389/fvets.2021.733357

Alvarado-Rybak M, Lepe-Lopez M, Peñafiel-Ricaurte A, Valenzuela-Sánchez A, Valdivia C, Mardones FO, Bacigalupe LD, Puschendorf R, Cunningham AA, Azat C (2021b) Bioclimatic and anthropogenic variables shape the occurrence of *Batrachochytrium dendrobatidis* over a large latitudinal gradient. Sci Rep 11:17383. https://doi.org/10.1038/s41598-021-96535-w

Andre SE, Parker J, Briggs CJ (2008) Effect of temperature on host response to *Batrachochytrium dendrobatidis* infection in the mountain yellow-legged frog (*Rana muscosa*). J Wildl Dis 44(3):716–720. https://doi.org/10.7589/0090-3558-44.3.716

Antwis RE, Preziosi RF, Harrison XA, Garner TWJ (2015) Amphibian symbiotic bacteria do not show a universal ability to inhibit growth of the global panzootic lineage of *Batrachochytrium dendrobatidis*. Appl Environ Microbiol 81(11):3706–3711. https://doi.org/10.1128/aem.00010-15

Azat C (2021) Not just a pathogen: the importance of recognizing genetic variability to mitigate a wildlife pandemic. Mol Ecol Resour 21(5):1410–1412. https://doi.org/10.1111/1755-0998.13348

Azat C, Valenzuela-Sánchez A, Delgado S, Cunningham AA, Alvarado-Rybak M, Bourke J, Briones R, Cabeza O, Castro-Carrasco C, Charrier A, Correa C, Crump ML, Cuevas CC, de la Maza M, Díaz-Vidal S, Flores E, Harding G, Lavilla EO, Mendez MA, Oberwemmer F, Ortiz JC, Pastore H, Peñafiel-Ricaurte A, Rojas-Salinas L, Serrano JM, Sepúlveda MA, Toledo V, Úbeda C, Uribe-Rivera DE, Valdivia C, Wren S, Angulo A (2021) A flagship for Austral temperate forest conservation: an action plan for Darwin's frogs bringing together key stakeholders. Oryx 55(3):356–363. https://doi.org/10.1017/S0030605319001236

Azat C, Alvarado-Rybak M, Solano-Iguaran JJ, Velasco A, Valenzuela-Sánchez A, Flechas SV, Peñafiel-Ricaurte A, Cunningham AA, Bacigalupe LD (2022) Synthesis of *Batrachochytrium dendrobatidis* infection in South America: amphibian species under risk and areas to focus research and disease mitigation. Ecography 2022(7):1–12. https://doi.org/10.1111/ecog.05977

Azevedo-Ramos C, Galatii U (2002) Patterns of amphibian diversity in Brazilian Amazonia: conservation implications. Biol Conserv 103(1):103–111. https://doi.org/10.1016/S0006-3207(01)00129-X

Bacigalupe L, Soto-Azat C, García-Vera C, Barría-Oyarzo I, Rezende EL (2017) Effects of amphibian phylogeny, climate and human impact on the occurrence of the amphibian-killing chytrid fungus. Glob Chang Biol 23(9):3543–3553. https://doi.org/10.1111/gcb.13610

Bacigalupe L, Vásquez IA, Estay SA, Valenzuela-Sánchez A, Alvarado-Rybak M, Peñafiel-Ricaurte A, Cunningham AA, Soto-Azat C (2019) *Batrachochytrium dendrobatidis* in a biodiversity hotspot: identifying and validating high-risk areas and refugia. Ecosphere 10(5):e02724. https://doi.org/10.1002/ecs2.2724

Baláž V, Vörös J, Civiš P, Vojar J, Hettyey A, Sós E, Dankovics R, Jehle R, Christiansen DG, Clare F, Fisher MC, Garner TWJ, Bielby J (2014) Assessing risk and guidance on monitoring of *Batrachochytrium dendrobatidis* in Europe through identification of taxonomic selectivity of infection. Conserv Biol 28(1):213–223. https://doi.org/10.1111/cobi.12128

Balseiro A, Dalton KP, Del Cerro A, Márquez I, Parra F, Prieto JM, Casais R (2010) Outbreak of common midwife toad virus in alpine newts (*Mesotriton alpestris cyreni*) and common midwife toads (*Alytes obstetricans*) in Northern Spain: a comparative pathological study of an emerging ranavirus. Vet J 186(2):256–258. https://doi.org/10.1016/j.tvjl.2009.07.038

Barrio Amorós CL, Costales M, Vieira J, Osterman E, Kaiser H, Arteaga A (2020) Back from extinction: rediscovery of the harlequin toad *Atelopus mindoensis* Peters, 1973 in Ecuador. Herpetol Notes 13:325–328

Barrionuevo S, Mangione S (2006) Chytridiomycosis in two species of *Telmatobius* (Anura: Leptodactylidae) from Argentina. Dis Aquat Org 73(2):171–174. https://doi.org/10.3354/dao073171

Barrionuevo S, Ponssa L (2008) Decline of three species of the genus *Telmatobius* (Anura: Leptodactylidae) from Tucumán Province, Argentina. Herpetologica 64(1):47–62. https://doi.org/10.1655/06-057.1

Becker CG, Rodriguez D, Lambertini C, Toledo LF, Haddad CFB (2016) Historical dynamics of *Batrachochytrium dendrobatidis* in Amazonia. Ecography 39(10):954–960. https://doi.org/10.1111/ecog.02055

Berger L, Speare R, Daszak P, Green DE, Cunningham AA, Goggin CL, Slocombe R, Ragan MA, Hyatt AD, McDonald KR, Hines HB, Lips KR, Marantelli G, Parkes H (1998) Chytridiomycosis causes amphibian mortality associated with population declines in the rain forests of Australia and Central America. Proc Natl Acad Sci U S A 95(15):9031–9036. https://doi.org/10.1073/pnas.95.15.9031

Berger L, Speare R, Hines H, Marantelli G, Hyatt AD, Skerratt LF, Olsen V, Clarke JM, Gillespie G, Mahony M, Sheppard N, Williams C, Tyler MJ (2004) Effect of season and temperature on mortality in amphibians due to chytridiomycosis. Aust Vet J 82(7):31–36. https://doi.org/10.1111/j.1751-0813.2004.tb11137.x

Berger L, Hyatt AD, Speare R, Longcore JE (2005) Life cycle stages of the amphibian chytrid *Batrachochytrium dendrobatidis*. Dis Aquat Org 68(1):51–63. https://doi.org/10.3354/dao068051

Berger L, Roberts AA, Voyles J, Longcore JE, Murray KA, Skerrat LF (2016) History and recent progress on chytridiomycosis in amphibians. Fungal Ecol 19:89–99. https://doi.org/10.1016/j.funeco.2015.09.007

Bielby J, Cooper N, Cunningham AA, Garner TWJ, Purvis A (2008) Predicting susceptibility to rapid declines in the world's frogs. Conserv Lett 1(2):82–90. https://doi.org/10.1111/j.1755-263X.2008.00015.x

Bletz MC, Loudon AH, Becker MH, Bell SC, Woodhams DC, Minbiole KPC, Harris RN (2013) Mitigating amphibian chytridiomycosis with bioaugmentation: characteristics of effective probiotics and strategies for their selection and use. Ecol Lett 16(6):807–820. https://doi.org/10.1111/ele.12099

Bollinger TK, Mao J, Schock D, Brigham RM, Chinchar VG (1999) Pathology, isolation, and preliminary molecular characterization of a novel iridovirus from tiger salamanders in Saskatchewan. J Wildl Dis 35(3):413–429. https://doi.org/10.7589/0090-3558-35.3.413

Borzée A, Kosch T, Kim M, Jang Y (2017) Introduced bullfrogs are associated with increased *Batrachochytrium dendrobatidis* prevalence and reduced occurrence of Korean treefrogs. PLoS One 12(12):e0190551. https://doi.org/10.1371/journal.pone.0177860

Bosch J, Sanchez-Tomé E, Fernández-Loras A, Oliver JA, Fisher MC, Garner TWJ (2020) Successful elimination of a lethal wildlife infectious disease in nature. Biol. Lett. 11:20150874. http://dx.doi.org/10.1098/rsbl.2015.087

Bower DS, Lips KR, Schwarzkopf L, Georges A, Clulow S (2017) Amphibians on the brink. Science 357(6350):454–455. https://doi.org/10.1126/science.aao0500

Brunner JL, Schock DM, Collins JP (2007) Transmission dynamics of the amphibian ranavirus *Ambystoma tigrinum* virus. Dis Aquat Org 77(2):87–95. https://doi.org/10.3354/dao01845

Brunner JL, Olson DH, Gray MJ, Miller DL, Duffus ALJ (2021) Global patterns of ranavirus detections. Facets 6(1):912–924. https://doi.org/10.1139/FACETS-2020-0013

Burrowes PA, De la Riva I (2017) Unravelling the historical prevalence of the invasive chytrid fungus in the Bolivian Andes: implications in recent amphibian declines. Biol Invasions 19(6):1781–1794. https://doi.org/10.1007/s10530-017-1390-8

Burrowes PA, James TY, Jenkinson TS, De la Riva I (2020) Genetic analysis of post-epizootic amphibian chytrid strains in Bolivia: Adding a piece to the puzzle. Transb. Emerg. Dis. 67(5):2163–2171. https://doi.org/10.1111/tbed.13568

Bustamante HM, Livo LJ, Carey C (2010) Effects of temperature and hydric environment on survival of the Panamanian golden frog infected with a pathogenic chytrid fungus. Integr Zool 5(2):143–153. https://doi.org/10.1111/j.1749-4877.2010.00197.x

Byrne AQ, Vredenburg VT, Martel AN, Pasmans F, Bell RC, Blackburn DC, Bletz MC, Bosch J, Briggs CJ, Brown RM, Catenazzi A, Familiar López M, Figueroa-Valenzuela R, Ghose SL, Jaeger JR, Jani AJ, Jirku M, Knapp RA, Muñoz A, Portik DM, Richards-Zawacki CL, Rockney H, Rovito SM, Stark T, Suleeman H, Tao NT, Voyles J, Waddle AW, Yuan Z, Rosenblum EB (2019) Cryptic diversity of a widespread global pathogen reveals expanded threats to amphibian conservation. Proc Natl Acad Sci U S A 116(41):20382–20387. https://doi.org/10.1073/pnas.1908289116

Candido M, Tavares LS, Alencar ALF, Ferreira CM, de Almeida Queiroz SR, Fernandes AM, de Sousa RLM (2019) Genome analysis of *Ranavirus frog virus 3* isolated from American Bullfrog (*Lithobates catesbeianus*) in South America. Sci Rep 9:171–135. https://doi.org/10.1038/s41598-019-53626-z

Carey C, Cohen N, Rollins-Smith L (1999) Amphibian declines: an immunological perspective. Dev Comp Immunol 23(6):459–472. https://doi.org/10.1016/s0145-305x(99)00028-2

Carvalho T, Becker CG, Toledo LF (2017) Historical amphibian declines and extinctions in Brazil linked to chytridiomycosis. Proc R Soc B 284(1848):20162254. https://doi.org/10.1098/rspb.2016.2254

Catenazzi A, Lehr E, Rodriguez LO, Vredenburg VT (2011) Batrachochytrium dendrobatidis and the collapse of anuran species richness and abundance in the upper Manu National Park, southeastern Peru. Conserv Biol 25(2):382–391. https://doi.org/10.1111/j.1523-1739.2010.01604.x

Collins JP, Crump ML (2009) Extinction in our times. Global amphibian decline. Oxford University Press, Oxford, p 273

Cullen BR, Owens L (2002) Experimental challenge and clinical cases of Bohle iridovirus (BIV) in native Australian anurans. Dis Aquat Org 49(2):83–92. https://doi.org/10.3354/dao049083

Cunningham AA, Langton TES, Bennett PM, Lewin JF, Drury SEN, Gough RE, MacGregor SK (1996) Pathological and microbiological findings from incidents of unusual mortality of the common frog (*Rana temporaria*). Philos Trans R Soc Lond B Biol Sci 351(1347):1539–1557. https://doi.org/10.1098/rstb.1996.0140

da Silva FR, Almeida-Neto M, Arena MVN (2014) Amphibian beta diversity in the Brazilian Atlantic Forest: contrasting the roles of historical events and contemporary conditions at different spatial scales. PLoS One 9(10):e109642. https://doi.org/10.1371/journal.pone.0109642

Daszak P, Berger L, Cunningham AA, Hyatt AD, Green DE, Speare R (1999) Emerging infectious diseases and amphibian population declines. Emerg Infect Dis 5(6):735–748. https://doi.org/10.3201/eid0506.990601

Drew A, Allen EJ, Allen LJ (2006) Analysis of climatic and geographic factors affecting the presence of chytridiomycosis in Australia. Dis Aquat Org 68(3):245–250. https://doi.org/10.3354/dao068245

Duffus ALJ, Cunningham AA (2010) Major disease threats to European amphibians. Herpetol J 20(3):117–127

Fijan N, Matasin Z, Petrinec Z, Valpotic I, Zwillenberg LO (1991) Isolation of an iridovirus-like agent from the green frog (*Rana esculenta* L.). Vet Arhiv 61:151–158

Fisher MC, Garner TWJ, Walker SF (2009) Global emergence of *Batrachochytrium dendrobatidis* and amphibian chytridiomycosis in space, time, and host. Annu Rev Microbiol 63:291–310. https://doi.org/10.1146/annurev.micro.091208.073435

Flechas SV, Medina EM, Crawford AJ, Sarmiento C, Cárdenas ME, Amézquita A, Restrepo S (2013) Characterization of the first *Batrachochytrium dendrobatidis* isolate from the Colombian Andes, an amphibian biodiversity hotspot. EcoHealth 10(1):72–76. https://doi.org/10.1007/s10393-013-0823-9

Flechas SV, Urbina J, Crawford AJ, Gutiérrez K, Corrales K, Castellanos LA, González MA, Cuervo AM, Catenazzi A (2023) First evidence of ranavirus in native and invasive amphibians in Colombia. Dis Aquat Org 153:51–58. https://doi.org/10.3354/dao03717

Foden WB, Butchart SHM, Stuart SN, Vié J-C, Akçakaya HR, Angulo A, DeVantier LM, Gutsche A, Turak E, Cao L, Donner SD, Katariya V, Bernard R, Holland RA, Hughes AF, O'Hanlon SE, Garnett ST, Şekercioğlu ÇH, Mace GM (2013) Identifying the world's most climate change vulnerable species: a systematic trait-based assessment of all birds, amphibians and corals. PLoS One 8(6):e65427. https://doi.org/10.1371/journal.pone.0065427

Fox SF, Greer AL, Torres-Cervantes R, Collins JP (2006) First case of ranavirus-associated morbidity and mortality in natural populations of the South American frog *Atelognathus patagonicus*. Dis Aquat Org 72(1):87–92. https://doi.org/10.3354/dao072087

Frost DR (2023) Amphibian species of the world: an online reference. Version 6.1. American Museum of Natural History, New York. https://doi.org/10.5531/db.vz.0001. Electronic Database accessible at https://amphibiansoftheworld.amnh.org/index.php. Accessed 28 Apr 2023

Galli L, Pereira A, Márquez A, Mazzoni R (2006) Ranavirus detection by PCR in cultured tadpoles (*Rana catesbeiana* Shaw, 1802) from South America. Aquaculture 257(1–4):78.82. https://doi.org/10.1016/j.aquaculture.2005.06.019

García-Rodríguez A, Basanta MD, García-Castillo MG, Zumbado-Ulate H, Neam K, Rovito S, Searle CL, Parra-Olea G (2022) Anticipating the potential impacts of *Batrachochytrium salamandrivorans* on Neotropical salamander diversity. Biotropica 54(1):157–169. https://doi.org/10.1111/btp.13042

Garner TWJ, Perkins MW, Govindarajulu P, Seglie D, Walker S, Cunningham AA, Fisher MC (2006) The emerging amphibian pathogen *Batrachochytrium dendrobatidis* globally infects introduced populations of the North American bullfrog, *Rana catesbeiana*. Biol Lett 2(3):455–459. https://doi.org/10.1098/rsbl.2006.0494

Garner TWJ, Schmidt BR, Martel A, Pasmans F, Muths E, Cunningham AA, Weldon C, Fisher MC, Bosch J (2016) Mitigating amphibian chytridiomycoses in nature. Philos Trans R Soc B 371(1709):20160207. https://doi.org/10.1098/rstb.2016.0207

Geng Y, Wang KY, Zhou ZY, Li CW, Wang J, He M, Yin ZQ, Lai WM (2011) First report of a ranavirus associated with morbidity and mortality in farmed Chinese giant salamanders (*Andrias davidianus*). J Comp Pathol 145(1):95–102. https://doi.org/10.1016/j.jcpa.2010.11.012

González-Del-Pliego P, Freckleton RP, Edwards DP, Koo MS, Scheffers BR, Pyron RA, Jetz W (2019) Phylogenetic and trait-based prediction of extinction risk for data-deficient amphibians. Curr Biol 29(9):1557–1563. https://doi.org/10.1016/j.cub.2019.04.005

Gower DJ, Doherty-Bone T, Loader SP, Wilkinson M, Kouete MT, Tapley B, Orton F, Daniel OZ, Wynne F, Flach E, Müller H, Menegon M, Stephen I, Browne RK, Fisher MC, Cunningham AA, Garner TWJ (2013) *Batrachochytrium dendrobatidis* infection and lethal chytridiomycosis in caecilian amphibians Gymnophiona. EcoHealth 10(2):173–183. https://doi.org/10.1007/s10393-013-0831-9

Granoff A, Came PE, Keen A, Rafferty KA Jr (1965) The isolation and properties of viruses from *Rana pipiens*: their possible relationship to the renal adenocarcinoma of the leopard frog. Ann N Y Acad Sci 126(1):237–255. https://doi.org/10.1111/j.1749-6632.1965.tb14278.x

Gray MJ, Miller DL, Schmutzer AC, Baldwin CA (2007) Frog virus 3 prevalence in tadpole populations inhabiting cattle-access and non-access wetlands in Tennessee, USA. Dis Aquat Org 77(2):97–103. https://doi.org/10.3354/dao01837

Gray MJ, Miller DL, Hoverman JT (2009) Ecology and pathology of amphibian ranaviruses. Dis Aquat Org 87(3):243–266. https://doi.org/10.3354/dao02138

Green DE, Converse KA, Schrader AK (2002) Epizootiology of sixty-four amphibian morbidity and mortality events in the USA, 1996-2001. Ann N Y Acad Sci 969:323–339. https://doi.org/10.1111/j.1749-6632.2002.tb04400.x

Greenspan SE, Lambertini C, Carvalho T, James TY, Toledo LF, Haddad CFB, Becker CG (2018) Hybrids of amphibian chytrid show high virulence in native hosts. Sci Rep 8:9600. https://doi.org/10.1038/s41598-018-27828-w

Greer AL, Berrill M, Wilson PJ (2005) Five amphibian mortality events associated with ranavirus infection in south central Ontario, Canada. Dis Aquat Org 67(1–2):9–14. https://doi.org/10.3354/dao067009

Haislip NA, Gray MJ, Hoverman JT, Miller DL (2011) Development and disease: how susceptibility to an emerging pathogen changes through anuran development. PLoS One 6(7):e22307. https://doi.org/10.1371/journal.pone.0022307

Harding G, Griffiths RA, Pavajeau L (2006) Developments in amphibian captive breeding and reintroduction programs. Conserv Biol 30(2):340–349. https://doi.org/10.1111/cobi.12612

He JG, Lü L, Deng M, He HH, Weng SP, Wang XH, Zhou SY, Long QX, Wang XZ, Chan SM (2002) Sequence analysis of the complete genome of an iridovirus isolated from the tiger frog. Virology 292(2):185–197. https://doi.org/10.1006/viro.2001.1245

Hoverman JT, Gray MJ, Miller DL (2010) Anuran susceptibilities to ranaviruses: role of species identity, exposure route, and a novel virus isolate. Dis Aquat Org 89(2):97–107. https://doi.org/10.3354/dao02200

Hudson M, Young RP, D'Urban Jackson J, Orozco-terWengel P, Martin L, James A, Sulton M, Garcia G, Griffiths RA, Thomas R, Magin C, Bruford MW, Cunningham AA (2016) Dynamics and genetics of a disease-driven species decline to near extinction: lessons for conservation. Sci Rep 6:1–13. https://doi.org/10.1038/srep30772

Hyatt AD, Gould AR, Zupanovic Z, Cunningham AA, Hengstberger S, Whittington RJ, Kattenbelt J, Coupar BE (2000) Comparative studies of piscine and amphibian iridoviruses. Arch Virol 145:301–331. https://doi.org/10.1007/s007050050025

IUCN (2023) The IUCN red list of threatened species. Version 2022-2. https://www.iucnredlist.org. Accessed 28 Apr 2023

James TY, Toledo LF, Rödder D, da Silva LD, Belasen AM, Betancourt-Román CM, Jenkinson TS, Soto-Azat C, Lambertini C, Longo AV, Ruggeri J, Collins JP, Burrowes PA, Lips KR,

Zamudio KR, Longcore JE (2015) Disentangling host, pathogen, and environmental determinants of a recently emerged wildlife disease: lessons from the first 15 years of amphibian chytridiomycosis. Ecol Evol 5(18):4079–4097. https://doi.org/10.1002/ece3.1672

Jancovich JK, Davidson EW, Morado JF, Jacobs BL, Collins JP (1997) Isolation of a lethal virus from the endangered tiger salamander *Ambystoma tigrinum stebbinsi*. Dis Aquat Org 31(3):161–167. https://doi.org/10.3354/dao031161

Jancovich JK, Mao J, Chinchar VG, Wyatt C, Case ST, Kumar S, Valente G, Subramanian S, Davidson EW, Collins JP, Jacobs BL (2003) Genomic sequence of a ranavirus (family Iridoviridae) associated with salamander mortalities in North America. Virology 316(1):90–103. https://doi.org/10.1016/j.virol.2003.08.001

Jenkins CN, Pimm SL, Joppa LN (2013) Global patterns of terrestrial vertebrate diversity and conservation. Proc Natl Acad Sci U S A 110(28):E2602–E2610. https://doi.org/10.1073/pnas.1302251110

Jenkinson TS, Betancourt Román CM, Lambertini C, Valencia-Aguilar A, Rodriguez D, Nunes-de-Almeida CHL, Ruggeri J, Belasen AM, da Silva Leite D, Zamudio KR, Longcore JE, Toledo LF, James TY (2016) Amphibian-killing chytrid in Brazil comprises both locally endemic and globally expanding populations. Mol Ecol 25(13):2978–2996. https://doi.org/10.1111/mec.13599

Johnson ML, Speare R (2003) Survival of *Batrachochytrium dendrobatidis* in water: quarantine and disease control implications. Emerg Infect Dis 9(8):922–925. https://doi.org/10.3201/eid0908.030145

Johnson AJ, Pessier AP, Wellehan JFX, Childress A, Norton TM, Stedman NL, Bloom DC, Belzer W, Titus VR, Wagner R, Brooks JW, Spratt J, Jacobson ER (2008) Ranavirus infection of free-ranging and captive box turtles and tortoises in the United States. J Wildl Dis 44(4):851–863. https://doi.org/10.7589/0090-3558-44.4.851

Kakoliris FP, Berkunsky I, Acosta JC, Acosta R, Agostini MG, Akmentis MS, Arellano ML, Azat C, Bach NC, Blanco GM, Calvo R, Charrier A, Corbalán V, Correa C, Cuello ME, Deutsch C, Di-Pierto D, Gastón MS, Gómez-Alés R, Kass C, Kass N, Lobos G, Martínez TA, Martínez-Aguirre T, Mora M, Nieva-Cocilio R, Pastore H, Pérez-Iglesias JM, Piaggio-Kokot L, Rabanal F, Rodriguez-Muñoz MJ, Sanchez LC, Tala C, Úbeda C, Vaira M, Velasco MA, Vidal M, Williams JD (2022) Current threats faced by amphibian populations in the southern cone of South America. J Nat Conserv 69:126254. https://doi.org/10.1016/j.jnc.2022.126254

Kik M, Martel A, Spitzen-van der Sluijs A, Pasmans F, Wohlsein P, Gröne A, Rijks JM (2011) Ranavirus-associated mass mortality in wild amphibians, The Netherlands, 2010: a first report. Vet J 190(2):284–286. https://doi.org/10.1016/j.tvjl.2011.08.031

Kriger KM, Pereoglou F, Hero JM (2007) Latitudinal variation in the prevalence and intensity of chytrid (*Batrachochytrium dendrobatidis*) infection in eastern Australia. Conserv Biol 21(5):1280–1290. https://doi.org/10.1111/j.1523-1739.2007.00777.x

Kueneman JG, Woodhams DC, Treuren WV, Archer HM, Knight R, McKenxzie VJ (2016) Inhibitory bacteria reduce fungi on early life stages of endangered Colorado boreal toads (*Anaxyrus boreas*). ISME J 10(4):934–944. https://doi.org/10.1038/ismej.2015.168

Küng D, Bigler L, Davis LR, Gratwicke B, Griffith E, Woodhams DC (2014) Stability of microbiota facilitated by host immune regulation: informing probiotic strategies to manage amphibian disease. PLoS One 9(1):e87101. https://doi.org/10.1371/journal.pone.0087101

La Marca E, Lips KR, Lötters S, Puschendorf R, Ibáñez R, Rueda-Almonacid JV, Schulte R, Marty C, Castro F, Manzanilla-Puppo J, García-Pérez JE, Bolaños F, Chaves G, Pounds JA, Toral E, Young BE (2005) Catastrophic population declines and extinctions in Neotropical harlequin frogs (Bufonidae: *Atelopus*). Biotropica 37(2):190–201. https://doi.org/10.1111/j.1744-7429.2005.00026.x

Lampo M, Rodríguez-Contreras A, La Marca E, Daszak P (2006) Chytridiomycosis epidemic and a severe dry season precede the disappearance of *Atelopus* species from the Venezuelan Andes. Herpetol J 16(4):395–402

Lampo M, Señaris C, García CZ (2017) Population dynamics of the critically endangered toad *Atelopus cruciger* and the fungal disease chytridiomycosis. PLoS One 12(6):e0179007. https://doi.org/10.1371/journal.pone.0179007

Lesbarrères D, Balseiro A, Brunner J, Chinchar VG, Duffus A, Kerby J, Miller DL, Robert J, Schock DM, Waltzek T, Gray MJ (2011) Ranavirus: past, present and future. Biol Lett 8(4):481–483. https://doi.org/10.1098/rsbl.2011.0951

Lips KR (2016) Overview of chytrid emergence and impacts on amphibians. Philos Trans R Soc Lond B Biol Sci 371(1709):20150465. https://doi.org/10.1098/rstb.2015.0465

Lips KR, Brem F, Brenes R, Reeve JD, Alford RA, Voyles J, Carey C, Livo L, Pessier AP, Collins JP (2006) Emerging infectious disease and the loss of biodiversity in a Neotropical amphibian community. Proc Natl Acad Sci U S A 103(9):3165–3170. https://doi.org/10.1073/pnas.0506889103

Lips KR, Diffendorfer J, Mendelson JR III, Sears MR (2008) Riding the wave: reconciling the roles of disease and climate change in amphibian declines. PLoS Biol 6(3):e72. https://doi.org/10.1371/journal.pbio.0060072

Longcore JE, Pessier AP, Nichols DK (1999) *Batrachochytrium dendrobatidis* gen. et sp. nov., a chytrid pathogenic to amphibians. Mycologia 91(2):219–227. https://doi.org/10.1080/00275514.1999.12061011

Martel A, Blooi M, Adriaensen C, Van Rooij P, Beukema W, Fisher MC, Farrer RA, Schmidt BR, Tobler U, Goka K, Lips KR, Muletz C, Zamudio KR, Bosch J, Lötters S, Wombwell E, Garner TWJ, Cunningham AA, Spitzen-van der Sluijs A, Salvidio S, Ducatelle R, Nishikawa K, Nguyen TT, Kolby JE, Van Bocxlaer I, Bossuyt F, Pasmans F (2014) Recent introduction of a chytrid fungus endangers Western Palearctic salamanders. Science 346(6209):630–631. https://doi.org/10.1126/science.1258268

Mavian C, López-Bueno A, Balseiro A, Casais R, Alcamí A, Alejo A (2012) The genome sequence of the emerging common midwife toad virus identifies and evolutionary intermediate within ranaviruses. J Virol 86(7):3617–3625. https://doi.org/10.1128/jvi.07108-11

Mazzoni R, de Mesquita AJ, Fleury LF, de Brito WM, Nunes IA, Robert J, Morales H, Coelho AS, Barthasson DL, Galli L, Catroxo MH (2009) Mass mortality associated with a frog virus 3-like ranavirus infection in farmed tadpoles *Rana catesbeiana* from Brazil. Dis Aquat Org 86(3):181–191. https://doi.org/10.3354/dao02096

McDonald KR, Méndez D, Müller R, Freeman AB, Speare R (2005) Decline in the prevalence of chytridiomycosis in upland frog populations in north Queensland, Australia. Pac Conserv Biol 11(2):114–120. https://doi.org/10.1071/PC050114

Mora M, Pons DJ, Peñafiel-Ricaurte A, Alvarado-Rybak M, Lebuy S, Soto-Azat C (2019) High abundance of invasive African clawed frog *Xenopus laevis* in Chile: challenges for their control and updated invasive distribution. Manage Biol Invasions 10(2):377–388. https://doi.org/10.3391/mbi.2019.10.2.11

Murphy PJ, St-Hilaire S, Corn PS (2011) Temperature, hydric environment, and prior pathogen exposure alter the experimental severity of chytridiomycosis in boreal toads. Dis Aquat Org 95(1):31–42. https://doi.org/10.3354/dao02336

Muths E, Gallant AL, Campbell EHC, Battaglin WA, Green DE, Staiger JS, Walls SC, Gunzburger MS, Kearney RF (2006) The amphibian research and monitoring initiative (ARMI): 5-year report. US Geological Survey Scientific Investigations Report 2006-5224

Nazir J, Spengler L, Marschang RE (2012) Environmental persistence of amphibian and reptilian ranaviruses. Dis Aquat Org 98(3):177–184. https://doi.org/10.3354/dao02443

Niederle MV, Bosch J, Ale CE, Nader-Macías ME, Aristimuño-Ficoseco MC, Toledo LF, Valenzuela-Sánchez A, Soto-Azat C, Pasteris SE (2019) Skin-associated lactic acid bacteria from American bullfrog as potential control agents of *Batrachochytrium dendrobatidis*. PLoS One 14(9):e0223020. https://doi.org/10.1371/journal.pone.0223020

O'Hanlon S, Rieux A, Farrer RA, Rosa GM, Waldman B, Bataille A, Kosch TA, Murray KA, Brankovics B, Fumagalli M, Martin MD, Wales N, Alvarado-Rybak M, Bates KA, Berger L, Böll S, Brookes L, Clare F, Courtois EA, Cunningham AA, Doherty-Bone TM, Ghosh P, Gower

DJ, Hintz WE, Höglund J, Jenkinson TS, Lin C-F, Laurila A, Loyau A, Martel A, Meurling S, Miaud C, Minting P, Pasmans F, Schmeller DS, Schmidt BR, Shelton JMG, Skerratt LF, Smith F, Soto-Azat C, Spagnoletti M, Tessa G, Toledo LF, Valenzuela-Sánchez A, Verster R, Vörös J, Webb RJ, Wierzbicki C, Wombwell E, Zamudio KR, Aanensen DM, James TY, Gilbert MTP, Weldon C, Bosch J, Balloux F, Garner TWJ, Fisher MC (2018) Recent Asian origin of chytrid fungi causing global amphibian declines. Science 360(6389):621–627. https://doi.org/10.1126/science.aar1965

Ouellet M, Mikaelian I, Pauli BD, Rodrigue J, Green DM (2005) Historical evidence of widespread chytrid infection in North American amphibian populations. Conserv Biol 19(5):1431–1440. https://doi.org/10.1111/j.1523-1739.2005.00108.x

Peñafiel-Ricaurte A, Price SJ, Lung WTM, Alvarado-Rybak M, Espinoza-Zambrano A, Valdivia C, Cunningham AA, Azat C (2023) Is *Xenopus laevis* introduction linked with *Ranavirus* incursion, persistence and spread in Chile? PeerJ 11:e14497. https://doi.org/10.7717/peerj.14497

Price SJ, Garner TWJ, Nichols RA, Balloux F, Ayres C, Mora-Cabello de Alba A, Bosch J (2014) Collapse of amphibian communities due to and introduced *Ranavirus*. Curr Biol 24(21):2586–2591. https://doi.org/10.1016/j.cub.2014.09.028

Price SJ, Leung WTM, Owen CJ, Puschendorf R, Sergeant C, Cunningham AA, Balloux F, Garner TWJ, Nichols RA (2019) Effects of historic and projected climate change on the range and impacts of an emerging wildlife disease. Glob Chang Biol 25(8):2648–2660. https://doi.org/10.1111/gcb.14651

Puschendorf R, Carnaval AC, VanDerWal J, Zumbado-Ulate H, Chaves G, Bolaños F, Alford RA (2009) Distribution models for the amphibian chytrid *Batrachochytrium dendrobatidis* in Costa Rica: proposing climatic refuges as a conservation tool. Divers Distrib 15(3):401–408. https://doi.org/10.1111/j.1472-4642.2008.00548.x

Rebollar EA, Simonetti SJ, Shoemaker WR, Harris RN (2016) Direct and indirect horizontal transmission of the antifungal probiotic bacterium *Janthinobacterium lividum* on green frog (*Lithobates clamitans*) tadpoles. Appl Environ Microbiol 82(8):2457–2466. https://doi.org/10.1128/aem.04147-15

Richmond JQ, Savage AE, Zamudio KR, Rosenblum EB (2009) Toward immunogenetic studies of amphibian chytridiomycosis: linking innate and acquired immunity. Bioscience 59(4):311–320. https://doi.org/10.1525/bio.2009.59.4.9

Robert J, Abramowitz L, Gantress J, Morales HD (2007) *Xenopus laevis*: a possible vector of ranavirus infection? J Wildl Dis 43(4):645–652. https://doi.org/10.7589/0090-3558-43.4.645

Rödder D, Kielgast J, Bielby J, Schmidtlein S, Bosch J, Garner TWJ, Veith M, Walker S, Fisher MC, Lötters S (2009) Global amphibian extinction risk assessment for the panzootic chytrid fungus. Diversity 1(1):52–66. https://doi.org/10.3390/d1010052

Rodriguez D, Becker CG, Pupin NC, Haddad CFB, Zamudio KR (2014) Long-term endemism of two highly divergent lineages of the amphibian-killing fungus in the Atlantic Forest of Brazil. Mol Ecol 23(4):774–787. https://doi.org/10.1111/mec.12615

Rojas S, Richards K, Jancovich JK, Davidson EW (2005) Influence of temperature on Ranavirus infection in larval salamanders *Ambystoma tigrinum*. Dis Aquat Org 63(2–3):95–100. https://doi.org/10.3354/dao063095

Rollins-Smith LA, Ramsey JP, Pask JD, Reinert LK, Woodhams DC (2011) Amphibian immune defenses against chytridiomycosis: impacts of changing environments. Integr Comp Biol 51(4):552–562. https://doi.org/10.1093/icb/icr095

Ron S, Duellman WE, Coloma LA, Bustamante MR (2003) Population decline of the Jambato toad *Atelopus ignescens* (Anura: Bufonidae) in the Andes of Ecuador. J Herpetol 37(1):116–126. https://doi.org/10.1670/0022-1511(2003)037[0116:PDOTJT]2.0.CO;2

Rosenblum EB, James TY, Zamudio KR, Poorten TJ, Ilut D, Rodriguez D, Eastman JM, Richards-Hrdlicka K, Joneson S, Jenkinson TS, Longcore JE, Parra Olea G, Toledo LF, Arellano ML, Medina EM, Restrepo S, Flechas SV, Berger L, Briggs CJ, Stajich JE (2013) Complex history of the amphibian-killing chytrid fungus revealed with genome resequencing data. Proc Natl Acad Sci U S A 110(23):9385–9390. https://doi.org/10.1073/pnas.1300130110

Ruggeri J, Ribeiro LP, Pontes MR, Toffolo C, Candido M, Carriero MM, Zanella N, Sousa RLM, Toledo LF (2019) Discovery of wild amphibians infected with ranavirus in Brazil. J Wildl Dis 55(4):897–902. https://doi.org/10.7589/2018-09-224

Savage AE, Sredl MJ, Zamudio KR (2011) Disease dynamics vary spatially and temporally in a North American amphibian. Biol Conserv 144(6):1910–1915. https://doi.org/10.1016/j.biocon.2011.03.018

Scheele BC, Hunter DA, Branelly LA, Skerrat LF, Driscoll DA (2017) Reservoir-host amplification of disease impact in an endangered amphibian. Conserv Biol 31(3):592–600. https://doi.org/10.1111/cobi.12830

Scheele BC, Pasmans F, Skerratt LF, Berger L, Martel A, Beukema W, Acevedo AA, Burrowes PA, Carvalho T, Catenazzi A, De la Riva I, Fisher MC, Flechas SV, Foster CN, Frías-Álvarez P, Garner TWJ, Gratwicke B, Guayasamin JM, Hirschfeld M, Kolby JE, Kosch TA, La Marca E, Lindenmayer DB, Lips KR, Longo AV, Maneyro R, McDonald CA, Mendelson J 3rd, Palacios-Rodriguez P, Parra-Olea G, Richards-Zawacki CL, Rödel M-O, Rovito SM, Soto-Azat C, Toledo LF, Voyles J, Weldon C, Whitfield SM, Wilkinson M, Zamudio KR, Canessa S (2019) Amphibian fungal panzootic causes catastrophic and ongoing loss of biodiversity. Science 363(6434):1459–1463. https://doi.org/10.1126/science.aav0379

Schloegel LM, Hero J-M, Berger L, Speare R, McDonald K, Daszak P (2006) The decline of the sharp-snouted day frog (*Taudactylus acutirostris*): the first documented case of extinction by infection in a free-ranging wildlife species? EcoHealth 3(1):35–40. https://doi.org/10.1007/s10393-005-0012-6

Schloegel LM, Daszak P, Cunningham AA, Speare R, Hill B (2010a) Two amphibian diseases, chytridiomycosis and ranaviral disease, are now globally notifiable to the World Organization for Animal Health (OIE): an assessment. Dis Aquat Org 92(2–3):101–108. https://doi.org/10.3354/dao02140

Schloegel LM, Ferreira CM, James TY, Hipolito M, Longcore JE, Hyatt AD, Yabsley M, Martins AMCRPF, Mazzoni R, Davies AJ, Daszak P (2010b) The North American bullfrog as a reservoir for the spread of *Batrachochytrium dendrobatidis*. Anim Conserv 13(1):53–61. https://doi.org/10.1111/j.1469-1795.2009.00307.x

Schloegel LM, Toledo LF, Longcore JE, Greenspan SE, Vieira CA, Lee M, Zhao S, Wangen C, Ferreira CM, Hipolito M, Davies AJ, Cuomo CA, Daszak P, James TY (2012) Novel, panzootic and hybrid genotypes of amphibian chytridiomycosis associated with the bullfrog trade. Mol Ecol 21(21):5162–5177. https://doi.org/10.1111/j.1365-294x.2012.05710.x

Seimon TA, Seimon A, Daszak P, Halloy SRP, Schloegel LM, Aguilar CA, Sowell P, Hyatt AD, Konecky B, Simmons JE (2007) Upward range extension of Andean anurans and chytridiomycosis to extreme elevations in response to tropical deglaciation. Glob Chang Biol 13(1):288–299. https://doi.org/10.1111/j.1365-2486.2006.01278.x

Sentenac H, Valenzuela-Sanchez A, Haddow-Brown N, Delgado S, Azat C, Cunningham AA (2023) Accounting for bias in prevalence estimation: the case of a globally emerging pathogen. J Appl Ecol 60:1–11

Soto-Azat C, Valenzuela-Sánchez A, Collen B, Rowcliffe JM, Veloso A, Cunningham AA (2013a) The population decline and extinction of Darwin's frogs. PLoS One 8(6):e66957. https://doi.org/10.1371/journal.pone.0066957

Soto-Azat C, Valenzuela-Sánchez A, Clarke BT, Busse K, Ortiz JC, Barrientos C, Cunningham AA (2013b) Is chytridiomycosis driving Darwin's frogs to extinction? PLoS One 8(11):e79862. https://doi.org/10.1371/journal.pone.0079862

Soto-Azat C, Valenzuela-Sánchez A, Ortiz JC, Díaz-Páez H, Castro C, Charrier A et al (2015) ASG Chile leads update of the extinction risk of Chilean amphibians for the IUCN red list of threatened species. FrogLog 23:6–7

Soto-Azat C, Peñafiel-Ricaurte A, Price SJ, Sallaberry-Pincheira N, García MP, Alvarado-Rybak M, Cunningham AA (2016) *Xenopus laevis* and emerging amphibian pathogens in Chile. EcoHealth 13(4):775–783. https://doi.org/10.1007/s10393-016-1186-9

Stöhr AC, Hoffmann A, Papp T, Robert N, Pruvost NBM, Reyer H-U, Marschang RE (2013) Long-term study of an infection with ranaviruses in a group of edible frogs (*Pelophylax kl. esculentus*) and partial characterization of two viruses based on four genomic regions. Vet J 197(2):238–244. https://doi.org/10.1016/j.tvjl.2013.02.014

Tapia EE, Coloma LA, Pazmiño-Otamendi G, Peñafiel N (2017) Rediscovery of the nearly extinct longnose harlequin frog *Atelopus longirostris* (Bufonidae) in Junín, Imbabura, Ecuador. Neotrop Biodivers 3(1):157–167. https://doi.org/10.1080/23766808.2017.1327000

Teacher AGF, Cunningham AA, Garner TWJ (2010) Assessing the long-term impact of *Ranavirus* infection in wild common frog populations. Anim Conserv 13(5):514–522. https://doi.org/10.1111/j.1469-1795.2010.00373.x

Une Y, Sakuma A, Matsueda H, Nakai K, Murakami M (2009) Ranavirus outbreak in North American bullfrogs (*Rana catesbeiana*), Japan. Emerg Infect Dis 15(7):1146–1147. https://doi.org/10.3201/eid1507.081636

Urgiles VL, Ramírez ER, Villalta CI, Siddons DC, Savage AE (2021) Three pathogens impact terrestrial frogs from a high-elevation tropical hotspot. EcoHealth 18:451–464. https://doi.org/10.1007/s10393-021-01570-8

Uribe-Rivera DE, Soto-Azat C, Valenzuela-Sánchez A, Bizama G, Simonetti JA, Pliscoff P (2017) Dispersal and extrapolation on the accuracy of temporal predictions from distribution models for the Darwin's frog. Ecol Appl 27(5):1633–1645. https://doi.org/10.1002/eap.1556

Valencia LM, Fonte LFM (2021) Harlequin toad (*Atelopus*) conservation action plan (2021–2041). Atelopus Survival Initiative, 52 pp. www.atelopus.org

Valenzuela-Sánchez A, Schmidt BR, Uribe-Rivera DE, Costas F, Cunningham AA, Soto-Azat C (2017) Cryptic disease-induced mortality may cause host extinction in an apparently stable host-parasite system. Proc R Soc B 284(1863):20171176. https://doi.org/10.1098/rspb.2017.1176

Valenzuela-Sánchez A, O'Hanlon SJ, Alvarado-Rybak M, Uribe-Rivera DE, Cunningham AA, Fisher MC, Soto-Azat C (2018) Genomic epidemiology of the emerging pathogen *Batrachochytrium dendrobatidis* from native and invasive amphibian species in Chile. Transbound Emerg Dis 65(2):309–314. https://doi.org/10.1111/tbed.12775

Valenzuela-Sánchez A, Azat C, Cunningham AA, Delgado S, Bacigalupe LD, Beltrand J, Serrano JM, Sentenac H, Haddow N, Toledo V, Schmidt BR, Cayuela H (2022) Interpopulation differences in male reproductive effort drive the population dynamics of a host exposed to an emerging fungal pathogen. J Anim Ecol 91(2):308–319. https://doi.org/10.1111/1365-2656.13603

Voyles J, Young S, Berger L, Campbell C, Voyles WF, Dinudom A, Cook D, Webb R, Alford RA, Skerrat LF, Speare R (2009) Pathogenesis of amphibian chytridiomycosis, a cause of catastrophic amphibian declines. Science 326(5952):582–585. https://doi.org/10.1126/science.1176765

Warne RW, La Bumbard B, La Grange S, Vredenburg VT, Catenazzi A (2016) Co-infection by chytrid fungus and ranaviruses in wild and harvested frogs in the tropical Andes. PLoS One 11(1):1–15. https://doi.org/10.1371/journal.pone.0145864

Williams T, Barbosa-Solomieu V, Chinchar VG (2005) A decade of advances in iridovirus research. Adv Virus Res 65:173–248. https://doi.org/10.1016/S0065-3527(05)65006-3

Woodhams DC, Bosch J, Briggs C, Cashins S, Davis LR, Lauer A, Muths E, Puschendorf R, Schmidt BR, Sheafor B, Voyles J (2011) Mitigating amphibian disease: strategies to maintain wild populations and control chytridiomycosis. Front Zool 8(1):8. https://doi.org/10.1186/1742-9994-8-8

Woodhams DC, Bletz M, Kueneman J, McKenzie V (2016) Managing amphibian disease with skin microbiota. Trends Microbiol 24(3):161–164. https://doi.org/10.1016/j.tim.2015.12.010

Young BE, Lips KR, Reaser JK, Ibáñez R, Salas AW, Celdeño JR, Coloma LA, Ron S, La Marca E, Meyer JR, Muñoz A, Bolaños F, Chaves G, Romo D (2001) Population declines and priorities for amphibian conservation in Latin America. Conserv Biol 15(5):1213–1223. https://doi.org/10.1111/j.1523-1739.2001.00218.x

Zhang Q-Y, Xiao F, Li Z-Q, Gui J-F, Mao J, Chinchar VG (2001) Characterization of an iridovirus from the cultured pig frog *Rana grylio* with lethal syndrome. Dis Aquat Org 48(1):27–36. https://doi.org/10.3354/dao048027

Zhou ZY, Geng Y, Liu XX, Ren SY, Zhou Y, Wang KY, Huang XL, Chen DF, Peng X, Lai WM (2013) Characterization of a ranavirus isolated from the Chinese giant salamander (*Andrias davidianus*, Blanchard, 1871) in China. Aquaculture 384–387:66–73. https://doi.org/10.1016/j.aquaculture.2012.12.018

Zupanovic Z, Lopez G, Hyatt AD, Green B, Bartran G, Parkes H, Whittington RJ, Speare R (1998a) Giant toads *Bufo marinus* in Australia and Venezuela have antibodies against "ranaviruses". Dis Aquat Org 32(1):1–8. https://doi.org/10.3354/dao032001

Zupanovic Z, Musso C, Lopez G, Louriero CL, Hyatt AD, Hengstberger S, Robinson AJ (1998b) Isolation and characterization of iridoviruses from the giant toad *Bufo marinus* in Venezuela. Dis Aquat Org 33(1):1–9. https://doi.org/10.3354/dao033001

Chapter 4
Disease Ecology in Terrestrial Reptiles from Latin America: A Call for Research

Randall Arguedas and Juan Carlos Troiano

4.1 Introduction

Global changes are driving many species to extinction, and most of the currently recognized threats include human intervention and activities, such as habitat conversion, fragmentation, destruction, climate change, overexploitation, pollution, invasive species, and diseases (Tingley et al. 2016). Reptiles represent a significant proportion of terrestrial biodiversity, with reported and suspected species declining globally, and approximately 19% of the world's reptiles are considered threatened (Hellebuyck et al. 2017). To date, only 45% of the approximately 10,400 recognized reptile species have been assessed by the IUCN, and those species that have been assigned a threatened Red List status have been primarily listed based on geographical range size due to insufficient data on population trends (Tingley et al. 2016).

Furthermore, much more effort has been made in studies of threatened reptiles on marine turtles than in terrestrial reptiles (Trimble and Van Aarde 2010); at the same time, vast information has been published on one of the most important diseases in marine turtles, the herpesvirus that causes fibropapillomatosis. In the case of terrestrial reptiles, only a few viral and mycotic infectious diseases have been documented as possible conservation threats to reptile populations, but all of them have been reported for North America and Europe; thus, the role of infectious diseases as threats to terrestrial reptile populations is less understood (Hellebuyck et al. 2017). In addition, disease ecology studies probably have more information on pathogens that affect human or domestic species when reptiles act as intermediate

R. Arguedas (✉)
Vetlab, División de Especies Exóticas y Fauna Silvestre, Curridabat, Costa Rica

Universidad Técnica Nacional, Sede Atenas, Alajuela, Costa Rica

J. C. Troiano
Universidad de Buenos Aires, Buenos Aires, Argentina

G. Acosta-Jamett, A. Chaves (eds.), *Ecology of Wildlife Diseases in the Neotropics*, https://doi.org/10.1007/978-3-031-50531-7_4

hosts for vector-borne diseases. For these reasons, this chapter is focused on Latin American terrestrial reptiles.

One of the widely recognized threats for reptiles that has implications for how they respond to diseases is climate change and global warming. Ectotherm physiological performance, including immunological status, is directly affected by temperature. This means that global warming may directly affect how reptiles can deal with diseases in their environment or with introduced pathogens due to human activities. For example, it is known that lizards cannot adapt, fast enough to follow climate change because of restraints coming from their genetic arrangement for thermal preferences (Sinervo et al. 2010), which suggests that this could be the same in terms of immune system performance and adaptation to parasites.

On the other hand, reptiles from Latin America are suffering from trade to regions such as North America and Europe, but many exotic reptiles are also incoming as pets (legally or illegally), enhancing chances of introducing known diseases that occur in captivity.

This chapter is aimed at covering known diseases affecting wild reptile populations, diseases that have been found in captivity in Latin America, and robust insight into reptile immunology and the effects of climate change on the immune system. Finally, the authors state the need for research on disease ecology and give an insight into what health studies have been done in this region.

4.2 Latin American Reptile Biodiversity and Conservation

Latin America (including the Caribbean) houses enormous biodiversity and has six of the world's most biodiverse countries: Brazil, Colombia, Ecuador, Mexico, Peru, and Venezuela. This region also harbors the most biodiverse habitat in the world, the Amazon Forest. It presents a very diverse ecosystem that reptiles have colonized, such as terrestrial habitats, including tropical forests, savannahs, grasslands, and xeric environments. Microhabitats include a variety of species, including aquatic and semiaquatic organisms, mainly turtles and crocodiles and also some squamates; terrestrial, semifossorial, and fossorial species; many arboreal lizards and snakes; and coastal species. Reptiles in Latin America are also found at almost all altitudes from sea level to the high Andean mountains, such as *Liolaemus* species, which can be found up to more than 5000 m.

On the other hand, reptiles from islands, such as the Caribbean and the Galapagos archipelago, represent a unique diversity and endemism that deserves the attention of researchers and conservationists.

Due to its vast habitats and ecosystems, Latin America has high reptile biodiversity and is one of the species richness hotspots of reptiles (Roll et al. 2017). Continental and insular reptiles in America are found in three orders: Testudines, with approximately nine families; Crocodylia, with two families; Squamata, with the suborder Sauria, with approximately 26 families; and Serpentes, with

Fig. 4.1 Number of reptile species described in Latin America up to 2020, separated by regions in Mexico and Central America, the Caribbean, and South America

approximately ten families[1] (Pyron et al. 2013; Vitt and Caldwell 2013). As of December 2020, 4457 reptile species have been described (Uetz 2021) (Fig. 4.1).

Nevertheless, there has been considerable loss of some of the habitats from the Latin American region. Habitat loss due to agriculture and pasture for livestock is the most important threat to biodiversity, and even though the rate of loss has decreased during the last decade, the total area transformed per year remains high (UNEP-WCMC 2016). This region as a whole presents a rising trend in all major pressures on biodiversity: land degradation and land use change; climate change; land-based pollution; unsustainable use of natural resources and invasive alien species. According to the IUCN Red List, there are 1993 reptile species (IUCN definition for an endemic species: native to, and restricted to, a particular geographical region) and a relevant number included in different threatened categories (Fig. 4.2) (IUCN 2021). Reptiles, particularly lizards and turtles, need to be better incorporated into conservation policies and decisions, and classic conservation focused strategies, such as tropical areas, should be extended to semiarid areas and islands (Roll et al. 2017).

[1] Taxonomic and phylogenetic classification will vary according to the author.

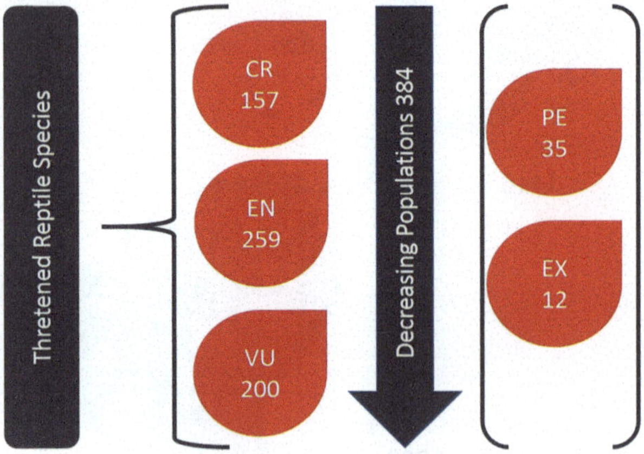

Fig. 4.2 The number of Latin American reptile species assessed by the International Union for Conservation of Nature (IUCN) in their Red List divided by the threatened categories: critically endangered (CR), endangered (EN), and vulnerable (VU), the number of reptile species of which populations are reported as decreasing and the species considered possibly extinct (PE) and extinct (EX)

4.3 An Overview of Global Knowledge of Continental Reptile Disease Ecology: Ranavirus, Herpesvirus and *Mycoplasma*, and Fungal Diseases

Continental and insular reptile parasite ecology is probably the least studied from all vertebrates around the world, and when compared to the other tetrapod taxa, there are scarce diseases that have been demonstrated to cause high morbidity and mortality in this paraphyletic group. This section describes important diseases that have been implicated in wild terrestrial reptile disease events.

4.3.1 Ranavirus

Ranavirus is an iridovirus that affects amphibians and fish but has been implicated in tortoise mortalities in the United States, specifically from Frog Virus 3 (FV3) (Price et al. 2017). This virus is a significant pathogen of chelonians, and it has been suggested that amphibians might serve as a source of infection for chelonians, or perhaps this virus exists at a low level of infection in chelonian populations that occasionally is induced into lethal expression by other stressors (Johnson et al. 2007). Fish and turtles can act as reservoirs for ranaviruses when amphibians are absent, and this interclass transmission can contribute to the persistence of these pathogens (Gray et al. 2009). In chelonians, ranavirus infections appear to be the result of a spillover from amphibian mortalities (Belzer and Seibert 2011).

Serological and PCR data in chelonian research in North America reflect a low incidence of infection, which is short-lived due to high virulence (Johnson et al. 2010; Allender et al. 2013) or because they are rapidly cleared from the host (Brunner et al. 2015).

Free-ranging gopher tortoise (*Gopherus polyphemus*), free-ranging eastern box turtle (*Terrapene carolina carolina*), and Florida box turtle (*Terrepene Carolina bauri*) mortalities have been reported from 2003 to 2005 in Georgia, Florida, and Pennsylvania (Johnson et al. 2007; Belzer and Seibert 2011), and recently the snapping turtle (*Chelydra serpentina*), which also occurs through Latin America, was reported (McKenzie et al. 2019a).

Ranavirus infection has also been described in Squamata, in the case of the suborder Serpentes, it has been reported from two families, Pythonidae: *Morelia viridis* (Hyatt et al. 2002) and *Python brongersmai* (Stöhr et al. 2015); and Viperidae: *Bothrops moojeni* (Johnsrude et al. 1997). In the case of the suborder Sauria, there are at least seven families: Agamidae: *Pogona vitticeps* (Stöhr et al. 2013) and *Japalura splendida* (Behncke et al. 2013); Anguidae: *Dopasia gracilis* (Stöhr et al. 2013); Dactyloidae: *Anolis sagrei* and *A. carolinensis* (Stöhr et al. 2013); Gekkonidae: *Uroplatus fimbriatus* (Marschang et al. 2005); Lacertidae: *Iberolacerta montícola* (Alves de Matos et al. 2011); Phrynosomatidae: *Sceloporus undulatus* (Goodman et al. 2018); and Iguanidae: *Iguana iguana* (Stöhr et al. 2013). These are mainly from captive animals brought for veterinary treatments or animals shipped for trade, and the only reports that come from wild animals are in *I. monticola* and *S. undulatus*.

Lesions found in animals affected by Ranavirus usually include ulcerative conjunctivitis and palpebral edema, nasal discharge, stomatitis with necrotizing yellow caseous plaques, aural abscesses, and splenomegaly. Histopathology usually reveals inclusion bodies in endothelial cells, tongue and hematopoietic cells, multisystemic fibrino necrotizing vasculitis, and severe fibrinous splenic necrosis (Johnson et al. 2008; McKenzie et al. 2019a).

4.3.2 Herpesvirus and **Mycoplasma** *in Chelonians*

Mycoplasma agassizi and testudinid herpesvirus (TeHV) are important causes of upper respiratory tract disease (URTD) in tortoises and are usually seen as a complex. Both agents were detected in wild desert tortoises, which supports the evidence that both *M. agassizii* and TeHV2 are agents of URTD in wild desert tortoises (Jacobson et al. 1991, 2012; Braun et al. 2014).

Clinical signs of mycoplasmosis are often a purulent nasal discharge (Jacobson et al. 1995), with occluded nares, crusts on the beak, dried mucus on the oral cavity, and eroded nostrils. Ocular signs include blepharoedema, periocular crusts, ocular discharge, protrusion of the globe, and mucus. Herpesvirus signs include plaques or ulcers in the oral cavity (Jacobson et al. 2012). This disease complex has also been found in Europe in different tortoises (Lecis et al. 2011; Kolesnik et al. 2017).

Mycoplasmas found in wild tortoises have been reported since the 1990s (Jacobson et al. 1991, 1995; Homer et al. 1998) in *Gopherus* species (desert tortoises) and have been implicated in disease. *Mycoplasma agassizii* in the desert tortoise *Gopherus agassizii* possibly contributes to declines in the abundance of this species in western North America observed in recent years (Brown et al. 1995, 2004). Another Mycoplasma, *M. testudineum*, has also been isolated from tortoises (Brown et al. 2004), but in a study in the Mojave Desert with *Gopherus agassizi*, it was not detected in wild tortoises, in contrast to *M. agassizi*, which represented 75% prevalence (Braun et al. 2014). In a study in Mexico, the authors could not find *Mycoplasma agassizi* or *M. testudineum* in *Gopherus morafkai* (Berry et al. 2015); this is the only study found in Latin America.

There are four known genetically distinct TeHV herpesviruses that can infect tortoises, referred to as TeHV1 through TeHV4. TeHV1, 2, and 3 have been found in individuals with severe stomatitis, glossitis, rhinitis, conjunctivitis, and hepatitis (Marschang 2019). In a study of *Gopherus* species, the survey revealed a 30.9% antibody prevalence for TeHV3 (Jacobson et al. 2012), but their findings found molecular evidence of TeHV2 infection in wild desert tortoises, supporting evidence of cross-reactivity between TeHV2 and TeHV3.

Detection of herpesviruses has also been performed in the family Emydidae. Infection in these animals has ranged from detection in clinically healthy animals to stomatitis, papillomatous skin lesions, rhinitis, and sudden death (Marschang 2019). The Emyididae family is one of the most distributed fresh turtle families in Latin America, and it represents a very important concern and research necessity.

4.3.3 Fungal Diseases

One of the most important diseases affecting wild terrestrial reptile populations is fungal disease, specifically snake fungal disease (SFD). Molecular analyses have revealed a genetically complex and diverse group of reptile fungal diseases, including different genera, particularly *Nannizziopsis*, *Ophidiomyces*, and *Paranannizziopsis*. Infections from these pathogens have been previously reported in several reptile species, including crocodilians, lizards, snakes, and tuataras, with special effects on wild snakes due to *Ophidiomyces* (Sigler et al. 2013; Woodburn et al. 2019).

SFD is caused by the fungus *Ophidiomyces ophiodiicola* and is an emerging infectious disease of wild snakes in North America and Europe (Lorch et al. 2016; Franklinos et al. 2017). Pathological effects initiate from an infected epidermis that becomes necrotic and thickened, producing yellow to brown crusts that are characteristic (Lorch et al. 2016). These crusts may break off, resulting in erosion or ulceration. Within necrotic skin, the fungus proliferates, and lesions may gradually expand in size (Lorch et al. 2016). Wild snakes often show several distinct lesions on various parts of the body, head, or tail but frequently cause mild infections, although severe morbidity and mortality with consequent population declines have also been observed in wild snakes, including in some threatened species snakes (Franklinos et al. 2017). To date, it is not known whether this fungus was an introduced pathogen or if it is a native pathogen in North America.

Interestingly, McKenzie et al. (2019b) found that clinical signs were a strong predictor of *O. ophiicola* presence in spring and summer seasons in Kentucky, USA, and that snakes occupying terrestrial environments had a lower overall probability of testing positive when compared to snakes occupying aquatic environments. This study indicates that both the clinical signs and the prevalence of the disease vary seasonally and are based on the habitat preferences of the host.

Moreover, Allender et al. (2015) observed activity at a range of temperatures and pH values, in addition to its ability to utilize complex carbon, nitrogen, and sulfur resources. These features allow this fungus to be present in a wide range of ecosystems, and consequently many snakes may be exposed to it.

Reported wild snake species in North America include family Colubridae, i.e., *Coluber constrictor* (Guthrie et al. 2016; Lorch et al. 2016), *Farancia abacura*, *Lampropeltis* spp. (Lorch et al. 2016), *Nerodia* spp. (Guthrie et al. 2016; Lorch et al. 2016), *Pantherophis* spp. (Lorch et al. 2016), *Thamnophis* spp. (Dolinski et al. 2014; Lorch et al. 2016), and *Virginia valeriae* (Lorch et al. 2016), as well as family Viperidae, i.e., *Sistrurus* spp. (Allender et al. 2011a, b; Lorch et al. 2016), *Crotalus horridus* (Smith et al. 2013; McBride et al. 2015; Lorch et al. 2016), and *Agkistrodon contortix* (Lorch et al. 2016). As seen, this fungus occurs on some snake genera that occur along Latin America, and due to the range of environmental conditions that the fungus tolerates, there is a strong possibility that this agent may be present.

In addition, there is a newly described fungal disease in freshwater turtles, an Onygenalean fungus associated with shell lesions (Woodburn et al. 2019). Based on morphological characteristics and the results of phylogenetic analysis, a new genus and species, *Emydomyces testavorans*, was proposed for these fungal pathogens isolated from chelonian carapace lesions (Woodburn et al. 2019). This disease and its effects on wild freshwater turtle populations are not yet known, but further research must be performed.

4.4 The Continuous Hazard of Captive Diseases Toward Wild Reptile Populations in Latin America: Paramyxovirus, Reptarenavirus, *Nanizziopsis*, and *Paranniziopsis*

Despite all the conservation and welfare issues due to illegal reptile extractions, there is a very important market for exotic and native reptiles kept as pets and in captive institutions. Many of these individuals have the potential to be resealed or escape to the wild, where they can become invasive or could potentially transmit diseases that are already reported in captive individuals, becoming a tangible epidemiological threat to wild reptiles. To date, there is very scarce information about common fatal diseases found in captive specimens affecting wild individuals, but due to climate change, there can be adaptations in those pathogens as well as immunologic impairment in reptiles that can lead to eventual outbreaks of known diseases.

Latin America has been identified as a source for legal and illegal reptile trade, mainly to the United States and Europe. Bush et al. (2014) found that of live individuals, 59.2% are wild sourced, and the only type of item predominantly sourced from captive operations is live eggs (13.1% wild sourced). In the case of reptiles (485 species), this group was the second most species-rich class reported in trade. Crocodylia is dominated by nonwild sources (90.6% nonwild), whereas Sauria (68.6% wild), Serpentes (46.6% wild), and Testudines (52.3% wild) individuals are more frequently wild sourced.

In contrast, many exotic reptile species are traded legally and illegally to Latin America, mainly for the pet market. Which lethal diseases have been found in captive reptiles in Latin America? There are several infectious diseases that have been implicated in reptile collection mortalities around the world. For most of them, the original source is not known, and at the same time, there is a large gap in information on how these pathogens may affect wild populations.

4.4.1 Paramyxovirus (Ferlavirus)

Reptile paramyxoviruses, known as ferlaviruses, have four genomic groups known as A, B, C, and tortoise (Marschang et al. 2009; La'Toya and Wellehan 2013; Kolesnik et al. 2019). This virus has been diagnosed in a wide variety of squamates, including reports in more than 25 species, more than 15 genera, and 4 different snake families (Viperidae, Colubridae, Boidae, and Pythonidae), as well as in saurians, such as caiman lizards (*Dracaena guianensis*) (Jacobson et al. 2001) and bearded dragons (*Pogona vitticeps*) (Abbas et al. 2012), *Xenosaurus* and *Abronia* (Marschang et al. 2002).

Infected animals present with severe respiratory, neurological, and immunosuppressive diseases (La'Toya and Wellehan 2013; Kolesnik et al. 2019). Ferlavirus can be transmitted by direct contact through oral, nasal, or cloacal discharges (Solis et al. 2017). All paramyxoviruses in snakes and other reptiles are associated with high mortality rates in captivity, and many reports have shown that ferlaviruses may cross species barriers (La'Toya and Wellehan 2013; Kolesnik et al. 2019). Since its initial discovery, ferlavirus infections in snakes have been implicated in disease outbreaks (Kolesnik et al. 2019), but they have not yet been implicated in wild events.

Ferlavirus has been reported in several wild and captive species in Latin America, either with antibody detection or PCR for native species (Table 4.1).

Table 4.1 Squamate species from Latin America that were positive for ferlavirus, country of origin, and reference

	Species	Country	Origin	Reference
Sauria	*Abronia graminea* (Anguidae)	Mexico	Wild	Marschang et al. (2002)
	Xenosaurus grandis (Xenosauridae)	Mexico	Wild	Marschang et al. (2002)
	Xenosaurus platyceps (Xenosauridae)	Mexico	Wild	Marschang et al. (2002)
	Ctenosaura similis (Iguanidae)	Honduras	Wild	Gravendyck et al. (1998)
	Ctenosaura bakeri (Iguanidae)	Honduras	Wild	Gravendyck et al. (1998)
	Iguana iguana (Iguanidae)	Honduras	Wild	Gravendyck et al. (1998)
Serpentes	*Boa constrictor* (Boidae)	Brazil	Captive	De Oliveira et al. (2019)
	Atropoides mexicanus (Viperidae)	Costa Rica	Captive	Solis et al. (2017)
	Atropoides picadoi (Viperidae)	Costa Rica	Captive	Solis et al. (2017)
	Bothriechis lateralis (Viperidae)	Costa Rica	Captive	Solis et al. (2017)
	Bothriechis schlegelli (Viperidae)	Costa Rica	Captive	Solis et al. (2017)
	Bothrops alternatus (Viperidae)	Brazil	Captive	Kolesnikovas et al. (2006); Nogueira et al. (2002)
	Bothrops asper (Viperidae)	Costa Rica	Captive	Solis et al. (2017)
	Bothrops jararaca (Viperidae)	Brazil	Captive	De Oliveira et al. (2019)
	Bothrops jararacussu (Viperidae)	Brazil	Captive	Nogueira et al. (2002)
	Crotalus durissus terrificus (Viperidae)	Brazil	Captive	Nogueira et al. (2002; Paiva et al. (2016)
	Crotalus simus (Viperidae)	Costa Rica	Captive	Solis et al. (2017)

4.4.2 Reptareanavirus (Boid Inclusion Body Disease)

Boid inclusion body disease (BIBD) is a disease that mainly affects Boidae and Pythonidae (Keller et al. 2017). BIBD has been associated with infection and, more recently, with coinfection by various reptarenavirus species (family Arenaviridae) (Hetlzel et al. 2013; Keller et al. 2017). BIBD has only been reported in captive snakes. The incubation period is not known, but transmission mechanisms are known to be vertical (Keller et al. 2017) and are thought to be primarily via contact with infected animals or contaminated materials, and it has been hypothesized that they may be transmitted by snake mites (La'Toya and Wellehan 2013). This disease was first described in the 1980s and 1990s, and one of the most important features was the presence of characteristic eosinophilic intracytoplasmic inclusions in neurons and epithelial cells of numerous tissues (Marschang 2020).

Clinically, BIBD is highly variable, particularly in boas, where affected animals can be free of clinical signs, die from secondary infections, or develop neurological signs, which are more pronounced in pythons (Keller et al. 2017). Central nervous system (CNS) signs are most often described (Fig. 4.3), but animals may develop anorexia, pneumonia, various skin lesions, mouth rot, and other problems (Marschang 2020).

The diagnosis of this disease can be performed by virus isolation, PCR, and light microscopy detection of IB in blood smears and liver biopsy specimens (Hetzel et al. 2013). Histologically, BIBD is characterized by typical eosinophilic to amphophilic, amorphous, intracytoplasmic IBs that are ultrastructurally electron dense and consist primarily of a protein that is found in most cell types (Hetzel et al. 2013; Keller et al. 2017).

Most snakes are infected with more than one virus strain, and these strains are undergoing recombination and reassortment. This situation was probably caused by mixing infected snakes in captivity with wild snakes, importation into several countries, and mixing of animals for breeding purposes (Stenglein et al. 2015; Marschang 2020). This may be an example of an anthropogenic cause of pathogen disruption (Stenglein et al. 2015; Marschang 2020).

There are only a few published reports of Reptarenavirus from Latin American countries, including *Boa constrictor* in Brazil (Argenta et al. 2020) and Costa Rica (Hetzel et al. 2013).

Fig. 4.3 *Boa constrictor* showing neurological signs and testing positive for BIBD after viral isolation. (Photo courtesy of Dr. Randall Arguedas)

4.4.3 **Nannizziopsis** *and* **Parannizziopsis** *(CANV Complex)*

The fungus *Chysosporium* anamorph *Nannizziopsis*, known as the CANV complex, was assigned to 16 species, either within *Nannizziopsis* or within the genera *Paranannizziopsis* and *Ophidiomyces*, and 14 of these species were newly described *Nannizziopsis*, which includes nine species associated with chamaeleonid, gekkonid, cordylid, teiid, agamid, and iguanid lizards and crocodiles. *Paranannizziopsis* includes four species that infect squamates and tuataras (Paré and Sigler 2016).

These fungi are known to be the primary etiologic agent of a deep granulomatous dermatomycosis commonly referred to as "yellow fungus disease", which initially appears as patchy yellowing of the skin that then turns dark and becomes necrotic and thickened. The crusts may come off, exposing the dermis, and the infection usually progresses to granulomatous inflammation of the deeper skin layers and is often lethal (Paré and Sigler 2016). Histopathologic findings include multifocal, chronic, ulcerative, and nodular pyogranulomatous dermatitis, with countless intralesional septate hyphae and arthroconidia (Le Donne et al. 2016).

These fungi have been reported widely in the literature from North America and Europe in captive individuals, and they have not been reported in the wild. For Latin America, there is only one report found at the moment of writing this chapter on *Iguana iguana* from Argentina, where the agent was classified as *Chysosporium* spp. (Delgado et al. 2017) (Fig. 4.4).

4.4.4 **Devrisea agamarum** *as an Example*

Devrisea agamarum is a bacterial disease reported in captive lizard collections characterized by severe dermatitis or granulomas that can sometimes result in septicemia and death, mainly in *Uromastyx* and other agamid species. Persistence of the disease within lizard collections is largely promoted by the presence of asymptomatic carriers (part of the oral microbiota) and long-term environmental survival leading to high morbidity and mortality. The presence of *D. agamarum* has been documented in healthy as well as clinically infected nondesert dwelling lizards belonging to the family Agamidae and the superfamily Iguanidae (La'Toya and Wellehan, 2013).

This bacterium has been reported to cause severe skin disease in free-ranging lesser antillean iguanas (*Iguana delicatissima*) on the French Caribbean island of Saint Barthélemy, where the authors identified *Devriesea agamarum* as the causative agent. The iguanas presented abscesses and infection in the oral cavity and cloaca. The presence of this bacterium was also demonstrated in healthy lizards (anoles) coinhabiting the island. All isolates from the iguanas corresponded to a single AFLP genotype that until now has exclusively been associated with infections in lizard species in captivity (Hellebuyck et al. 2017).

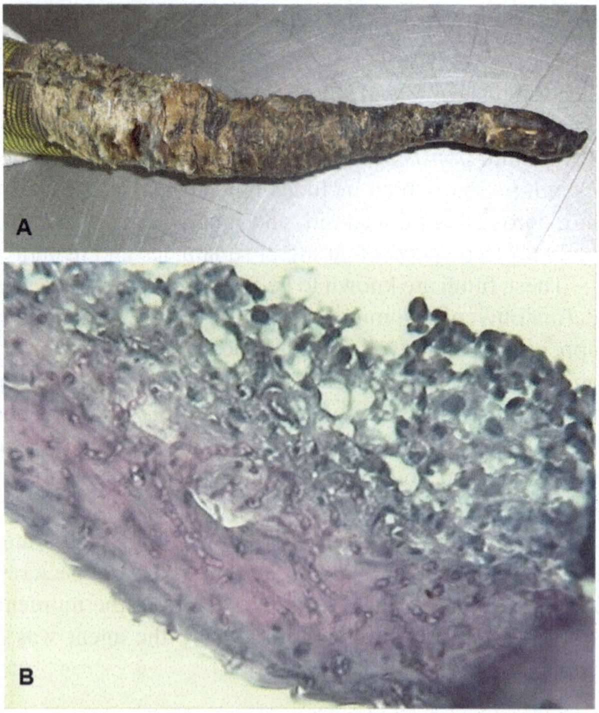

Fig. 4.4 (**a**) Green iguana (*Iguana iguana*) with skin lesions caused by *Chysosporium* spp. (**b**) Histological appearance of the skin of a green iguana (*I. iguana*) with *Chysosporium* spp. (HandE 40×). (Photo courtesy: Dr. Juan Carlos Troiano)

4.5 Reptiles as Disease Intermediate Hosts: Equine Encephalitis Virus and West Nile Virus

Reptiles have been implicated in several zoonotic vector-borne diseases, including *Leishmania* and *Trypanosoma*, *Anaplasma*, *Borrelia*, *Coxiella*, *Ehrlichia*, and *Rickettsia* (Mendoza-Roldan et al. 2021). Reptiles are also important reservoirs for arboviruses since they have low specificity for mosquitoes and flies that affect humans and other vertebrates (Mendoza-Roldan et al. 2021).

Viruses include Alphaviruses (EEV, family Togaviridae), West Nile Virus (Flaviviridae), and more recently, Chikungunya virus (Togaviridae) (Bosco-Lauth et al. 2018).

4.5.1 West Nile Virus (WNV)

WNV is a flavivirus with zoonotic implications and is the cause of a serious and lethal neurodegenerative disease in humans, horses, birds, and several wildlife species (Marschang 2011). This virus has been extensively investigated in natural hosts and various animal models, such as rodents, lagomorphs, birds, and reptiles (Sbrana et al. 2005; Ariel 2011). Mosquitoes from the genus *Culex* are the most important vector, but some other forms of transmission, including ingestion, aerosol, and direct contact, have been reported under experimental conditions (Sbrana et al. 2005). In humans and recently in breeding alligators and crocodiles, the intrauterine/oviductal route was described (Jacobson et al. 2005; Colpitts et al. 2012).

Despite the amount of generated information, its pathogenesis is still controversial, and many aspects remain to be discovered. Increasing globalization and climate change are important predisposing factors in the appearance and reappearance of the virus and disease (Kilpatrick 2011). Briefly, the life cycle of WNV involves several virus reservoirs (mainly birds without signs of clinical disease), mosquitoes as vectors, and incidental hosts. On the other hand, the final hosts are generally dead-end hosts, with the exception of crocodiles, which, unlike other final hosts, also amplify the virus (Habarugira et al. 2020). After ingestion of blood, the virus reaches the gut of the mosquito, where it is amplified and spreads to the salivary glands before infecting the final host during subsequent blood ingestion by the mosquito (Colpitts et al. 2012). Although the *Culex pipiens* mosquito is the main intermediary host, other species of mosquitoes, such as *Culex interrogator* and *C. nigripalpus,* have transmitted the virus in Mexico and other Latin American countries (Farfan-Ale et al. 2006). The role of other blood-sucking arthropods should also be highlighted, i.e., hard ticks (*Hyalomma marginatum* and *Rhipicephalus sanguineus*), soft ticks (*Ornithodoros maritimus* and *Argas hermanni*), blood-sucking bugs (*Oeciacus hirundinis*), and bird ticks (*Ornithonyssus sylviarum*) (Sardelis and Turell 2001).

The virus infects cells by means of cellular receptor-mediated endocytosis mechanisms as well as other mechanisms, such as dendritic cell-specific adhesive molecules (Davis et al. 2006). Once the virus penetrates the host cell endosomal vesicles, the E viral protein acidifies and triggers the conformational changes necessary for viral and cell membrane fusion, and viral RNA is released into the host cell cytoplasm to initiate replication. After replication in the cytoplasm of the infected cell, the new virus particles acquire a lipid envelope (Davis et al. 2006).

The virus has complex ecoepidemiological characteristics, where birds act as reservoirs and hosts, but the role of reptiles, especially in wild populations, is still poorly understood. WNV infection in terrestrial and aquatic reptiles is rare, and except for a few epidemiological studies, most of the available reports are of experimental infections. A sero-epidemiological study in Mexico found a prevalence of the virus of 41% in natural populations of *Crocodylus moreletii* compared to 30% in captive populations (Machain-Williams et al. 2013). Another study reported a prevalence of 86% in crocodiles raised in Mexico (Farfan-Ale et al. 2006). In Israel, Steinman et al. (2003) reported a prevalence as high as 70% in Nile crocodiles (*C. niloticus*). All of these reports were based on the detection of clinically healthy crocodiles. The first clinical illness due to WNV infection in crocodile species was reported in captive American alligators (*Alligator mississippiensis*), and a severe outbreak between 2001 and 2003 resulted in the death of approximately 2000 alligators (Miller et al. 2003). Clinical signs included anorexia, weakness, swimming in circles, bloody diarrhea, and scoliosis. Pathological lesions, including pulmonary congestion, hemorrhagic intestines and trachea, and hydropericardium, were found in *Crocodylus niloticus* deaths in Zambia (Simulundu et al. 2020). Other studies suggest that both American alligators and saltwater crocodiles (*C. porosus*) are also amplifiers of the virus (Klenk et al. 2004).

Cutaneous lesions were found in crocodiles; they appeared 4 or 5 weeks after the acute infection. Histologically, the lesions are a cutaneous lymphocytic proliferative syndrome. These lesions are also known as "pix" ("Pix" is defined as areas of approximately 1–2 mm in crocodile skin, usually detected during skin processing). These lesions are commonly found in the ventral region extending from the ventral caudal part of the neck to the ventral pelvis (Nevarez 2007).

In garter snake (*Thamnophis sirtalis sirtalis*) experimental infections, approximately 56% of the animals developed viremia and seroconversion, which makes these amplifying host snakes competent. On the other hand, experimentally infected snakes showed a variety of clinical forms, from sudden death after an asymptomatic infection to clinical signs such as paralysis, weakness, loss of appetite, and cachexia. The lethargic state of these diseased animals did not allow them to survive natural predation, thus potentially becoming the source of infection for their predators. At necropsy, the virus was isolated from various organs, such as the liver, kidney, heart, intestine, skin, and skeletal muscle. The main lesions observed were histiocytic hepatitis and multifocal splenic inflammation. The presence of the viral antigen was also detected in the cytoplasm of macrophages of the liver and spleen (Steinman et al. 2006). Free-ranging *T. s. sirtalis* tested positive with PCR, which confirms the presence of the virus in wild populations (Dahlin et al. 2016).

It is speculated that reptiles develop a slower humoral immune response to WNV infection due to their slow metabolism and lack of lymph nodes. This implies that it takes several weeks or even months for the antibodies to reach a peak (Habarugira et al. 2020).

WNV is an arbovirus with a complex life cycle that requires an interaction between vectors, vertebrates acting as reservoirs, and the final host, all of which interact in the viral life cycle. Climatic conditions, particularly environmental factors (temperature and rain), are predominant factors in the mosquito life cycle and the WNV amplification and dynamics of infection (Sardelis and Turell 2001). The possibility that WNV can be introduced in nonendemic areas is a close scenario since mosquitoes are transported by different means, including bird migration and trade, other vertebrates, and human movements. On the other hand, the alteration of the natural habitat of virus reservoirs by anthropogenic factors such as land use, urbanization, and agriculture also contributes to dispersal to and from natural populations.

4.5.2 Equine Encephalitis Virus (EEV)

Reptiles are well-known reservoirs for Alphaviruses, and experimental and wild species surveillance has been performed. Experimental inoculation with EEV resulted in prolonged viremia, as reported in garter snakes (Thomas et al. 1958) and Texas tortoises (*Gopherus berlandieri*) following inoculation with the virus (Bowen 1977). It has been found that the viral titers achieved in snakes were lower than those observed in birds but for a longer time and that snakes can even remain viremic during hibernation (Thomas et al. 1980; White et al. 2011).

Anole lizards (*Anolis carolinensis*) also show susceptibility to viral infection but with low viremia. In addition, other reptile species are mentioned as species that can harbor EEV antibodies (*Coluber constrictor*, *Sternotherus odoratus*, *Trachemys scripta*, *Kinosternon subrubrum*, and *Storeria dekayi*) (Graham et al. 2012). These differences could be due to many factors, including capture site, differences in the total number of sampled specimens, and differences in the assay-specific sensitivity. However, a seropositive finding is only indicative of virus exposure and not necessarily a latent infection (Graham et al. 2012).

In Latin America, outbreaks of Venezuelan equine encephalitis (VEE) in humans and equines in Venezuela suggested a sylvatic focus of an epizootic subtype of VEE virus that includes reptiles (Walder et al. 1984).

Mosquito species are also a fundamental component in viral propagation. At this point, some researchers mention that *Culex peccator* and *C. erraticus* feed primordially in ectothermic hosts and prefer snakes before amphibians (Graham et al. 2012; Klenk et al. 2004). It is known that the feeding strategies of the different mosquito species act as a limitation to exposure and that ectothermic animals can be overlooked reservoirs and perhaps an important winter reservoir for EEV (Kilpatrick 2011).

4.6 An Ecological View of the Reptile Immune System Facing Climate Change and Its Implications for Disease Vulnerability

4.6.1 Receptor Immune System

All multicellular organisms have some type of innate immunity, from simple antimicrobial peptides to complex phagocytic cells, and knowledge of the immune response and the immunological dynamics that occur in reptiles is important, especially in regard to diseases caused by micro- and macroparasites.

It is necessary to highlight the importance of temperature influence on the physiological processes and immune responses of reptiles. In addition, the animal's health status (preexisting parasite infection and nutritional status), population dynamics, seasonal influence (linked to environmental temperature), and type of antigen must be considered.

The reptile immune system is organized in a similar way as other vertebrates, lymphoid organs, and tissues; innate immunity comprises cells and molecules (e.g., lysozymes or cytokines) and adaptive immunity with antibodies and lymphocytes (Fig. 4.5).

The most important lymphoid tissues that intervene in the reptile immune response are the thymus, spleen, and lymphoid tissues associated with the intestinal submucosa and bone marrow (Sherif and El Ridi 1992; Buchmann 2014).

The thymus is morphologically similar to that of higher vertebrates (avian and mammalian species). It is slightly differentiated by the absence of Hassal corpuscles and by the presence of a large number of myoid cells in adult reptiles. This organ presents, with the exception of members of the order Sphenodontia (tuataras), important characteristics such as the existence of periodic involution phases according to temperature, which generates variation in the reptilian immune response (Muñoz et al. 2000; Origgi 2007). The spleen is also microscopically similar to that observed in higher vertebrates and presents seasonal variations. The white pulp presents lymphocyte-bordering arteriole areas, and the red pulp is separated from the previous pulp by reticular flat cells (Muñoz et al. 2000).

Reptiles do not present lymph nodes; they are called lymphoid accumulations and are associated with the intestinal submucosa in crocodiles, turtles, lizards, and snakes (Riera Romo et al. 2016). Some authors correlate these accumulations with intestinal Peyer's patches (Origgi 2007). These accumulations are associated with the *lamina propria* and the submucosa of the digestive tract in the ileus, ileocecal junction, colon, and cloaca (GALT). The predominant cells in these structures are

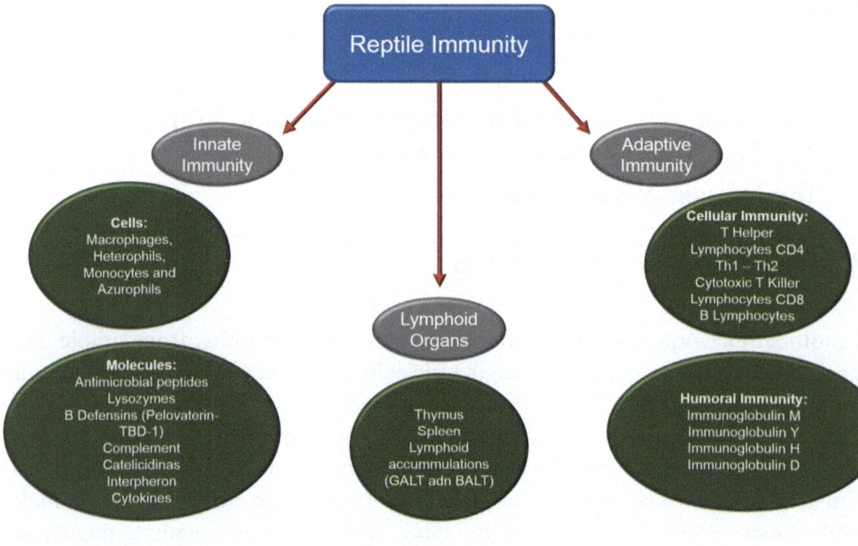

Fig. 4.5 Schematic summary of the reptile immune system

lymphocytes and plasma cells. Outside of the location in the digestive tract, similar structures are found in various organs, such as the lungs (snakes and turtles), kidneys (turtles), pancreas (lizards), axillary region (turtles and lizards), tonsils (snakes), urinary bladder wall (turtles), and testes (lizards) (Origgi 2007).

The innate immune system comprises a variety of molecules and cells that act as the first barrier against infection. Some of these molecules are antimicrobial peptides (lysozymes), the complement system, and nonspecific phagocytic cells (Zimmerman et al. 2010).

Lysozymes are enzymes that promote bacterial lysis by hydrolysis of their cell wall and have been isolated from Asian aquatic turtles of *Tryonix*, *Pelodiscus*, and *Amyda* species. Additionally, antimicrobial proteins similar to defensins with a cysteine-rich domain and six disulfuric bridges, the β-defensins, have been described. Peptides or substances such as hydrophobic and anionic pelovaterin with bactericidal action on Gram-negative bacteria have been isolated from the eggs of some species of marine turtles (Van Hoek 2014).

Other peptides isolated from the different tissues of reptiles are cathelicidins, linear proteins without disulfide bridges such as cathelicidin-BF isolated from elapid *Bungarus fasciatus* venom, and an anionic protein isolated from *Crocodylus siamensis* with antimicrobial activity against strains of *Salmonella*, *Klebsiella*, and *Pseudomonas* (Van Hoek 2014).

The complement system is also part of innate immunity, and in reptiles, the activity of the classical pathway and the alternative pathway that produces opsonization and bacterial lysis has been demonstrated (Buchmann 2014).

Nonspecific leukocyte cells that are involved in phagocytosis phenomena include macrophages, monocytes, heterophiles, basophils, eosinophils, and even thrombocytes (Stacy et al. 2011). An important function of the innate immune system is the response to infections via the inflammatory response. It is known that there is an increase in total leukocyte counts during higher temperature seasons (i.e., summer), which highlights the importance of thermoregulation in reptile immunity (Troiano et al. 2008).

The cellular response of reptiles is different from that of mammals; extracellular pathogens induce the formation of granulomas, with accumulation of degranulated heterophiles and subsequent necrosis, while intracellular pathogens form histiocytic granulomas (Origgi 2007).

The inflammatory response is driven by a series of cytokines and interleukins, including tumor necrosis factor α, interleukin-6, and interleukin-1. Another different class of cytokines is interferons, which are known for their interference in viral replication and have been sequenced in terrestrial turtle species (Montali 1988; Terio 2004).

Adaptive immunity includes immunity mediated by T lymphocytes, which regulate the production of antibodies. These cells are found in all reptiles and are capable of differentiating into two cell types, the Tc (or cytotoxic) lymphocyte and the Th (helper) lymphocyte, responsible for the regulation of the immune response. The processes of cellular immunity in reptiles are influenced by seasonal cycle, age, sex, and the presence of environmental pollutants such as chlorinated compounds and mercury (Riera Romo et al. 2016).

The presence of the major histocompatibility antigen has also been described, which is highly dependent on temperature and health, especially nutrition. In terrestrial turtles, it has been reported that allograft rejection occurs between 20 and 40 days with temperatures between 20 and 25 °C, and at a temperature of 10 °C, the response is null (Glinski and Buczek 1999).

The cell-mediated immunological response is similar to that found in mammalian species. Examples of these phenomena are the presence of a delayed skin reaction of hypersensitivity to tuberculin and alloantigens, rejection of allografts and xenografts, and phagocytosis (Glinski and Buczek 1999). As occurs with mammals, the immune-mediated response by cells can be measured by in vitro cell proliferation assays using whole blood or peripheral blood mononuclear cells cultured with phytohemagglutinin, concanavalin A, or soluble egg lysozyme (Work et al. 2000). In saurians, cases of graft rejection are mentioned within 2 months at a temperature of 24 °C and 245 days at 20 °C, where encapsulated cells such as lymphocytes, monocytes, macrophages, and plasma cells are found in the reactive area.

The humoral immune response is regulated by B lymphocytes and is also highly influenced by environmental temperature. In addition, the antigen concentration, route of administration, type of antigen, and adjuvant must be considered. Antibody molecules are composed of two identical heavy chains and two identical light chains linked by sulfur bridges. Both the heavy chain and the light chain contain a constant region and a variable region. Together, the variable regions of both chains form the antigen-binding site. The constant region determines the immunoglobulin isotype, antibodies are divided into classes based on the heavy chain isotype, and each class has different functions. Reptiles mainly produce two classes of immunoglobulins, IgM and IgY. The first is produced in response to Gram-negative bacteria and is a lytic antibody with a half-life of 10 days, while IgY has a shorter life span, is produced in large quantities in response to infections, and is transmitted from the mother to the embryo through the egg yolk. IgY is often considered the functional equivalent of mammalian IgG and is expressed in the liver and spleen (Pettinello and Dooley 2014).

In chelonians, a very rapid and very high primary and secondary response is found in summer and a slower and lower response is found in winter. For example, at 25–27 °C, an antibody response occurs at 27–30 days, while at 10 °C, it is suppressed. In *Sphenodon* spp., antibodies against *Salmonella* antigens are found in 60–80 days at a temperature of 20 °C. It is very important to remember that the preferred optimum temperature for this animal is 10–12 °C (Pettinello and Dooley 2014).

In saurians exposed to equine encephalomyelitis virus, there is no immune response at 4 °C; at 20 °C, the response occurs in 50 days; at 30 °C, it takes only 20 days; and at 37 °C, the primary and secondary immune responses are produced more quickly. On the other hand, it is known that certain species of lizards of the genus *Egernia* do not produce an immune response at a temperature of 20 °C, considering that this species has an optimal temperature of 36 °C (Wright and Schapiro 1973).

In the case of snakes, a similar behavior was found in terms of the production of antibodies according to the ambient temperature. The higher the temperature is, the faster and higher the antibody titer, for example, against mammalian erythrocytes (Pettinello and Dooley 2014). In crocodilians, the maximum formation of antibodies occurs after 25 days at a temperature of 37 °C, while at 20 °C, the response is lower and has a lower titer (Origgi 2007).

4.6.2 Facing Climate Change and Its Implications for Disease Vulnerability

Organisms may exhibit different responses to global climate change. Given sufficient time and dispersal abilities, specy distributions may shift to more favorable thermal environments; another possibility is that organisms may adjust to the new local climate either by acclimation plasticity or adaptive evolutionary responses. However, there can also occur an undesirable response, which is a failure to address the new climatic conditions, leading to the extirpation of local populations and possibly the extinction of a given species (Piantoni et al. 2016; Sinervo et al. 2018).

Due to their marked dependence on environmental factors (mainly temperature), reptiles are particularly vulnerable to thermal fluctuations that directly influence their physiology and behavior, which affect their performance and fitness. For example, body temperature fluctuations in lizards affect several physiological and self-maintenance cornerstones, including their digestion, metabolism, growth reproduction, and susceptibility to diseases (Laspiur et al. 2021). Among the most insidious results of climate change and its effect on reptile populations are some sublethal effects that are capable of indirectly causing an increase in morbidity and mortality rates due to diseases, which may lead to a gradual and progressive decline in natural reptile populations.

Immunological conditions are crucial to survival, as they directly influence the physiology and health status of the animal. Extreme weather events such as floods and droughts cause prolonged glucocorticoid elevation (Martin et al. 2010; Refsnider et al. 2015), and glucocorticoid elevation often correlates with depressed immune function (Millet et al. 2007). In addition, shifts in the thermal environment demand increased time–energy allocation to thermoregulation, and food resources are limited. Ectotherms may have to decrease immune function to maintain a positive energy balance (McCoy et al. 2017).

Consequently, populations exposed to prolonged periods of environmental disturbance, such as higher temperatures and a higher frequency of extreme weather events associated with global warming (Solomon et al. 2007), are likely to experience physiological stress in the form of increased glucocorticoid levels and tend to have significantly higher levels of corticosterone than those living in more benign environments (Saad and El Ridi 1988; Munoz and la Fuente 2001; Martin et al. 2010).

High glucocorticoid levels can lead to decreased immune function and can increase mortality in vulnerable populations (Jessop et al. 2013). The effects on corticosterone dynamics and immune function impairment have been demonstrated in freshwater turtles (*Macroclemyys temmincki*) and were negatively correlated with bactericidal capacity in birds (Millet et al. 2007) and body condition in snakes (Moore et al. 2000).

The importance of environmental temperature in reptiles has been demonstrated in the role of incubation temperature and how these variables affect the immune function of soft-shell turtle *Pelodiscus sinensis* neonates. When this species was exposed to bacterial infections, the offspring that hatched at 24 °C had lower cumulative mortality (55%; therefore, greater immunocompetence) than those incubated at 28 or 32 °C (100%). Likewise, the incubation temperature had higher IgM, IgD, and CD3γ expression than their counterparts at the other two higher incubation temperatures. In contrast, the activity of enzymes related to immunity did not coincide with the difference between temperatures in immune function. Specifically, enzyme activity, such as alkaline phosphatase, was higher at intermediate temperatures, but other enzymes, such as acid phosphatase or lysozymes, were not affected by incubation temperature. These results indicate that the reduced immunity induced by high development temperatures may increase the vulnerability of reptiles to outbreaks of diseases under global warming scenarios (Martin et al. 2010).

In another study carried out in Argentina with wild populations of *Liolaemus sarmentoi*, it was analyzed how body temperature, speed to escape, locomotor resistance, and body condition can be affected by immune status by analyzing the leukocyte profiles by estimating the percentages of cells such as heterophiles, eosinophils, basophils, lymphocytes, and monocytes. In particular, they studied how lizards' immune status induces a variation in thermoregulation capacity and a reduction in locomotor performance. The conclusions were that juveniles of *Liolaemus sarmentoi* with a high percentage of heterophils showed less locomotor resistance, and adult lizards and females with eggs did not show differences in locomotor performance in relation to their leukocyte profile (Duran et al. 2019).

This study represents a starting point in the eco-immunology of a wild lizard population of *Liolaemus* in the cold and temperate environments of Patagonia, where the southward movement in the geographical ranges of pathogen populations due to global warming represents a threat to the natural and wild populations. One of the most important conclusions of this experiment is that when reptiles assign physiological resources to immune processes, they can suffer a general decrease in their body condition (Refsnider et al. 2015), which generally affects the performance and interactions of the individual with the environment (Duran et al. 2019).

Lizard populations with low body condition have also been studied, which may be limited in their ability to increase their temperature to counteract pathogens due to the energy costs of maintaining a high temperature. This is the case for juvenile green iguanas (*Iguana iguana*) and green anoles (*Anolis carolinensis*) with low energy reserves (poor body condition) that developed hypothermia as a defense strategy when they were experimentally infected with *Escherichia coli* lipopolysaccharide cell walls in order to conserve the energy reserves of the individual (Merchant et al. 2008; Voogdt et al. 2016).

Infected or unhealthy states not only lead to a change in the blood cell profile (Zamora-Camacho et al. 2015) but could also affect much of the general physiology of a given individual, as evidenced by studies with *Podarcis lilfordi* lizards, which exhibit fast speed and the best corporal condition when compared with similar populations showing high parasitemia by hemoparasites (Zamora-Camacho et al. 2015). Another study with snake fungal disease in *Sistrurus miliaris* demonstrated that the severity of clinical signs varies seasonally and in relationship to seasonal variation in air temperature and the mean energetic status of the individuals (McCoy et al. 2017).

In summary, the effect of environmental changes on the physiological ability of reptiles to deal with pathogens is an ongoing menace that exerts strong selection pressure on these groups. It is expected, according to the previous studies mentioned, that reptile populations could be affected by the cost of reduction in thermoregulation performance resulting in an impaired immunological and physical response to disease.

4.7 The Need for Terrestrial Reptile Disease Ecology Studies in Latin America: A Call for Research

There is an important account of literature reporting or describing macroparasites of Reptilia in Latin America, mainly Helminths, Arthropods, and Protozoans. This information is very valuable and helpful for understanding what exists in the wild but does not evaluate parasite ecology or pathogen–host dynamics.

Research that evaluates reptile health in a more integral way is also found for neotropical reptiles and provides better insight into the health status of a given population based on physiological parameters, which finally show a picture of important ecological aspects such as the cellular immune system, morphology, and other physiological traits that can be affected when ecological disruption occurs. Health assessments are valuable when baseline data on normal health parameters are available when an apparently normal population is available (Arguedas et al. 2021), especially when such assessments offer the only available data for a species (Innis 2014). This information helps to identify potential effects of disease or other changing environmental conditions that would be difficult to understand without knowledge of normal species-specific variations in physiological variables (Smyth et al.

2014; Lewbart et al. 2015) and represents a way to understand how wild animals are impacted by and respond to these environmental stressors (Altizer et al. 2013) and to monitor the health and resilience of wild populations (Stacy et al. 2011; Maceda-Veiga et al. 2015).

There are several native species that have been included in health status studies in Latin America (Table 4.2). Many of them have restricted geographical distribution. Noticeably, most of the studies have been performed with saurians.

Although this kind of study is very valuable, there is a lack of research that evaluates wild species morbidity and mortality or studies that at least look for known diseases that affect reptiles at other latitudes, such as the ones explained above.

Table 4.2 Health assessment performed in the wild reptile population from Latin America, the country where the study was performed, the geographical distribution of the studied species and the reference

Order	Family	Species	Country of the study	Distribution	Reference
Squamata (Sauria)	Iguanidae	*Amblyrhynchus cristatus*	Ecuador	Galápagos	Lewbart et al. (2015)
	Iguanidae	*Conolophus subcristatus*	Ecuador	Galápagos	Lewbart et al. (2019)
	Iguanidae	*Conolophus pallidus*	Ecuador	Galápagos	Lewbart et al. (2019)
	Iguanidae	*Cyclura cychlura inornata*	Bahamas	Northern Exuma Cays	James et al. (2006)
	Iguanidae	*Cyclura ricordii*	Dominican Republic	Hispaniola Island	Maria et al. (2007)
	Iguanidae	*Cyclura rileyi rileyi*	Bahamas	San Salvador Island	Kishbaugh et al. (2020)
	Iguanidae	*Cyclura lewisi*	Cayman Islands	Grand Cayman	Rainwater et al. (2021)
	Corytophanidae	*Basiliscus plumifrons*	Costa Rica	Honduras, Nicaragua Costa Rica, Panama	Dallwig et al. (2011)
	Polychrotidae	*Polychrus gutturosus*	Costa Rica	Honduras, Nicaragua, Costa Rica, Panama, W Ecuador, E Colombia	Arguedas et al. (2021)
	Tropiduridae	*Microlophus bivittatus*	Ecuador	San Cristobal Island	Arguedas et al. (2018)
	Helodermatidae	*Heloderma horridum*	Mexico	Mexico	Espinosa-Avilés et al. (2008)
Crocodylidae	Alligatoridae	*Caiman crocodilus*	Peru	Mexico through Brazil and Peru	Carlos et al. (2017)

In addition, there is no information about known captive diseases affecting reptiles, i.e., it is not yet known whether the viruses (Paramyxovirus or Reptarenavirus) are circulating in the wild or if they are endemic (it has been suggested), which always bears the possibility of evolving into more pathogenic forms due to changes in the host immune system because of climate change or pollution.

4.7.1 Proposed Model for Latin American Reptile Disease Ecology

As a first step of research, disease monitoring of reptiles that are legally imported or exported can be beneficial to understand what is happening with disease introductions. At the same time, a quest for known diseases in wild reptile populations that affect reptiles at other latitudes as well as diseases that affect captive reptiles must be made.

Since information on pathogen dynamics in the population will be assessed, it is important to generate experiments or models that examine the initial host–pathogen process in a particular environment or community and the way it will affect population dynamics.

As has been exposed, all the agents that have been implicated in reptiles infect multiple species that may coexist in a geographical area; thus, pathogens can be maintained in a population and in a community. Therefore, one important goal in the research process is to assess whether each infected species is infected via a dead-end process or is contributing to maintenance by ongoing transmission.

As explored above, immunity and population health status are very important in reptiles due to their susceptibility to environmental changes; hence, in a given reptile population, immune responses may focus on a specific pathogen or may act on an entire community, and these interactions may be restricted to specific immunological responses (i.e., cellular, humoral, or both). Thus, the type of immune response may modulate host population assembly and host–pathogen dynamics (Fig. 4.6).

4.8 Conclusions

It is evident that the vast and biodiverse region of Latin America lacks studies in reptile disease ecology and that only isolated efforts that can give a clear window of what is happening in wild reptile communities have to be addressed. This chapter places the need to conduct research on wild Latin American reptile disease ecology since climate change is affecting the capacity of reptiles to deal with and adapt to

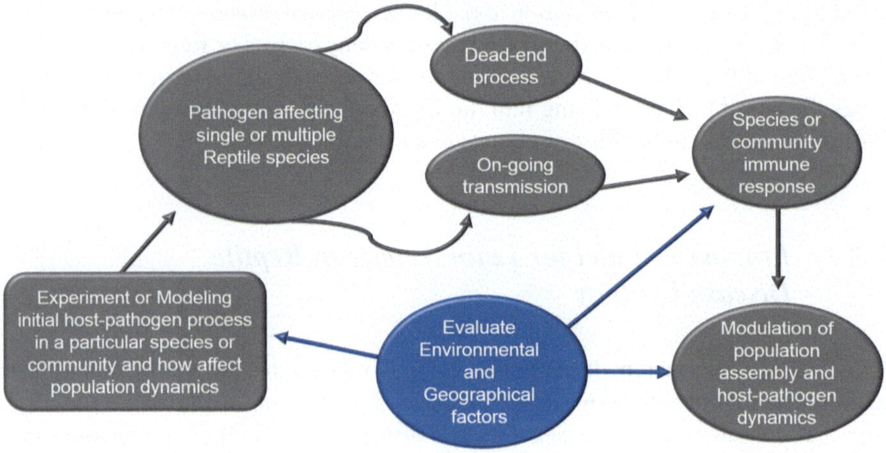

Fig. 4.6 Proposed research model for disease ecology studies in Latin American wild reptile populations

rising disease threats that the scientific community fails to understand; thus, remedial actions and conservation efforts cannot be proposed if a fatal event occurs in a threatened or endemic species.

References

Abbas MD, Ball I, Ruckova Z, Öfner S, Stöhr AC, Marschang RE (2012) Virological screening of bearded dragons (Pogona vitticeps) and the first detection of paramyxoviruses in this species. J Herpetol Med Surg 22(3–4):86–90

Allender MC, Dreslik M, Wylie S, Phillips C, Wylie DB, Maddox C, Delaney MA, Kinsel MJ (2011a) Chrysosporium sp. infection in eastern massasauga rattlesnakes. Emerg Infect Dis 17(12):2383

Allender MC, Dreslik MJ, Wylie SJ, Phillips CA, Wylie D, Delaney MA, Kinsel M (2011b) An unusual mortality event associated with Chrysosporium in the Eastern Massasauga (Sistrurus catenatus catenatus). Emerg Infect Dis 17:2383–2384

Allender MC, Mitchell MA, Torres T, Sekowska J, Driskell EA (2013) Pathogenicity of frog virus 3-like virus in red-eared slider turtles (Trachemys scripta elegans) at two environmental temperatures. J Comp Pathol 149(2–3):356–367

Allender MC, Raudabaugh DB, Gleason FH, Miller AN (2015) The natural history, ecology, and epidemiology of Ophidiomyces ophiodiicola and its potential impact on free-ranging snake populations. Fungal Ecol 17:187–196

Altizer S, Ostfeld RS, Johnson PT, Kutz S, Harvell CD (2013) Climate change and infectious diseases: from evidence to a predictive framework. Science 341(6145):514–519

Alves de Matos PA, da Silva Trabucho MFA, Papp T, Matos BADCA, Correia ACL, Marschang RE (2011) New viruses from Lacerta monticola (Serra da Estrela, Portugal): further evidence for a new group of nucleo-cytoplasmic large deoxyriboviruses. Microsc Microanal 17(1):101–108

Argenta FF, Hepojoki J, Smura T, Szirovicza L, Hammerschmitt ME, Driemeier D, Kipar A, Hetzel U (2020) Identification of reptarenaviruses, hartmaniviruses, and a novel chuvirus in captive native Brazilian boa constrictors with boid inclusion body disease. J Virol 94(11):e00001-20

Arguedas R, Steinberg D, Lewbart GA, Deresienski D, Lohmann KJ, Muñoz-Pérez JP, Valle CA (2018) Hematology and biochemistry of the San Cristóbal Lava Lizard (Microlophus bivittatus). Conserv Physiol 6(1):coy046

Arguedas R, Ovares L, Arguedas VP, Vargas R, Barquero MD (2021) Health status of Polychrus gutturosus based on physical examination, hematology and biochemistry parameters in Costa Rica. PeerJ 9:e10649

Ariel E (2011) Viruses in reptiles. Vet Res 42(1):1–12

Behncke H, Stöhr AC, Heckers KO, Ball I, Marschang RE (2013) Mass-mortality in green striped tree dragons (Japalura splendida) associated with multiple viral infections. Vet Rec 173(10):248–248

Belzer WR, Seibert S (2011) A natural history of Ranavirus in an eastern box turtle population. Turtle Tortoise Newsl 15:18–25

Berry KH, Brown MB, Vaughn M, Gowan TA, Hasskamp MA, Torres MCM (2015) Mycoplasma agassizii in Morafka's desert tortoise (Gopherus morafkai) in Mexico. J Wildl Dis 51(1):89–100

Bosco-Lauth AM, Hartwig AE, Bowen RA (2018) Reptiles and amphibians as potential reservoir hosts of Chikungunya virus. Am J Trop Med Hyg 98(3):841

Bowen GS (1977) Prolonged western equine encephalitis viremia in the Texas tortoise (Gopherus berlandieri). Am J Trop Med Hyg 26(1):171–175

Braun J, Schrenzel M, Witte C, Gokool L, Burchell J, Rideout BA (2014) Molecular methods to detect Mycoplasma spp. and Testudinid herpesvirus 2 in desert tortoises (Gopherus agassizii) and implications for disease management. J Wildl Dis 50(4):757–766

Brown DR, Crenshaw BC, McLaughlin GS, Schumacher IM, McKenna CE, Klein PA, Jacobson ER, Brown MB (1995) Taxonomic analysis of the tortoise mycoplasmas Mycoplasma agassizii and Mycoplasma testudinis by 16S rRNA gene sequence comparison. Int J Syst Evol Microbiol 45(2):348–350

Brown DR, Merritt JL, Jacobson ER, Klein PA, Tully JG, Brown MB (2004) Mycoplasma testudineum sp. nov., from a desert tortoise (Gopherus agassizii) with upper respiratory tract disease. Int J Syst Evol Microbiol 54(5):1527–1529

Brunner JL, Storfer A, Gray MJ, Hoverman JT (2015) Ranavirus ecology and evolution: from epidemiology to extinction. In: Ranaviruses. Springer, Cham, pp 71–104

Buchmann K (2014) Evolution of innate immunity: clues from invertebrates via fish to mammals. Front Immunol 5:459

Bush ER, Baker SE, Macdonald DW (2014) Global trade in exotic pets 2006–2012. Conserv Biol 28(3):663–676

Carlos N, Nuñez del Prado Y, Quispe D, Capuñay C (2017) Valores hematológicos y bioquímica sérica del Caimán Blanco (Caiman crocodilus) de vida libre en Madre de Dios, Perú. Rev Invest Vet Perú 28(1):189–194

Colpitts TM, Conway MJ, Montgomery RR, Fikrig E (2012) West Nile virus: biology, transmission, and human infection. Clin Microbiol Rev 25(4):635–648

Dahlin CR, Hughes DF, Meshaka WE Jr, Coleman C, Henning JD (2016) Wild snakes harbor West Nile virus. One Health 2:136–138

Dallwig RK, Paul-Murphy J, Thomas C, Medlin S, Vaughan C, Sullivan L, Sladky KK, Ramirez O, Herrera G (2011) Hematology and clinical chemistry values of free-ranging basilisk lizards (Basiliscus plumifrons) in Costa Rica. J Zoo Wildl Med 42(2):205–213

Davis CW, Nguyen HY, Hanna SL, Sánchez MD, Doms RW, Pierson TC (2006) West Nile virus discriminates between DC-SIGN and DC-SIGNR for cellular attachment and infection. J Virol 80(3):1290–1301

De Oliveira CC, Melo L, Ullmann L, Malossi C, Ferreira RS, Barraviera B, Araujo JP, Dos Santos LD (2019) Ophidian paramyxovirus: a serum epidemiological study in captive snakes. Toxicon 168:S40

Delgado AR, Lorente JA, Troiano JC, Rejf PK (2017) Isolamento de Chrysosporium spp em dermatite em Iguana iguana-primeiro relato na Argentina. Clín Vet 22 (127):82–85

Dolinski AC, Allender MC, Hsiao V, Maddox CW (2014) Systemic Ophidiomyces ophiodiicola infection in a free-ranging plains garter snake (Thamnophis radix). J Herpetol Med Surg 24(1–2):7–10

Duran F, Boretto JM, Fernandez JB, Molina MI, Medina MS, Ibargüengoytía NR (2019) Impact of immunological state on eco-physiological variables in one of the southernmost lizards in the world. An Acad Bras Cienc 91:e20190055

Espinosa-Avilés D, Salomón-Soto VM, Morales-Martínez S (2008) Hematology, blood chemistry, and bacteriology of the free-ranging Mexican beaded lizard (Heloderma horridum). J Zoo Wildl Med 39(1):21–27

Farfan-Ale JA, Blitvich BJ, Marlenee NL, Loroño-Pino MA, Puerto-Manzano F, García-Rejón JE, Rosado-Paredes EP, Flores-Flores LF, Ortega-Salazar A, Chávez-Medina J, Cremieux-Grimaldi JC (2006) Antibodies to West Nile virus in asymptomatic mammals, birds, and reptiles in the Yucatan Peninsula of Mexico. Am J Trop Med Hyg 74(5):908–914

Franklinos LH, Lorch JM, Bohuski E, Fernandez JRR, Wright ON, Fitzpatrick L, Petrovan S, Durrant C, Linton C, Baláž V, Cunningham AA (2017) Emerging fungal pathogen Ophidiomyces ophiodiicola in wild European snakes. Sci Rep 7(1):1–7

Glinski Z, Buczek J (1999) Aspects of reptile immunity. Med Weter 55:574–578

Goodman RM, Hargadon KM, Carter ED (2018) Detection of ranavirus in eastern fence lizards and eastern box turtles in Central Virginia. Northeast Nat 25(3):391–398

Graham SP, Hassan HK, Chapman T, White G, Guyer C, Unnasch TR (2012) Serosurveillance of eastern equine encephalitis virus in amphibians and reptiles from Alabama, USA. Am J Trop Med Hyg 86(3):540

Gravendyck M, Ammermann P, Marschang RE, Kaleta EF (1998) Paramyxoviral and reoviral infections of iguanas on Honduran Islands. J Wildl Dis 34(1):33–38

Gray MJ, Miller DL, Hoverman JT (2009) Ecology and pathology of amphibian ranaviruses. Dis Aquat Org 87(3):243–266

Guthrie AL, Knowles S, Ballmann AE, Lorch JM (2016) Detection of snake fungal disease due to Ophidiomyces ophiodiicola in Virginia, USA. J Wildl Dis 52(1):143–149

Habarugira G, Moran J, Colmant AM, Davis SS, O'Brien CA, Hall-Mendelin S, McMahon J, Hewitson G, Nair N, Barcelon J, Suen WW (2020) Mosquito-independent transmission of West Nile virus in farmed saltwater crocodiles (Crocodylus porosus). Viruses 12(2):198

Hellebuyck T, Questel K, Pasmans F, Van Brantegem L, Philip P, Martel A (2017) A virulent clone of Devriesea agamarum affects endangered Lesser Antillean iguanas (Iguana delicatissima). Sci Rep 7(1):1–6

Hetzel U, Sironen T, Laurinmäki P, Liljeroos L, Patjas A, Henttonen H, Vaheri A, Artelt A, Kipar A, Butcher SJ, Vapalahti O. (2013) Isolation, identification, and characterization of novel arenaviruses, the etiological agents of boid inclusion body disease. J Virol 15;87(20):10918–10935

Homer BL, Berry KH, Brown MB, Ellis G, Jacobson ER (1998) Pathology of diseases in wild desert tortoises from California. J Wildl Dis 34(3):508–523

Hyatt AD, Williamson M, Coupar BEH, Middleton D, Hengstberger SG, Gould AR, Selleck P, Wise TG, Kattenbelt J, Cunningham AA, Lee J (2002) First identification of a ranavirus from green pythons (Chondropython viridis). J Wildl Dis 38(2):239–252

Innis CJ (2014) Conservation issues. In: Mader DR, Divers SJ (eds) Current therapy in reptile medicine and surgery. Saunders, St. Louis, pp 296–303

IUCN (2021) The IUCN red list of threatened species. Version 2021-1. https://www.iucnredlist.org. Downloaded on 5 Aug 2021

Jacobson ER, Gaskin JM, Brown MB, Harris RK, Gardiner CH, LaPointe JL, Adams HP, Reggiardo C (1991) Chronic upper respiratory tract disease of free-ranging desert tortoises (Xerobates agassizii). J Wildl Dis 27(2):296–316

Jacobson ER, Brown MB, Schumacher IM, Collins BR, Harris RK, Klein PA (1995) Mycoplasmosis and the desert tortoise (Gopherus agassizii) in Las Vegas Valley, Nevada. Chelonian Conserv Biol 1(4):279–284

Jacobson ER, Origgi F, Pessier AP, Lamirande EW, Walker I, Whitaker B, Stalis IH, Nordhausen R, Owens JW, Nichols DK, Heard D (2001) Paramyxovirus infection in caiman lizards (Draecena guianensis). J Vet Diagn Investig 13(2):143–151

Jacobson ER, Ginn PE, Troutman JM, Farina L, Stark L, Klenk K, Burkhalter KL, Komar N (2005) West Nile virus infection in farmed American alligators (Alligator mississippiensis) in Florida. J Wildl Dis 41(1):96–106

Jacobson ER, Berry KH, Wellehan JF Jr, Origgi F, Childress AL, Braun J, Schrenzel M, Yee J, Rideout B (2012) Serologic and molecular evidence for Testudinid herpesvirus 2 infection in wild Agassiz's desert tortoises, Gopherus agassizii. J Wildl Dis 48(3):747–757

James SB, Iverson J, Greco V, Raphael BL (2006) Health assessment of Allen Cays rock iguana, Cyclura cychlura inornata. J Herpetol Med Surg 16(3):93–98

Jessop TS, Woodford R, Symonds MR (2013) Macrostress: do large-scale ecological patterns exist in the glucocorticoid stress response of vertebrates?. Funct Ecol 27(1):120–130

Johnson AJ, Pessier AP, Jacobson ER (2007) Experimental transmission and induction of ranaviral disease in western ornate box turtles (Terrapene ornata ornata) and red-eared sliders (Trachemys scripta elegans). Vet Pathol 44(3):285–297

Johnson AJ, Pessier AP, Wellehan JF, Childress A, Norton TM, Stedman NL, Bloom DC, Belzer W, Titus VR, Wagner R, Brooks JW. (2008). Ranavirus infection of free-ranging and captive box turtles and tortoises in the United States. J Wildl Dis 1;44(4):851–863

Johnson AJ, Wendland L, Norton TM, Belzer B, Jacobson ER (2010) Development and use of an indirect enzyme-linked immunosorbent assay for detection of iridovirus exposure in gopher tortoises (Gopherus polyphemus) and eastern box turtles (Terrapene carolina carolina). Vet Microbiol 142(3–4):160–167

Johnsrude JD, Raskin RE, Hoge AYA, Erdos GW (1997) Intraerythrocytic inclusions associated with iridoviral infection in a fer de lance (Bothrops moojeni) snake. Vet Pathol 34(3):235–238

Keller S, Hetzel U, Sironen T, Korzyukov Y, Vapalahti O, Kipar A, Hepojoki J (2017) Coinfecting reptarenaviruses can be vertically transmitted in boa constrictor. PLoS Pathog 13(1):e1006179

Kilpatrick AM (2011) Globalization, land use, and the invasion of West Nile virus. Science 334(6054):323–327

Kishbaugh J, Tunseth D, Lung NP, Hope K, Hayes WK, Hayek LAC, Murray S (2020) Hematologic and biochemical ranges and health assessments of free-ranging San Salvador iguanas (Cyclura rileyi rileyi) in a translocation program. J Herpetol Med Surg 30(1):28–37

Klenk K, Snow J, Morgan K, Bowen R, Stephens M, Foster F, Gordy P, Beckett S, Komar N, Gubler D, Bunning M (2004) Alligators as West Nile virus amplifiers. Emerg Infect Dis 10(12):2150

Kolesnik E, Obiegala A, Marschang RE (2017) Detection of Mycoplasma spp., herpesviruses, topiviruses, and ferlaviruses in samples from chelonians in Europe. J Vet Diagn Investig 29(6):820–832

Kolesnik E, Hyndman TH, Müller E, Pees M, Marschang RE (2019) Comparison of three different PCR protocols for the detection of ferlaviruses. BMC Vet Res 15(1):1–8

Kolesnikovas CKM, Grego KF, De Albuquerque LR, Jacobson ER, Monezi TA, Mehnert DU, Catão-Dias JL (2006) Ophidian paramyxovirus in Brazilian vipers (Bothrops alternatus). Vet Rec 159(12):390

La'Toya VL, Wellehan J (2013) Selected emerging infectious diseases of squamata. Vet Clin North Am Exot Anim Pract 16(2):319–338

Laspiur A, Santos JC, Medina SM, Pizarro JE, Sanabria EA, Sinervo B, Ibargüengoytía NR (2021) Vulnerability to climate change of a microendemic lizard species from the central Andes. Sci Rep 11(1):1–14

Le Donne V, Crossland N, Brandão J, Sokolova Y, Fowlkes N, Nevarez JG, Langohr IM, Gaunt SD (2016) Nannizziopsis guarroi infection in 2 inland bearded dragons (Pogona vitticeps): clinical, cytologic, histologic, and ultrastructural aspects. Vet Clin Pathol 45(2):368–375

Lecis R, Paglietti B, Rubino S, Are BM, Muzzeddu M, Berlinguer F, Chessa B, Pittau M, Alberti A (2011) Detection and characterization of Mycoplasma spp. and Salmonella spp. in free-living European tortoises (Testudo hermanni, Testudo graeca, and Testudo marginata). J Wildl Dis 47(3):717–724

Lewbart GA, Hirschfeld M, Brothers JR, Muñoz-Pérez JP, Denkinger J, Vinueza L, García J, Lohmann KJ (2015) Blood gases, biochemistry and hematology of Galápagos marine iguanas (Amblyrhynchus cristatus). Conserv Physiol 3(1):cov034

Lewbart GA, Grijalva CJ, Calle PP, Ingerman K, Muñoz-Pérez JP, Quezada G, Vera CA, Gentile G, Valle CA (2019) Health assessment of Conolophus subcristatus, Conolophus pallidus, and C. subcristatus X Amblyrhynchus cristatus hybrid (Galápagos land iguanas). PLoS One 14(10):e0222884

Lorch JM, Knowles S, Lankton JS, Michell K, Edwards JL, Kapfer JM, Staffen RA, Wild ER, Schmidt KZ, Ballmann AE, Blodgett D (2016) Snake fungal disease: an emerging threat to wild snakes. Philos Trans R Soc B Biol Sci 371(1709):20150457

Maceda-Veiga A, Figuerola J, Martínez-Silvestre A, Viscor G, Ferrari N, Pacheco M (2015) Inside the Redbox: applications of hematology in wildlife monitoring and ecosystem health assessment. Sci Total Environ 514:322–332

Machain-Williams C, Padilla-Paz SE, Weber M, Cetina-Trejo R, Juarez-Ordaz JA, Loroño-Pino MA, Ulloa A, Wang C, Garcia-Rejon J, Blitvich ABJ (2013) Antibodies to West Nile virus in wild and farmed crocodiles in southeastern Mexico. J Wildl Dis 49(3):690–693

Maria R, Ramer J, Reichard T, Tolson PJ, Christopher MM (2007) Biochemical reference intervals and intestinal microflora of free-ranging Ricord's iguanas (Cyclura ricordii). J Zoo Wildl Med 38(3):414–419

Marschang RE (2011) Viruses infecting reptiles. Viruses 3(11):2087–2126

Marschang RE (2019) Emerging reptile viruses. In: Miller RE, Lamberski N, Calle P (eds) Fowler's zoo and wild animal medicine current therapy, vol 9. Saunders, St. Louis, pp 267–273

Marschang RE, Donahoe S, Manvell R, Lemos-Espinal J. (2002). Paramyxovirus and reovirus infections in wild-caught Mexican lizards (Xenosaurus and Abronia spp). J Zoo Wildl Med 33(4):317–321

Marschang RE, Braun S, Becher P (2005) Isolation of a ranavirus from a gecko (Uroplatus fimbriatus). J Zoo Wildl Med 36(2):295–300

Marschang RE, Papp T, Frost JW (2009) Comparison of paramyxovirus isolates from snakes, lizards and a tortoise. Virus Res 144(1–2):272–279

Marschang RE, Bogan J, Bradford CM. (2020) What's New in the Scientific Literature? Infectious Diseases of Reptiles and Amphibians: Peer-reviewed publications, July 2019–December 2019. J Herpetol Med Surg 30(2):54–62

Martin LB, Hopkins WA, Mydlarz LD, Rohr JR (2010) The effects of anthropogenic global changes on immune functions and disease resistance. Ann N Y Acad Sci 1195(1):129–148

McBride MP, Wojick KB, Georoff TA, Kimbro J, Garner MM, Wang X, Childress AL, Wellehan JF (2015) Ophidiomyces ophiodiicola dermatitis in eight free-ranging timber rattlesnakes (Crotalus horridus) from Massachusetts. J Zoo Wildl Med 46(1):86–94

McCoy CM, Lind CM, Farrell TM (2017) Environmental and physiological correlates of the severity of clinical signs of snake fungal disease in a population of pigmy rattlesnakes, Sistrurus miliarius. Conserv Physiol 5(1):cow077

McKenzie CM, Piczak ML, Snyman HN, Joseph T, Theijin C, Chow-Fraser P, Jardine CM (2019a) First report of ranavirus mortality in a common snapping turtle Chelydra serpentina. Dis Aquat Org 132(3):221–227

McKenzie JM, Price SJ, Fleckenstein JL, Drayer AN, Connette GM, Bohuski E, Lorch JM (2019b) Field diagnostics and seasonality of Ophidiomyces ophiodiicola in wild snake populations. EcoHealth 16(1):141–150

Mendoza-Roldan JA, Mendoza-Roldan MA, Otranto D (2021) Reptile vector-borne diseases of zoonotic concern. Int J Parasitol Parasites Wildl 15:132–142

Merchant M, Fleury L, Rutherford R, Paulissen M (2008) Effects of bacterial lipopolysaccharide on thermoregulation in green anole lizards (Anolis carolinensis). Vet Immunol Immunopathol 125(1–2):176–181

Miller DL, Mauel MJ, Baldwin C, Burtle G, Ingram D, Hines ME (2003) West Nile virus in farmed alligators. Emerg Infect Dis 9(7):794

Millet S, Bennett J, Lee KA, Hau M, Klasing KC (2007) Quantifying and comparing constitutive immunity across avian species. Dev Comp Immunol 31(2):188–201

Montali RJ (1988) Comparative pathology of inflammation in the higher vertebrates (reptiles, birds and mammals). J Comp Pathol 99(1):1–26

Moore IT, Lerner JP, Lerner DT, Mason RT (2000) Relationships between annual cycles of testosterone, corticosterone, and body condition in male red-spotted garter snakes, Thamnophis sirtalis concinnus. Physiol Biochem Zool 73(3):307–312

Munoz FJ, la Fuente MD (2001) The effect of the seasonal cycle on the splenic leukocyte functions in the turtle Mauremys caspica. Physiol Biochem Zool 74(5):660–667

Muñoz FJ, Galván A, Lerma M, De la Fuente M (2000) Seasonal changes in peripheral blood leukocyte functions of the turtle Mauremys caspica and their relationship with corticosterone, 17-β-estradiol and testosterone serum levels. Vet Immunol Immunopathol 77(1–2):27–42

Nevarez JG (2007) Lymphohistiocytic proliferative syndrome of alligators (Alligator mississippiensis): a cutaneous manifestation of West Nile virus. Louisiana State University and Agricultural and Mechanical College

Nogueira MF, Barrella TH, Silva RD, Lopes CAM, Araujo Junior JP (2002) Isolation of an Ophidian Paramyxovirus (OPMV) in a captive rattlesnake (Crotalus durissus terrificus) from Botucatu, São Paulo State, Brazil. J Venomous Anim Toxins 8(1):168–173

Origgi F (2007) Reptile immunology. In: Jacobson E (ed) Infectious diseases and pathology of reptiles: color atlas and text. CRC Press Taylor and Francis Group, Boca Raton, pp 131–166

Paiva MIS, Araújo Junior JP, Barraviera B (2016) Inquérito sorológico para paramixovirus ofídico em serpentes cativas. Rev Educ Contin Med Vet Zootec CRMV-SP 14(1):22–27

Paré JA, Sigler L (2016) An overview of reptile fungal pathogens in the genera Nannizziopsis, Paranannizziopsis, and Ophidiomyces. J Herpetol Med Surg 26(1–2):46–53

Pettinello R, Dooley H (2014) The immunoglobulins of cold-blooded vertebrates. Biomolecules 4(4):1045–1069

Piantoni C, Navas CA, Ibargüengoytía NR (2016) Vulnerability to climate warming of four genera of New World iguanians based on their thermal ecology. Anim Conserv 19(4):391–400

Price SJ, Ariel E, Maclaine A, Rosa GM, Gray MJ, Brunner JL, Garner TW (2017) From fish to frogs and beyond: impact and host range of emergent ranaviruses. Virology 511:272–279

Pyron RA, Burbrink FT, Wiens JJ. (2013). A phylogeny and revised classification of Squamata, including 4161 species of lizards and snakes. BMC Evol Biol 13:1–54

Rainwater KL, McClave C, Raphael BL, Cray C, Franklin AD, Powell DM, Burton FJ, Calle PP (2021) Hematology, plasma biochemistry, and plasma protein electrophoresis reference intervals for blue iguanas (Cyclura lewisi) from Grand Cayman Island. J Zoo Wildl Med 51(4):933–947

Refsnider JM, Palacios MG, Reding DM, Bronikowski AM (2015) Effects of a novel climate on stress response and immune function in painted turtles (Chrysemys picta). J Exp Zool A Ecol Genet Physiol 323(3):160–168

Riera Romo M, Pérez-Martínez D, Castillo Ferrer C (2016) Innate immunity in vertebrates: an overview. Immunology 148(2):125–139

Roll U, Feldman A, Novosolov M, Allison A, Bauer AM, Bernard R, Böhm M, Castro-Herrera F, Chirio L, Collen B, Colli GR (2017) The global distribution of tetrapods reveals a need for targeted reptile conservation. Nat Ecol Evol 1(11):1677–1682

Saad AH, El Ridi R (1988) Endogenous corticosteroids mediate seasonal cyclic changes in immunity of lizards. Immunobiology 177(4–5):390–403

Sardelis MR, Turell MJ (2001) Ochlerotatus j. japonicus in Frederick County, Maryland: discovery, distribution, and vector competence for West Nile virus. J Am Mosq Control Assoc 17(2):137–141

Sbrana E, Tonry JH, Xiao SY, Da Rosa APT, Higgs S, Tesh RB (2005) Oral transmission of West Nile virus in a hamster model. Am J Trop Med Hyg 72(3):325–329

Sherif M, El Ridi R (1992) Natural cytotoxic cell activity in the snake Psammophis sibilans. Immunobiology 184(4–5):348–358

Sigler L, Hambleton S, Paré JA (2013) Molecular characterization of reptile pathogens currently known as members of the Chrysosporium anamorph of Nannizziopsis vriesii complex and relationship with some human-associated isolates. J Clin Microbiol 51(10):3338–3357

Simulundu E, Ndashe K, Chambaro HM, Squarre D, Reilly PM, Chitanga S, Changula K, Mukubesa AN, Ndebe J, Tembo J, Kapata N (2020) West Nile virus in farmed crocodiles, Zambia, 2019. Emerg Infect Dis 26(4):811

Sinervo B, Mendez-De-La-Cruz F, Miles DB, Heulin B, Bastiaans E, Villagrán-Santa Cruz M, Lara-Resendiz R, Martínez-Méndez N, Calderón-Espinosa ML, Meza-Lázaro RN, Gadsden H (2010) Erosion of lizard diversity by climate change and altered thermal niches. Science 328(5980):894–899

Sinervo B, Miles DB, Wu Y, Méndez-De La Cruz FR, Kirchhof S, Qi Y (2018) Climate change, thermal niches, extinction risk and maternal-effect rescue of toad-headed lizards, Phrynocephalus, in thermal extremes of the Arabian Peninsula to the Qinghai—Tibetan Plateau. Integr Zool 13(4):450–470

Smith CE, Edwards J, Lorch JM (2013) Crotalus horridus (timber rattlesnake) fungal pathogens. Herpetol Rev 44(519):e520

Smyth AK, Smee E, Godfrey SS, Crowther M, Phalen D (2014) The use of body condition and hematology to detect widespread threatening processes in sleepy lizards (Tiliqua rugosa) in two agricultural environments. R Soc Open Sci 1(4):140257

Solis C, Arguedas R, Baldi M, Piche M, Jimenez C (2017) Seroprevalence and molecular characterization of ferlavirus in captive vipers of Costa Rica. J Zoo Wildl Med 48(2):420–430

Solomon S, Qin D, Manning M, Chen Z, Marquis M, Averyt KB, Tignor M, Miller HL (2007) Climate change 2007: synthesis report. Contribution of Working Group I, II and III to the Fourth Assessment Report of the Intergovernmental Panel on Climate Change. Summary for Policymakers. Cambridge University Press, Cambridge

Stacy NI, Alleman AR, Sayler KA (2011) Diagnostic hematology of reptiles. Clin Lab Med 31(1):87–108

Steinman A, Banet-Noach C, Tal S, Levi O, Simanov L, Perk S, Malkinson M, Shpigel N. (2003). West Nile virus infection in crocodiles. Emerg Infect Dis 9(7):887

Steinman A, Banet-Noach C, Simanov L, Grinfeld N, Aizenberg Z, Levi O, Lahav DAN, Malkinson M, Perk S, Shpigel NY (2006) Experimental infection of common garter snakes (Thamnophis sirtalis) with West Nile virus. Vector Borne Zoonotic Dis 6(4):361–368

Stenglein MD, Jacobson ER, Chang LW, Sanders C, Hawkins MG, Guzman DS, Drazenovich T, Dunker F, Kamaka EK, Fisher D, Reavill DR (2015) Widespread recombination, reassortment, and transmission of unbalanced compound viral genotypes in natural arenavirus infections. PLoS Pathog 11(5):e1004900

Stöhr AC, Blahak S, Heckers KO, Wiechert J, Behncke H, Mathes K, Günther P, Zwart P, Ball I, Rüschoff B, Marschang RE (2013) Ranavirus infections associated with skin lesions in lizards. Vet Res 44(1):1–10

Stöhr AC, López-Bueno A, Blahak S, Caeiro MF, Rosa GM, Alves de Matos AP, Martel A, Alejo A, Marschang RE (2015) Phylogeny and differentiation of reptilian and amphibian ranaviruses detected in Europe. PLoS One 10(2):e0118633

Terio KA (2004) Comparative inflammatory responses of non-mammalian vertebrates. In: 55th annual meeting of the American College of Veterinary Pathologists (ACVP) and 39th annual meeting of the American Society of Clinical Pathology (ASVCP), Orlando, FL, USA

Thomas LA, Eklund CM, Rush WA (1958) Susceptibility of garter snakes (Thamnophis spp.) to western equine encephalomyelitis virus. Proc Soc Exp Biol Med 99(3):698–700

Thomas LA, Patzer ER, Cory JC, Coe JE (1980) Antibody development in garter snakes (Thamnophis spp.) experimentally infected with western equine encephalitis virus. Am J Trop Med Hyg 29(1):112–117

Tingley R, Meiri S, Chapple DG (2016) Addressing knowledge gaps in reptile conservation. Biol Conserv 204:1–5

Trimble MJ, Van Aarde RJ (2010) Species inequality in scientific study. Conserv Biol 24(3):886–890

Troiano JC, Gould EG, Gould I (2008) Hematological reference intervals in argentine lizard Tupinambis merianae (Sauria—Teiidae). Comp Clin Pathol 17(2):93–97

Uetz P (ed) (2021) The Reptile Database. http://www.reptile-database.org. Downloaded on 2 Aug 2021

UNEP-WCMC (2016) The State of Biodiversity in Latin America and the Caribbean: a mid-term review of progress toward the Aichi Biodiversity Targets. UNEP-WCMC, Cambridge

Van Hoek ML (2014) Antimicrobial peptides in reptiles. Pharmaceuticals 7(6):723–753

Vitt LJ, Caldwell JP. (2013). Herpetology: an introductory biology of amphibians and reptiles. Academic press

Voogdt CG, Bouwman LI, Kik MJ, Wagenaar JA, Van Putten JP (2016) Reptile Toll-like receptor 5 unveils adaptive evolution of bacterial flagellin recognition. Sci Rep 6(1):1–10

Walder R, Suarez OM, Calisher CH (1984) Arbovirus studies in the Guajira region of Venezuela: activities of eastern equine encephalitis and Venezuelan equine encephalitis viruses during an interepizootic period. Am J Trop Med Hyg 33(4):699–707

White G, Ottendorfer C, Graham S, Unnasch TR (2011) Competency of reptiles and amphibians for eastern equine encephalitis virus. Am J Trop Med Hyg 85(3):421

Woodburn DB, Miller AN, Allender MC, Maddox CW, Terio KA (2019) Emydomyces testavorans, a new genus and species of onygenalean fungus isolated from shell lesions of freshwater aquatic turtles. J Clin Microbiol 57(2):e00628-18

Work TM, Balazs GH, Rameyer RA, Chang SP, Berestecky J (2000) Assessing humoral and cell-mediated immune response in Hawaiian green turtles, Chelonia mydas. Vet Immunol Immunopathol 74(3–4):179–194

Wright RK, Schapiro HC (1973) Primary and secondary immune responses of the desert iguana, Dipsosaurus dorsalis. Herpetologica 29:275–280

Zamora-Camacho FJ, Reguera S, Rubiño-Hispán MV, Moreno-Rueda G (2015) Eliciting an immune response reduces sprint speed in a lizard. Behav Ecol 26(1):115–120

Zimmerman LM, Vogel LA, Bowden RM (2010) Understanding the vertebrate immune system: insights from the reptilian perspective. J Exp Biol 213(5):661–671

Chapter 5
Avian Disease Ecology in the Neotropics

Paulina Álvarez-Mendizábal, María José Tolsa, Octavio Rojas-Soto, Ian MacGregor-Fors, and Diego Santiago-Alarcon

5.1 Introduction

Globally, birds are the best studied taxonomic group of vertebrates. Over 10,000 bird species have radiated and spread across the Earth's biomes (Avibase; avibase.bsc-eoc.org), and ornithologists have mapped and dated the diversification of all extant bird species worldwide (Jetz et al. 2012). Through ecological studies, ornithologists have uncovered the high taxonomic and functional diversity of birds, their biogeographical patterns and the drivers of these patterns across different regions and scales (Rahbek and Graves 2001; Herzog et al. 2005; Moura et al. 2016; Matuoka et al. 2020; Sol et al. 2020). However, parasite species affecting bird populations and communities have received fewer research efforts, hindering our capacity to manage and conserve bird species. Bird species play a critical role as hosts of a large array of parasite species that affect animal and human health (Dobson et al.

P. Álvarez-Mendizábal (✉)
Biología y Conservación de Vertebrados, Instituto de Ecología, A.C.-CONACYT, Xalapa, Veracruz, México

Facultad de Medicina Veterinaria y Zootecnia, Universidad Nacional Autónoma de México, Ciudad de México, México

M. J. Tolsa (✉)
Institut de Recherche pour le Développement, México, Ciudad de México, México

O. Rojas-Soto
Biología Evolutiva, Instituto de Ecología, A.C.-CONACYT, Xalapa, Veracruz, México

I. MacGregor-Fors
Ecosystems and Environment Research Programme, Faculty of Biological and Environmental Sciences, University of Helsinki, Lahti, Finland

D. Santiago-Alarcon
Department of Integrative Biology, University of South Florida, Tampa, FL, USA

G. Acosta-Jamett, A. Chaves (eds.), *Ecology of Wildlife Diseases in the Neotropics*, https://doi.org/10.1007/978-3-031-50531-7_5

2008; Nabi et al. 2021). Birds have been reported to harbor as many zoonoses as those found in flying mammals, such as in bats (Order Chiroptera; Mollentze and Streicker 2020). Moreover, birds latitudinal and altitudinal migration movements between wintering and breeding areas are implicated in the geographical spread of major parasitic species (e.g., influenza viruses, West Nile virus; Feare 2010; Winker and Gibson 2010; Lee et al. 2015; Morin et al. 2018; Mine et al. 2019). For example, Avian Influenza viruses (AIVs) are known to successfully complete reassortment processes (i.e., the processes in which gene segments are exchanged between different influenza viruses) in geographical hotspots for viral transport and mixing, where breeding areas are shared between North American and Asian migrant bird species (Morin et al. 2018). In the Beringia region, phylogeographic analyses have shown that viral reassortment has likely occurred between highly pathogenic influenza strains originating in China (e.g., Asian H5N8) and North American influenza lineages that have ultimately spread across the continental United States (Saito et al. 2015).

Avian parasites can have negative effects on their host populations and ecosystems (Møller 2005). Parasites can significantly affect host fitness by changing their behavior and sexual selection and can regulate avian populations by negatively impacting host reproduction (i.e., affecting the fecundity or survival of individuals) (May and Anderson 1978; Møller 2005). Parasites can also mediate intraspecific competition, predator–prey interactions, and food web stability and determine host community structure (Møller and Erritzøe 2000; Navarro 2004; Lafferty et al. 2006; Wood et al. 2007). Moreover, they can exert evolutionary pressures on their hosts' immune systems, promoting greater investment in immune function that produces fitter and more viable host populations (Møller and Erritzøe 2002). For example, avian malaria (genus *Plasmodium*) has produced epizootic die-offs in immunologically naive Hawaiian insular avifauna (LaPointe et al. 2012), causing population declines and significant altitudinal shifts in the distribution of susceptible forest birds (e.g., Hawaii amakihi; *Hemignathus virens*) (Samuel et al. 2011). While such population declines are commonly associated with acute malaria infections that result in increased predation or starvation of avian hosts, chronic infections in continental avifauna produce long-term subclinical effects that can indirectly impair reproductive success (e.g., *Plasmodium*-infected male songbirds have been observed to reduce song frequency, potentially impacting sexual selection) (Gilman et al. 2007; Meza-Montes et al. 2023).

Although parasitic species such as avian malaria and related haemosporidians are constrained to affect avian populations, others have had critical effects on both avian and human populations. Such is the case for the neurotropic and zoonotic West Nile virus (WNV), which has produced massive bird die-offs in highly susceptible hosts (e.g., American crows; *Corvus brachyrhynchos*), and neurological outbreaks of disease in humans across Europe and the United States (LaDeau et al. 2011; George et al. 2015; Hadfield et al. 2019). WNV is capable of infecting over 608 species of birds (Tolsá et al. 2018); however, not all bird species are amplifying hosts for maintaining and transmitting the infection (Komar et al. 2003). Similarly, not all infected avian species are affected by neurologic illness and death (Steele et al. 2000). The effects of WNV on birds have affected only a few susceptible host species (Kilpatrick and Wheeler 2019), but WNV-associated declines of avian populations have had indirect ecological and health effects on other species. For instance, abundance variation in susceptible WNV species such as American robins promoted shifts in mosquito feeding preferences to humans and other avian species, increasing the prevalence of human disease (Kilpatrick et al. 2006).

Avian hosts can also produce important effects on parasite populations and communities. For example, when infected by vector-borne viruses, bird species can exert selection pressures over parasite genetic populations, removing variants of lower fitness and affecting intrahost viral diversity (Jerzak et al. 2005). This regulation has been observed for WNV, where strong processes of purifying selection can evolutionarily constrain its genetic diversity in bird hosts (Jerzak et al. 2008). On a much broader scale, avian host ecological traits (e.g., body mass, geographic range size, and relative abundance) can influence a parasite's colonization success, reproduction, and contact with susceptible hosts, affecting parasite diversity and community composition (Poulin and Valtonen 2001; Dáttilo et al. 2020). Moreover, the phylogenetic relatedness of hosts can drive the phylogenetic diversity of parasites across spatial scales (Clark and Clegg 2017).

Few study systems are well understood in relation to the effects of the ecology of avian parasites on the ecology of avian host populations (and vice versa). Additionally, the potential for bird parasites to produce spillover events and emerge as novel zoonoses remains unclear. This stems from the fact that, in general, very little is known about the global parasitic faunas of birds. Most studies on the ecology of birds and their parasites have been performed in temperate regions (i.e., Nearctic and Palearctic) (Titley et al. 2017). Accordingly, a gap in knowledge on avian disease ecology remains for other biogeographical regions, such as the Neotropics. This scenario is worrying, as the Neotropical region is home to approximately 30% of the global avifauna and a high number of endemic species (Reboreda et al. 2019). The high diversity of Neotropical bird species provides a diverse set of ecological niches for parasites (Hudson et al. 2006; Poulin 2014). Furthermore, host species diversity has been shown to be a positive driver of parasite species diversity (Dobson et al. 2008; Poulin 2014; Kamiya et al. 2014). Consequently, the Neotropical region is predicted to harbor not only a great proportion of the world's avifauna but also an extraordinary amount of parasite species (Dobson et al. 2008; Santiago-Alarcon and Rojas Soto 2021).

In the context of global change, parasites are most likely to jump across species and disproportionately affect endangered and naïve native avian populations (e.g., Ortega-Guzmán et al. 2022). Yet, some parasite species are also predicted to decline in abundance and geographical range size due to climate change and habitat destruction (Dobson et al. 2008). This will be contingent on the complexity of a parasites' life cycle, their dependence on host population density, their type of transmission, their host specificity, and the geographical range size of their hosts, among other factors (Dobson et al. 2008). Unfortunately, Neotropical species tend to have smaller range sizes than those inhabiting temperate or polar regions (Jetz and Rahbek 2002); hence, it is likely that their population declines will lead to the decline of their parasitic faunas (Dobson et al. 2008). This is predicted to impact several ecosystem services that parasite species provide, including regulation of host abundance, increase in food web connectance, concentration and removal of pollutants from their hosts, among others (Sures 2004; Møller 2005; Lafferty et al. 2006). Because parasites represent fundamental components of biodiversity, understanding their ecological and evolutionary interactions with their host species will aid to maintain animal, human, and ecosystem health.

In this chapter, we conducted an extensive search in the Web of Science Core Collection for current information regarding parasite ecology in birds across the Neotropical region from 2000 to 2021. The Web of Science Core Collection is a selective and balanced database that uses relevant information from multiple sources (Birkle et al. 2020). Our first search aimed to retrieve studies focused on parasite groups (i.e., virus, bacteria, protozoa, fungi, and disease), birds (i.e., birds, avian, aves, and avifauna), and the Neotropical region (i.e., Neotropic, South America, and Latin America). We excluded all scientific papers that reported avian parasites in captive bird species, review papers, laboratory/experimental studies, and museum collection studies.

Our second search aimed to retrieve studies that show how anthropogenic (i.e., fragmentation, land use, landscape, urban, and habitat perturbation) and ecological (i.e., temperature, latitude, altitude, precipitation, humidity, and climate change) factors can be associated with the prevalence and diversity of each parasite group. According to prior knowledge, anthropogenic and ecological keywords were chosen regarding fundamental factors that influence the prevalence and diversity of parasite species in their hosts, environment, and space. Last, we conducted a third search seeking to retrieve studies that explicitly assessed the relation between anthropogenic/ecological factors and the prevalence, diversity and/or richness of specific parasite genera or species within each parasite group. We selected one parasite species (or genus) per parasite group (i.e., the parasite species or genus that had the highest number of observations in our first search). An avian parasite record was considered to be an independent observation of a parasite genus or species in a particular Neotropical country, locality, and year. Boolean codes for all three searches can be found in Appendix Table 5.4.

5.2 Results

In our first search, we retrieved and assessed 3064 studies. Only papers that contained parasite group and/or species, avian order and/or species and country were selected, leaving a total of 1338 studies. Only 11 of 26 were conducted in Neotropical countries (i.e., Brazil, Peru, Argentina, Colombia, Chile, Ecuador, Mexico, Venezuela, Costa Rica, Dominican Republic, French Guiana, Panama, Guatemala, Bolivia, and Uruguay), and 15 Neotropical countries were not represented (i.e., Belize, El Salvador, Honduras, Nicaragua, Paraguay, Guyana, Suriname, Cuba, Jamaica, Haiti, and Puerto Rico). We found that Brazil is the Neotropical country that contains the highest number of avian parasite records (553 observations), followed by Colombia (268 observations), Peru (252 observations), Argentina (86 observations), Chile (53 observations), Mexico (28 observations), and Ecuador (27 observations) (Fig. 5.1). In terms of groups of parasites, Brazil and Mexico were the only countries that had parasite records of all five groups considered in this chapter: Virus, Bacteria, Fungi, Protista, and Animalia (i.e., parasitic insects, worms, cestodes, and rotifers) (Fig. 5.1). Brazil mainly had records of parasitic insects (i.e., Arthropoda, Fig. 5.2), but it also had a significant number of protozoa of the phylum Mizozoa (subphylum Apicomplexa) (e.g., avian haemosporidian parasites) (Fig. 5.3) and of viral families such as Orthomyxoviridae, Coronaviridae, and Bornaviridae (e.g., avian coronaviruses, avian influenza, and avian bornaviruses) (Fig. 5.4). Mexico mainly contained records of viral families such as Flaviviridae (e.g., West Nile virus) (Fig. 5.4), platyhelminthes and rotifers (Fig. 5.2), and Apicomplexa protozoa (i.e., avian haemosporidians) (Fig. 5.3).

Chile followed Brazil and Mexico in number of groups of parasites, showing records of parasitic arthropods, nematodes, and platyhelminthes (Fig. 5.2) but also containing records of Apicomplexa protozoa (Fig. 5.3), and of Orthomyxoviridae and Poxviridae viral families (e.g., Avian Pox and Influenza viruses) (Fig. 5.4). Argentina mainly contained records of parasitic platyhelminthes (Fig. 5.2) and of the Orthomyxoviridae, Adenoviridae, Coronaviridae, Paramyxoviridae, and Bunyaviridae families (e.g., avian influenza virus, avian adenovirus, avian infectious bronchitis virus, and avian paramyxovirus) (Fig. 5.4). After Brazil, Colombia contained the most records of parasitic arthropods (Fig. 5.2) and an important number of records of parasitic protozoa (i.e., both Mizozoa and Euglenozoa phyla) (Fig. 5.3). Following Colombia, Peru also contained an important number of records of parasitic arthropods (Fig. 5.2) and showed records of viral families such as Orthomyxoviridae and Adenoviridae (e.g., avian influenza, Siadenoviruses, and Mastadenoviruses) (Fig. 5.4). Brazil and Argentina contained most studies focusing on avian bacteria (Fig. 5.1), mainly the phyla Proteobacteria (e.g., *Escherichia coli*, *Vibrio cholerae*, *Salmonella* spp), Firmicutes (e.g., *Bacillus* spp., *Clostridium* spp.), and Chlamydiota (e.g., *Chlamydia psittaci*) (Fig. 5.5).

Brazil also represented the country with the highest parasite species richness ($S = 243$), followed by Argentina ($S = 58$), Colombia ($S = 48$), Peru ($S = 43$), and Chile ($S = 44$) (Fig. 5.6). These five countries contained most of the studies that we were able to retrieve (Fig. 5.6). Additionally, our first search revealed that almost all avian parasites included in our study (over 800 observations out of 1338 total parasite observations) are being recorded and studied in bird species of the order Passeriformes ($n = 72$ studies), followed by birds of the order Columbiformes ($n = 27$ studies), Charadriiformes ($n = 18$ studies), and Psittaciformes ($n = 17$ studies) (Fig. 5.7). Most records observed in the order Passeriformes belong to parasitic insects of the phylum Arthropoda (Fig. 5.8), followed by Apicomplexa protozoa (Fig. 5.9), and by viral species of the Poxviridae, Flaviviridae, and Bunyaviridae families (Fig. 5.10). Studies that focused on avian bacteria were mostly observed in the orders Psittaciformes, Charadriiformes, and Cathartiformes (Fig. 5.11).

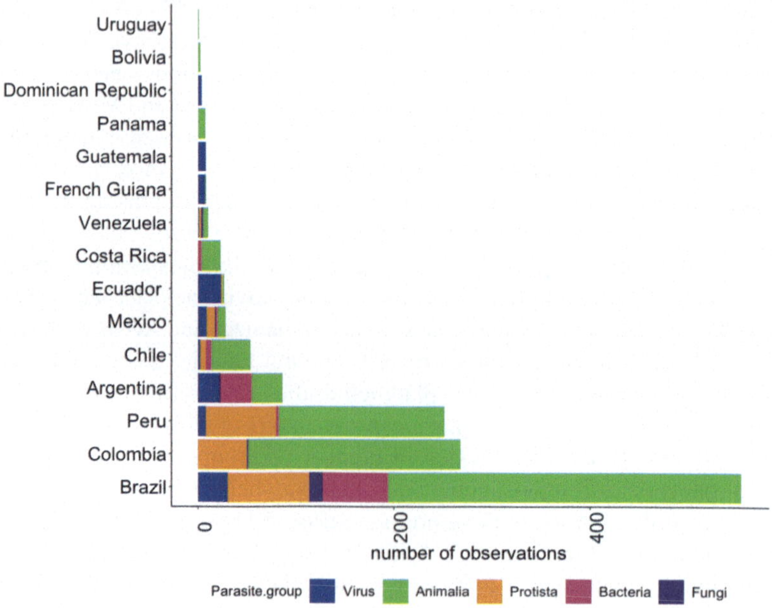

Fig. 5.1 Number of parasite records belonging to the kingdoms Virus, Animalia (i.e., Arthropoda, Nematoda, Platyhelminthes, Rotifera), Protista, Bacteria, and Fungi per studied country in the Neotropical region

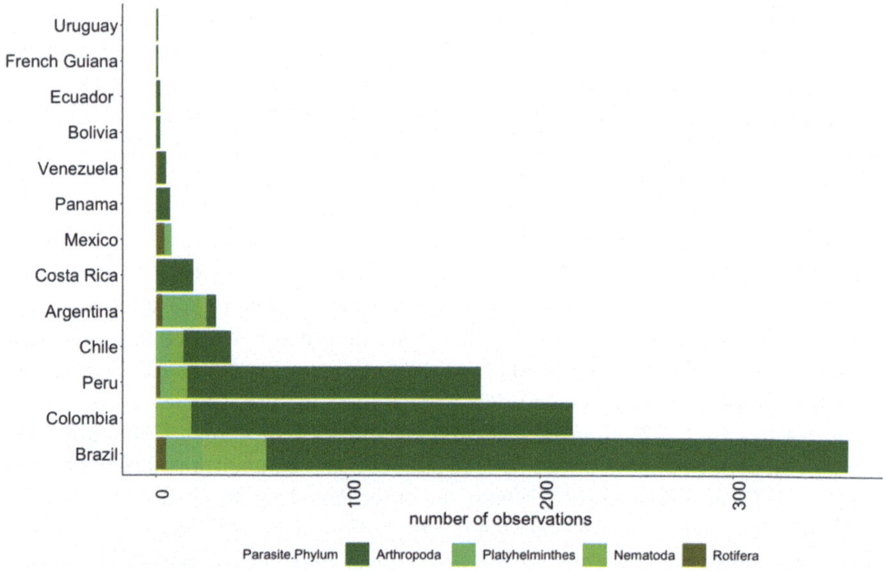

Fig. 5.2 Number of records of avian parasitic arthropods, nematodes, platyhelminthes and rotifers per studied country in the Neotropical region

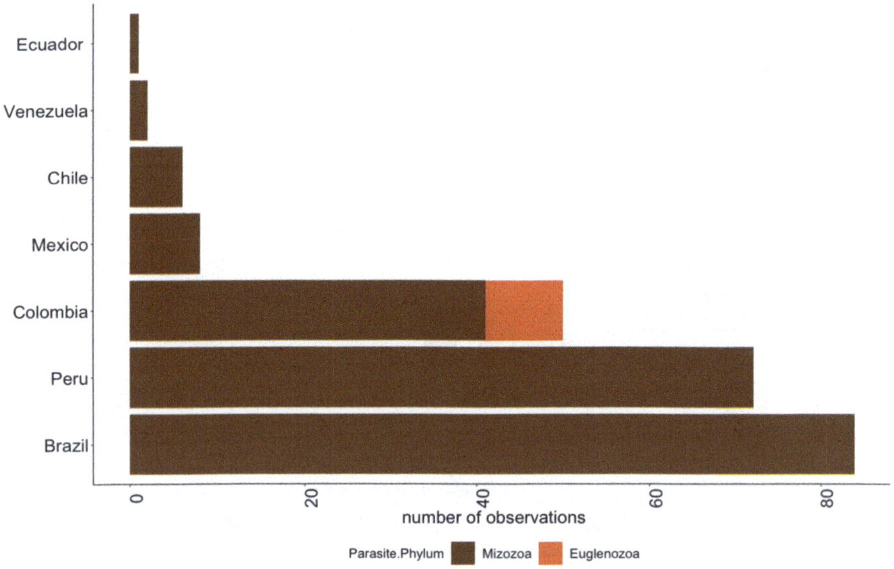

Fig. 5.3 Number of records of avian protozoa per studied country in the Neotropical region

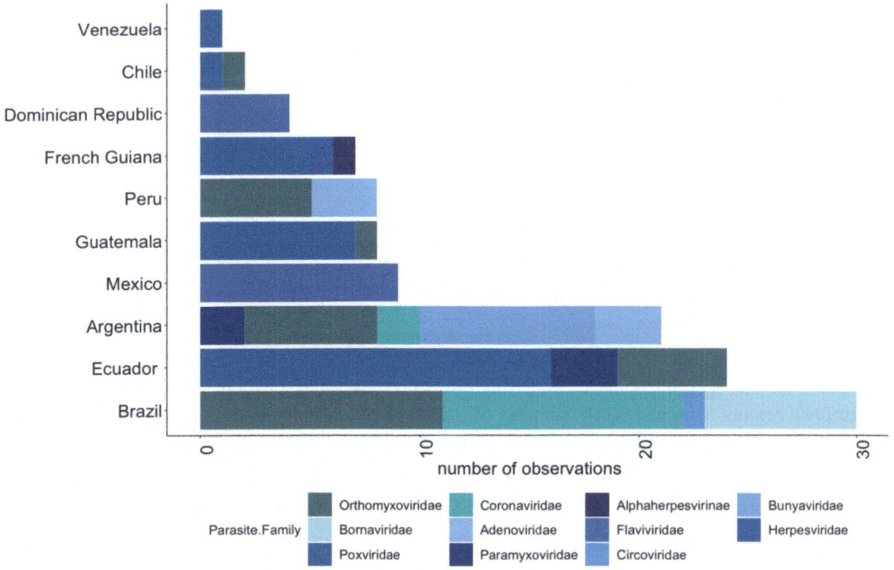

Fig. 5.4 Number of records of viral families per studied country in the Neotropical region

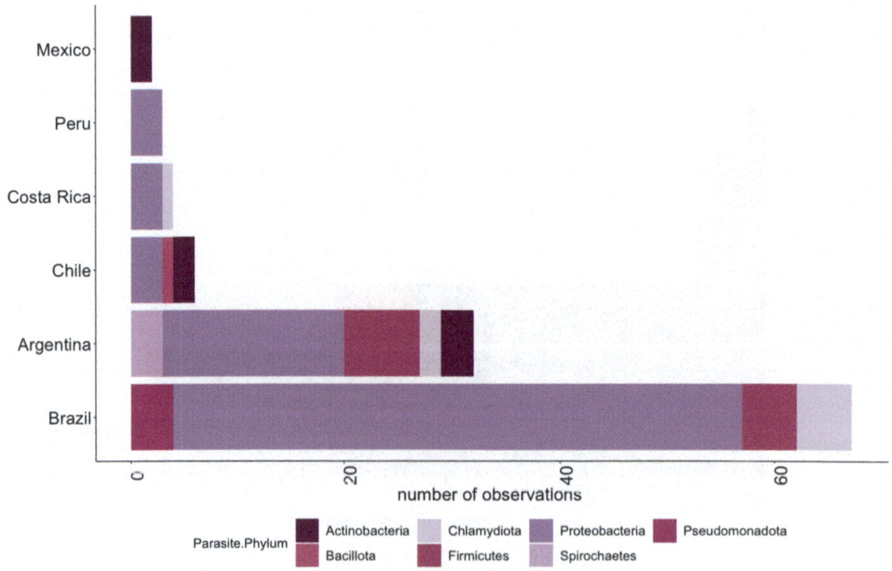

Fig. 5.5 Number of records of bacterial phyla per studied country in the Neotropical region

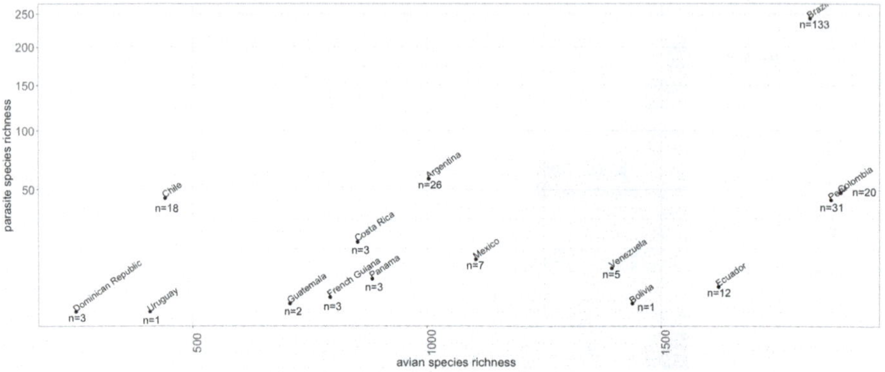

Fig. 5.6 Parasite species richness as a function of avian species richness as analyzed in the Neotropical region. *N* represents the number of studies retrieved from each country. The Y axis was transformed to Sqrt (square root)

In our second search, we retrieved 1100 and 268 studies where anthropogenic and ecological factors, respectively, were studied concerning the prevalence and diversity of each parasite group. However, only in eight studies, parasite prevalence and/or diversity were explicitly assessed (i.e., using mathematical modeling) in relation to anthropogenic (e.g., land use type, distance to urban sites, forest fragment size, etc.) and ecological (i.e., temperature, precipitation, latitude, etc.) factors (Table 5.1). The parasite groups assessed in these eight studies included Apicomplexa protozoa (i.e., avian haemosporidians), avian viruses (i.e., West Nile virus, St Louis

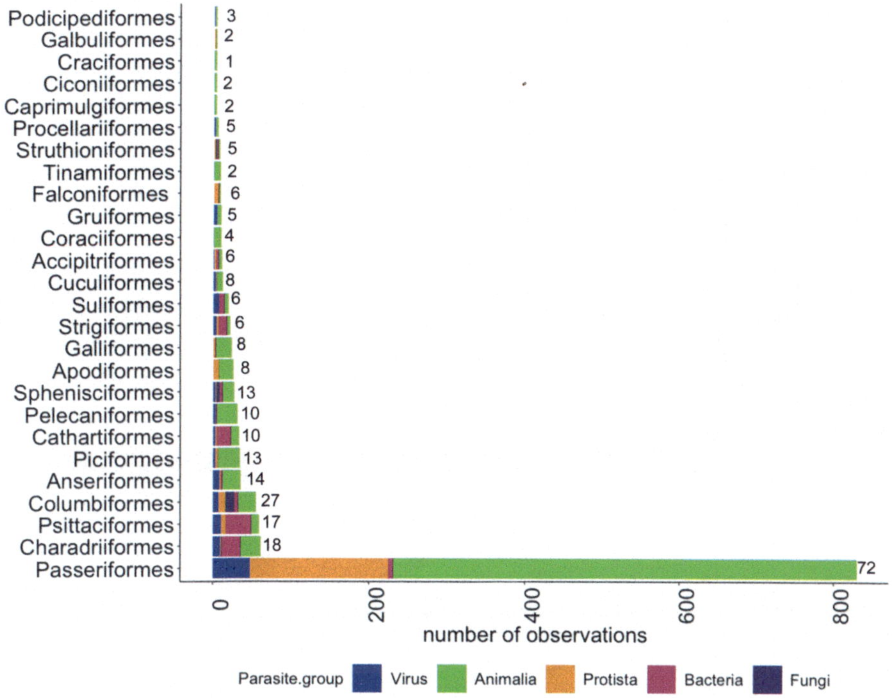

Fig. 5.7 Number of parasite records belonging to the kingdoms Virus, Animalia (i.e., Arthropoda, Nematoda, Platyhelminthes, Rotifera), Protista, Bacteria, and Fungi, per taxonomic bird order. Numbers next to bars represent the number of studies found per bird order

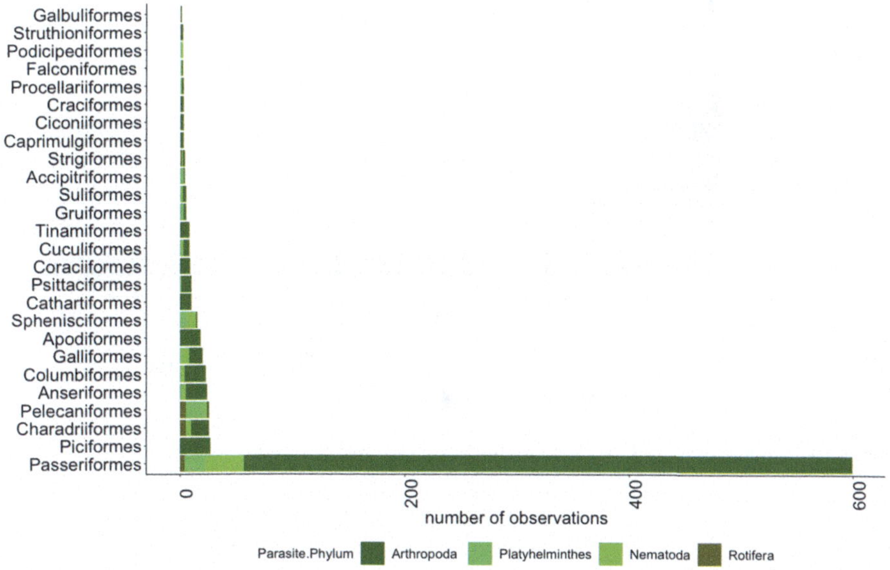

Fig. 5.8 Number of records of avian parasitic arthropods, platyhelminthes, nematodes, and rotifers per taxonomic bird order

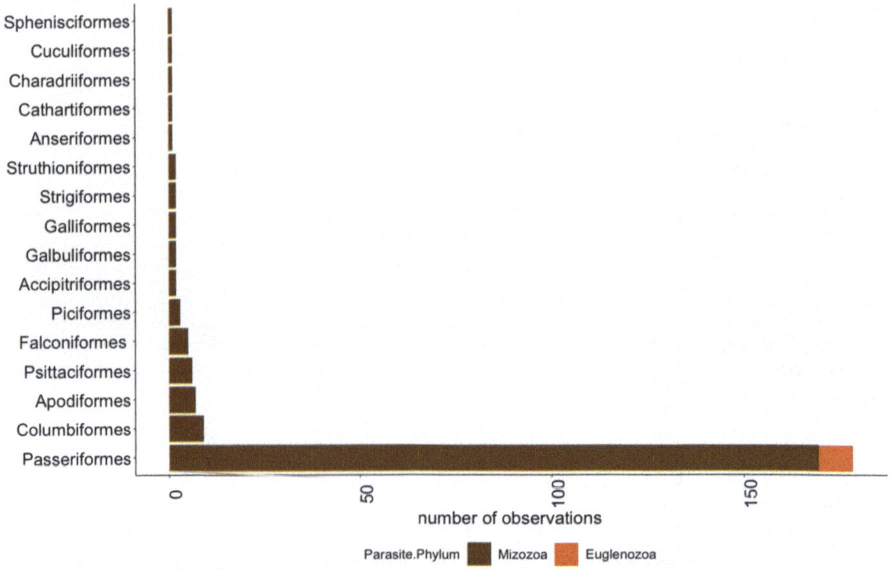

Fig. 5.9 Number of records of avian protozoa per taxonomic bird order

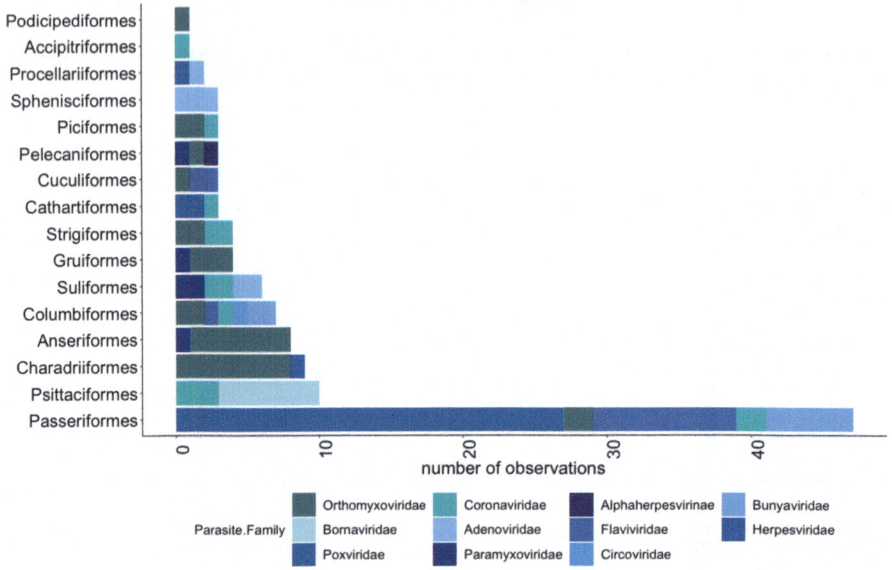

Fig. 5.10 Number of records of viral families per taxonomic bird order

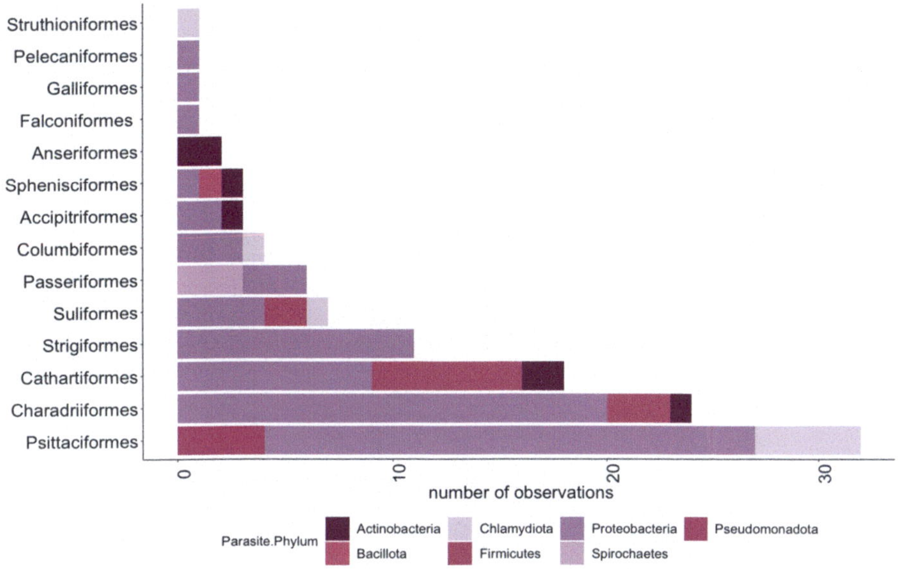

Fig. 5.11 Number of records of bacterial phyla per taxonomic bird order

Encephalitis virus, Avian pox virus), parasitic arthropods (i.e., *Amblyomma* spp.), and helminth communities (Table 5.1).

Finally, our third search retrieved 287 and 62 studies for anthropogenic and ecological factors, respectively. Of these, 12 (Table 5.2) and 17 (Table 5.3) were studies where the prevalence and/or diversity of each parasite genus/species was explicitly assessed for anthropogenic and ecological factors. All studies analyzed avian haemosporidian parasites.

5.3 Discussion

Parasite ecology in birds has gained greater attention in the last two decades. Animal and public health concerns linked to avian disease outbreaks (e.g., highly pathogenic avian flu and West Nile fever) underscore the importance of investigating the ecological and evolutionary relationships between avian parasites, their host communities, and their environments (e.g., Robles-Fernández et al. 2022). Despite this, our research suggests that avian parasite diversity surveys remain rudimentary in most countries of the Neotropical region, and therefore, we have a poor understanding of disease dynamics as a function of ecological and anthropogenic factors.

Our extensive search revealed that within the Neotropics, most avian parasites are being detected in Brazil, Colombia, and Peru. This is not surprising, as these countries represent the three most bird-rich countries in the world (birdlife.org). Several studies have shown that parasite species richness follows host species

Table 5.1 Summary of the selected (second search) studies that explore the association between anthropogenic/ecological factors and parasite prevalence/diversity in Neotropical avifauna

Type of factor	Country	Parasite studied	Avian order	Explanatory variable(s)	Response variable (s)	Effect type	References
Anthropogenic	Brazil	Arthropoda: *Amblyomma* spp.	Passeriformes Columbiformes Coraciiformes	Forest fragment size	Prevalence	Predictive	Ogrzewalska et al. (2011)
	Brazil	Protozoa (Apicomplexa): Haemosporidia (*Plasmodium* sp., *Haemoproteus* sp.)	Passeriformes	Native vegetation cover Distance to urban areas Number of water bodies	Infection probability (molecular prevalence)	Predictive native vegetation increases infection probability	Fecchio et al. (2020)
	Argentina	Virus (Flaviviridae): West Nile virus St. Louis Encephalitis	Passeriformes Piciformes Columbiformes Cuculiformes Accipitriformes	Agricultural area Tree coverage Urban sites Distance to water bodies	Seroprevalence	Predictive	Mansilla et al. (2022)
	Ecuador	Virus (Poxviridae): Avian Pox virus	Passeriformes	Habitat type: Urban Agricultural Undeveloped	Molecular prevalence	Predictive agriculture increases prevalence	Zylberberg et al. (2013)

Ecological	Argentina	Animalia: (Helminths) Cestoda, Digenea, Nematoda, Acanthocephala	Charadriiformes	Used environments across migratory flyways: Continental Marine	Parasite prevalence, richness and evenness	At component community: predictive At infracommunity: nonpredictive	Capasso et al. (2022)
	Brazil	Protozoa (Apicomplexa): Haemosporidia (*Plasmodium* sp., *Haemoproteus* sp.)	Passeriformes	Temperature Precipitation	Alpha and beta parasite diversity	Predictive	Lópes et al. (2020)
	Puerto Rico	Protozoa (Apicomplexa): Haemosporidia (*Plasmodium* sp., *Haemoproteus* sp.)	Passeriformes	1- and 10-year time intervals	Lineage composition	1-year interval: predictive negative 10-year interval: predictive positive	Fallon et al. (2004)
	Ecuador	Protozoa (Apicomplexa): Haemosporidia (*Plasmodium* sp., *Haemoproteus* sp.)	Passeriformes	Elevation	Prevalence	Predictive	Escallon et al. (2016)

Table 5.2 Summary of the selected (third search) studies that explore the association between anthropogenic factors and parasite prevalence/diversity in Neotropical avifauna

Country	Avian order	Explanatory variables	Response variables	Effect type	References
Ecuador	NA	Forest fragment size	Prevalence (*Plasmodium*)	Nonpredictive	Sebaio et al. (2010)
		Forest fragment size	Prevalence (*Haemoproteus*)	Nonpredictive	
		Forest dependence	Prevalence (*Plasmodium*)	Predictive	
		Forest dependence	Prevalence (*Haemoproteus*)	Nonpredictive	
Ecuador	Passeriformes	Fragment area	Infection status (Haemosporidian)	Nonpredictive	Rivero de Aguilar et al. (2018)
		Forest quality	Infection status (Haemosporidian)	Nonpredictive	
		Tree cover around fragments	Infection status (Haemosporidian)	Nonpredictive	
		Mist-net location (edge or interior)	Infection status (Haemosporidian)	Nonpredictive	
		Date	Infection status (*Plasmodium*)	Predictive	
		Cover	Infection status (*Haemoproteus*)	Predictive	
Brazil	Passeriformes	Proportion of native vegetation	Probability of infection (Haemosporidian)	Predictive	Fecchio et al. (2021)
		Distance to urban regions	Probability of infection (Haemosporidian)	Nonpredictive	
Brazil	NA	Habitat type (disturbed and undisturbed)	prevalence (Haemosporidian)	Predictive	Fecchio et al. (2021)
		Habitat type (disturbed and undisturbed)	prevalence (*Haemoproteus*)	Predictive	
		Habitat type (disturbed and undisturbed)	Prevalence (*Plasmodium*)	Predictive	
		Habitat type (disturbed and undisturbed)	Diversity (*Plasmodium*)	Predictive	
		Habitat type (disturbed and undisturbed)	Diversity (Haemosporidian)	Predictive	

(continued)

Table 5.2 (continued)

Country	Avian order	Explanatory variables	Response variables	Effect type	References
México	Passeriformes	Land use types	Lineage richness (Haemosporidian)	Predictive	Hernandez-Lara et al. (2020)
		Land use types	Prevalence (Haemosporidian)	Predictive	
		Land use types	Parasitemia (Haemosporidian)	Nonpredictive	
		Land use types	Aggregation (*Plasmodium*)	Nonpredictive	
Brazil	Passeriformes	Habitat type qué tipos?	Prevalence (Haemosporidian)	Predictive	Belo et al. (2011)
			Diversity (Haemosporidian)	Predictive	
Argentina, Uruguay, Bolivia, Mexico, Brazil, Peru, and Guatemala	Passeriformes	Host species diversity	Prevalence (*Haemoproteus*)	Predictive	Doussang et al. (2021)
Argentina, Uruguay, Bolivia, Mexico, Brazil, Peru, and Guatemala	Passeriformes	Host species diversity	Prevalence (*Plasmodium*)	Predictive	
Guadeloupe and Martinique	Passeriformes	Habitat loss	Prevalence (Haemosporidian)	Predictive	Perez-Rodriguez et al. (2018)
		Habitat fragmentation	Prevalence (Haemosporidian)	Predictive	

(continued)

Table 5.2 (continued)

Country	Avian order	Explanatory variables	Response variables	Effect type	References
Brazil	NA	Proportion native forest	Prevalence (*Haemoproteus*)	Nonpredictive	Fecchio et al. (2021)
			Taxonomic diversity (*Haemoproteus*)	Nonpredictive	
			Phylogenetic diversity (*Haemoproteus*)	Nonpredictive	
		Host functional diversity	Prevalence (*Haemoproteus*)	Nonpredictive	
			Taxonomic diversity (*Haemoproteus*)	Nonpredictive	
			Phylogenetic diversity (*Haemoproteus*)	Nonpredictive	
		Host phylogenetic diversity	Prevalence (*Haemoproteus*)	Nonpredictive	
			Taxonomic diversity (*Haemoproteus*)	Nonpredictive	
			Phylogenetic diversity (*Haemoproteus*)	Nonpredictive	
		Host taxonomic diversity	Prevalence (*Haemoproteus*)	Nonpredictive	
			Taxonomic diversity (*Haemoproteus*)	Predictive	
			Phylogenetic diversity (*Haemoproteus*)	Nonpredictive	
		Proportion native forest	Prevalence (*Plasmodium*)	Predictive	
			Taxonomic diversity (*Plasmodium*)	Nonpredictive	
			Phylogenetic diversity (*Plasmodium*)	Nonpredictive	

(continued)

Table 5.2 (continued)

Country	Avian order	Explanatory variables	Response variables	Effect type	References
		Host functional diversity	Prevalence (*Plasmodium*)	Predictive	
			Taxonomic diversity (*Plasmodium*)	Predictive	
			Phylogenetic diversity (*Plasmodium*)	Nonpredictive	
		Host phylogenetic diversity	Prevalence (*Plasmodium*)	Nonpredictive	
			Taxonomic diversity (*Plasmodium*)	Nonpredictive	
			Phylogenetic diversity (*Plasmodium*)	Nonpredictive	
		Host taxonomic diversity	Prevalence (*Plasmodium*)	Nonpredictive	
			Taxonomic diversity (*Plasmodium*)	Predictive	
			Phylogenetic diversity (*Plasmodium*)	Predictive	
		Null model (without any predictor variables)	Prevalence (*Haemoproteus*)	Nonpredictive	
		Null model (without any predictor variables)	Taxonomic diversity (*Haemoproteus*)	Nonpredictive	
		Null model (without any predictor variables)	Phylogenetic diversity (*Haemoproteus*)	Nonpredictive	
Mexico	Passeriformes	Land use types	Prevalence (Haemosporidian)	Predictive	Hernandez-Lara et al. (2017)
		Land use types	Parasitaemia (Haemosporidian)	Predictive	
		Land use types	Aggregation (Haemosporidian)	Nonpredictive	
Brazil	Passeriformes and Columbiformes	Succession stage	Prevalence (Haemosporidian)	Predictive	Ferreira et al. (2017)
			Diversity (Haemosporidian)	Predictive	

Table 5.3 Summary of the selected (third search) studies that explore the association between ecological factors and parasite prevalence/diversity in Neotropical avifauna

Country	Avian order	Explanatory variables	Response variables	Effect type	References
Chile, Costa Rica, Bolivia, Peru, and Argentina	Passeriformes	Altitude	Prevalence (*Haemoproteus*)	Predictive	Doussang. et al. (2021)
			Prevalence (*Plasmodium*)	Predictive	
			Genetic diversity (*Haemoproteus*)	Nonpredictive	
			Genetic diversity (*Plasmodium*)	Nonpredictive	
		Latitude	Prevalence (*Haemoproteus*)	Predictive	
			Prevalence (*Plasmodium*)	Predictive	
			Genetic diversity (*Haemoproteus*)	Nonpredictive	
			Genetic diversity (*Plasmodium*)	Nonpredictive	
Brazil	Apodiformes, Passeriformes, Columbiformes, Accipitriformes, Galbuliformes, Psittaciformes	Humidity	Infection (Haemosporidian)	Predictive	Ferreira de Souza et al. (2020)
Peru	Passeriformes	Altitude	Prevalence (*Plasmodium*)	Predictive	Jones et al. (2013)
Chile	Passeriformes	Latitude	Prevalence (*Plasmodium*)	Predictive	Cuevas et al. (2020)
			Diversity (*Plasmodium*)	Predictive	
			Prevalence (*Leucocytozoon*)	Predictive	
			Diversity (*Leucocytozoon*)	Predictive	
			Prevalence (*Haemoproteus*)	Predictive	
			Diversity (*Haemoproteus*)	Nonpredictive	

(continued)

Table 5.3 (continued)

Country	Avian order	Explanatory variables	Response variables	Effect type	References
Chile	NA	Latitude	Prevalence (*Leucocytozoon*)	Predictive	Merino et al. (2008)
			Prevalence (*Plasmodium*)	Predictive	
			Prevalence (*Haemoproteus*)	Predictive	
			Prevalence (*Haemosporidian*)	Predictive	
			Parasite richness (*Leucocytozoon*)	Nonpredictive	
			Parasite richness (*Plasmodium*)	Nonpredictive	
			Parasite richness (*Haemoproteus*)	Nonpredictive	
			Parasite richness (Haemosporidian)	Nonpredictive	
Ecuador	Passeriformes	Precipitation	Prevalence (Haemosporidian)	Predictive	Cadena-Ortiz et al. (2019)
		Site	Parasitemia (Haemosporidian)	Predictive	
Colombia	Passeriformes	Precipitation	Prevalence (Haemosporidian) *Eucometis penicillata*	Predictive	Pulgarin-R et al. (2018)
			Prevalence (Haemosporidian) *Manacus manacus*	Predictive	
			Diversity (Haemosporidian)	Nonpredictive	
Brazil	Passeriformes and Columbiformes	Temperature	Prevalence (*Plasmodium*)	Predictive	Rodriguez et al. (2021)
			Prevalence (*Haemoproteus*)	Nonpredictive	
		Rainfall	Prevalence (*Plasmodium*)	Nonpredictive	
			Prevalence (*Haemoproteus*)	Nonpredictive	

(continued)

Table 5.3 (continued)

Country	Avian order	Explanatory variables	Response variables	Effect type	References
Peru	Passeriformes, Caprimulgiformes, Apodiformes, Strigiformes, Cuculiformes, Galbuliformes, Piciformes, Tinamiformes, Trogoniformes, and Columbiformes	Precipitation	Parasite turnover (Haemosporidian)	Predictive	McNew et al. (2021)
		Temperature	Parasite turnover (Haemosporidian)	Nonpredictive	
		Elevation	Parasite turnover (Haemosporidian)	Predictive	
		Net primary production	Parasite turnover (Haemosporidian)	Predictive	
		Host turnover	Parasite turnover (Haemosporidian)	Predictive	
		Precipitation	Taxonomic diversity	Nonpredictive	
		Temperature	Taxonomic diversity	Nonpredictive	
		Elevation	Taxonomic diversity	Predictive	
		Net primary production	Taxonomic diversity	Predictive	
		Host turnover	Taxonomic diversity	Predictive	
Mexico	Columbiformes, Cuculiformes, Apodiformes, Strigiformes, Coraciiformes, Piciformes, Passeriformes	Elevation	Prevalence (*Plasmodium*)	Predictive	Rodríguez-Hernandez et al. (2021)
			Prevalence (*Haemoproteus*)	Predictive	
			Prevalence (*Leucocytozoon*)	Predictive	
			Parasitemia (*Plasmodium*)	Predictive	
			Parasitemia (*Haemoproteus*)	Predictive	
			Parasitemia (*Leucocytozoon*)	Predictive	
			Aggregation (*Plasmodium*)	Predictive	
			Aggregation (*Haemoproteus*)	Predictive	
			Aggregation (*Plasmodium*)	Predictive	
			Prevalence (Haemosporidia)	Predictive	
			Parasitemia (Haemosporidia)	Predictive	
			Aggregation (Haemosporidia)	Predictive	

(continued)

Table 5.3 (continued)

Country	Avian order	Explanatory variables	Response variables	Effect type	References
Mexico	Columbiformes, Cuculiformes, Apodiformes, Strigiformes, Coraciiformes, Piciformes, Passeriformes	Temperature	Lineage community composition	Predictive	Alvarez-Medizabal et al. (2021)
		Precipitation	Lineage community composition	Predictive	
Brazil	Passeriformes	Number of water bodies	Probability of infection (Haemosporidian)	Nonpredictive	Fecchio et al. (2021)
		Precipitation	Probability of infection (Haemosporidian)	Predictive	
		Temperature	Probability of infection (Haemosporidian)	Predictive	
Mexico	Passeriformes	Seasonality	Prevalence (Haemosporidia)	Predictive	Hernandez-Lara et al. (2020)
			Prevalence (*Haemoproteus*)	Predictive	
			Prevalence (*Plasmodium*)	Predictive	
			Parasitemia (Haemosporidia)	Predictive	
			Aggregation (Haemosporidia)	Predictive	
Brazil	NA	Temperature	Prevalence (*Plasmodium*)	Nonpredictive	Fecchio et al. (2021)
			Taxonomic diversity (*Plasmodium*)	Nonpredictive	
			Phylogenetic diversity (*Plasmodium*)	Nonpredictive	
			Prevalence (*Haemoproteus*)	Nonpredictive	
			Taxonomic diversity (*Haemoproteus*)	Nonpredictive	
			Phylogenetic diversity (*Haemoproteus*)	Nonpredictive	
Mexico	Passeriformes	Seasonality	Prevalence	Predictive	Hernandez-Lara et al. (2017)
			Parasitaemia	Predictive	
			Aggregation	Nonpredictive	

(continued)

Table 5.3 (continued)

Country	Avian order	Explanatory variables	Response variables	Effect type	References
Brazil	Passeriformes and Columbiformes	Seasonality	Prevalence	Nonpredictive	Ferrerira et al. (2017)
Ecuador	Passeriformes	Seasonality	Infection status (*Plasmodium*)	Nonpredictive	Rivero de Aguilar et al. (2018)

richness (Dobson et al. 2008; Poulin 2014; Kamiya et al. 2014). Consequently, these three bird-rich countries offer a high diversity of hosts for parasites to occupy, increasing the likelihood of parasite detection. Most observations for Brazil, Colombia, and Peru are of parasitic Arthropoda, which are mainly represented by chewing lice of the genus *Myrsidea* sp. Avian chewing lice are ubiquitous and highly diverse ectoparasites of birds in Neotropical regions (Soto-Madrid et al. 2020). Apart from being studied for their relevance as members of Neotropical communities, arthropods such as chewing lice are methodologically more straightforward to detect than other groups of parasites (e.g., viruses). We suggest that their vast diversity, direct detection, and usefulness as parasite models to understand ecological interactions explains the high number of arthropod records observed in Brazil, Colombia, and Peru.

Following arthropod insects, avian protozoa comprised the next best-studied parasite group in the Neotropics. Brazil, Peru, Colombia, Mexico, and Chile contained most avian protozoa records. Phylum Mizozoa (subphylum Apicomplexa) represented most of the observations, mainly including avian haemosporidian parasites (*Plasmodium* sp., *Haemoproteus* sp., and *Leucocytozoon* sp.). Apicomplexan parasites such as avian haemosporidians have become a model system for ecological and evolutionary studies of bird–parasite interactions (e.g., see chapters in Santiago-Alarcon and Marzal 2020b). Their importance for avian conservation, worldwide distribution, and high probability of detection in blood smears and PCR molecular samples has contributed to more scientists dedicating their research to this highly diverse parasite system (Valkiūnas and Atkinson 2020; Bensch and Hellgren 2020). Although much effort to detect and understand avian haemosporidian diversity in tropical settings has taken place in the last decade, most of their studies have been performed in temperate regions (Santiago-Alarcon and Marzal 2020a; Chapa-Vargas et al. 2020). Moreover, haemosporidian haplotype sequence divergence has suggested that haemosporidian genetic richness could match worldwide avian species richness (i.e., 10,000 bird species) (Bensch et al. 2004; Bensch and Hellgren 2020). Consequently, detection of avian haemosporidian diversity will help to understand complex antagonistic interactions – particularly of vector-borne parasites – and should be prioritized within Neotropical settings considered hotspots for host and parasite biodiversity (Santiago-Alarcon and Rojas Soto 2021).

After avian protozoa, avian viruses were the next best-studied group. In particular, we found that the Poxviridae, Orthomyxoviridae, Flaviviridae, and Coronaviridae families have been the most studied avian viruses in Neotropical countries. Ecuador, French Guiana, Guatemala, Chile, and Venezuela contained all studies focusing on avian pox virus. Avian pox viruses have been observed in over 350 species of birds (Williams et al. 2021). Because their incidence has increased over the last 10 years, avian pox has been considered an emerging viral disease (Alehegn et al. 2014), mainly affecting songbirds, birds of prey, marine birds, and poultry within tropical regions. A high prevalence of the disease has been observed in remote populations (e.g., Galapagos Islands), suggesting a threat to naive endemic insular avifauna in the Neotropical region (Williams et al. 2021). In wild continental birds, pox prevalence has been highly variable, and information regarding host specificity, strain virulence, and strain diversity remains elusive (Williams et al. 2021). In addition to finding many observations on avian pox viruses, our results showed avian influenza viruses (AIV) as the second most studied avian viruses in Neotropical countries. Brazil, Argentina, Peru, Ecuador, Chile, and Guatemala contained most AIV observations. Research efforts have been conducted worldwide to better understand the ecology of AIVs in wild birds (Wille and Barr 2022). Today, we know that highly pathogenic AIVs (HPAIVs) circulate on all continents except Australia and Antarctica (Willie and Barr 2022). Moreover, HPAIVs have caused recent outbreaks in South America, killing over 22,000 wild birds, mainly Peruvian pelicans (*Pelecanus thagusI*) and boobies (*Sula variegata*) (Gamarra-Toledo et al. 2023). Apart from these species, HPAIV has affected the near-threatened Guanay cormorant (*Leucocarbo bougainvillii*) across the coastline of Peru and has been suggested to be a potential threat to scavenger species such as the Andean condor (*Vultur gryphus*) (Gamarra-Toledo et al. 2023). Researchers are calling for epidemiological surveillance to better understand the extent to which such outbreaks could progress.

Following avian influenza, West Nile virus was the third most studied avian virus in the Neotropical region. Mexico and the Dominican Republic contained all of the records. WNV belongs to the Flavivirus genus, and its transmission cycle involves mosquitoes as vectors and wild birds as their vertebrate hosts. For more than 20 years, this virus has represented a threat to bird populations with persistent impacts on some wild bird species in North America (Passeriformes comprises the order with more susceptible avian species for this virus) and on human populations producing outbreaks of neurological disease (LaDeau et al. 2007; George et al. 2015). However, despite its importance for bird conservation, human health, and evidence suggesting migratory birds are important carriers of this virus, WNV is poorly known in the Neotropical region. A worldwide study showed that Mexico, Guatemala, and Argentina are countries with serological and molecular evidence in wild birds (Tolsá et al. 2018). However, WNV has already been classified as a neglected tropical disease because funding and research efforts have declined over time (Ronca et al. 2021).

Regarding bacteria, we found few records, even though birds are recognized as hosts of many bacterial species relevant to avian and human health (Reed 2002; Benskin et al. 2009). *Escherichia coli* was the best-represented bacterial species in our search and was studied in Brazil, Chile, Costa Rica, and Argentina mainly in Charadriiformes, Psittaciformes, and Sphenisciformes hosts. In the Neotropics, the study of *E. coli* should be a priority because it has been associated with mass die-offs of various wild bird species, which are suggested to act as carriers of multidrug-resistant *E. coli* (Reed 2002; Smith et al. 2020). The genus *Salmonella* was the second most studied bacteria in the Neotropics, mainly in Peru and Argentina in the Charadriiformes and Passeriformes. Various studies have shown that this genus, specifically *Salmonella typhimurium,* is commonly isolated in intestinal samples of wild birds and has been associated with mass mortality events (Smith et al. 2020).

Fungi was the least studied parasite group in the Neotropics (n = 19). The best-represented species were *Cryptococcus neoformans* and *Aspergillus* spp. *Cryptococcus neoformans* was identified infecting the order Columbiformes (pigeons and doves) in Brazil, Venezuela, and Colombia. *Cryptococcus* spp. have been found in avian orders such as Passeriformes, Anseriformes, Accipitriformes, Apterygiformes, Psittaciformes, and Columbiformes, with wide variability in susceptibility between bird species (Johnston et al. 2016). Additionally, *Aspergillus* spp. were identified in samples of Sphenisciformes, Suliformes, and Charadriiformes in Brazil, Argentina, and Mexico, respectively. At the parasite species level, we also recorded *Aspergillus fumigatus*, a common parasite among wild and domestic birds, with a prevalence of approximately 95% (Arné et al. 2021).

5.3.1 Parasite Dynamics as a Function of Anthropogenic and Ecological Factors

Understanding parasite infection patterns, their diversity and distribution in relation to environmental, spatial, and anthropogenic factors remains a challenge for infectious disease ecology. Our synthesis found that the main anthropogenic variables associated to prevalence and diversity of avian parasites are land-use or habitat type, habitat fragmentation or loss, the proportion of native vegetation, and habitat-associated host diversity in its taxonomic, phylogenetic, and functional forms. These variables often relate differently to each parasite system, sometimes producing opposing trends even within the same parasite taxa. For example, the prevalence of avian pox viruses has increased with agricultural intensification (Zylberberg et al. 2013). In contrast, the probability of haemosporidian parasite infection has increased with higher proportion of native vegetation cover (Fecchio et al. 2020) or decrease with advanced stages of forest succession (Ferreira Junior et al. 2017). In the case of avian haemosporidians, parasite genera can produce these contrasting results because each Haemosporidian genus is transmitted by different Diptera families and species whose reproduction and development depend on different microclimatic conditions (Rodríguez-Hernández et al. 2021). For instance, the *Plasmodium* genus might be favored by anthropogenic transformation, while the *Haemoproteus* genus might be more prevalent in preserved forests (Hernández-Lara et al. 2020).

This shows how the response of parasites to anthropogenic drivers of prevalence varies spatially and may be contingent on the host–parasite system.

Habitat-associated host diversity has also been related to contrasting infection patterns and parasite diversity in systems such as avian haemosporidians. Increased host functional diversity in preserved remnants of the Brazilian Atlantic Forest decreased *Plasmodium* lineage diversity (Fecchio et al. 2021). However, high host taxonomic diversity within the same locations increased *Plasmodium* taxonomic lineage diversity but decreased *Plasmodium* phylogenetic lineage diversity (i.e., more diverse avian communities in preserved remnants harbored more taxonomically diverse but phylogenetically clustered parasite assemblages). Apparently, different facets of avian host diversity (e.g., functional, taxonomic, and phylogenetic) can produce differential effects over the taxonomic and phylogenetic components of haemosporidian parasite diversity, underscoring the complexity and relevance of assessing the spatial patterns of host diversity in all of its facets as drivers of parasite diversity and distribution.

Beyond the diverse effects of anthropogenic factors, the prevalence and diversity of avian parasites can be driven by ecological factors (i.e., environmental and spatial factors). We found that latitude, altitude, temperature, precipitation, humidity, and seasonality are the most studied ecological drivers of parasite prevalence and diversity in the Neotropics. For instance, the inverse relationship between altitude and temperature has been associated with the differential distribution of avian haemosporidian genera across elevational gradients in different settings (Zamora-Vilchis et al. 2012; van Rooyen et al. 2013; Álvarez-Mendizábal et al. 2021). Similarly, an inverse latitudinal diversity gradient has explained the infection probability and phylogenetic diversity of *Leucocytozoon* parasites in New World birds (Fecchio et al. 2019). Moreover, within Neotropical countries such as Mexico and Brazil, haemosporidian parasite prevalence, diversity, and community composition have been explained by temperature, precipitation, humidity, and seasonality (Hernández-Lara et al. 2017; Hernández-Lara et al. 2020; Ferreira de Souza et al. 2020). In the Caatinga of Brazil, there was a high frequency of hemoparasites in birds due to high levels of humidity and rainfall (Ferreira de Souza et al. 2020). Similarly, contrasting patterns in prevalence among bird species were reported in Colombia (e.g., gray-headed tanagers, *Eucometis penicillata*, exhibited higher prevalence rates in locations with less precipitation, while white-bearded manakins, *Manacus manacus,* showed higher prevalence in locations with high precipitation) (Pulgarín-R et al. 2018). In Mexico, increased infection rates have been associated with wet seasons, where there is an increase in the abundance of mosquito vectors, and coincide with avian host breeding seasons, where reproduction generates a trade-off between immune function and reproductive output (Hernández-Lara et al. 2020).

Environmental and spatial heterogeneity can structure the distribution and diversity of avian parasites and their hosts and vectors. However, a wide variation in responses to the heterogeneity in ecological factors can be observed for most host–parasite systems. The relative importance of environmental versus spatial drivers of parasite infection and diversity can also depend on the scale of the study (e.g., microclimatic variables can explain parasite distribution patterns at local-landscape scales, while geographic barriers or latitude could be more important drivers at regional and global scales). Consequently, it is fundamental to continue studying avian host–parasite systems in regions that provide this type of environmental heterogeneity (i.e., tropical and subtropical) at different spatiotemporal scales to better understand the relative importance of such drivers for specific parasite systems.

5.4 Conclusions and Synthesis

Our study showed that arthropod insects, protozoa, and viruses are the main parasite groups being studied in birds across the Neotropical region. Although bacteria and fungi have been less studied, the potential for these parasites to produce diseases in avian populations should not be overlooked. Most of the published information regarding parasite ecology in Neotropical birds has been directed to the order Passeriformes, mainly because of its high number of species and because there is a sampling bias for passerine birds when using understory mist nets. This has allowed in-depth knowledge of parasite dynamics within some passerine populations (e.g., West Nile virus infecting American robins); however, the diversity and distribution of avian parasites among nonpasserine birds remains scarce, and it is an area of opportunity to advance the field. Our data showed that parasite species richness was higher in Brazil, Argentina, Colombia, Peru, and Chile than in the other analyzed Neotropical countries. This was certainly associated with the high bird species richness that these countries harbor, and it should also be associated with the high number of studies that we retrieved for those countries. The low parasite and bird species richness that was observed for countries, such as the Dominican Republic, Guatemala, Bolivia, and Uruguay, is likely explained by the few studies that we found, suggesting that parasite surveillance and scientific publication in these countries is insufficient. The lack of studies focusing on avian disease ecology in Neotropical islands (e.g., Cuba, Jamaica, Haiti, Puerto Rico; but see (Parker 2018) for synthesis of disease ecology research in the Galápagos Islands) underscores the need to focus our scientific research on parasite dynamics of endemic insular avifauna.

We identified several anthropogenic and ecological factors affecting parasite prevalence and diversity across space. However, our synthesis did not include all relevant factors that could potentially drive parasite diversity and distribution (e.g., host-associated factors such as life-history traits). Although we showed that the diversity and distribution of some Neotropical avian parasites have been thoroughly assessed concerning anthropogenic and ecological drivers (e.g., avian haemosporidians), most parasite systems remain understudied in the Neotropical region (i.e., their presence in avian hosts is not assessed as a function of relevant drivers of disease), and their diversity is yet to be uncovered (Fernandez Correa et al. 2019). We underscore the large knowledge gap when considering the temporal aspect of disease ecology (e.g., Rubio et al. 2017). Future research efforts should focus on determining how the temporal component (e.g., variation across the day and across seasons or months) (Castro-Arellano et al. 2010) affects the ecological dynamics of host-parasite systems.

Appendix A: Supplementary Data

Table 5.4 Boolean codes used for our first, second, and third searches

First search	
General overview of parasite groups	((virus* OR bacteria* OR protozoa* OR *parasite* OR fungi* OR disease*) AND (birds OR avian OR aves OR avifauna) AND (neotropic* OR South America OR Latin America) NOT (review* OR laboratory OR experiment*) NOT (zoos OR captivity OR domestic OR poultry))
Second search	
Anthropogenic factors/ parasite groups	(birds OR avian OR aves OR avifauna) AND (neotropic* OR South America OR Latin America) AND (fragmentation OR land use OR landscape OR urban*OR habitat perturbation) NOT (review* OR laboratory OR experiment*) NOT (zoos OR captivity OR domestic OR poultry) AND (**virus***)
	(birds OR avian OR aves OR avifauna) AND (neotropic* OR South America OR Latin America) AND (fragmentation OR land use OR landscape OR urban*OR habitat perturbation) NOT (review* OR laboratory OR experiment*) NOT (zoos OR captivity OR domestic OR poultry) AND (**bacteria***)
	(birds OR avian OR aves OR avifauna) AND (neotropic* OR South America OR Latin America) AND (fragmentation OR land use OR landscape OR urban*OR habitat perturbation) NOT (review* OR laboratory OR experiment*) NOT (zoos OR captivity OR domestic OR poultry) AND (**fungi*** OR **fungus**)
	(birds OR avian OR aves OR avifauna) AND (neotropic* OR South America OR Latin America) AND (fragmentation OR land use OR landscape OR urban*OR habitat perturbation) NOT (review* OR laboratory OR experiment*) NOT (zoos OR captivity OR domestic OR poultry) AND (**protista** OR **protozoa***)
	(birds OR avian OR aves OR avifauna) AND (neotropic* OR South America OR Latin America) AND (fragmentation OR land use OR landscape OR urban*OR habitat perturbation) NOT (review* OR laboratory OR experiment*) NOT (zoos OR captivity OR domestic OR poultry) AND (**Arthropoda** OR **Platyhelminthes** OR **Nematoda** OR **Rotifera**)

(continued)

Table 5.4 (continued)

Ecological factors/ parasite groups	(birds OR avian OR aves OR avifauna) AND (neotropic* OR South America OR Latin America) AND (temperature OR latitude OR altitude OR precipitation OR humidity OR climate change) NOT (review* OR laboratory OR experiment*) NOT (zoos OR captivity OR domestic OR poultry) AND (**virus***)
	(birds OR avian OR aves OR avifauna) AND (neotropic* OR South America OR Latin America) AND (temperature OR latitude OR altitude OR precipitation OR humidity OR climate change) NOT (review* OR laboratory OR experiment*) NOT (zoos OR captivity OR domestic OR poultry) AND (**bacteria***)
	(birds OR avian OR aves OR avifauna) AND (neotropic* OR South America OR Latin America) AND (temperature OR latitude OR altitude OR precipitation OR humidity OR climate change) NOT (review* OR laboratory OR experiment*) NOT (zoos OR captivity OR domestic OR poultry) AND (**fungi*** OR **fungus**)
	(birds OR avian OR aves OR avifauna) AND (neotropic* OR South America OR Latin America) AND (temperature OR latitude OR altitude OR precipitation OR humidity OR climate change) NOT (review* OR laboratory OR experiment*) NOT (zoos OR captivity OR domestic OR poultry) AND (**protista** OR **protozoa***)
	(birds OR avian OR aves OR avifauna) AND (neotropic* OR South America OR Latin America) AND (temperature OR latitude OR altitude OR precipitation OR humidity OR climate change) NOT (review* OR laboratory OR experiment*) NOT (zoos OR captivity OR domestic OR poultry) AND (**Arthropoda** OR **Platyhelminthes** OR **Nematoda** OR **Rotifera**)
Third search	
Anthropogenic factors/ prevalence and diversity/ parasite genus or species	(birds OR avian OR aves OR avifauna) AND (neotropic* OR South America OR Latin America) AND (fragmentation OR land use OR landscape OR urban*OR habitat perturbation) AND (**presence** OR **prevalence** OR **diversity** OR **richness**) AND (**Avian pox virus**)
	(birds OR avian OR aves OR avifauna) AND (neotropic* OR South America OR Latin America) AND (fragmentation OR land use OR landscape OR urban*OR habitat perturbation) AND (**presence** OR **prevalence** OR **diversity** OR **richness**) AND (**Escherichia coli**)
	(birds OR avian OR aves OR avifauna) AND (neotropic* OR South America OR Latin America) AND (fragmentation OR land use OR landscape OR urban*OR habitat perturbation) AND (**presence** OR **prevalence** OR **diversity** OR **richness**) AND (**Cryptococcus neoformans**)
	(birds OR avian OR aves OR avifauna) AND (neotropic* OR South America OR Latin America) AND (fragmentation OR land use OR landscape OR urban*OR habitat perturbation) AND (presence OR prevalence OR diversity OR richness) AND (**Plasmodium sp**)
	(birds OR avian OR aves OR avifauna) AND (neotropic* OR South America OR Latin America) AND (fragmentation OR land use OR landscape OR urban*OR habitat perturbation) AND (presence OR prevalence OR diversity OR richness) AND (**Myrsidea sp**)

(continued)

Table 5.4 (continued)

Ecological factors/ prevalence and diversity/ parasite genus or species	(birds OR avian OR aves OR avifauna) AND (neotropic* OR South America OR Latin America) AND (temperature OR latitude OR altitude OR precipitation OR humidity OR climate change) AND (presence OR prevalence OR diversity OR richness) AND (**Avian pox virus**)
	(birds OR avian OR aves OR avifauna) AND (neotropic* OR South America OR Latin America) AND (temperature OR latitude OR altitude OR precipitation OR humidity OR climate change) AND (presence OR prevalence OR diversity OR richness) AND (**Escherichia coli**)
	(birds OR avian OR aves OR avifauna) AND (neotropic* OR South America OR Latin America) AND (temperature OR latitude OR altitude OR precipitation OR humidity OR climate change) AND (presence OR prevalence OR diversity OR richness) AND (**Cryptococcus neoformans**)
	(birds OR avian OR aves OR avifauna) AND (neotropic* OR South America OR Latin America) AND (temperature OR latitude OR altitude OR precipitation OR humidity OR climate change) AND (presence OR prevalence OR diversity OR richness) AND (**Plasmodium sp**)
	(birds OR avian OR aves OR avifauna) AND (neotropic* OR South America OR Latin America) AND (temperature OR latitude OR altitude OR precipitation OR humidity OR climate change) AND (presence OR prevalence OR diversity OR richness) AND (**Myrsidea sp**)

References

Alehegn E, Mersha C, Mengesha D (2014) A systematic review of serological and clinicopathological features and associated risk factors of avian pox. Br J Poult Sci 3(3):78–87. https://doi.org/10.20372/NADRE:1547201106.08

Álvarez-Mendizábal P, Villalobos F, Rodríguez-Hernández K, Hernández-Lara C, Rico-Chávez O, Suzán G, Chapa-Vargas L, Santiago-Alarcon D (2021) Metacommunity structure reveals that temperature affects the landscape compositional patterns of avian malaria and related haemosporidian parasites across elevations. Acta Oecol 113(November):103789. https://doi.org/10.1016/j.actao.2021.103789

Arné P, Risco-Castillo V, Jouvion G, Le Barzic C, Guillot J (2021) Aspergillosis in Wild Birds. J Fungi (Basel). 7(3):241. https://doi.org/10.3390/jof7030241. PMID: 33807065; PMCID: PMC8004873.

Belo NO, Pinheiro RT, Reis ES, Ricklefs RE, Braga ÉM (2011) Correction: Prevalence and lineage diversity of avian haemosporidians from three distinct cerrado habitats in Brazil. PLoS ONE 6(4). https://doi.org/10.1371/annotation/993592b8-5719-45b1-814e-2052318532e7

Bensch S, Hellgren O (2020) The use of molecular methods in studies of avian haemosporidians. In: Santiago-Alarcon D, Marzal A (eds) Avian malaria and related parasites in the tropics. Springer, Cham, pp 113–135. https://doi.org/10.1007/978-3-030-51633-8_4

Bensch S, Péarez-Tris J, Waldenströum J, Hellgren O (2004) Linkage between nuclear and mitochondrial dna sequences in avian malaria parasites: multiple cases of cryptic speciation? Evolution 58(7):1617–1621. https://doi.org/10.1111/j.0014-3820.2004.tb01742.x

Benskin CM, Wilson K, Jones K, Hartley IR (2009) Bacterial pathogens in wild birds: a review of the frequency and effects of infection. Biol Rev Camb Philos Soc 84(3):349–373. https://doi.org/10.1111/j.1469-185X.2008.00076.x. Epub Apr 29. PMID: 19438430.

Birkle C, Pendlebury DA, Schnell J, Adams J (2020) Web of science as a data source for research on scientific and scholarly activity. Quant Sci Stud 1(1):363–376. https://doi.org/10.1162/qss_a_00018

Cadena-Ortiz H, Mantilla JS, de Aguilar JR, Flores D, Bahamonde D, Matta NE, Bonaccorso E (2019) Avian haemosporidian infections in rufous-collared sparrows in an Andean dry forest: diversity and factors related to prevalence and parasitaemia. Parasitology. 146(6):765–773. https://doi.org/10.1017/S0031182018002081. Epub 2018 Dec 26. Erratum in: Parasitology. 2022 Sep;149(11):1521–1522. PMID: 30585147.

Capasso S, D'Amico VL, Minardi G et al (2022) The helminth community structure as an indicator of habitat use: the case of the Baird's Sandpiper (Calidris bairdii) and the White-Rumped Sandpiper (Calidris fuscicollis) on their non-breeding grounds in Patagonia. Polar Biol 45, 297–307 https://doi.org/10.1007/s00300-021-02990-6

Castro-Arellano I, Lacher TE, Willig MR, Rangel TF (2010) Assessment of assemblage-wide temporal niche segregation using null models: assemblage-wide temporal niche overlap assessment. Methods Ecol Evol 1(3):311–318. https://doi.org/10.1111/j.2041-210X.2010.00031.x

Chapa-Vargas L, Matta NE, Merino S (2020) Effects of ecological gradients on tropical avian hemoparasites. In: Santiago-Alarcon D, Marzal A (eds) Avian malaria and related parasites in the tropics. Springer, Cham, pp 349–377. https://doi.org/10.1007/978-3-030-51633-8_10

Clark NJ, Clegg SM (2017) Integrating phylogenetic and ecological distances reveals new insights into parasite host specificity. Mol Ecol 26(11):3074–3086. https://doi.org/10.1111/mec.14101

Cuevas E, Vianna JA, Botero-Delgadillo E, Doussang D, González-Acuña D, Barroso O, Rozzi R, Vásquez RA, Quirici V (2020) Latitudinal gradients of haemosporidian parasites: prevalence, diversity and drivers of infection in the Thorn-tailed Rayadito (Aphrastura spinicauda). Int J Parasitol Parasites Wildl. 11:1–1. https://doi.org/10.1016/j.ijppaw.2019.11.002

Dáttilo W, Barrozo-Chávez N, Lira-Noriega A, Guevara R, Villalobos F, Santiago-Alarcon D, Neves FS, Izzo T, Ribeiro SP (2020) Species-level drivers of mammalian ectoparasite faunas. Edited by Jenny Dunn. J Anim Ecol 89(8):1754–1765. https://doi.org/10.1111/1365-2656.13216

de Souza F, Carolina R, Teixeira BLB, Gusmão CLS, Fernandes AM (2020) Humidity effects on avian blood parasites in the Caatinga of Brazil. Ornithol Res 28(2):98–104. https://doi.org/10.1007/s43388-020-00009-y

Dobson A, Lafferty KD, Kuris AM, Hechinger RF, Jetz W (2008) Homage to linnaeus: how many parasites? How many hosts? Proc Natl Acad Sci 105(Supplement 1):11482–11489. https://doi.org/10.1073/pnas.0803232105

Doussang D, Sallaberry-Pincheira N, Cabanne GS, Lijtmaer DA, González-Acuña D, Vianna JA (2021) Specialist versus generalist parasites: the interactions between host diversity, environment and geographic barriers in avian malaria. Int J Parasitol 51(11):899–911. https://doi.org/10.1016/j.ijpara.2021.04.003

Escallón C, Weinstein NM, Tallant JA, Wojtenek W, Rodríguez-Saltos CA, Bonaccorso E, Moore IT (2016) Testosterone and haemosporidian parasites along a tropical elevational gradient in rufous-collared sparrows (Zonotrichia capensis). J Exp Zool 325A:501–510.

Fallon SM, Ricklefs RE, Latta SC, Bermingham E (2004) Temporal stability of insular avian malarial parasite communities. Proc Biol Sci 271(1538):493–500. https://doi.org/10.1098/rspb.2003.2621. PMID: 15129959; PMCID: PMC1691613.

Feare CJ (2010) Role of wild birds in the spread of highly pathogenic avian influenza virus H5N1 and implications for global surveillance. Avian Dis 54(s1):201–212. https://doi.org/10.1637/8766-033109-ResNote.1

Fecchio A, Bell JA, Bosholn M, Vaughan JA, Tkach VV, Lutz HL, Cueto VR et al (2019) An inverse latitudinal gradient in infection probability and phylogenetic diversity for *Leucocytozoon* blood parasites in new world birds. Edited by David Richardson. J Anim Ecol. https://doi.org/10.1111/1365-2656.13117

Fecchio A, Ribeiro RM, Ferreira FC, de Angeli D, Dutra GT-P, Alquezar RD, Khan AU et al (2020) Higher infection probability of haemosporidian parasites in Blue-black Grassquits (*Volatinia Jacarina*) inhabiting native vegetation across Brazil. Parasitol Int 80(February):102204. https://doi.org/10.1016/j.parint.2020.102204

Fecchio A, Lima MR, Bell JA, Schunck F, Corrêa AH, Beco R, Jahn AE et al (2021) Loss of forest cover and host functional diversity increases prevalence of avian malaria parasites in the Atlantic forest. Int J Parasitol 51(9):719–728. https://doi.org/10.1016/j.ijpara.2021.01.001

Ferreira Junior FC, Rodrigues RA, Ellis VA, Leite LO, Borges MAZ, Braga ÉM (2017) Habitat modification and seasonality influence avian haemosporidian parasite distributions in Southeastern Brazil. Edited by Petr Heneberg. PLoS One 12(6):e0178791. https://doi.org/10.1371/journal.pone.0178791

Fernández-Correa I, Truchado DA, Gomez-Lucia E, et al (2019) A novel group of avian astroviruses from Neotropical passerine birds broaden the diversity and host range of Astroviridae. Sci Rep 9:9513. https://doi.org/10.1038/s41598-019-45889-3

Gamarra-Toledo V, Plaza PI, Gutiérrez R, Luyo P, Hernani L, Angulo F, Lambertucci SA (2023) Avian flu threatens neotropical birds. Science 379(6629):246–246. https://doi.org/10.1126/science.adg2271

George TL, Harrigan RJ, LaManna JA, DeSante DF, Saracco JF, Smith TB (2015) Persistent impacts of West Nile virus on North American bird populations. Proc Natl Acad Sci 112(46):14290–14294. https://doi.org/10.1073/pnas.1507747112

Gilman S, Blumstein DT, Foufopoulos J (2007) The effect of hemosporidian infections on white-crowned sparrow singing behavior. Ethology 113(5):437–445. https://doi.org/10.1111/j.1439-0310.2006.01341.x

Hadfield J, Brito AF, Swetnam DM, Vogels CBF, Tokarz RE, Andersen KG, Smith RC, Bedford T, Grubaugh ND (2019) Twenty years of West Nile virus spread and evolution in the Americas visualized by Nextstrain. Edited by Tom C. Hobman. PLoS Pathog 15(10):e1008042. https://doi.org/10.1371/journal.ppat.1008042

Hernández-Lara C, González-García F, Santiago-Alarcon D (2017) Spatial and seasonal variation of avian malaria infections in five different land use types within a neotropical montane forest matrix. Landsc Urban Plan 157(January):151–160. https://doi.org/10.1016/j.landurbplan.2016.05.025

Hernández-Lara C, Carbó-Ramírez P, Santiago-Alarcon D (2020) Effects of land use change (rural-urban) on the diversity and epizootiological parameters of avian haemosporida in a widespread neotropical bird. Acta Trop 209(September):105542. https://doi.org/10.1016/j.actatropica.2020.105542

Herzog SK, Kessler M, Bach K (2005) The elevational gradient in Andean bird species richness at the local scale: a foothill peak and a high-elevation plateau. Ecography 28(2):209–222. https://doi.org/10.1111/j.0906-7590.2005.03935.x

Hudson PJ, Dobson AP, Lafferty KD (2006) Is a healthy ecosystem one that is rich in parasites? Trends Ecol Evol 21(7):381–385. https://doi.org/10.1016/j.tree.2006.04.007

Jerzak G, Bernard KA, Kramer LD, Ebel GD (2005) Genetic variation in West Nile virus from naturally infected mosquitoes and birds suggests quasispecies structure and strong purifying selection. J Gen Virol 86(8):2175–2183. https://doi.org/10.1099/vir.0.81015-0

Jerzak GVS, Brown I, Shi P-Y, Kramer LD, Ebel GD (2008) Genetic diversity and purifying selection in West Nile virus populations are maintained during host switching. Virology 374(2):256–260. https://doi.org/10.1016/j.virol.2008.02.032

Jetz W, Rahbek C (2002) Geographic range size and determinants of avian species richness. Science 297(5586):1548–1551. https://doi.org/10.1126/science.1072779

Jetz GH, Thomas JB, Joy KH, Mooers AO (2012) The global diversity of birds in space and time. Nature 491(7424):444–448. https://doi.org/10.1038/nature11631

Jones MR, Cheviron ZA, Carling MD (2013) Spatial patterns of avian malaria prevalence in Zonotrichia capensis on the western slope of the Peruvian Andes. J Parasitol 99(5):903–905. https://doi.org/10.1645/12-147.1

Johnston SA, Voelz K, May RC (2016) Cryptococcus neoformans Thermotolerance to Avian Body Temperature Is Sufficient For Extracellular Growth But Not Intracellular Survival In Macrophages. Sci Rep 6:20977. https://doi.org/10.1038/srep20977. PMID: 26883088; PMCID: PMC4756366.

Kamiya T, O'Dwyer K, Nakagawa S, Poulin R (2014) Host diversity drives parasite diversity: meta-analytical insights into patterns and causal mechanisms. Ecography 37(7):689–697. https://doi.org/10.1111/j.1600-0587.2013.00571.x

Kilpatrick AM, Wheeler SS (2019) Impact of West Nile virus on bird populations: limited lasting effects, evidence for recovery, and gaps in our understanding of impacts on ecosystems. Edited by William Reisen. J Med Entomol 56(6):1491–1497. https://doi.org/10.1093/jme/tjz149

Kilpatrick AM, Kramer LD, Jones MJ, Marra PP, Daszak P (2006) West Nile virus epidemics in North America are driven by shifts in mosquito feeding behavior. Edited by Rick Ostfeld. PLoS Biol 4(4):e82. https://doi.org/10.1371/journal.pbio.0040082

Komar N, Langevin S, Hinten S, Nemeth N, Edwards E, Hettler D, Davis B, Bowen R, Bunning M (2003) Experimental infection of North American birds with the New York 1999 strain of West Nile virus. Emerg Infect Dis 9(3):311–322

LaDeau S, Kilpatrick A, Marra P (2007) West Nile virus emergence and large-scale declines of North American bird populations. Nature 447:710–713. https://doi.org/10.1038/nature05829

LaDeau SL, Calder CA, Doran PJ, Marra PP (2011) West Nile virus impacts in American crow populations are associated with human land use and climate. Ecol Res 26(5):909–916. https://doi.org/10.1007/s11284-010-0725-z

Lafferty KD, Dobson AP, Kuris AM (2006) Parasites dominate food web links. Proc Natl Acad Sci 103(30):11211–11216. https://doi.org/10.1073/pnas.0604755103

LaPointe DA, Atkinson CT, Samuel MD (2012) Ecology and conservation biology of avian malaria. Ann N Y Acad Sci 1249(1):211–226. https://doi.org/10.1111/j.1749-6632.2011.06431.x

Lee D-H, Torchetti MK, Winker K, Ip HS, Song C-S, Swayne DE (2015) Intercontinental spread of Asian-origin H5N8 to North America through Beringia by migratory birds. Edited by A. García-Sastre. J Virol 89(12):6521–6524. https://doi.org/10.1128/JVI.00728-15

Lopes VL, Costa FV, Rodrigues RA et al (2020) High fidelity defines the temporal consistency of host-parasite interactions in a tropical coastal ecosystem. Sci Rep 10:16839. https://doi.org/10.1038/s41598-020-73563-6

Mansilla AP, Grande JM, Diaz A (2022) Effect of Agroecosystems on Seroprevalence of St. Louis Encephalitis and West Nile Viruses in Birds, La Pampa, Argentina, 2017–2019. Emerg Infect Dis 28(7):1393–1402. https://doi.org/10.3201/eid2807.211485. PMID: 35731160; PMCID: PMC9239869.

Matuoka MA, Benchimol M, de Almeida-Rocha JM, Morante-Filho JC (2020) Effects of anthropogenic disturbances on bird functional diversity: a global meta-analysis. Ecol Indic 116(September):106471. https://doi.org/10.1016/j.ecolind.2020.106471

May RM, Anderson RM (1978) Regulation and stability of host-parasite population interactions: II. destabilizing processes. J Anim Ecol 47(1):249. https://doi.org/10.2307/3934

McNew SM, Barrow LN, Williamson JL, Galen SC, Skeen HR, DuBay SG, Gaffney AM, Johnson AB, Bautista E, Ordoñez P, Schmitt CJ (2021) Contrasting drivers of diversity in hosts and parasites across the tropical Andes. Proceedings of the National Academy of Sciences. 118(12):e2010714118. https://doi.org/10.1073/pnas.2010714118

Merino S, Moreno J, Vasquez RA, Martinez J, Sánchez-Monsávez IN, Estades CF, Ippi S, Sabat P, Rozzi R, Mcgehee S (2008) Haematozoa in forest birds from southern Chile: latitudinal gradients in prevalence and parasite lineage richness. Austral Ecol 33(3):329–340. https://doi.org/10.1111/j.1442-9993.2008.01820.x

Meza-Montes E, Fernández-Gómez RA, Llanes-Quevedo A, Navarro-Sigüenza AG, Santiago-Alarcon D, Roberto Sosa-López J (2023) Vocal behaviour, parasitic infection, chronic stress and body condition in Rufous-naped Wrens (*Campylorhynchus Rufinucha*). Ibis 165(2):676–684. https://doi.org/10.1111/ibi.13130

Mine J, Uchida Y, Sharshov K, Sobolev I, Shestopalov A, Saito T (2019) Phylogeographic evidence for the inter- and intracontinental dissemination of avian influenza viruses via migration flyways. Edited by Charles J. Russell. PLoS One 14(6):e0218506. https://doi.org/10.1371/journal.pone.0218506

Mollentze N, Streicker DG (2020) Viral zoonotic risk is homogenous among taxonomic orders of mammalian and avian reservoir hosts. Proc Natl Acad Sci 117(17):9423–9430. https://doi.org/10.1073/pnas.1919176117

Møller AP (2005) Parasitism and the regulation of host populations. In: Thomas F, Renaud F, Guegan J-F (eds) Parasitism and ecosystems. Oxford University Press, pp 43–53. https://doi.org/10.1093/acprof:oso/9780198529873.003.0004

Møller AP, Erritzøe J (2000) Predation against birds with low immunocompetence. Oecologia 122(4):500–504. https://doi.org/10.1007/s004420050972

Møller AP, Erritzøe J (2002) Coevolution of host immune defence and parasite-induced mortality: relative spleen size and mortality in altricial birds. Oikos 99(1):95–100. https://doi.org/10.1034/j.1600-0706.2002.990110.x

Morin CW, Stoner-Duncan B, Winker K, Scotch M, Hess JJ, Meschke JS, Ebi KL, Rabinowitz PM (2018) Avian influenza virus ecology and evolution through a climatic lens. Environ Int 119(October):241–249. https://doi.org/10.1016/j.envint.2018.06.018

Moura MR, Villalobos F, Costa GC, Garcia PCA (2016) Disentangling the role of climate, topography and vegetation in species richness gradients. Edited by Maura (Gee) Geraldine Chapman. PLoS One 11(3):e0152468. https://doi.org/10.1371/journal.pone.0152468

Nabi G, Wang Y, Lü L, Jiang C, Ahmad S, Yuefeng W, Li D (2021) Bats and birds as viral reservoirs: a physiological and ecological perspective. Sci Total Environ 754(February):142372. https://doi.org/10.1016/j.scitotenv.2020.142372

Navarro C (2004) Predation risk, host immune response, and parasitism. Behav Ecol 15(4):629–635. https://doi.org/10.1093/beheco/arh054

Ogrzewalska M, Uezu A, Jenkins CN, Labruna MB (2011) Effect of forest fragmentation on tick infestations of birds and tick infection rates by rickettsia in the Atlantic forest of Brazil. Ecohealth 8(3):320–331. https://doi.org/10.1007/s10393-011-0726-6. Epub 2011 Dec 16. PMID: 22173291.

Ortega-Guzmán L, Rojas-Soto O, Santiago-Alarcon D, Huber-Sannwald E, Chapa-Vargas L (2022) Climate predictors and climate change projections for avian haemosporidian prevalence in Mexico. Parasitology 149(8):1129–1144. https://doi.org/10.1017/S0031182022000683

Parker PG (ed) (2018) Disease ecology: Galapagos birds and their parasites, Social and ecological interactions in the Galapagos Islands. Springer, Cham. https://doi.org/10.1007/978-3-319-65909-1

Pérez-Rodríguez A, Khimoun A, Ollivier A, Eraud C, Faivre B, Garnier S (2018) Habitat fragmentation, not habitat loss, drives the prevalence of blood parasites in a Caribbean passerine. Ecography, 41:1835–1849. https://doi.org/10.1111/ecog.03189

Poulin R (2014) Parasite biodiversity revisited: frontiers and constraints. Int J Parasitol 44(9):581–589. https://doi.org/10.1016/j.ijpara.2014.02.003

Poulin R, Tellervo Valtonen E (2001) Nested assemblages resulting from host size variation: the case of endoparasite communities in fish hosts. Int J Parasitol 31(11):1194–1204. https://doi.org/10.1016/S0020-7519(01)00262-4

Pulgarín-R PC, Gómez JP, Robinson S, Ricklefs RE, Cadena CD (2018) Host Species, and not environment, predicts variation in blood parasite prevalence, distribution, and diversity along a humidity gradient in Northern South America. Ecol Evol 8(8):3800–3814. https://doi.org/10.1002/ece3.3785

Rahbek C, Graves GR (2001) Multiscale assessment of patterns of avian species richness. Proc Natl Acad Sci 98(8):4534–4539. https://doi.org/10.1073/pnas.071034898

Reboreda, J. C., Fiorini, V. D., Tuero, D. T. (Eds.) (2019). Behavioral Ecology of Neotropical Birds. Springer International Publishing. ISBNs 978-3-03-014279-7. https://doi.org/10.1007/978-3-030-14280-3

Reed KD (2002) Laboratory testing for Lyme disease: possibilities and practicalities. J Clin Microbiol. 40(2):319–324. https://doi.org/10.1128/JCM.40.2.319-324.2002. PMID: 11825936; PMCID: PMC153420.
Rivero de Aguilar J, Castillo F, Moreno A, Peñafiel N, Browne L, Walter ST, Karubian J, Bonaccorso E (2018) Patterns of avian haemosporidian infections vary with time, but not habitat, in a fragmented Neotropical landscape. PLoS One. 13(10):e0206493. https://doi.org/10.1371/journal.pone.0206493. PMID: 30379912; PMCID: PMC6209335.
Robles-Fernández ÁL, Santiago-Alarcon D, Lira-Noriega A (2022) Wildlife susceptibility to infectious diseases at global scales. Proc Natl Acad Sci 119(35):e2122851119. https://doi.org/10.1073/pnas.2122851119
Rodríguez-Hernández K, Álvarez-Mendizábal P, Chapa-Vargas L, Escobar F, González-García F, Santiago-Alarcon D (2021) Haemosporidian prevalence, parasitaemia and aggregation in relation to avian assemblage life history traits at different elevations. Int J Parasitol 51(5):365–378. https://doi.org/10.1016/j.ijpara.2020.10.006
Ronca SE, Ruff JC, Murray KO (2021) A 20-year historical review of West Nile virus since its initial emergence in North America: has West Nile virus become a neglected tropical disease? Edited by Pedro F. C. Vasconcelos. PLoS Negl Trop Dis 15(5):e0009190. https://doi.org/10.1371/journal.pntd.0009190
Rubio AV, Castro-Arellano I, Mills JN, List R, Ávila-Flores R, Suzán G (2017) Is species richness driving intra- and interspecific interactions and temporal activity overlap of a hantavirus host? An experimental test. Edited by W. F. de Boer. PLoS One 12(11):e0188060. https://doi.org/10.1371/journal.pone.0188060
Saito T, Tanikawa T, Uchida Y, Takemae N, Kanehira K, Tsunekuni R (2015) Intracontinental and intercontinental dissemination of Asian H5 highly pathogenic avian influenza virus (Clade 2.3.4.4) in the winter of 2014-2015: dissemination of clade 2.3.4.4 HPAIV in 2014-2015. Rev Med Virol 25(6):388–405. https://doi.org/10.1002/rmv.1857
Samuel MD, Hobbelen PHF, DeCastro F, Ahumada JA, LaPointe DA, Atkinson CT, Woodworth BL, Hart PJ, Duffy DC (2011) The dynamics, transmission, and population impacts of avian malaria in native Hawaiian birds: a modeling approach. Ecol Appl 21(8):2960–2973. https://doi.org/10.1890/10-1311.1
Santiago-Alarcon D, Marzal A (2020a) Research on avian haemosporidian parasites in the tropics before the year 2000. In: Santiago-Alarcon D, Marzal A (eds) Avian malaria and related parasites in the tropics. Springer, Cham, pp 1–44. https://doi.org/10.1007/978-3-030-51633-8_1
Santiago-Alarcon D, Marzal A (2020b) Avian malaria and related parasites in the tropics – ecology, evolution and systematics. Springer
Santiago-Alarcon D, Soto OR (2021) Discovering parasite biodiversity in the tropics: a framework based on multi-host-species ecological niche models. Ecotropicos 33(May):1–9. https://doi.org/10.53157/ecotropicos.33e0016
Sebaio F, Braga É, Branquinho F, Manica L, Marini M (2010) Blood parasites in Brazilian atlantic forest birds: effects of fragment size and habitat dependency. Bird Conserv Int 20(4):432–439. https://doi.org/10.1017/S0959270910000110
Smith OM, Snyder WE, Owen JP (2020) Are we overestimating risk of enteric pathogen spillover from wild birds to humans?. Biol Rev 95(3):652–679. https://doi.org/10.1111/brv.12581
Sol D, Trisos C, Múrria C, Jeliazkov A, González-Lagos C, Pigot AL, Ricotta C, Swan CM, Tobias JA, Pavoine S (2020) The worldwide impact of urbanisation on avian functional diversity. Edited by David Storch. Ecol Lett 23(6):962–972. https://doi.org/10.1111/ele.13495
Soto-Madrid R, Sychra O, Benedick S, Edwards DP, Efeykin BD, Fandrem M, Haugaasen T, Teterina A, Tomassi S, Tolstenkov O (2020) Diversity and host associations of Myrsidea chewing lice (Phthiraptera: Menoponidae) in the tropical rainforest of Malaysian Borneo. Int J Parasitol Parasites Wildl 13(December):231–247. https://doi.org/10.1016/j.ijppaw.2020.10.011
Steele KE, Linn MJ, Schoepp RJ, Komar N, Geisbert TW, Manduca RM, Calle PP et al (2000) Pathology of fatal West Nile virus infections in native and exotic birds during the 1999 outbreak in New York City, New York. Vet Pathol 37(3):208–224. https://doi.org/10.1354/vp.37-3-208

Sures B (2004) Environmental parasitology: relevancy of parasites in monitoring environmental pollution. Trends Parasitol 20(4):170–177. https://doi.org/10.1016/j.pt.2004.01.014

Titley MA, Snaddon JL, Turner EC (2017) Scientific research on animal biodiversity is systematically biased towards vertebrates and temperate regions. Edited by Bernd Schierwater. PLoS One 12(12):e0189577. https://doi.org/10.1371/journal.pone.0189577

Tolsá MJ, García-Peña GE, Rico-Chávez O, Roche B, Suzán G (2018) Macroecology of birds potentially susceptible to West Nile virus. Proc R Soc B Biol Sci 285(1893):20182178. https://doi.org/10.1098/rspb.2018.2178

Valkiūnas G, Atkinson CT (2020) Introduction to life cycles, taxonomy, distribution, and basic research techniques. In: Santiago-Alarcon D, Marzal A (eds) Avian malaria and related parasites in the tropics. Springer, Cham, pp 45–80. https://doi.org/10.1007/978-3-030-51633-8_2

van Rooyen J, Lalubin F, Glaizot O, Christe P (2013) Altitudinal variation in haemosporidian parasite distribution in great tit populations. Parasit Vectors 6(1):139. https://doi.org/10.1186/1756-3305-6-139

Wille M, Barr IG (2022) Resurgence of avian influenza virus. Science 376(6592):459–460. https://doi.org/10.1126/science.abo1232

Williams RAJ, Truchado DA, Benitez L (2021) A review on the prevalence of poxvirus disease in free-living and captive wild birds. Microbiol Res 12(2):403–418. https://doi.org/10.3390/microbiolres12020028

Winker K, Gibson DD (2010) The Asia-to-America influx of avian influenza wild bird hosts is large. Avian Dis 54(s1):477–482. https://doi.org/10.1637/8741-032509-Reg.1

Wood CL, Byers JE, Cottingham KL, Altman I, Donahue MJ, Blakeslee AMH (2007) Parasites alter community structure. Proc Natl Acad Sci 104(22):9335–9339. https://doi.org/10.1073/pnas.0700062104

Zamora-Vilchis I, Williams SE, Johnson CN (2012) Environmental temperature affects prevalence of blood parasites of birds on an elevation gradient: implications for disease in a warming climate. Edited by Justin David Brown. PLoS One 7(6):e39208. https://doi.org/10.1371/journal.pone.0039208

Zylberberg M, Lee KA, Klasing KC, Wikelski M (2013) Variation with land use of immune function and prevalence of avian pox in Galapagos finches: avian pox, immunity, and land use. Conserv Biol 27(1):103–112. https://doi.org/10.1111/j.1523-1739.2012.01944.x

Chapter 6
Marine Bird of Neotropics, What We Know, and We Should Know of Diseases in a Changing World

Galaxia Cortés-Hinojosa

6.1 State of Knowledge Regarding Infectious Diseases in Neotropical Marine Birds: Penguins, Albatrosses, Cormorants, and Shorebirds

Marine birds are sentinels of the health of ecosystems due to their characteristics as top predators, long life, and diverse lifestyles, from coastal residents to intercontinental travelers. Their health can be an indicator of the health of the environment at different scales. Infectious diseases are one of the major causes of wild population extinctions (Smith et al. 2006). The effect of climate change on the distribution of infectious diseases represents a great concern for the research community working in aquatic ecosystems because of the effect of the temperature shift on the distribution of potential pathogens and their consequent interaction with naïve populations of wildlife (Marcogliese 2008). Pathogens, contaminants, and stress can cause lethal and sublethal effects on their hosts (Grilo et al. 2016; Sebastiano et al. 2022; Asghar et al. 2015), which can be assessed and documented on physiological alterations, such as changes in growth and body condition, reduction in reproductive success, or, in more dramatic cases, direct mortalities (Best et al. 2010). Coinfection by opportunistic pathogens can occur and has been reported in marine animals. In addition, subclinical infections could reduce their ability to endure environmental changes (Beldomenico and Begon 2015).

The Humboldt upwelling system is one of the most productive marine ecosystems in the world, bringing cold nutrient-rich water from the Antarctic to the Pacific coast of South America (Gutierrez et al. 2016). This system supports diverse marine

G. Cortés-Hinojosa (✉)
Escuela de Medicina Veterinaria, Facultad de Agronomía e Ingeniería Forestal, Facultad de Ciencias Biológicas y Facultad de Medicina, Pontificia Universidad Católica de Chile, Santiago, Chile
e-mail: galaxia.cortes@uc.cl

G. Acosta-Jamett, A. Chaves (eds.), *Ecology of Wildlife Diseases in the Neotropics*, https://doi.org/10.1007/978-3-031-50531-7_6

fauna in Peru and northern-central Chile (Alheit and Niquen 2004). Approximately every 4–7 years, the El Niño Southern Oscillation (ENSO) cycle results in oceanic conditions that prevent the nutrient-rich cold water of the Humboldt Current from reaching the euphotic layer in the water column where photosynthesis occurs. This significantly reduces primary productivity and fish stocks, causing migration or mortality in higher trophic marine predator species, such as seabirds, especially juveniles (Soto et al. 2004, 2006). In addition, climate change can cause unexpected changes in the complex ocean–atmosphere interactions needed for upwelling to occur.

The diversity and genomic features of pathogens of South American marine species are shaped by the complex interaction between environmental factors and host genetics. By understanding the effect of the environment on the immune response and pathogen dynamics in natural populations, we can comprehend the possible impact of climate change on the ecology of infectious diseases to better monitor and conserve vulnerable wildlife populations. Multicellular organisms are exposed to multiple pathogens belonging to different taxonomic groups; consequently, the rapid response of the body against potential threats is essential. The immune system is a complex intercommunicated machinery formed of cells, molecules, large tissues, and organs that support the balance between potential pathogens and multicellular organisms (Tizard 2019). In tetrapods, the immune response is divided into two major groups: innate and adaptive immune responses. Innate immunity is composed mainly of myeloid-derived cell (heterophil, eosinophil, monocyte, and other) complement systems and inflammatory responses that are key to preventing the initial invasion and proliferation of pathogens. These two systems are interconnected, and innate immunity is the driver for the initiation of adaptive immunity (Tizard 2019). Adaptive immunity relies mainly on the ability to present foraging material to specialized lymphoid-derived cells, such as T cells (cellular mediate response) and B cells (humoral response). The presentation is carried out for specialized proteins in the major histocompatibility complex (MHC) family. Of the four members of this gene family in mammalian organisms, I and II are considered "classical MHC genes" (Tizard 2019).

For the development of a proper adaptive immune response against intracellular pathogens, MHC-I is essential. MHCs bind to peptides from the pathogen to present molecules on the cell surface to be recognized by other components of the adaptive immune response. For extracellular organisms, the leading player is the MHC-II gene (Engelhard 1994). MHC II molecules are mainly expressed by professional presenting cells, such as macrophages, dendritic cells, and B lymphocytes. Immunogenetic studies are primarily based on MHC-II studies. Some authors recognize the need to expand our understanding of the immune response in wildlife, expanding the research horizon from MHC genes (Acevedo-Whitehouse and Cunningham 2006). However, few immunogenetic studies have been carried out in marine birds, including research in immunogenomics and how pathogens can shape the immune response and from an evolutionary and conservation genetic perspective (Sallaberry-Pincheira et al. 2016; Levy et al. 2020).

Most of our knowledge related to immunology in birds is related to domestic animals, with an enormous limit to understanding how environmental change, new stressors and contamination can affect the response to pathogens in marine birds. However, a recent publication showed a possible interaction between contaminants and their effect on the immune system and herpesvirus in frigatebirds from French Guiana (Sebastiano et al. 2022).

6.2 Most Relevant Diseases in Birds and Seabirds

To implement an appropriate literature search in such a vast topic, we use the following search criteria in PUBMED. For example, "seabird virus (Brazil OR Chile OR Argentina OR Mexico OR South America or LATAM)." Then, the word Virus was replaced according to the search by other keywords such as bacteria, parasite, plasmodium, etc. (plasmodium seabirds (Brazil OR Chile OR Mexico OR Argentina OR South America OR LATAM)).

6.2.1 RNA Viruses

Coronaviridae Coronaviruses are large (120–160 nm), round, toroidal, or bacilliform, positive-sense, single-stranded envelope viruses with intracytoplasmic replication. They belong to the order Nidovirales. The family *Coronaviridae* has three subfamilies. The subfamily *Orthocoronaviridae* has four genera: *Alphacoronavirus*, *Betacoronavirus*, *Deltacoronavirus*, and *Gammacoronavirus*. Betacoronaviruses are the best-known clade for containing the severe acute respiratory syndrome (SARS) virus (Woo et al. 2023). Avian infectious bronchitis (Coronaviridae, Orthocoronavirinae, genus *Gammacoronavirus*) is a disease of great relevance in poultry and is an world organisation for animal health (WOAH) reportable disease; consequently, investigations have been carried out in neotropic marine birds. Researchers have detected coronaviruses in two species, neotropical cormorant (*Phalacrocorax brasilianus*) (Verdugo et al. 2019) from Chile and magellanic penguins (*Spheniscus magellanicus*) (Uhart et al. 2020) from Argentina. In both cases, the discovery was part of surveillance and not associated with clinical diseases.

Influenza Virus *Orthomyxoviruses* are medium size (80–120 nm), segmented genome, pleomorphic, enveloped, negative-sense single-stranded viruses with intranuclear and intracytoplasmic replication. The family *Orthomyxoviridae* has seven genera: *Influenza virus A, Influenza virus B, Influenza virus C, Influenza virus D, Thogotovirus*, *Quaranjavirus*, and *Isavirus*. A relevant feature of segmented viruses is the possibility of reassortment (antigenic shift) between homologous segments of different viruses that coinfect a particular host (Ma 2022). This ability, in combination with mutations in specific genes used as antibody binding sites for the host (antigenic drift), helps these viruses jump between hosts and avoid immunosurveillance (Ma 2022). Influenza A is a major zoonotic pathogen and WOAH reportable disease. Their main reservoirs are aquatic birds, such as waterfowl and shorebirds. This virus causes major diseases in domestic animals; nonetheless, clinical cases in the wildlife population of marine birds were historically less common. However, since 2001, H5N1 has caused the death of several species of wild birds worldwide, and the most recent outbreak has resulted in great mortality for marine birds in the neotropics (see more in Sect. 6.5). Aquatic birds are the reservoir of influenza of high relevance to aquatic systems, and many studies on the circulation of influenza virus have a major role in surveillance. Nevertheless, many studies have focused on aquatic birds and, less commonly, marine birds until recently.

Newcastle Disease Virus *Paramyxoviruses* are 150–300 nm, pleomorphic, enveloped, negative-sense single-stranded viruses with intracytoplasmic replication. The family *Paramyxoviridae* is part of the order *Mononegavirales* and has 4 subfamilies and 14 genera (Rima et al. 2019). In birds, one of the most studied paramyxoviruses is Newcastle disease virus (NDV; avian paramyxovirus 1, genus Orthoavula, subfamily Avulavirinae), which is a worldwide disease of importance for domestic and wild birds and is an WOAH reportable disease. Newcastle disease virus shows great genetic diversity, which leads to the presence of two clades of viruses (classes I and II). Of those, the most divers are clade II, which has been detected in cormorants among other wild domestic birds. Clinical signs may include respiratory distress, depression, diarrhea, decreased egg production, neurologic signs, torticollis, and death (Dimitrov et al. 2016)). Cormorants in the family *Phalacrocoracidae* have been indicated to be at risk of contracting this virus, and outbreaks related to this virus have occurred in North America and South America. In LATAM, outbreaks have been reported by the local government from Chile (SAG final report, (Moreira 2009)), and this document indicates that 382 bird deaths are associated with this outbreak, including penguins (*Spheniscus magellanicus*), cormorants (*Phalacrocorax*), and boobies (*Sula variegata*); 25% of the animals sampled alive tested positive for the virus. In peer review literature, studies carried out in healthy cormorants from Chile showed no positive animals (0/104) (Verdugo et al. 2019). Other paramyxoviruses (APV-II and -X) have been detected in *Spheniscus magellanicus* from Brazil, with no clinical relevance (Fornells et al. 2012).

6.2.2 DNA Viruses

Adenoviruses are nonenveloped, double-stranded DNA viruses with a medium-sized genome of 26–48 kbp. Viruses in the family Adenoviridae are classified into six genera (*Mastadenovirus*, *Aviadenovirus*, *Atadenovirus*, *Siadenovirus*, *Testadenovirus* and *Ichtadenovirus*) (Benkő et al. 2022) The genera *Aviadenovirus*, *Atadenovirus*, and *Siadenovirus* all utilize avian hosts. Persistent infections are not uncommon. Multiple novel adenoviruses have been discovered in Humboldt penguins, including aviadenoviruses, siadenoviruses, and mastadenoviruses. Preliminary data obtained in systems with high resource variability and high population densities of animals of different species, such as Punta San Juan, are an ideal situation to evaluate adenoviral diversity and the ability to jump hosts in a natural system. Preliminary data include aviadenoviruses in a mammalian host, and this finding drove us to request samples from marine birds in the area. We received historical samples of Humboldt penguin (HP, *Spheniscus humboldti*) to complement the pinniped study (Cortés-Hinojosa et al. 2021). Avian adenovirus was also detected. These viruses have not been associated with clinical diseases in wild or managed populations.

Herpesviruses are large (160–300 nm), icosahedral, enveloped, double-stranded viruses with intranuclear replication and high host fidelity; interestingly, they remain latent for prolonged periods in many animals. Most of the characterized avian herpesviruses are members of the subfamily Alphaherpesvirinae, with two subfamilies recognized genera relevant to avian hosts: *Iltovirus* and *Mardivirus* (Gatherer et al. 2021). Herpesviruses have been widely reported in poultry and wild animals, including waterfowl, gamebirds, and raptors. Members of the genus *Iltovirus*, which include clinically relevant herpesviruses in poultry, such as Gallid Herpesvirus 1 (GaHV1, infectious laryngotracheitis), an WOAH reportable disease, have been isolated from a large variety of psittacine species (Gatherer et al. 2021). In aquatic birds, Gaviid Herpesvirus 1 (GavHV1) from common loons (*Gavia immer*) has also been shown to cluster in the genus *Iltovirus*. Reactivation of latent herpesvirus infections is associated with stress and immune suppression, such as rehousing and chemical pollutants (Hughes et al. 1989; Goldberg et al. 1990), and has been associated with depression, respiratory distress, hemorrhagic lesions, and sudden death. The stress of captivity may induce GavHV1 reactivation in loons, and this may be one factor that predisposes them to respiratory infection with *Aspergillus* sp. Penguins, such as loons, are also susceptible to respiratory infection in captivity. In penguins, few viruses have been reported, and a recent work published by Niemeyer et al. detected herpesvirus in Magellanic penguins in rehabilitation facilities. They found a 38% prevalence in animals in rehabilitation and a 5.6% prevalence in free-range animals in apparently healthy conditions (Niemeyer et al. 2017), which is consistent with the reactivation of latent viruses. Cases of death related to herpesvirus affected two collections of banded penguins in Germany; Humboldt (*Spheniscus humboldti*) and African (*Spheniscus demersus*) chicks were affected. The cause of the deaths was a novel alphaherpesvirus, SpAHV-1. The authors of the paper call for more investigation on the topic to identify the reservoir of this novel virus in natural populations (Pfaff et al. 2017). African penguin (*Spheniscus demersus*) chicks were affected. This virus was detected in Humboldt penguins from Punta San Juan without clinical significance in healthy animals (the author, (Angles et al. 2023)). Since 2005, mortalities due to herpesvirus have been reported in the magnificent frigatebird (*Fregata magnificens*) from French Guiana. This herpesvirus is related to a high mortality of hatchlings, with up to 90% mortalities of chicks every year since the initial report (Sebastiano et al. 2022). In this case, the author indicated that the detrimental effect of viral infection could be related to high levels of contaminants, particularly Hg, which could lead to immunologic impairments (Sebastiano et al. 2022).

Polyomaviruses are 40–45 nm, icosahedral, nonenveloped, circular, double-stranded DNA viruses with intranuclear replication. The family *Polyomaviridae* has recently been split into six genera, with the genus Gammapolyomavirus infecting only birds. Currently, nine recognized avian polyomaviruses (AVVs) have been reported (Moens et al. 2017) and can cause inflammatory disease and, in some cases, acute diseases. Avian polyomaviruses (APVs), formally referred to as budgerigar fledgling disease (BFD), have great importance in cage birds and impact parrot worldwide. This virus can cause diseases in young animals. Polyomavirus has been detected by next-generation sequencing on Adelie penguins from Antarctica without clinical significance (Varsani et al. 2015). However, they have not been reported in peer review literature of polyomavirus from neotropic marine birds.

Poxvirus Poxviruses are large (140–260 × 220–450 nm), brick or ovoid-shaped, enveloped double-stranded viruses with intracytoplasmic replication (Delhon 2022). The family *Poxviridae* has 2 subfamilies and 22 genera. The subfamily *Chordopoxvirinae* includes viruses that infect chordates, and the subfamily *Entomopoxvirinae* includes viruses that infect invertebrates. Viruses in the subfamily *Chordopoxvirinae* can be transmitted directly, indirectly, and by vectors. In general, viruses in this subfamily cause proliferative skin disease in vertebrates. Poxviruses are of great relevance to domestic and terrestrial birds and have been reported in two clinical presentations. In seabirds, this pathogen has been reported in several species across the globe, including Antarctic penguins and northern royal albatross (*Diomedea sanfordi*) (Gyuranecz et al. 2013; Parsons et al. 2018). In LATAM, penguins have been reported in Chile and Ecuador (including Galapagos), and in Magellanic penguins, reports have occurred in Argentina and Brazil, Humboldt penguins and Galapagos penguins. The lesion is described from a wart-like lesion in the skin and mucosal membrane to the less common systemic diphtheric form. Cases of poxvirus on penguins have been described in wild animals and rehabilitation facilities; however, cases are more common in chicks than in adults. In addition, the most concerning cases of mortality in Waved Albatross from Galapagos were 14 nestlings with pox-like lesions and a mortality of 8%; in this case, the most common presentation was the cutaneous or "dry" form (Tompkins et al. 2017).

6.2.3 Bacteria and Fungi

Rare bacterial and fungal organisms are the primary cause of diseases in wild seabirds (Woods et al. 2009). Studies from lower latitudes have shown that migratory birds can act as reservoirs or vehicles of human pathogenic bacteria (Abulreesh et al. 2007). Studies of human pathogens such as Salmonella and Campylobacter have been carried out; however, the role of seabirds remains a topic that needs more research (González-Acuña and Llanos-Soto 2020). One of the concerns for wildlife is the increase in antibiotic-resistant bacteria (AMR). Some studies on this topic have been carried out in LATAM, and it is important to consider that marine animals can be sentinels for these organisms in ecosystems (Ewbank et al. 2021b).

Lyme disease could be a concern in the population of seabirds because of the presence of ticks; however, this disease has not been reported in seabirds in LATAM.

Aspergillosis Aspergillus is a genus of ubiquitous soil fungi and is considered an opportunistic mycosis of relevance in veterinary medicine. In penguins, it is considered a primary fungal disease (Wallace 2014). Within the genus Aspergillus, most cases of aspergillosis are related to *A. fumigatus*; however, *A. flatus* has been reported in avian species (Leishangthem et al. 2015). In a study in Magallanes penguins among zoological institutions, the author considered aspergillosis to be the most relevant cause of death in captive settings (Krol et al. 2020). In addition, a recent report showed that Aspergillus affects *Thalassarche melanophris* under rehabilitation in Brazil, and the author indicated that this pathogen could play a major role in the failure of their rehabilitation (Melo et al. 2020a, b). Aspergillus currently does not represent a major issue in wild populations of penguins in LATAM; however, a report from Brazil indicated that Aspergillus was found in 3.7% of the necropsies of Magellanic penguins (*Spheniscus magellanicus*) found dead in Sao Paulo, Brazil (Ewbank et al. 2021a). Additional reports of Aspergillus in two sick wild birds have occurred in Brazil, brown-hooded gull (*Chroicocephalus maculipennis*) and white-chinned petrel (*Procellaria aequinoctialis*), both of which die before reaching the rehabilitation facility (Melo et al. 2020b). The effect on breeding colonies has not been described; however, we do not know the effect of climate change on seabird health and disease dynamics and whether this pathogen can pose a major threat in the future in species if we consider the data presented by Ewbank et al. recently.

6.2.4 Parasites

Ectoparasites are the major parasites reported in marine birds, *and* Ornithodoros ticks have been reported in guano burrows of the Humboldt penguin from Peru. One of the most relevant issues related to external parasites is the transmission of other viruses, bacteria, or internal parasites.

Plasmodium This is a worldwide spread of mosquito-borne intracellular protozoan parasites. This parasite can have low mortalities in an endemic host, but in a naïve population of birds, it can be highly susceptible. It is particularly important to understand the cases of Plasmodium in African penguins, *Spheniscus demersus*, where this parasite is decimated in wild populations. The problem is particularly severe in Western Cape in South Africa, where 35% of penguins admitted to rehabilitation facilities are positive for *Plasmodium* spp. (Parsons and Underhill

2005). Several species of penguins appear susceptible to plasmodium, and reports in captive or rehabilitation facility settings have occurred in South America (Grilo et al. 2016). In the case of free-living neotropical marine birds, the first report of the parasite occurred on the Galapagos penguins (*Spheniscus mendiculus*). Records that indicate the arrival of *Plasmodium* spp. in 2003 at the Galapagos Islands but not clinical disease (Levin et al. 2009), previous research carried out in 1996 indicated that this species was free from this pathogen (Levin et al. 2009; Miller et al. 2001). The presence of the pathogen is critical for an endangered species with a small population size that was severely affected by El Niño in 1982 and 1996, with an extreme decline in its populations. A study carried out with samples from 2010 to 2013 from Peru, Chile, and Argentina indicated that this pathogen is only present in Humboldt penguins from Peru but not in Chilean populations or Magellanic penguins from the localities under study (Sallaberry-Pincheira et al. 2015). However, a report from the local government from Chile indicates that it is possible that this pathogen is present in *Spheniscus magellanicus* (Carvajal and Alvarado 2009). Therefore, it's imperative to conduct additional reserch and monitoring of this pathogen in naïve populations.

6.3 Main Environmental and Anthropogenic Impacts on Seabird Disease Ecology on the Pacific Coast

Anthropogenic pressures can directly or indirectly impact neotropical fauna, and some impacts include eutrophication, overfishing, bycatch, plastic and organic pollution, and climate change (Abelson et al. 2020).

6.3.1 Extensive Extraction of Guano

Historically, guano extraction has had a great impact on marine birds. Seabird associated with Humboldt upwelling nested in areas of enormous guano accumulations. Species such as Cormorants and Penguins build their burros in the guano. The role of the guano is so relevant in the construction of nests of some species that the reduction in Guanay cormorant (*Phalacrocorax bougainvillii*) is linked to the extraction of Guano in combination with the effects of the El Niño Southern Oscillation (ENSO) and El Niño Event. In addition, research has proposed that the lack of protection provided by guano makes this species more susceptible to habitat destruction and increases the chance of nest desertions. Other species highly affected by the harvested guano are Humboldt penguins, which are vulnerable species. Due to the lack of their preferred type of nest, they have been forced to use other less optimal nesting options.

6.3.2 *Overfishing and Bycatch*

The impact of large-scale industrial fisheries is a serious concern for most global marine ecosystems. Fisheries can impact the health of marine vertebrates through direct interactions during fishing activities (e.g., Bycatch) or by overfishing key resources for top predators. The neotropics contain the largest single ecosystem producer of seafood on the planet, the Humboldt Current, because it is one of the highest productivity systems of the glove, which has encountered historical overfishing, leading to the depletion of key resources for marine mammals (Gutierrez et al. 2016; Lima et al. 2020). Several articles have indicated the detrimental effect of bycatch on marine birds at the global scale and in the neotropics. An increased number of reports indicates the relevance of incorporating new techniques that reduce bycatch. International treatments for conservation have been adopted for countries in the neotropics, with evidence of a reduction in bycatch. However, it is still difficult to qualify all the events that occurred. This threat may not directly affect interactions with infectious diseases, but if we think of populations that are undergoing a reduction in population size because of multiple threats, it is a concern that their genetic variability suffers and consequently is not able to respond properly to novel threats. For example, a new variant of viruses has expanded the range of old pathogens due to climate change and more hospice habitats for intermediate hosts.

6.3.3 *Contaminants*

Contaminants can bioaccumulate and biomagnify in the trophic web, and marine birds, as top predators, can be a good sentinel for a wide range of organisms as well as an indicator of a lack of food resources and, sadly, have been recognized as indicators of plastic contamination in the oceans. Marine pollution is a serious concern for biodiversity and health in neotropical regions. Although it is not fully understood how many of the organic and inorganic pollutants affect sentinel species, the evidence for some of the best studied substances is alarming. It is well known that polychlorinated biphenyls and heavy metals have been associated with cancer, immunosuppression and higher susceptibility to infectious diseases in marine mammals worldwide (Desforges et al. 2016). In the neotropics, research on the impact of these contaminants is accumulating, although concentrations of organic pollutants are usually lower compared to more developed regions of the planet (Durante et al. 2016; Alava et al. 2020). Environmental contaminants can interfere with the immune system of aquatic organisms, especially in fish-eating animals such as seabirds. This contamination can interfere in different ways with the normal function of the immune system, consequently increasing the potential risk of disease outbreaks. Studies on contaminants and their interaction with infectious diseases in seabirds are rare in general. However, the information offered by researchers in French

Guiana (Sebastiano et al. 2022) must be taken into careful consideration since it indicates a possible connection between Hg contamination and clinical cases of herpesvirus-associated chick death in the magnificent frigatebird (*Fregata magnificens*).

6.3.4 El Niño–Southern Oscillation (ENSO)

Seabirds that inhibit Humboldt upwelling are adapted to an environment with high productivity and ENSO variation. They live in physiological equilibrium with their environment, which is more limited during the breeding season. For that reason, researchers are able to establish an optimal range of feeding during breeding (Culik and Luna-Jorquera 1997; Culik et al. 1998); consequently, when prey is unavailable, the parent will abandon the nest. This has been documented during El Niño events for several species but with more dramatic effects for endemic species such as the *Galápagos* (*Spheniscus mendiculus*) and flightless cormorant (*Phalacrocorax harrisi*) during the 1982–1983 event (Valle and Coulter 1987). Changes in the frequency and severity of ENSO can be devastating for the equilibrium of populations, and the combination with other threats, such as infectious diseases, can be distressing for any species.

6.4 Next Step, in Which Areas Should We Focus Our Research Efforts to Better Understand the Ecology of Diseases in Neotropical Birds?

Ongoing efforts at two sites are great examples of how collaboration between veterinarians and biologists can contribute to the understanding of marine bird ecology, establishment of health baselines and monitoring of the prevalence of infectious diseases. In Punta San Juan and Islas Galapagos, several papers have been published on these topics. Thanks to the collaborative effort, it was possible to obtain evidence for the introduction of Plasmodium on the penguins.

To better understand the effect of diseases and the environmental impact of the Anthropocene. In addition, there is a direct impact of human activities. It is necessary to obtain more information on the life cycle of the species and be able to generate prediction models that help to develop conservation measures based on evidence.

6.5 2022–2023 Outbreak HPAI Influenza in LATAM

6.5.1 History of Influenza in LATAM

Influenza was reported in early 2000 in LATAM from wild birds in Bolivia and since then has been reported in several countries from South America. The North American flyaway has been indicated as the main migratory route that connects birds on the American continent. Birds migrate from North America to South America, with some of them having annual migration routes that connect Alaska with South Patagonia and Chile (Olsen et al. 2006). In addition, the flow of pathogens such as influenza is quite relevant for understanding disease dynamics. In the case of this virus, such as influenza, this large interchange of variants of potential hosts leads to a logically large diversity of strains. Low pathogenic influenza virus circulating in LATAM, some of them apparently coevolve in South America, and others have a North American origin (Jiménez-Bluhm et al. 2018). The origins and prevalence of North American strains have changed between countries in South America since the late 2000s and after the outbreak in poultry from Chile. An increase in surveillance in South America is helping to better understand the epidemiology of local strains (Jiménez-Bluhm et al. 2018).

6.5.2 HPIV 2.3.4.4 Outbreak in LATAM 2022

During the elaboration of this chapter, we encounter the influenza virus outbreak LATAM. This influenza A virus belongs to the 2.3.4.4 clade first reported in 2021; it is the largest outbreak of influenza reported thus far (Shi et al. 2023; WOAH 2023) and reaches the level of a panzootic. Reports from the scientific community indicate that this outbreak has been devastating to several species across the globe, and the consequences for conservation must be carefully examined case by case (Cunningham et al. 2022). In North America, it has affected several birds, but apparently more cases have been reported in aquatic birds, raptors, and vultures. In the case of LATAM, perhaps because of the promiscuous nature of the virus in combination with a vast diversity and richness of naïve hosts that inhabit one of highest productive areas of the world's oceans, a large diversity of seabirds have been affected (Table 6.1, Fig. 6.1). The first mortalities occurred in Peru, with cases on the order of thousands of pelicans reported in the news (https://www.reuters.com/business/healthcare-pharmaceuticals/bird-flu-kills-sea-lions-thousands-pelicans-perus-protected-areas-2023-02-21/). Peer review publications in press indicate that several birds from Peru have been reported to be positive for Peruvian pelican (*Pelecanus thagus*), Guanay cormorant (*Phalacrocorax bougainvillii*), Peruvian booby (*Sula variegata*), and Humboldt penguin (*Spheniscus humboldti*) (Leguia et al. 2023). In Peru, more than 22,000 wild birds die by the end of 2022, mainly pelicans and boobies (Gamarra-Toledo et al. 2023a, b). However, the impact of the mortalities on the population and the consequences for the conservation of the species under the IUCN red list and those heavily impacted by this virus remain unclear. In Chile, the trend was similar, with more cases in Pelicans followed by boobies and cormorants. Pelicans have 60% positivity in Chile according to local authorities. Data provided by local Chilean authorities, SAG and SERNAPESCA, indicate that more than 16,000 seabirds have stranded or died from the beginning of this panzootic (SERNAPESCA 2023; SAG 2023). SERNAPESCA indicates 1037 Humboldt penguins are reported stranded, most of them death stranding, with a low positivity to H5N1 with the current testing protocol (SERNAPESCA 2023) (Table 6.2). However, they reported a more than 1000% increase in strandings over the 90 reported animals on average between 2009 and 2022 (Table 6.3). Additionally, more than 3000 marine birds have been tested for influenza (SAG 2023). To evaluate the situation on LATAM, we base our information on the information provided to WAHIS (WAHIS 2023) to compare among countries; however, the information available on the platform depended on information provided by local authorities, and we identified discrepancies with local information and peer review data. Consequently, it is important to take the information provided here as a reference and need to be re-evaluated in light of new peer review data. According to these available data, the most impacted countries are Chile and Peru, and Chile shows a larger diversity of affected seabirds (Table 6.1, Fig. 6.1). It is relevant to note that this virus has spread to several species of marine mammals (see Chap. 6).

It is challenging to talk about the realities of all countries in LATAM, but at least to our knowledge, this outbreak has been a great challenge for wildlife, since LATAM does have the capabilities and facilities that are available in the Northern Hemisphere, and the impact on the rescue and rehabilitation of all birds needs to be evaluated carefully. In the case of Chile, many rehabilitation centers are closed, and many wild birds have been euthanized because of the zoonotic potential of this virus. Local associations called for the implementation of emergency quarantine facilities and requests for equipment for local rescue centers.

Finally, this panzootic is a call to work together in a true One Health approach to understand the effect of different stressors and anthropogenic impacts on the wildlife of LATAM and across the globe. We are looking for data to predict the effect of this panzootic on the wildlife population, and the information needed is lacking for even the most loved and well-studied species. We need information on virus genetics and evolution in real time and data on the ecology and population biology of the species to build a life table and then be able to generate prediction models of the effect of different stressors in combination with diseases on the long-term survival of the species.

Table 6.1 List of affected seabirds reported to be positive for HPIV H5N1 in LATAM (data collected on April 4, 2023, from WAHIS)

	Scientific name	Conservation status	Chile	Colombia	Costa Rica	Ecuador	Guatemala	Honduras	Panama	Peru	Venezuela	Total/specie
American Oystercatcher	*Haematopus palliatus*	Least concern	1									1
Band-tailed gull	*Larus belcheri*	Least concern	19									19
Black Skimmer	*Rynchops niger*	Least concern	3									3
Black-crowned night-heron	*Nycticorax nycticorax*	Least concern	1									1
Black-headed gull	*Chroicocephalus ridibundus*	Least concern				1						1
Blue-footed booby	*Sula nebouxii*	Least concern				1				2		3
Brown pelican	*Pelecanus occidentalis*	Least concern		302	7		11	324	11			655
Elegant tern	*Thalasseus elegans*	Near threatened	6									6
Franklin's gull	*Leucophaeus pipixcan*	Least concern	2									2
Fregatidae (unidentified)		NA				3						3
Gray gull	*Leucophaeus modestus*	Least concern	17									17
Gray-headed gull	*Chroicocephalus cirrocephalus*	Least concern								2		2
Guanay cormorant	*Leucocarbo bougainvillii*	Near threatened	69							6		75
Humboldt penguin	*Spheniscus humboldti*	Vulnerable	2									2

(continued)

Table 6.1 (continued)

	Scientific name	Conservation status	Chile	Colombia	Costa Rica	Ecuador	Guatemala	Honduras	Panama	Peru	Venezuela	Total/specie
Inca tern	*Larosterna inca*	Near threatened	1									1
Kelp gull	*Larus dominicanus*	Least concern	29									29
Neotropic cormorant	*Phalacrocorax brasilianus*	Least concern	5									5
Peruvian booby	*Sula variegata*	Least concern	67							51		118
Peruvian pelican	*Pelecanus thagus*	Near threatened	153							505	172	830
Red-legged cormorant	*Poikilocarbo gaimardi*	Near threatened	12							1		13
Rock shag	*Leucocarbo magellanicus*	Least concern	1									1
Sanderling	*Calidris alba*	Least concern	2									2
South American tern	*Sterna hirundinacea*	Least concern	1									1
Total			**391**	**302**	**7**	**5**	**11**	**324**	**11**	**567**	**172**	**1790**
Number of species reported			18	1	1	3	1	1	1	6	1	23

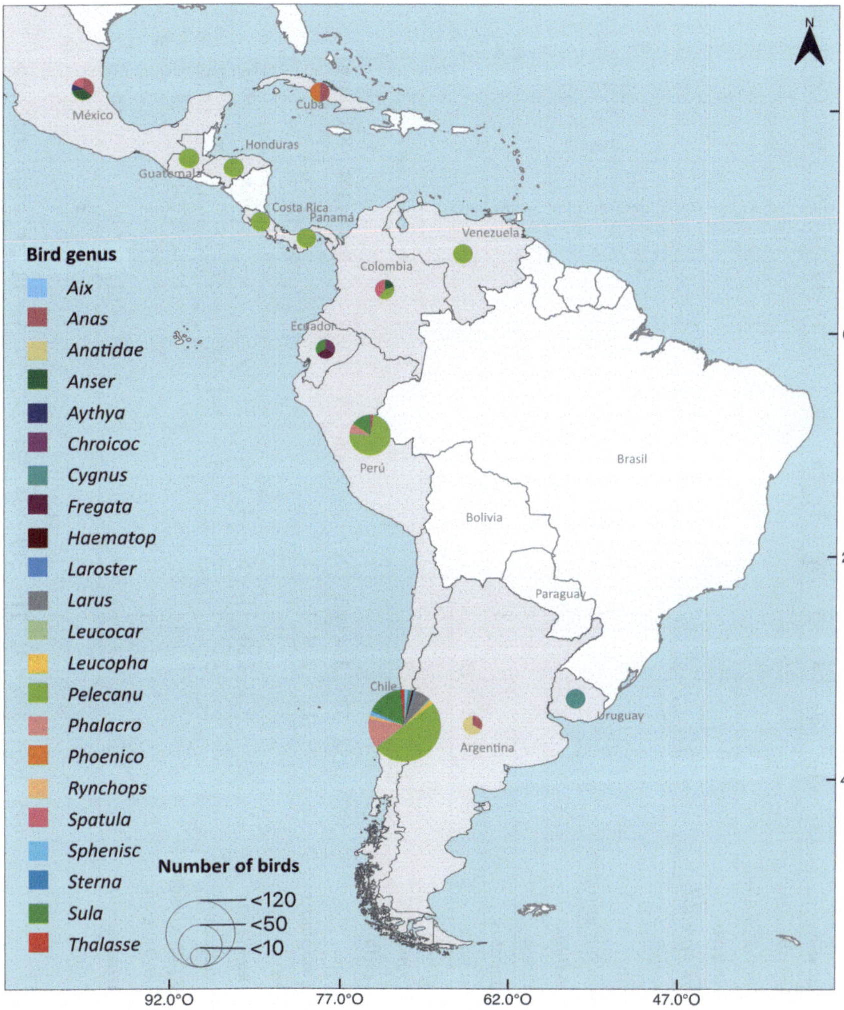

Fig. 6.1 Genus of aquatic birds affected by HPIV H5N1 in LATAM. (Data reported to WAHIS April 4, 2023. Map generated by Nicole Atero, One Health Initiative, UC)

Table 6.2 Stranded marine mammals and penguins in Chile during 2023, data from Sernapesca (2023)

Common name	Scientific name	Chilean states represented by their assigned number															Grand total
		1	2	3	4	5	6	7	8	9	10	11	12	14	15	16	
Ballena de Aleta	*Balaenoptera physalus*								1								1
Ballena Jorobada	*Megaptera novaeangliae*							1	2								3
Ballena Picuda de Cuvier	*Ziphius cavirostris*						1										1
Ballena Sei	*Balaenoptera borealis*											3					3
Cachalote Enano de Cabeza Corta	*Kogia breviceps*			1													1
Chungungo	*Lontra felina*	3		2	5	7			1						3		21
Delfín Chileno	*Cephalorhynchus eutropia*					2		8								3	13
Delfín Común	*Delphinus delphis*							1									1
Delfin Gris	*Grampus griseus*					1											1
Delfín Naríz de Botella	*Tursiops truncatus*					1									2		3
Elefante Marino	*Mirounga leonina*			2	1		1		1					1	1		7
Foca Leopardo	*Hydrurga leptonyx*				1												1
Lobo Fino Austral	*Arctophoca australis*		1		2						1	1					5
Lobo Fino de Juan Fernández	*Arctophoca philippii*					3	1	8								1	13
Lobo Marino Común	*Otaria bryonia*	599	1038	528	324	82	207	79	306	15	47	1	5	15	1391	393	5030
Marsopa Espinosa	*Phocoena spinipinnis*		1	1	2	1	1	1							9		16
Orca	*Orcinus orca*												1				1
Pingüino de Humboldt	*Spheniscus humboldti*	43	23	330	210	11	5	4	6	33	9				363		1037
Pingüino de Magallanes	*Spheniscus magellanicus*			2	5	7	3		7	194	21		4	22		7	272
Pingüino Rey	*Aptenodytes patagonicus*												1				1
Tortuga Olivácea	*Lepidochelys olivacea*					1									1		2
Zifio de Arnoux	*Berardius arnuxii*												1				1
Tortuga Verde	*Chelonia mydas*				1			1							9		11
Unknow	*NA*							1			1				1		3
Grand Total		645	1063	866	551	116	219	104	324	242	79	5	12	38	1780	404	6448

Not all animals tested positive for H5N1. In orange, marked species were more affected

Table 6.3 Historical data of stranded Humboldt penguins, 2009–2022, data from Sernapesca, Strandings (http://www.sernapesca.cl/informacion-utilidad/registro-de-varamientos)

Spheniscus humboldi strandings	Years														
Regions	2009	2010	2011	2012	2013	2014	2015	2016	2017	2018	2019	2020	2021	2022	Total
Antofagasta	1					4	1	1	1	2	3	8	12	14	47
Araucanía								12	1	1	4		5	1	24
Arica and Parinacota	2			2	2	5	2	2		1		3	1	4	24
Atacama		1					8	7	3	2	8	3	2	8	42
Aysén del General Carlos Ibáñez del Campo				1			1								2
BIO-BIO	19	41	9	9	15	6	17	11	25	11	7	9	3	8	190
Coquimbo		12	12	7	27	9	8	31	10	1	11	17	25	35	205
Libertador General Bernardo O'Higgins					1	2	2	6	5	4	3	15	5	9	52
Los Lagos	3	1	1	1	3	3	12	16	48	12	10	10	7	19	146
Los Ríos						1		13	1	4	7	3	1	1	31
Maule	11	5	1	4	3	5	5	4	5	2	4	4	15	10	78
Ñuble											2	1			3
Tarapacá		1	2	5	2			1		62		4		2	79
Valparaíso	14	19	11	59	56	11	14	24	24	15	37	18	15	26	343
Total per year	50	80	36	88	109	46	70	128	123	117	96	95	91	137	1266

References

Abulreesh HH, Goulder R, Scott GW (2007) Wild birds and human pathogens in the context of ringing and migration. Ringing Migr 23(4):193–200

Abelson A, Reed DC, Edgar GJ, Smith CS, Kendrick GA, Orth RJ, Airoldi L et al (2020) Challenges for Restoration of Coastal Marine Ecosystems in the Anthropocene. Front Mar Sci 7:544105

Acevedo-Whitehouse K, Cunningham AA (2006) Is MHC enough for understanding wildlife immunogenetics? Trends Ecol Evol 21(8):433–438. https://doi.org/10.1016/j.tree.2006.05.010

Alheit J, Niquen M (2004) Regime shifts in the Humboldt current ecosystem. Prog Oceanogr 60(2–4):201–222. https://doi.org/10.1016/j.pocean.2004.02.006

Alava JJ, Calle P, Tirapé A, Biedenbach G, Cadena OA, Maruya K, Lao W et al (2020) Persistent Organic Pollutants and Mercury in Genetically Identified Inner Estuary Bottlenose Dolphin (Tursiops truncatus) Residents of the Guayaquil Gulf, Ecuador: Ecotoxicological Science in Support of Pollutant Management and Cetacean Conservation. Front Mar Sci 7:122

Angles R, Adkesson MJ, Cárdenas-Alayza S, Adamovicz L, Allender MC (2023) Detection and prevalence of Spheniscid alpha-herpesvirus-1 (SpAHV-1) in a sample of Humboldt penguins (*Spheniscus humboldti*) AT Punta San Juan, Peru. J Zoo Wildl Med 54(1):159–163

Asghar M, Hasselquist D, Hansson B, Zehtindjiev P, Westerdahl H, Bensch S (2015) Chronic infection. Hidden costs of infection: chronic malaria accelerates telomere degradation and senescence in wild birds. Science 347(6220):436–438. https://doi.org/10.1126/science.1261121. https://www.ncbi.nlm.nih.gov/pubmed/25613889

Beldomenico PM, Begon M (2015) Stress-host-parasite interactions: a vicious triangle? Rev FAVE Secc Cienc Vet 14(1–2):6–19

Benkő M, Aoki K, Arnberg N, Davison AJ, Echavarría M, Hess M, Jones MS, Kaján GL, Kajon AE, Mittal SK (2022) ICTV virus taxonomy profile: Adenoviridae 2022. J Gen Virol 103(3):001721

Best A, White A, Boots M (2010) Resistance is futile but tolerance can explain why parasites do not always castrate their hosts. Evolution 64(2):348–357

Carvajal ER, Alvarado PM (2009) Reporte de caso Pesquisa de Plasmodium spp. en pingüinos de Magallanes (Spheniscus magellanicus) de la Región de los Ríos Malaria aviar como nueva patología de interés en la avifauna local. Boletin Veterinario Oficial 10(2)

Cortés-Hinojosa G, Adkesson MJ, Cárdenas-Alayza S, Seguel M, Pavés H, Wellehan JFX Jr (2021) Adenovirus diversity in fur seal and penguin colonies of South America. J Wildl Dis 57(4):964–969

Culik BM, Luna-Jorquera G (1997) Satellite tracking of Humboldt penguins (*Spheniscus humboldti*) in northern Chile. Mar Biol 128:547–556

Culik BM, Luna-Jorquera G, Oyarzo H, Correa H (1998) Humboldt penguins monitored via VHF telemetry. Mar Ecol Prog Ser 162:279–286

Cunningham EJA, Gamble A, Hart T, Humphreys EM, Philip E, Tyler G, Wood MJ (2022) The incursion of Highly Pathogenic Avian Influenza (HPAI) into North Atlantic seabird populations: an interim report from the 15th International Seabird Group conference. Seabird 34

Desforges, Jean-Pierre W et al (2016) "Immunotoxic effects of environmental pollutants in marine mammals." *Environment international 86*: 126–139

Durante CA, Santos-Neto EB, Azevedo A, Crespo EA, Lailson-Brito J (2016) POPs in the South Latin America: Bioaccumulation of DDT, PCB, HCB, HCH and Mirex in blubber of common dolphin (Delphinus delphis) and Fraser's dolphin (Lagenodelphis hosei) from Argentina. Science of The Total Environment 572:352–360

Delhon G (2022) Poxviridae. Vet Microbiol:522–532

Dimitrov KM, Ramey AM, Qiu X, Bahl J, Afonso CL (2016) Temporal, geographic, and host distribution of avian paramyxovirus 1 (Newcastle disease virus). Infect Genet Evol 39:22–34

Engelhard VH (1994) Structure of peptides associated with Class I and Class II MHC molecules. https://doi.org/10.1146/annurev.iy.12.040194.001145. https://www.annualreviews.org/doi/abs/10.1146/annurev.iy.12.040194.001145

Ewbank AC, Duarte-Benvenuto A, Zamana-Ramblas R, Navas-Suárez PE, Gattamorta MA, Carla P, dos Santos-Costa J, Catão-Dias L, Sacristán C (2021a) Case report of respiratory aspergillosis and candidiasis in wild Magellanic penguins (*Spheniscus magellanicus*), Brazil. Braz J Microbiol 52:967–975

Ewbank AC, Esperón F, Sacristán C, Sacristán I, Neves E, Costa-Silva S, Antonelli M, Lorenço JR, Kolesnikovas CKM, Catão-Dias JL (2021b) Occurrence and quantification of antimicrobial resistance genes in the gastrointestinal microbiome of two wild seabird species with contrasting behaviors. Front Vet Sci 8:651781

Fornells LAMG, Silva TF, Bianchi I, Travassos CEPF, Liberal MHT, Andrade CM, Petrucci MP, Veiga VF, Vaslin MFS, Couceiro JNSS (2012) Detection of paramyxoviruses in Magellanic penguins (*Spheniscus magellanicus*) on the Brazilian tropical coast. Vet Microbiol 156(3–4):429–433

Gamarra-Toledo V, Plaza PI, Gutiérrez R, Inga-Diaz G, Saravia-Guevara P, Pereyra-Meza O, Coronado-Flores E, Calderón-Cerrón A, Quiroz-Jiménez G, Martinez P (2023a) Mass mortality of marine mammals associated with highly pathogenic influenza virus (H5N1) in South America. bioRxiv:2023.02.08.527769

Gamarra-Toledo V, Plaza PI, Gutiérrez R, Luyo P, Hernani L, Angulo F, Lambertucci SA (2023b) Avian flu threatens Neotropical birds. Science 379(6629):246–246

Gatherer D, Depledge DP, Hartley CA, Szpara ML, Vaz PK, Benkő M, Brandt CR, Bryant NA, Dastjerdi A, Doszpoly A (2021) ICTV virus taxonomy profile: herpesviridae 2021. J Gen Virol 102(10):001673

Goldberg DR, Yuill TM, Burgess EC (1990) Mortality from duck plague virus in immunosuppressed adult mallard ducks. J Wildl Dis 26(3):299–306

González-Acuña D, Llanos-Soto S (2020) Una revisión sistemática de los patógenos virales y bacterianos de aves silvestres en Chile. Rev Chil Infectol 37(4):422–442

Grilo ML, Vanstreels RET, Wallace R, García-Párraga D, Braga ÉM, Chitty J, Catão-Dias JL, Madeira LM, de Carvalho. (2016) Malaria in penguins–current perceptions. Avian Pathol 45(4):393–407

Gutierrez O, Panario D, Nagy GJ, Bidegain M, Montes C (2016) Climate teleconnections and indicators of coastal systems response. Ocean Coast Manag 122:64–76. https://doi.org/10.1016/j.ocecoaman.2016.01.009

Gyuranecz M, Foster JT, Dán Á, Ip HS, Egstad KF, Parker PG, Higashiguchi JM, Skinner MA, Höfle U, Kreizinger Z (2013) Worldwide phylogenetic relationship of avian poxviruses. J Virol 87(9):4938–4951

Hughes CS, Gaskell RM, Jones RC, Bradbury JM, Jordan FTW (1989) Effects of certain stress factors on the re-excretion of infectious laryngotracheitis virus from latently infected carrier birds. Res Vet Sci 46(2):274–276

Jiménez-Bluhm P, Karlsson EA, Freiden P, Sharp B, Di Pillo F, Osorio JE, Hamilton-West C, Schultz-Cherry S (2018) Wild birds in Chile Harbor diverse avian influenza A viruses. Emerg Microbes Infect 7(1):1–4

Krol L, Moore RP, Mutlow AG, Brady SM, Dorsa D (2020) A retrospective analysis of mortality in captive Magellanic penguins (*Spheniscus magellanicus*) in the United States, 2008–2018. Zoo Biol 39(6):405–410

Lima M, Canales TM, Wiff R, Montero J (2020) The Interaction Between Stock Dynamics, Fishing and Climate Caused the Collapse of the Jack Mackerel Stock at Humboldt Current Ecosystem. Front Mar Sci 7:123

Leguia M, Garcia-Glaessner A, Munoz-Saavedra B, Juarez D, Barrera P, Calvo-Mac C, Jara J, Silva W, Ploog K, Amaro L (2023) Highly pathogenic avian influenza A (H5N1) in marine mammals and seabirds in Peru. bioRxiv:2023.03.03.531008

Leishangthem GD, Singh ND, Brar RS, Banga HS (2015) Aspergillosis in avian species: a review. J Poult Sci Technol 3(1):1–14

Levin II, Outlaw DC, Hernán Vargas F, Parker PG (2009) Plasmodium blood parasite found in endangered Galapagos penguins (*Spheniscus mendiculus*). Biol Conserv 142(12):3191–3195

Levy H, Fiddaman SR, Vianna JA, Noll D, Clucas GV, Sidhu JKH, Polito MJ, Bost CA, Phillips RA, Crofts S, Miller GD, Pistorius P, Bonnadonna F, Le Bohec C, Barbosa A, Trathan P, Raya Rey A, Frantz LAF, Hart T, Smith AL (2020) Evidence of pathogen-induced immunogenetic selection across the large geographic range of a wild seabird. Mol Biol Evol 37(6):1708–1726. https://doi.org/10.1093/molbev/msaa040. https://www.ncbi.nlm.nih.gov/pubmed/32096861

Ma W (2022) Orthomyxoviridae. Vet Microbiol:573–588

Marcogliese DJ (2008) The impact of climate change on the parasites and infectious diseases of aquatic animals. Rev Sci Tech 27(2):467–484. https://www.ncbi.nlm.nih.gov/pubmed/18819673

Melo AM, Silva Filho RPD, Poester VR, Fernandes CG, von Groll A, Stevens DA, Sabino R, Xavier MO (2020a) Aspergillosis in albatrosses. Med Mycol 58(6):852–855. https://doi.org/10.1093/mmy/myz122. https://www.ncbi.nlm.nih.gov/pubmed/31782484

Melo AM, Stevens DA, Tell LA, Verissimo C, Sabino R, Xavier MO (2020b) Aspergillosis, avian species and the one health perspective: the possible importance of birds in azole resistance. Microorganisms 8(12). https://doi.org/10.3390/microorganisms8122037. https://www.ncbi.nlm.nih.gov/pubmed/33352774

Miller GD, Hofkin BV, Snell H, Hahn A, Miller RD (2001) Avian malaria and Marek's disease: potential threats to Galapagos penguins Spheniscus mendiculus. Mar Ornithol 29:43–46

Moens U, Calvignac-Spencer S, Lauber C, Ramqvist T, Feltkamp MCW, Daugherty MD, Verschoor EJ, Ehlers B, ICTV Report Consortium (2017) ICTV virus taxonomy profile: Polyomaviridae. J Gen Virol 98(6):1159–1160

Moreira R (2009) Informe epidemiológico final: detección de un brote de la enfermedad de Newcastle (ENC) en aves marinas, en la zona costera de Constitución, Región del Maule, Chile 2007

Niemeyer C, Favero CM, Shivaprasad HL, Uhart M, Musso CM, Rago MV, Silva-Filho RP, Canabarro PL, Craig MI, Olivera V (2017) Genetically diverse herpesviruses in South American Atlantic coast seabirds. PLoS One 12(6):e0178811

Olsen B, Munster VJ, Wallensten A, Waldenström J, Osterhaus ADME, Fouchier RAM (2006) Global patterns of influenza A virus in wild birds. Science 312(5772):384–388

Pfaff, Florian et al (2017) "A novel alphaherpesvirus associated with fatal diseases in banded Penguins." *Journal of General Virology 98*.1: 89–95

Parsons NJ, Underhill LG (2005) Oiled and injured African penguins Spheniscus demersus and other seabirds admitted for rehabilitation in the Western Cape, South Africa, 2001 and 2002. Afr J Mar Sci 27(1):289–296

Parsons NJ, Gous TA, Cranfield MR, Cheng LI, Schultz A, Horne E, Last RP, Lampen F, Ludynia K, Bousfield B (2018) Novel vagrant records and occurrence of vector-borne pathogens in King Penguins (Aptenodytes patagonicus) in South Africa. Polar Biol 41:79–86

Rima B, Balkema-Buschmann A, Dundon WG, Duprex P, Easton A, Fouchier R, Kurath G, Lamb R, Lee B, Rota P (2019) ICTV virus taxonomy profile: Paramyxoviridae. J Gen Virol 100(12):1593

SAG (2023) Servicio Agrícola y Ganadero, Ministerio de Agricultura de Chile. Plataforma de influenza aviar. Accessed and Downloaded 30 Apr 2023. https://www.sag.gob.cl/ia

Sallaberry-Pincheira N, Gonzalez-Acuña D, Herrera-Tello Y, Dantas GPM, Luna-Jorquera G, Frere E, Valdés-Velasquez A, Simeone A, Vianna JA (2015) Molecular epidemiology of avian malaria in wild breeding colonies of Humboldt and Magellanic penguins in South America. EcoHealth 12:267–277

Sallaberry-Pincheira N, González-Acuña D, Padilla P, Dantas GPM, Luna-Jorquera G, Frere E, Valdés-Velásquez A, Vianna JA (2016) Contrasting patterns of selection between MHC I and II across populations of Humboldt and Magellanic penguins. Ecol Evol 6(20):7498–7510

Sebastiano M, Costantini D, Eens M, Pineau K, Bustamante P, Chastel O (2022) Possible interaction between exposure to environmental contaminants and nutritional stress in promoting disease occurrence in seabirds from French Guiana: a review. Reg Environ Chang 22(2):63

SERNAPESCA (2023) Servicio Nacional de Pesca. Plataforma de influenza aviar. Accessed and Downloaded 30 Apr 2023. http://www.sernapesca.cl/influenza-aviar

Shi J, Zeng X, Cui P, Yan C, Chen H (2023) Alarming situation of emerging H5 and H7 avian influenza and effective control strategies. Emerg Microbes Infect 12(1):2155072

Smith KF, Sax DF, Lafferty KD (2006) Evidence for the role of infectious disease in species extinction and endangerment. Conserv Biol 20(5):1349–1357. https://doi.org/10.1111/j.1523-1739.2006.00524.x. https://www.ncbi.nlm.nih.gov/pubmed/17002752

Soto KH, Trites AW, Arias-Schreiber M (2004) The effects of prey availability on pup mortality and the timing of birth of South American sea lions (Otaria flavescens) in Peru. J Zool 264:419–428. https://doi.org/10.1017/S0952836904005965

Soto KH, Trites AW, Arias-Schreiber M (2006) Changes in diet and maternal attendance of South American sea lions indicate changes in the marine environment and prey abundance. Mar Ecol Prog Ser 312:277–290. https://doi.org/10.3354/meps312277

Tizard I (2019) Veterinary immunology, 10th edn. Sauders

Tompkins EM, Anderson DJ, Pabilonia KL, Huyvaert KP (2017) Avian Pox discovered in the critically endangered waved Albatross (Phoebastria irrorata) from the Galápagos Islands, Ecuador. J Wildl Dis 53(4):891–895

Uhart M, Vanstreels RET, Gallo L, Cook RA, Karesh WB (2020) Serological survey for select infectious agents in wild Magellanic penguins (Spheniscus magellanicus) in Argentina, 1994–2008. J Wildl Dis 56(1):66–81

Valle CA, Coulter MC (1987) Present status of the flightless cormorant, Galapagos penguin and greater flamingo populations in the Galapagos Islands, Ecuador, after the 1982-83 El Nino. Condor 89:276–281

Varsani A, Porzig EL, Jennings S, Kraberger S, Farkas K, Julian L, Massaro M, Ballard G, Ainley DG (2015) Identification of an avian polyomavirus associated with Adelie penguins (Pygoscelis adeliae). J Gen Virol 96(4):851–857

Verdugo C, Pinto A, Ariyama N, Moroni M, Hernandez C (2019) Molecular identification of avian viruses in Neotropic cormorants (Phalacrocorax brasilianus) in Chile. J Wildl Dis 55(1):105–112

WAHIS (2023) World Organization for Animal Health, World animal health information system. Accessed and Downloaded 4 Apr 2023. https://wahis.woah.org/#/home

Wallace RS (2014) Sphenisciformes (penguins). Fowler's Zoo Wild Anim Med 8:82–88

WOAH (2023) Controlling the surge of avian influenza cases in Central and South America. https://www.woah.org/en/controlling-the-surge-of-avian-influenza-cases-in-central-and-south-america/

Woo PCY, de Groot RJ, Haagmans B, Lau SKP, Neuman BW, Perlman S, Sola I, van der Hoek L, Wong ACP, Yeh S-H (2023) ICTV virus taxonomy profile: Coronaviridae 2023. J Gen Virol 104(4):001843

Woods R, Jones HI, Watts J, Miller GD, Shellam GR (2009) Diseases of Antarctic seabirds. In: Health of Antarctic wildlife: a challenge for science and policy, pp 35–55. https://doi.org/10.1007/978-3-540-93923-8_3

Chapter 7
Rodents as Key Hosts of Zoonotic Pathogens and Parasites in the Neotropics

Carlos N. Ibarra-Cerdeña, César R. Rodríguez-Luna, Eduardo E. Palomo-Arjona, Martha P. Ibarra-López, Margarida F. Barber, and Rodolfo Dirzo

7.1 Introduction

Zoonotic diseases are infections that can be naturally transmitted between vertebrate animals and humans (WHO 2012). It is estimated that approximately 60% of emerging human infections are zoonotic (Rahman et al. 2020; Taylor et al. 2001). Zoonoses represent the most important threat among infectious diseases emerging worldwide (Morens et al. 2004; Woolhouse 2006; Karesh et al. 2012) because of their high impact on human health and the global economy (Jones et al. 2008; Rahman et al. 2020).

In the twentieth century, influenza A virus was responsible for pandemic episodes such as the Spanish Flu (1918–1919), Asian Flu (1957–1958), and Hong Kong Flu (1968), which together claimed the lives of 20–50 million people

Supplementary Information The online version contains supplementary material available at https://doi.org/10.1007/978-3-031-50531-7_7.

C. N. Ibarra-Cerdeña (✉) · C. R. Rodríguez-Luna · E. E. Palomo-Arjona · M. F. Barber
Departamento de Ecología Humana, Centro de Investigación y de Estudios Avanzados del Instituto Politécnico Nacional (Cinvestav), Mérida, Mexico
e-mail: cibarra@cinvestav.mx

M. P. Ibarra-López
Departamento de Ecología Humana, Centro de Investigación y de Estudios Avanzados del Instituto Politécnico Nacional (Cinvestav), Mérida, Mexico

Departamento de Recursos Naturales, Centro Universitario de la Costa Sur, Universidad de Guadalajara, Autlán, Mexico

R. Dirzo
Departments of Biology and Earth Systems Science, Stanford University, Stanford, CA, USA

G. Acosta-Jamett, A. Chaves (eds.), *Ecology of Wildlife Diseases in the Neotropics*, https://doi.org/10.1007/978-3-031-50531-7_7

worldwide. In the twenty-first century, the 2009 H1N1 influenza pandemic has already caused approximately 18,000 human deaths (Rewar et al. 2015). In economic terms, the 2014 Ebola virus disease outbreak in West Africa cost the global economy more than USD 53 billion (Hurber et al. 2018), and diseases transmitted by *Aedes aegypti* and *Aedes albopictus* mosquitoes accumulated a cost of more than USD 87 billion in the period 1975–2020, according to the most conservative estimate (Roiz et al. 2023). However, the most notable and recent zoonotic phenomenon is undoubtedly the crisis generated by the new SARS-CoV-2 coronavirus disease (COVID-19). While there is ongoing debate regarding the intermediate host species responsible for the initiation of human infections, it is most likely that the spillover occurred in the wet markets of Wuhan (Lytras et al. 2021) or as a result of interactions between wildlife and the human inhabitants of rural populations (Li et al. 2020; Córdoba-Aguilar et al. 2021). Globally, there have been 765,903,278 confirmed cases of COVID-19, including 6,927,378 deaths, reported to the WHO (https://covid19.who.int; consulted in May 2023), in addition to the macroeconomic impact it represents since more than 90% of the global economy experienced a contraction of the gross domestic product per capita (Yeyati and Filippini 2021). Thus, the significance of the impact of pandemic episodes caused by zoonotic diseases on the human population and development is clear.

The role played by mammals as zoonotic reservoirs is notable since most emerging human diseases originate from this group of animals (Cleaveland et al. 2001; Morse et al. 2012; Han et al. 2016). Within the terrestrial mammalian taxonomic group of Orders, there are wide variations in the proportion of species that have been identified as zoonotic hosts. Generally, the number of species that are zoonotic hosts increases in direct proportion to the total species richness of the order, and thus, the most species-rich orders present the greatest zoonotic diversity (Han et al. 2016; Mollentzea and Streickera 2020). In a review of global patterns of zoonotic diseases in mammals, Han et al. (2016) highlighted the orders Rodentia, Chiroptera, Soricomorpha, Primates, Carnivora, and Artiodactyla as the most species-rich groups of mammals harboring the greatest diversity of zoonoses.

The order Rodentia includes 2552 species grouped into 33 families and 513 genera, which places rodents as the most diverse group, with 39.30% of the total number of species among mammals (Burgin et al. 2018). Rodents are widely distributed since they have proven capable of colonizing nearly every terrestrial ecosystem worldwide (with the exception of a few isolated islands and Antarctica; Wilson and Reeder 2005). In general, rodents are small animals that reach sexual maturity at an early age, presenting high reproductive rates and large litters (Krebs et al. 1973). Due to their high abundance, low immunological investment, and high tolerance to anthropogenic environments, these species with short life spans tend to harbor and propagate more zoonotic pathogens than species with long life cycles (Albery and Becker 2021; Estavillo et al. 2022). These characteristics, coupled with their prevalence and range of impacts they have on agriculture, urban areas, natural ecosystems, and public health (Singleton et al. 1999; Capizzi et al. 2014), make rodents a group of special attention for research focused on zoonoses.

Rodents are hosts of a great diversity of pathogens, endo- and ectoparasites, including some of considerable public health importance (Mills and Childs 1998; Meerburg et al. 2009; Luis et al. 2013; Capizzi et al. 2014; Han et al. 2015; Albery et al. 2020). Salmonellosis, plague, leptospirosis, leishmaniasis, toxoplasmosis, Lassa fever, Hantavirus hemorrhagic fever, bartonellosis, listeriosis, and Lyme disease, among many others, are some examples of human diseases, the etiological agents of which can be transmitted by rodents (Meerburg et al. 2009; Battersby 2015; Dahmana et al. 2020).

Rodents comprise the largest number of zoonotic hosts compared to any other order of terrestrial mammals, as ~10% of rodent species are zoonotic hosts (Han et al. 2015, 2016). Currently, 244 rodent species have been identified as reservoirs of 66 zoonotic diseases caused by viruses, bacteria, fungi, helminths, and protozoa (Meerburg et al. 2009; Luis et al. 2013; Battersby 2015; Han et al. 2015, 2016). The geographic distribution of rodent hosts worldwide is heterogeneous, and it is evident that the areas with the highest concentration of host species are found in the high latitudes of the globe (~40° N; see Han et al. 2015). Nevertheless, the intertropical regions of the planet also contain a high number of rodent hosts (Han et al. 2015). These are areas in which biodiverse nations are still undergoing economic-social development and where zoonotic diseases pose a greater risk to humans as they experience ever increasing contact with wildlife (Jones et al. 2008; Han et al. 2016; Albery et al. 2020). In the Americas, the intertropical region is one of the areas with the highest concentration of rodent host species, particularly central Mexico and the Atlantic coast region of South America where a high potential exists for the emergence of new host species (Han et al. 2015). Given its importance as an area of high potential for the presence and emergence of zoonoses, in this chapter, we review the role of rodents as zoonotic reservoirs in the Neotropical biogeographic region in America, the latitudinal range of which approximately covers between 31° N and 44° S and extends from north-central Mexico, with the state of Sonora being the northernmost limit, to south-central Argentina and Chile in South America (Morrone 2014; Morrone et al. 2022).

Here we examine the general role played by rodents as hosts of zoonotic pathogens and parasites in the Neotropics. To this end, we conducted a systematic literature search from September to December 2022, following the recommendations of the PRISMA guide (Moher et al. 2010). We reviewed a total of 236 published studies that refer to the detection of zoonotic pathogens in Neotropical rodent hosts in 24 countries in the Americas, covering the period 1950–2022 (Fig. 7.1). We classified the zoonotic role of Neotropical rodents into two groups: (1) hosts of pathogens that cause diseases in humans through direct (nonvectorial) transmission, such as Machupo virus–hemorrhagic fever, Hantavirus–pulmonary syndrome, and leptospirosis, among others, and (2) hosts of ectoparasites that transmit disease-causing pathogens in humans, domestic animals, or livestock through indirect (vectorial) transmission, such as Chagas disease, leishmaniasis, and spotted fever, among others. Given the prevalence of anthropogenic impact on ecosystems, we also reviewed two land use scenarios in which rodents can be important propagators of parasites: fragmented landscapes and agricultural areas. Since rodents can persist in highly

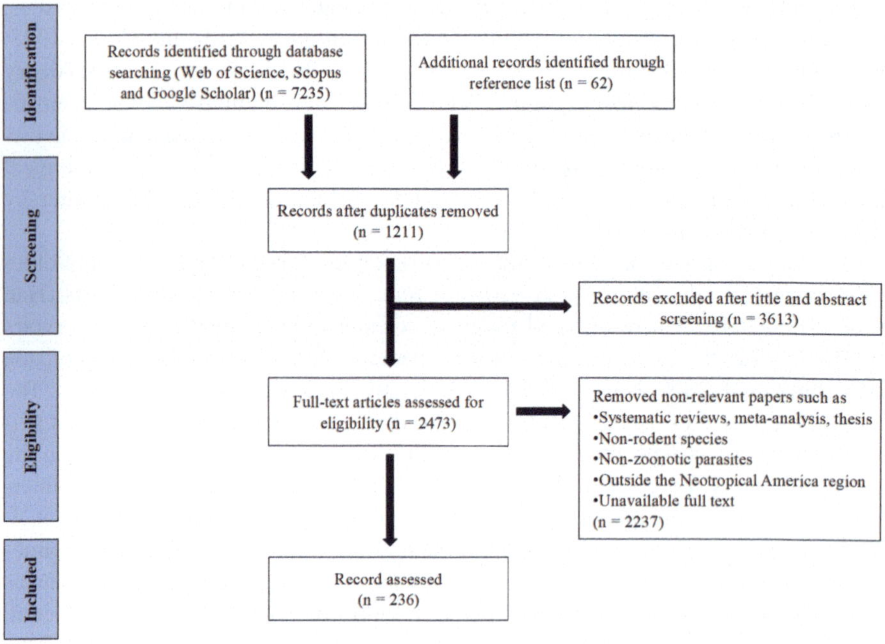

Fig. 7.1 Flow chart of article selection according to the PRISMA protocol

fragmented landscapes (Pires and Gallet 2022) and can be very damaging as crop pests (Stenseth et al. 2003), we revised the significance of rodents as hosts of zoonotic pathogens with public health implications and analyzed the evidence suggesting that these patterns may exhibit predictability across time and geographical regions.

The search included the following databases: Web of Science, Scopus, and Google Scholar, including studies published in the period 1950–2022. The search strategy included words related to rodents (rodent* OR rat OR mice), zoonotic diseases (pathogen OR parasite OR zoonotic OR disease OR infection OR virus OR bacteria OR protozoa OR helminth), and Neotropical location (America OR Neotropical OR Latin America OR South America). All articles published in English, Spanish, and Portuguese were included. The selection criteria for the articles were as follows: (1) full text provided; (2) the study was conducted in the Neotropical region of the Americas; (3) geographic location provided, including specific information of the province or country; and (4) the study reported zoonotic parasites or those with zoonotic potential. Systematic review articles and meta-analyses were excluded. For data extraction, location variables (i.e., country, study site, habitat type), rodent-specific data (i.e., family, genus, species), and zoonotic pathogen-specific data (i.e., pathogen type, etiologic agent, associated disease) were selected.

7.2 Rodents as Hosts of Pathogens and Zoonotic Parasites in the Neotropics

Our study covers 24 countries and 245 reports of the detection of zoonotic vectoring pathogens belonging to four classes (i.e., helminths, bacteria, protozoa, and viruses) in at least 130 species of Neotropical rodents. Based on the number of reports per nation, the following countries stood out: Brazil with 33.20% of the reports ($n = 81$), Mexico with 13.93% ($n = 34$), Argentina with 11.89% ($n = 29$), Colombia with 5.33% ($n = 13$), and Peru with 4.51% ($n = 11$). Cuba, Guatemala, Jamaica, and Trinidad and Tobago had one report each (0.41%). The lowest number of reports corresponded to Central America and the Caribbean Islands or Insular America (Fig. 7.2). Strikingly, there is no published information on the detection of zoonotic pathogens in rodents for Nicaragua, Guyana, Haiti, or other Antillean Island nations.

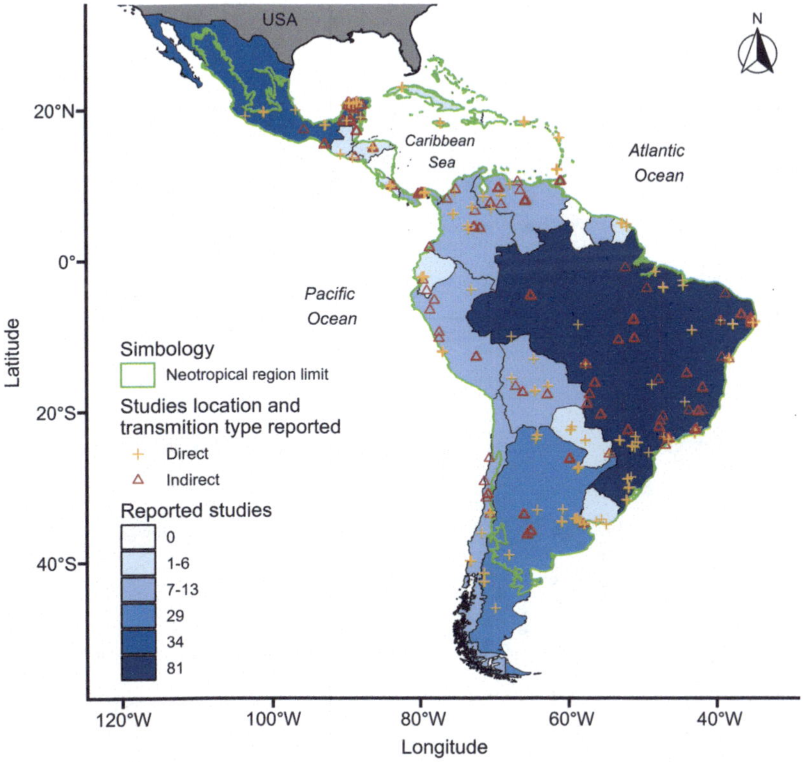

Fig. 7.2 Approximate locations and study frequency by country reporting zoonotic pathogens. Pathogens of direct transmission (plus sign) and indirect transmission (triangles) in the Neotropical region (gray line) of America

This distribution pattern of the published reports on the detection of zoonotic pathogens is maintained on consideration of the type of etiological agent causing the zoonosis since for pathogens such as helminths, bacteria, and protozoa, the largest number of records are found in the Yucatan Peninsula of Mexico and the southern cone of South America and in Brazil and Argentina. However, for zoonotic diseases in which the etiological agents are viruses, the vast majority of published reports correspond to South American countries (Fig. 7.3).

In the Neotropics, 130 rodent species, represented by 56 genera and 12 families, have been identified as confirmed hosts of zoonotic diseases (Table 7.1 and Fig. 7.4). The families with the highest representation were Cricetidae (63.85%; n = 83), Echimyidae (10.00%; n = 13), Heteromyidae (4.58%; n = 6), and Sciuridae (4. 58%; n = 6), while the most represented genera were *Akodon* (6.15%; n = 8), *Calomys*

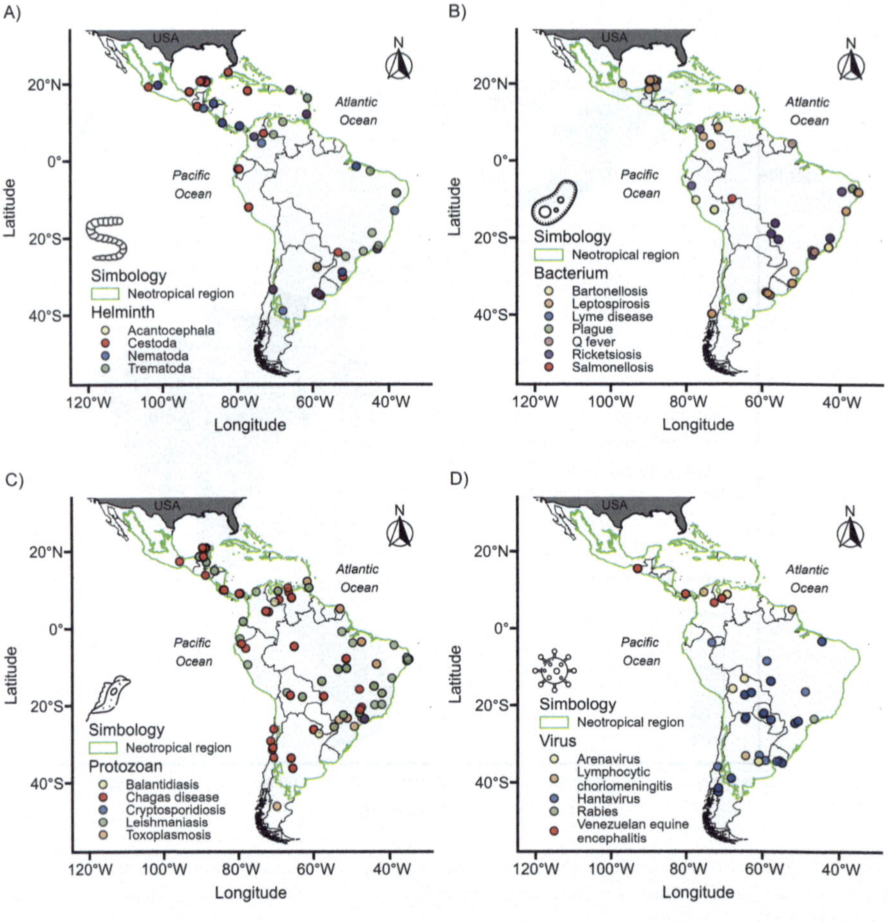

Fig. 7.3 Approximate locations of publication records of zoonotic parasites and pathogens by group: (**a**) helminth, (**b**) bacteria, (**c**) protozoa, and (**d**) virus in the Neotropical region (blue line)

Table 7.1 Etiological agents and classification criteria for zoonotic diseases detected in Neotropical rodent hosts based on 236 publications

Transmission	Pathogen	Disease	Etiological agent	Reservoir/host specie	Geographical distribution	References
Direct	Virus	Rabies	Rabies virus	*Hydrochoerus hydrochaeris*	Brazil	[31]
Direct	Virus	Hantavirus cardiopulmonary syndrome	Hantavirus: Anajatuba, Andes, Araucaria, Araraquara, Bermejo, Castelo dos Sonhos, Choclo, Itapua, Juquitiba, Laguna Negra, Lechiguanas, Oran, Rio Mamore, Rio Mearim	*Abrothrix longipilis, Abrothrix olivaceus, Akodon montensis, Akodon paranaensis, Calomys callidus, Calomys callosus, Calomys laucha, Holochilus sciureus, Necromys lasiurus, Oligoryzomys chacoensis, Oligoryzomys flavescens, Oligoryzomys fornesi, Oligoryzomys fulvescens, Oligoryzomys longicaudatus, Oligoryzomys microtis, Oligoryzomys nigripes, Oxymycterus nasutus, Oxymycterus quaestor*	Argentina, Bolivia, Brazil, Chile, Panama, Paraguay, Peru, Uruguay	[14, 28, 33, 63, 64, 76, 98, 106, 111, 112, 125, 126, 158, 169, 170, 173, 177, 182, 198, 200, 201, 212–214, 228]
Direct	Virus	Hemorrhagic fevers caused by Arenaviruses	Arenavirus: Chapare, Guanarito, Junin, Machupo, Sabia	*Akodon azarae, Calomys callosus, Calomys laucha, Calomys musculinius, Necromys obscurus, Oligoryzomys microtis, Sigmodon alstoni, Zygodontomys brevicauda*	Argentina, Bolivia, Brazil, Venezuela	[36, 85, 113, 130, 165, 185]
Direct	Virus	Lymphocytic choriomeningitis	Lymphocytic choriomeningitis virus	*Mus musculus*	Argentina, Colombia, French Guiana	[30, 123, 178]
Direct	Bacteria	Leptospirosis	*Leptospira interrogans, Leptospira borgpetersenii, Leptospira kirschneri, Leptospira noguchii, Leptospira santarosai*	*Abrothrix longipilis, Abrothrix olivacea, Akodon azarae, Cavia aperea, Heteromys gaumeri, Holochilus sciureus, Hydrochoerus hydrochaeris, Mus musculus, Oligoryzomys longicaudatus, Oligoryzomys nigripes, Oligoryzomys* sp., *Ototylomys phyllotis, Rattus norvegicus, Rattus rattus, Scapteromys aquaticus, Zygodontomys brevicauda*	Argentina, Brazil, Chile, Colombia, Mexico, Puerto Rico, Venezuela	[3, 9, 21, 38, 47, 50, 52, 69, 77, 99, 114, 148, 151, 159, 180, 186, 191, 207, 209, 221, 223, 234]
Direct	Bacteria	Q fever	*Coxiella burnetii*	*Akodon cursor, Hydrochoerus hydrochaeris, Mus musculus, Oligoryzomys nigripes, Oxymycterus dasytrichus, Proechimys* sp., *Wiedomys pyrrhorhinos*	Brazil, French Guiana	[32, 58, 87, 183]
Direct	Bacteria	Salmonellosis	*Salmonella enterica, Salmonella* spp.	*Hydrochoerus hydrochaeris, Rattus norvegicus*	Brazil	[72, 195]

(continued)

Table 7.1 (continued)

Transmission	Pathogen	Disease	Etiological agent	Reservoir/host specie	Geographical distribution	References
Direct	Protozoan	Toxoplasmosis	*Toxoplasma gondii*	*Akodon cursor, Akodon* sp., *Coendou prehensilis, Cuniculus paca, Dasyprocta agouti, Dasyprocta* sp., *Galea spixii, Hydrochoerus hydrochaeris, Mus musculus, Myocastor coypus, Myoprocta acouchy, Nectomys rattus, Oligoryzomys nigripes, Rattus norvegicus, Rattus rattus, Thrichomys laurentius*	Argentina, Brazil, Costa Rica, Grenada, French Guiana, Mexico, Panama	[2, 25–27, 31, 37, 46, 49–50, 60, 67, 83–84, 88, 107, 138, 143, 153, 184, 194, 208, 217, 219, 232]
Direct	Protozoan	Cryptosporidiosis	*Cryptosporidium parvum*	*Hydrochoerus hydrochaeris*	Brazil	[140]
Direct	Protozoan	Balantidiasis	*Balantidium coli*	*Hydrochoerus hydrochaeris*	Argentina, Colombia	[44, 220]
Direct	Helminth	Fasciolosis	*Fasciola hepatica*	*Hydrochoerus hydrochaeris*	Argentina, Brazil	[7, 119, 199]
Direct	Helminth	Schistosomiasis	*Schistosoma mansoni*	*Holochilus sciureus, Holochilus* sp., *Nectomys squamipes, Rattus norvegicus, Rattus rattus*	Brazil, Guadeloupe, Venezuela	[8, 39, 66, 91, 94, 146]
Direct	Helminth	Unknown	*Plagiorchis muris*	*Hydrochoerus hydrochaeris*	Colombia	[220]
Direct	Helminth	Taeniasis	*Hydatigera taeniaeformis*	*Akodon azarae, Mus musculus, Rattus rattus, Rattus norvegicus, Sciurus yucatanensis, Sigmodon toltecus*	Argentina, Colombia, Costa Rica, Ecuador, Grenada, Mexico, Panama, Puerto Rico	[13, 22, 42, 55, 68, 79, 89, 100–101, 139, 145, 160–163, 181, 189, 197, 229]
Direct	Helminth	Capillariasis	*Capillaria hepatica*	*Rattus norvegicus*	Argentina, Brazil, Colombia, Puerto Rico	[55, 68, 86, 101]

Direct	Helminth	Himenolepiasis	*Hymenolepis diminuta, Rodentolepis microstoma, Rodentolepis nana*	*Akodon azarae, Hodomys alleni, Mus musculus, Rattus rattus, Rattus norvegicus*	Argentina, Brazil Chile, Colombia, Costa Rica, Cuba, Ecuador, Grenada, Guatemala, Jamaica, Mexico, Panama, Peru, Puerto Rico	[1, 4, 13, 22, 34–35, 40, 42–43, 48, 55, 59, 78, 90, 93, 100–102, 108, 128, 135, 144, 147, 159–160, 163, 171, 197, 203, 229–230]
Direct	Helminth	Raillietiniasis	*Raillietina celebensis, Raillietina demerariensis, Raillietina* sp.	*Handleyomys* sp., *Rattus rattus, Rattus norvegicus*	Brazil, Jamaica, Peru, Mexico	[1, 57, 59, 80, 137, 230]
Direct	Helminth	Acanthocephaliasis	*Moniliformis moniliformis*	*Rattus rattus, Rattus norvegicus*	Argentina, Brazil, Costa Rica, Ecuador, Grenada, Guatemala, Jamaica, Mexico, Panama, Peru, Puerto Rico	[1, 22, 34, 42–43, 55, 59, 89, 100, 102, 128, 135, 197, 229–230]
Direct	Helminth	Trichinosis	*Trichinella spiralis*	*Rattus rattus, Rattus norvegicus*	Argentina, Chile, Mexico	[70, 176, 188, 203]
Direct	Helminth	Eosinophilic meningitis	*Angiostrongylus cantonensis*	*Rattus rattus, Rattus norvegicus*	Brazil	[128, 150, 193]
Direct	Helminth	Abdominal angiostrongyliasis	*Angiostrongylus costaricensis*	*Heteromys adspersus, Melanomys caliginosus, Mus musculus, Oligoryzomys fulvescens, Oligoryzomys nigripes, Peromyscus* sp., *Rattus norvegicus, Rattus rattus, Sigmodon hispidus, Sooretamys angouya, Zygodontomys brevicauda*	Brazil, Colombia, Costa Rica, El Salvador, Grenada, Guadeloupe, Honduras, Panama	[92, 115–116, 133, 152, 187, 205]
Indirect by free-living vector	Virus	Venezuelan equine encephalitis	Venezuelan equine encephalitis virus	*Handleyomys alfaroi, Oryzomys couesi, Proechimys chrysaeolus, Proechimys guairae, Proechimys semispinosus, Sigmodon hispidus*	Colombia, Mexico, Panama, Venezuela	[12, 61, 71, 95, 154]

(continued)

Table 7.1 (continued)

Transmission	Pathogen	Disease	Etiological agent	Reservoir/host specie	Geographical distribution	References
Indirect by ectoparasites	Bacteria	Bartonella Illnesses	*Bartonella* spp.	*Akodon cursor, Akodon montensis, Delomys dorsalis, Euryoryzomys russatus, Hylaeamys perenensis, Nectomys squamipes, Oecomys* sp., *Oxymycterus dasytrichus, Rattus norvegicus, Rattus rattus*	Brazil, Peru	[16, 47, 136, 183]
Indirect by ectoparasites	Bacteria	Lyme disease	*Borrelia burgdorferi*	*Heteromys gaumeri, Mus musculus, Rattus rattus*	Mexico	[180, 196]
Indirect by ectoparasites	Bacteria	Murine typhus	*Rickettsia typhi*	*Mus musculus, Rattus rattus*	Mexico	[166, 211]
Indirect by ectoparasites	Bacteria	Flea-borne spotted fever	*Rickettsia felis*	*Heteromys anomalus, Heteromys gaumeri, Hydrochoerus hydrochaeris, Mus musculus, Oligoryzomys* sp., *Ototylomys phyllotis, Oxymycterus rufus, Peromyscus yucatanicus, Rattus norvegicus, Sigmodon toltecus*	Argentina, Brazil, Colombia, Mexico, Peru	[17, 41, 132, 141, 164, 166]
Indirect by ectoparasites	Bacteria	Tick-borne spotted fever	*Rickettsia parkeri*	*Akodon* sp., *Cerradomys maracajuensis, Cerradomys subflavus, Euryoryzomys russatus, Hydrochoerus hydrochaeris, Necromys* sp., *Nectomys squamipes, Rattus rattus*	Brazil	[15, 132, 142, 157, 202]
Indirect by ectoparasites	Bacteria	Brazilian spotted fever	*Rickettsia rickettsii*	*Akodon* sp., *Cerradomys maracajuensis, Cerradomys scotti, Cerradomys subflavus, Euryoryzomys russatus, Hydrochoerus hydrochaeris, Necromys* sp., *Nectomys squamipes, Rattus rattus*	Brazil	[15, 58, 118, 132, 142, 157, 202]
Indirect by ectoparasites	Bacteria	Plague	*Yersinia pestis*	*Akodon dolores, Calomys callosus, Calomys laucha, Calomys musculinus, Calomys venustus, Cavia aperea, Cerradomys subflavus, Eligmodontia moreni, Galea musteloides, Galea spixii, Graomys griseoflavus, Holochilus chacarius, Holochilus sciureus, Hylaeamys perenensis, Lagostomus maximus, Microcavia australis, Necromys lasiurus, Oecomys* sp., *Oligoryzomys flavescens, Oligoryzomys nigripes, Phyllotis darwini, Rattus rattus, Thrichomys laurentius*	Argentina, Brazil, Peru	[5, 6, 54, 136]

Indirect by free-living vector	Protozoan	Chagas disease	*Trypanozoma cruzi*	*Abrothrix olivaceus, Akodon boliviensis, Akodon cursor, Akodon molinae, Akodon montensis, Akodon toba, Baiomys musculus, Calomys callosus, Calomys expulsus, Calomys laucha, Calomys musculinus, Calomys tener, Cavia aperea, Cerradomys subflavus, Clyomys laticeps, Coendou insidiosus, Coendou mexicanus, Coendou prehensilis, Coendou quichua, Coendou vestitus, Cuniculus paca, Dasyprocta agouti, Dasyprocta azarae, Dasyprocta fuliginosa, Dasyprocta punctata, Dasyprocta* sp., *Delomys dorsalis, Diplomys labilis, Galea spixii, Graomys chacoensis, Graomys griseoflavus, Heteromys desmarestianus, Heteromys gaumeri, Heteromys irroratus, Heteromys pictus, Hydrochoerus hydrochaeris, Hylaeamys megacephalus, Mus musculus, Necromys lactens, Necromys lasiurus, Nectomys squamipes, Neotoma mexicana, Octodon degus, Oecomys concolor, Oecomys mamorae, Oligoryzomys chacoensis, Oligoryzomys nigripes, Oligoryzomys* sp., *Oryzomys couesi, Otospermophilus variegatus, Ototylomys phyllotis, Oxymycterus hispidus, Pattonomys semivillosus, Peromyscus leucopus, Peromyscus maniculatus, Peromyscus melanophrys, Peromyscus mexicanus, Peromyscus yucatanicus, Phyllotis darwini, Phyllotis osilae, Proechimys guyannensis, Proechimys semispinosus, Rattus rattus, Reithrodontomys fulvescens, Rhipidomys macrurus, Sciurus aestuans, Sciurus granatensis, Sciurus ignitus, Sciurus igniventris, Sigmodon hispidus, Sigmodon mascotensis, Sigmodon toltecus, Thalpomys lasiotis, Thaptomys nigrita, Thrichomys apereoides, Thrichomys inermis, Thrichomys laurentius, Thrichomys pachyurus, Thrichomys* sp., *Tylomys panamensis, Wiedomys pyrrhorhinos*	Argentina, Bolivia, Brazil, Chile, Colombia, Costa Rica, Ecuador, El Salvador, Mexico, Panama, Peru, Venezuela	[11, 18, 20, 23, 45, 74–75, 97, 103, 105, 109–110, 131, 149, 156, 159, 168, 174–175, 210, 218, 222, 227, 231, 233, 235]

(continued)

Table 7.1 (continued)

Transmission	Pathogen	Disease	Etiological agent	Reservoir/host specie	Geographical distribution	References
Indirect by free-living vector	Protozoan	Leishmaniasis	*Leishmania* spp.	*Agouti paca, Akodon montensis, Akodon* sp., *Cerradomys subflavus, Clyomys laticeps, Coendou prehensilis, Coendou* sp., *Cuniculus paca, Dasyprocta azarae, Dasyprocta* sp., *Euryoryzomys nitidus, Euryoryzomys russatus, Handleyomys melanotis, Heteromys anomalus, Heteromys desmarestianus, Heteromys gaumeri, Heteromys* sp., *Holochilus sciureus, Hydrochoerus hydrochaeris, Hylaeamys acritus, Hylaeamys megacephalus, Kannabateomys amblyonyx, Mus musculus, Neacomys spinosus, Necromys lasiurus, Nectomys squamipes, Nyctomys sumichrasti, Oecomys concolor, Oligoryzomys nigripes, Oligoryzomys* sp., *Ototylomys phyllotis, Peromyscus yucatanicus, Phyllotis andinum, Proechimys canicollis, Proechimys guyannensis, Proechimys semispinosus, Proechimys* sp., *Rattus norvegicus, Rattus rattus, Reithrodontomys gracilis, Rhipidomys mastacalis, Sciurus granatensis, Sigmodon hispidus, Sigmodon toltecus, Thrichomys apereoides, Thrichomys inermis, Thrichomys laurentius, Thrichomys pachyurus*	Argentina, Belize, Bolivia, Brazil, Colombia, Costa Rica, Ecuador, French Guiana, Honduras, Mexico, Panama, Peru, Trinidad and Tobago, Venezuela	[10, 19, 24, 29, 50–51, 53, 56, 62, 65, 73, 81–82, 96, 104, 117, 120–122, 124, 127, 129, 134, 155, 167, 172, 179, 190, 192, 204, 206, 215–216, 224–226, 236]

(5.38%; *n* = 7), *Oligoryzomys* (5.38%; *n* = 7), *Heteromys* (4.62%; *n* = 6; Fig. 7.5a), and *Necromys* (4.62%; *n* = 6; Table 7.1). Considering the tolerance of these species to fragmentation of their natural habitat, 76 (58.46%) Neotropical rodents are wild species with affinity to the unmodified natural environment, 51 (39.23%) species have affinity to agricultural and/or fragmented landscapes, and three (2.31%) species are considered synanthropic (Fig. 7.6).

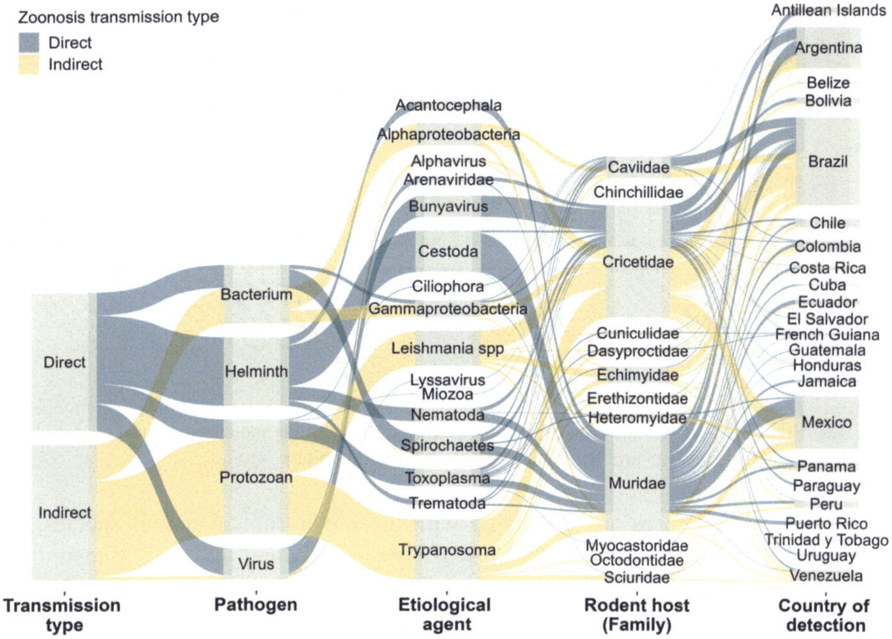

Fig. 7.4 Distribution flow of transmission type of zoonosis by parasite and pathogen group, host, and place of occurrence. The width of the flow bars is proportional to the frequency of records from the 245 reports of zoonotic parasite and pathogen detections from 236 publications

Fig. 7.5 (**a**) *Heteromys gaumeri* (Heteromydae) captured in a degraded semievergreen tropical forest of the Calakmul region, and (**b**) *Mus musculus* (Muridae) captured in a traditional agricultural field of the south of Yucatan—credits to Marga Barber

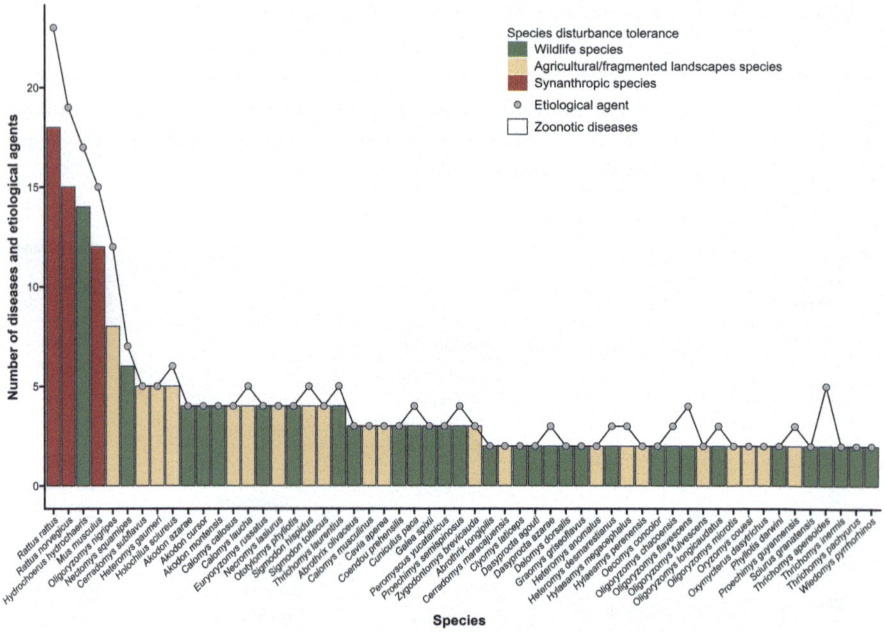

Fig. 7.6 Frequency of zoonotic diseases (bars) and etiological agents (gray points) reported by multipathogen rodent hosts and their tolerance to disturbed habitats

At least 31 zoonoses are associated with these Neotropical rodent species, of which 67.74% ($n = 21$) and 32.26% ($n = 10$) are directly and indirectly transmitted by vectors, respectively (Table 7.1; Figs. 7.2 and 7.4). Parasitic zoonoses due to helminths were the most frequently recorded, with 35.48% ($n = 11$), followed by bacterial zoonoses (32.26%; $n = 10$), while parasitic zoonoses due to protozoa and those of viral origin each represented 16.13% ($n = 5$) of the zoonoses identified (Table 7.1). The highest numbers of etiologic agents of these zoonoses are viruses ($n = 22$), followed by 16 species of both helminths and protozoa and 15 species of bacteria (Table 7.1 and Fig. 7.7).

The zoonoses with the highest number of Neotropical rodent hosts are diseases transmitted indirectly by vectors: Chagas disease (Trypanosomiasis), which has been detected in 78 rodent species and reported in 12 Neotropical countries; leishmaniasis, detected in 42 rodent species and reported in 14 countries; and plague, detected in 23 rodent species and three countries (Table 7.1 and Fig. 7.5). Direct transmission zoonoses such as Hantavirus cardiopulmonary syndrome, leptospirosis, and toxoplasmosis have been reported in seven countries and detected in 18, 15, and 14 rodent species, respectively (Fig. 7.8).

Chagas disease, caused by the protozoan *Trypanosoma cruzi*, is endemic to the Americas and infects hundreds of species of wild and synanthropic mammals (Jansen et al. 2018). This disease is vector-borne by dozens of species of triatomine arthropods (Ibarra-Cerdeña et al. 2017), with rodents comprising the mammalian

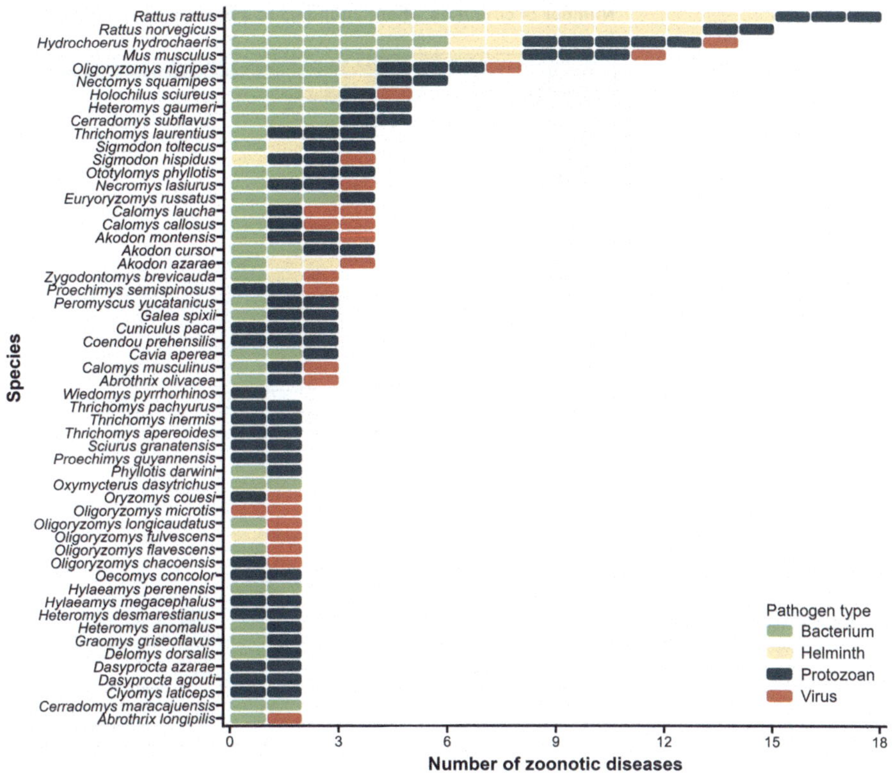

Fig. 7.7 Frequency of zoonotic diseases by the pathogen in multipathogen rodent host

group with the second highest number of *T. cruzi* records, behind only the order Didelphimorphia (Brenière et al. 2016). Leishmaniasis is distributed in regions of Asia, Africa, and America; its etiological agents are protozoa of the genus *Leishmania*, and in the Neotropical region of America, it is transmitted by the bite of phlebotomine sandflies of the genus *Lutzomyia* (Sharma and Singh 2008), with domestic dogs and small mammals being the main reservoirs (Brandão-Filho et al. 2003; Quaresma et al. 2011). On the other hand, the disease known as plague is caused by infection with the bacterium *Yersinia pestis*, which is transmitted mainly by the bite of fleas associated with rodents (Barbieri et al. 2020). It is worth noting that in countries classified as endemic for plague (such as Peru, Brazil, Bolivia, and Ecuador), where human cases continue to occur, susceptible rodents and fleas play a crucial role in maintaining plague transmission in the natural environment (Schneider et al. 2014), highlighting the importance of monitoring rodent populations in regions prone to outbreaks in the Neotropics.

Among the Neotropical rodent hosts, 77 species (58.78%) have been identified as being associated with a single pathogen and 54 species (41.22%) as multi-pathogen reservoirs (Table 7.1 and Fig. 7.6). The Neotropical multi-pathogen

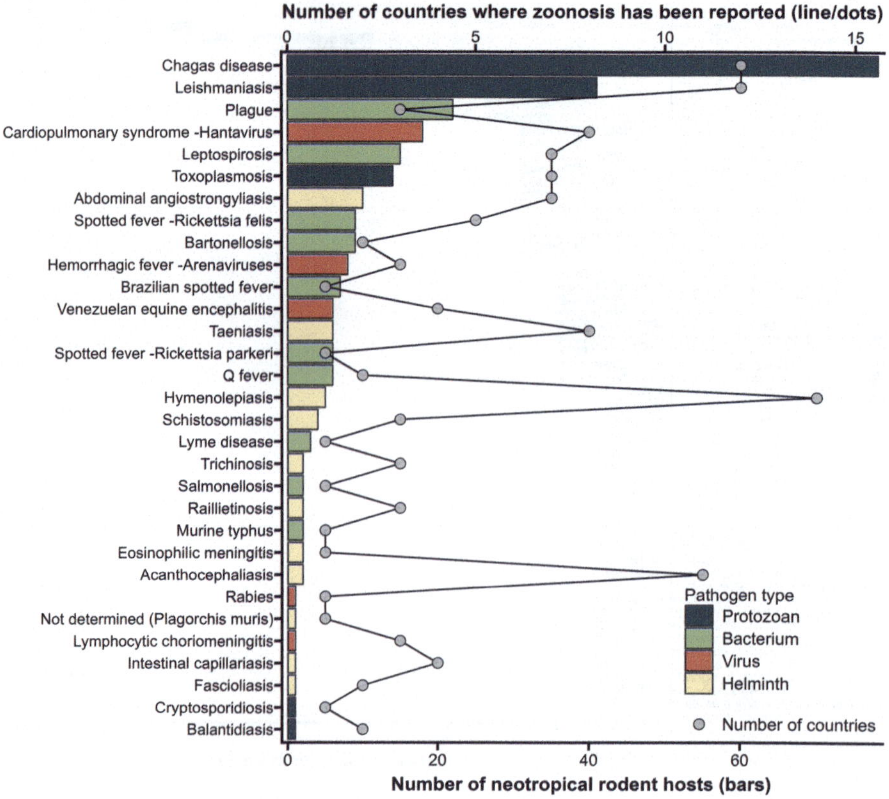

Fig. 7.8 Multihost zoonotic diseases ($n = 31$) by pathogen type, rodent host frequency (bars), and countries of occurrence (gray points)

rodents associated with the highest number of zoonoses (*z*) and etiologic agents (ea) are *Rattus rattus* (*z* = 18, ea = 23), *R. norvegicus* (*z* = 15, ea = 19), *Hydrochoerus hydrochaeris* (*z* = 14, ea = 17), *Mus musculus* (*z* = 12, ea = 15; Fig. 7.5b), *Oligoryzomys nigripes* (*z* = 8, ea = 12), *Nectomys squamipes* (*z* = 6, ea = 7), and *Holochilus sciureus* (*z* = 5, ea = 6). Prominent among the multi-pathogen rodents are the three synanthropic species (*R. rattus*, *R. norvegicus*, and *M. musculus*), one associated with wild settings (*H. hydrochaeris*), and one associated with agricultural/fragmented landscapes (*O. nigripens*; Fig. 7.6).

Synanthropic rodents play a key role as hyper reservoirs in the Neotropics. Collectively, *R. rattus*, *R. norvegicus*, and *M. musculus* are multi-pathogen species that can transmit 100% ($n = 10$) of bacteria-caused zoonoses in the Neotropics. Even when only *R. rattus* and *R. norvegicus* species are considered, they can together transmit 90% ($n = 9$) of the zoonoses caused by bacteria and 81.82% ($n = 9$) of those caused by parasitic helminths; *R. rattus* and *M. musculus* can be reservoirs for 85.71% ($n = 6$) and 57. 10% ($n = 4$), respectively, of infections transmitted by ectoparasites. In addition, *M. musculus* can be a reservoir for 50% ($n = 5$) of zoonoses caused by bacteria and is the only reservoir species for the virus that causes lymphocytic choriomeningitis in the Neotropics (Table 7.1). These rodent species are distributed in temperate and tropical zones worldwide, are closely related to human environments (Macdonald et al. 2015), and are considered pest species given the threat they pose to agricultural production, damage to infrastructure, and impact on public health and ecosystems (Almeida et al. 2013; Capizzi et al. 2014; Rabiee et al. 2018). These rodents can play an important role in the transmission of diseases to humans since they are hosts of a high number of zoonotic parasites and have a close relationship with human populations (Rabiee et al. 2018). Thus, it is important to study the effect of control of these rodents on the transmission of zoonotic diseases, especially in biodiverse areas where humans increasingly experience contact with wildlife (Jones et al. 2008; Albery et al. 2020), such as the Neotropical region of the Americas. Clearly, this is a research topic that warrants considerable effort going forward (Zeppelini et al. 2022).

On the other hand, capybara (*H. hydrochaeris*) is another important multipathogen species in the Neotropics. It is a wild rodent that is widely distributed in South America (Mones and Ojasti 1986) and is a reservoir and host of at least 14 zoonotic diseases: six caused by bacteria, five by helminths, two by protozoa, and one by a virus (Table 7.1 and Fig. 7.7). This large rodent represents a frequent and important source of protein for indigenous and rural communities (Mones and Ojasti 1986; Ali and Jones 2020), as well as an ecotourism attraction (Herrera and Barreto 2013), and consequently, its relationship with human populations is extremely prominent (Verdade and Ferraz 2006; Serra-Medeiros et al. 2021). Given these characteristics, it is important to consider the potential risk to public health posed by rodents, and it is therefore necessary to monitor these animals in anthropogenic environments and use them in an appropriate manner to prevent contagion and zoonotic outbreaks.

Next, we examine the variety of rodent-borne diseases and discuss the different modes of zoonotic disease transmission in humans that can result from pathogens in the Neotropics.

7.3 Direct Transmission: Rodents as Hosts of Pathogens That Cause Human Diseases in the Neotropics

The direct transmission of zoonotic pathogens occurs via the participation of two agents: the host of the pathogen (in this case, rodents) and humans. For these, contagion occurs through contact with rodent fluids and wastes such as saliva, excreta, urine, or even through ingestion of contaminated food and water (Rahman et al. 2020). Of the 31 zoonoses associated with Neotropical rodents, 21 (67.74%) are directly transmitted (Table 7.1 and Fig. 7.4). Among the directly transmitted zoonoses, 11 (52.38%) were caused by helminths, four (19.05%) by viruses, three (14.29%) by bacteria, and three (14.29%) by protozoa (Table 7.1). Synanthropic rodents are involved in the transmission of 13 directly transmitted zoonotic diseases (Table 7.1).

Zoonotic diseases can also be classified according to the environments in which transmission occurs: synanthropic and exoanthropic (Bedi et al. 2022). The most common direct rodent-borne diseases in the Neotropics are synanthropic zoonotic diseases, that is, the circle of transmission occurs in the urban and peri-urban environment, involving domestic fauna and wildlife adapted to human-developed environments (Rahman et al. 2020; Bedi et al. 2022). Within this categorization, helminth zoonoses are the most diverse in the Neotropics, and the most widely distributed helminth in Neotropical countries ($n = 11$) is *Moniliformis moniliformis* (Dhaliwal and Juyal 2013). This helminth causes human acanthocephaliasis, which is transmitted among rodents only by species of the genus *Rattus* (*R. rattus* and *R. norvegicus*; Table 7.1), while humans are accidental hosts (Dhaliwal and Juyal 2013; Pan American Health Organization 2003). Another helminth involved in zoonoses transmitted by synanthropic rodents is *Angiostrongylus costaricensis*, which causes abdominal angiostrongyliasis and for which rodents are the definitive host (Graeff-Teixeira et al. 1990). In the first instance, the parasite establishes itself in the pulmonary arteries of the rodent and then travels in larval form to its feces, where fecal deposition can contaminate food and thus produce transmission to humans (Dhaliwal and Juyal 2013). In the Neotropics, 11 rodents of the families Cricetidae and Muridae have been confirmed as host species of *A. costaricensis* (Table 7.1). Another parasite that is transmitted by food contamination is the helminth *Trichinella spiralis*, which is responsible for trichinosis and for which *Rattus* rats are a reservoir (Panti-May et al. 2021). This rodent genus represents an important reservoir of pathogenic microorganisms due to their wide geographic distribution and ease of establishment and reproduction in urban areas (Runge et al. 2013). They are also responsible for the transmission of *Rodentolepis nana*, which is the most common parasitic cestode in humans, with a prevalence of 4% worldwide (Bogitsh et al. 2012), and which, together with *Hymenolepis diminuta* and *Rodentolepis microstoma*, causes the disease known as hymenolepiasis. Another synanthropic zoonosis directly transmitted by Neotropical rodents is salmonellosis, which is caused by gram-negative bacteria of the Enterobacteriaceae family and is considered one of the most important foodborne pathogens worldwide (Meerburg and Kijlstra 2007). Transmission of salmonellosis in humans occurs by consuming food or water sources that have previously been contaminated by Salmonella hosts (Conover and Vail 2015). Although it has been shown that rodents can become infected with Salmonella (Bastiaan and Aize 2007) through contact with feces from diseased animals and other wildlife, the role of rodents in the transmission of salmonellosis is still unclear (Battersby 2015; Raufu et al. 2019). Rodents can be long-term sources of infection since their droppings can be contaminated by the pathogen for up to 3 months (Davies and Wray 1995), and they are also considered amplifiers of infection on busy food animal production farms (Meerburg and Kijlstra 2007; Meerburg et al. 2009), where prevalence values of up to 24% have been reported (Henzler and Opitz 1992). In general, salmonellosis is considered a global public health problem (Sabour et al. 2022) since, at least until 2010, 93.8 million cases and

155 thousand deaths per year were estimated (Majowicz et al. 2010). There are considerable effects in the Americas: in Mexico, an estimated 92,000 people are infected per year (DGE 2021), and in the USA, the economic costs of salmonellosis have been estimated at USD 1.4 billion per year (Roberts 1988).

There are also exoanthropic diseases in the Neotropics. These are transmitted mainly by wildlife and occur in nonanthropic environments. They include rabies (virus) and cryptosporidiosis, caused by the protozoan *Cryptosporidium parvum*. These zoonoses have been detected in Brazil, with the capybara (*Hydrochoerus hydrochaeris*) being the rodent intermediate host, although most cases of transmission of these diseases are due to other hosts (e.g., *Canis lupus familiaris*) in anthropogenic environments (Sing 2015; Garg 2014), causing 3.7 million cases of rabies and 59,000 deaths per year worldwide, with economic losses valued at USD 8.6 billion annually (Hampson et al. 2015).

However, most zoonotic diseases circulate in both natural and anthropogenic environments (Marquardt 2004; Rahman et al. 2020). In particular, there is a record of three directly transmitted diseases characterized by a large number of reservoirs or hosts and a wide Neotropical distribution: Hantavirus cardiopulmonary syndrome (Hantavirus), leptospirosis (*Leptospira* sp. bacteria), and toxoplasmosis caused by the protozoan *Toxoplasma gondii* (Table 7.1).

Hantavirus family viruses can be transmitted to humans and cause two diseases: hemorrhagic fever with renal syndrome (HFRS), present in Eurasia, and Hantavirus cardiopulmonary syndrome (HCPS) in the Americas (Heyman et al. 2012). The first indications of Hantavirus hemorrhagic fever date back over 900 years when it was described in Chinese writings, and the first description and isolation of the etiological agent subsequently occurred in 1978 (Smadel 1953; Lee et al. 1978; Avšic-Zupanc et al. 2016). Transmission occurs through inhalation of aerosols from excreta and the secretions of infected rodents (Jonsson et al. 2010; Pinto-Junior et al. 2014). Regarding its reservoirs, several species of the Muridae and Cricetidae families are natural hosts of the virus but do not develop the disease (Conover and Vail 2015). The incidence of HCPS is therefore strongly correlated with the population dynamics of the hosts (Heyman et al. 2012; Conover and Vail 2015). In Latin America, at least 14 Hantaviruses are of zoonotic potential, with each of these viruses varying in terms of distribution and rodent reservoir species: Anajatuba by *Oligoryzomys fornesi*; Araucaria by *Oligoryzomys nigripes*; Araraquara by *Necromys lasiurus* in Brazil; Andes by *Oligoryzomys longicaudatus* in Argentina; Bermejo by *Oligoryzomys chacoensis* in Argentina and Paraguay; Castelo dos Sonhos by *O. nigripes* in Brazil; Choclo by *Oligoryzomys fulvescens* in Panama; Itapua by *O. nigripes* in Paraguay; Juquitiba by *O. nigripes* in Brazil; Laguna Negra by *Calomys callosus* and *Calomys laucha* in Argentina, Bolivia, and Paraguay; Lechiguanas by *Oligoryzomys flavescens* in Argentina and Uruguay; Orán by *O. longicaudatus* in Argentina; Rio Mamore by *Oligoryzomys microtis* in Bolivia and Peru; and Rio Mearim by *Holochilus sciureus* in Brazil (Rosa et al. 2005; Firth et al. 2012; Figueiredo et al. 2014; Avšic-Zupanc et al. 2016). Comparatively, HFRS affects a larger number of people, although it is less lethal. Between 150,000 and 200,000 cases are reported annually, of which 10,000 are estimated in Europe, while most occur in Asia (Schmaljohn 2009; Heyman et al. 2012; D'Souza and Patel 2020). In the case of HCPS, the prevalence is lower, but mortality is greater than 50% in South America (Bedi et al. 2022). In the Neotropics, approximately 280 cases of HCPS are reported per year and attributed to the Andes, Brazil, Araraquara, and Juquitiba Hantaviruses, mainly in Argentina and Chile. The highest case fatality rate is for the Araraquara virus at 44.5%, followed by the Andes virus at 21.4–35.9% and the Juquitiba virus at 32.5%, while for the Choclo and Laguna Negra viruses, the rates are less than 15% (Vial et al. 2023).

The Arenaviridae family of viruses is a diverse group of RNA viruses that are etiologic agents of several emerging zoonoses characterized by high case fatality rates (Peters 2002; Ma et al. 2021). The first identification of Arenavirus occurred in 1933 when the virus was isolated from lymphocytic choriomeningitis (Armstrong and Lillie 1934). In the Neotropical region, the first Arenavirus identified were the Junin virus in Argentina in the 1950s and the Machupo virus in Bolivia in the 1960s (Peters 2002). These etiological agents are responsible for hemorrhagic fevers and are transmitted by inhalation of excreta or contact with the urine, feces, or saliva of infected rodents (Mills 2006). At least six Arenaviruses are of zoonotic potential in the South American region: Guanarito, causing Venezuelan hemorrhagic fever and transmitted by *Zygodontomys brevicauda*; Junín, causing Argentine hemorrhagic fever and transmitted by *Calomys musculinus*; Machupo, an agent of Bolivian hemorrhagic fever caused by *Calomys callosus*; Chapare, causing Chapare hemorrhagic fever in Bolivia caused by *O. microtis*; Sabiá, which causes Brazilian hemorrhagic fever, the reservoir of which remains unknown; and lymphocytic choriomeningitis caused by lymphocytic choriomeningitis virus and transmitted by *Mus musculus* (Peters 2002; Frank et al. 2021; Loayza-Mafayle et al. 2022). In the case of lymphocytic choriomeningitis virus, seroprevalences of less than 5% have been reported in humans in Argentina, while the seroprevalence in rodents was 20% (Vilibic-Cavlek et al. 2021). Among the most important Arenaviruses, the Junin virus causes annual outbreaks in the central region of Argentina, where approximately five million people are at risk of infection, and the disease has a fatality rate of approximately 20% (Enria et al. 2008). Machupo virus causes one of the most devastating diseases, with a mortality rate of 25–35%, and since 2005, a re-emergence of cases has taken place in Bolivia (Patterson et al. 2014).

Leptospirosis is the most common zoonotic disease in the world within the animal kingdom (Guerrant et al. 2006). It is one of the first zoonotic diseases to be attributed to rodents, being first described in 1886 (Weil 1886). It is caused by bacteria of the genus *Leptospira* that chronically infect rodents by establishing themselves in the renal system, with subsequent transfer in large quantities through the urine (Ellis 2015). Humans can contract the disease by direct contact with urine or through contaminated water (Cosson et al. 2014; Haake and Levett 2015; Sabour et al. 2022). Leptospirosis is one of the zoonotic diseases with the highest morbidity and mortality since the number of people affected by leptospirosis worldwide is estimated at 1.03 million per year, with a mortality of 58,900 per year (Costa et al. 2015). In the Neotropics, approximately 77,000 cases of leptospirosis are estimated annually, with a mortality of 100–1600 deaths, mainly in Central American countries (Costa et al. 2015).

Toxoplasmosis, caused by the protozoan *Toxoplasma gondii*, is another globally distributed disease that is transmitted in both synanthropic and exoanthropic environments and has been estimated to affect six billion people worldwide (Klaren and Kijlstra 2002). It is most prevalent in areas with humid and temperate climates at low elevations, and its prevalence therefore varies according to geographic region (Walton et al. 1966; Schwartzman and Maguire 2011; de Lima Bessa et al. 2021). In the United States of America and United Kingdom, prevalence values of 23% have been documented (Joynson 1992; Holland 2003), while higher values have been reported in Latin American countries, even reaching approximately 50% (Bigna et al. 2020). The most common form of transmission is through consumption of contaminated food and water or by ingestion of meat with *T. gondii* oocysts (Schwartzman and Maguire 2011).

Q fever is caused by the bacterium *Coxiella burnetii* and is also one of the zoonoses present in both urban and natural areas. There are reports of Q fever cases worldwide (except in New Zealand), and both the disease and the pathogen were first identified in Australia in the 1930s (Conover and Vail 2015). The reservoirs of *C. burnetii* comprise a wide variety of vertebrates, including domestic and wild animals, such as rodents (Parker et al. 2006). Transmission to humans occurs primarily through inhalation of aerosols from ruminating, parturient, or slaughtered animals (Parker et al. 2006; Angelakis and Raoult 2010). In the Neotropics, human cases have been described in several Latin American countries, but the country with the highest incidence is French Guiana (Epelboin et al. 2021). Although rodents are not the main reservoirs of *C. burnetii*, their presence in anthropic environments represents a risk of transmission of the pathogen to domestic animals (Parker et al. 2006). In synanthropic environments in the Neotropics, the rodent reservoir of *C. burnetii* is *M. musculus* (Rozental et al. 2017), although species of the genus *Rattus* have also been reported as hosts and potential transmitters of the pathogen in other parts of the world (Sabour et al. 2022; Izquierdo-Rodríguez et al. 2019). In exoanthropic environments, the pathogen has been detected in *Hydrochoerus hydrochaeris* (Caviidae; Christen et al. 2020) and in rodents of the family Cricetidae and the genus *Proechimys* of the family Echimyidae (Gardon et al. 2001; Rozental et al. 2017; de Oliveira et al. 2020).

7.4 Vectorial Transmission: Rodents Are Hosts of Endo- and Ectoparasitic Vectors of Pathogens That Cause Disease in Humans, Domestic Animals, and Livestock

In the Neotropical region, seven zoonotic diseases have been identified that are transmitted through the bite of ectoparasitic vectors in rodents (Table 7.1). All of these zoonoses are caused by bacteria, of which 57.14% ($n = 4$) belong to the genus

Rickettsia: murine typhus, flea-borne spotted fever, tick-borne spotted fever, and Brazilian spotted fever. The remaining three diseases are human bartonellosis, the etiological agent of which is bacteria of the genus *Bartonella*; Lyme disease, caused by *Borrelia burgdorferi*; and plague, caused by *Yersinia pestis*. Together, these diseases have been detected in 39 rodent species in the Neotropics.

Bartonellosis is an infectious disease caused by gram-negative bacteria of the genus *Bartonella*, which are transmitted mainly by arthropod vectors such as lice, fleas, sandflies, and ticks (Billeter et al. 2008; Chomel et al. 2009). These bacteria infect a large number of mammals, including rodents, which act as natural reservoirs (Gutiérrez et al. 2015). At least ten species of rodent reservoirs of *Bartonella* have been identified in the Neotropics (Table 7.1). The first etiological agent described was *Bartonella bacilliformis* causing Carrion's disease (Bartonellosis) in 1905 during an outbreak in a mining town in Los Andes, Peru (Bass et al. 1997; Maco et al. 2004); however, DNA evidence of another bacterial species of the same genus, *B. quintana*, has been found in human samples from more than 4000 years ago (Drancourt et al. 2005). At least 13 species of *Bartonella* are pathogenic to humans (Chomel et al. 2009). The diseases they cause include Carrion's disease, trench fever, cat scratch disease, and clinical manifestations such as bacillary angiomatosis, septicemia, endocarditis, chronic lymphadenopathy, and neurological disorders (Maurin et al. 1997). Although it is estimated that more than 12,000 cases of bartonellosis occur annually in the United States (Nelson et al. 2018), the incidence of confirmed cases of bartonellosis associated with rodent transmission is very low, with only 24 confirmed cases identified worldwide (Krügel et al. 2022). Cases of human bartonellosis have been described in Brazil and Peru (Maco et al. 2004; Favacho et al. 2014), although there is no evidence that these infections were caused by or associated with rodents. Krügel et al. (2022) explain that the absence of case reports of rodent-associated *Bartonella* infections may be the result of several factors, including misdiagnosis, lack of alertness, and lack of sophisticated detection equipment. It is therefore necessary to continue research aimed at this zoonosis in the Neotropics.

Lyme disease is caused by bacteria of the phylum Spirochaetes and the genus *Borrelia*. At least five species cause this disease in humans: *Borrelia afzelii*, *B. bavariensi*, *B. burgdorferi*, *B. garinii*, and *B. spielmanii* (Tilly et al. 2008; Stanek et al. 2012). The most frequent clinical manifestation of this disease is erythema migrans, which are skin lesions that appear at the site of the tick bite, while the most severe manifestations include affectations of the skin, nervous system, joints, and even the heart (Stanek et al. 2012; Shapiro 2014). Lyme disease is distributed in Europe, Asia, and the Americas (Schmid 1985). In North America, the only species responsible for infection is *B. burgdorferi* (Stanek et al. 2012; Shapiro 2014). The reservoirs of *B. burgdorferi* include mammals, birds, and reptiles, with rodents being the most frequently studied of these (Wolcott et al. 2021). Lyme disease is transmitted incidentally through the bites of ticks of the genus *Ixodes*; in North America, the main species are *Ixodes scapularis* and *I. pacificus* (Stanek et al. 2012; Shapiro 2014). The first description of Lyme disease in America occurred in 1977 in 51 residents of Connecticut, USA (Steere et al. 1977). In 1982, the etiologic agent was described by identifying the presence of the bacterium in *I. dammini* ticks (Burgdorfer et al. 1982). In the Americas, an estimated 476,000 cases are reported annually in the United States (Kugeler et al. 2021), while in the Neotropical region of Mexico, the number of cases has been low, and no association between human cases and reservoirs or vectors has been demonstrated (Colunga-Salas et al. 2020), although the presence of *B. burgdorferi* has been identified in three rodent species in Mexico: *Heteromys gaumeri*, *M. musculus*, and *R. rattus* (Solís-Hernández et al. 2016; Rodríguez-Rojas et al. 2020). In Latin America, the countries that have reported human cases are Costa Rica, Colombia, Brazil, Peru, Bolivia, Chile, and Argentina; however, these cases have been scarce, and there are no records of rodents or other mammals as reservoirs (Robles et al. 2018).

Murine typhus is an acute febrile disease caused by *Rickettsia typhi*, a gram-negative, obligate intracellular bacterium with worldwide distribution. It was first described in 1926 (Azad 1990; Quintal 1996; Tsioutis et al. 2017). Its reservoirs are mammals, mainly rats (*R. norvegicus* and *R. rattus*), and it is therefore associated with urban areas where these synanthropic rodents are abundant (Civen and Ngo 2008; Peniche-Lara et al. 2012). Fleas, mainly of the species *Xenopsylla cheopis*, function as vectors, and transmission to humans occurs when flea bites and feces are inoculated at the site of the bite (Azad 1990). In general, murine typhus is considered a mild disease with a relatively low mortality rate (1–4%). The characteristic symptoms are fever, musculoskeletal pain, and a skin rash (Bolaños et al. 2004). In the last two decades in the Neotropics, autochthonous human cases of murine typhus have been reported in Mexico, Honduras, Nicaragua, Colombia, and Brazil, and cases have been confirmed with serological evidence in countries such as Panama, Peru, Chile, and Argentina (Faccini-Martínez et al. 2021). Recently, molecular evidence of *R. typhi* infection has been obtained for the synanthropic rodents *M. musculus* and *R. rattus* in Yucatán, Mexico (Peniche-Lara et al. 2015; Torres-Castro et al. 2018).

Flea-borne spotted fever is a rickettsiosis caused by the intracellular bacterium *Rickettsia felis*, the hosts of which can be mammals, including humans, and transmission is mainly by the vector *Ctenocephalides felis*, the cat flea (Pérez-Osorio et al. 2008; Reif and Macaluso 2009; Brown and Macaluso 2016). Its distribution is worldwide, and molecular evidence of *R. felis* has been found in different arthropods, including fleas, mites, and ticks, on five different continents (Pérez-Osorio et al. 2008; Reif and Macaluso 2009). It was first described in 1990 and found in the cytoplasm of *C. felis* flea cells (Adams et al. 1990). In 1994, the first human case of *R. felis* infection was identified in the United States, demonstrating its zoonotic potential (Schriefer et al. 1994). The clinical symptoms include fever, skin rash, eschar, and, in some cases, neurological affectations (Zavala-Velazquez et al. 2006; Parola 2011). Human cases in the Neotropics have been rare, occurring only in Mexico (Zavala-Velázquez et al. 2000, 2006) and Brazil (Galvão et al. 2006). However, in a large number of countries in the region, the presence of *R. felis* has been confirmed in arthropod vectors, as well as in domestic and wild mammals (Labruna et al. 2011; Bermúdez and Troyo 2018). In the Neotropical region, 10 rodents have been confirmed as reservoirs of *R. felis*, present in five countries:

Mexico, Colombia, Peru, Brazil, and Argentina (Table 7.1). Evidence of this *Rickettsia* has been found in ectoparasites collected from rodents, including *C. felis* in the rodent *Peromyscus yucatanicus* and *Polygenis odiosus* in *Ototylomys phyllotis* in Mexico (Peniche-Lara et al. 2015) and in *P. axius* in *Oxymycterus rufus* in Argentina (Melis et al. 2020). The presence of this pathogen throughout the Neotropical region, together with the growing number of studies reporting infection in various ectoparasitic arthropods, highlights the potential public health risk posed by *R. felis*.

Tick-borne spotted fever is a rickettsiosis that in America is caused by the intracellular bacterium *Rickettsia parkeri*. Its reservoirs are mammals, including rodents, and it can be transmitted to humans by tick vectors of the genus *Amblyomma* (Paddock et al. 2004; Silva-Ramos et al. 2021; Scott et al. 2022). In 1937, *R. parkeri* was first isolated from the *Amblyomma maculatum* tick in Texas, the USA (Parker et al. 1939). However, it was not until 2004 that the zoonotic potential of the pathogen was identified, with an infection described in a patient in the United States (Paddock et al. 2004). It is considered a relatively nonaggressive disease with varied symptoms, including fever, skin rash, myalgia, headache, and regional adenopathy, among others (Paddock et al. 2004; Silva-Ramos et al. 2021). The cases reported in Latin America are from Mexico, Colombia, Brazil, Argentina, and Uruguay (Silva-Ramos et al. 2021; Torres-Castro et al. 2022). Rodents have only been identified as reservoirs in Brazil in at least nine species: *Akodon* sp., *Cerradomys maracajuensis*, *C. scotti*, *C. subflavus*, *Euryoryzomys russatus*, *H. hydrochaeris*, *Necromys* sp., *Nectomys squamipes*, and *R. rattus* (Pacheco et al. 2007; Milagres et al. 2013; Szabó et al. 2013; Binder et al. 2016; Luz et al. 2019).

Rocky Mountain spotted fever or Brazilian spotted fever is caused by the bacterium *Rickettsia rickettsii* (Labruna 2009). This species is a gram-negative, obligate intracellular bacterium that is transmitted to vertebrates and humans through the bite of infected ticks (Dantas-Torres 2007; Greca et al. 2008). The tick species *A. cajennense* and *A. aureolatum* are considered the main vectors in South America (Thorner et al. 1998; Labruna 2009). The first case of Rocky Mountain spotted fever was recorded in Idaho, USA, and published in 1896 (Thorner et al. 1998). In South America, the first case described was in Brazil in 1929 (Del Guercio et al. 1997). The recurrent symptoms of this disease are fever, headache, vomiting, diarrhea, and skin rashes (Estripeaut et al. 2023). Among the rickettsioses associated with human infections, those caused by *R. rickettsii* are predominant and account for 42.2% of the reports (Zhang et al. 2022). In Latin America, the disease is endemic in Mexico, Panama, Costa Rica, Colombia, Argentina, and Brazil (Dantas-Torres 2007; Dzul-Rosado et al. 2019). This rickettsiosis is considered to have the highest case fatality rate in Brazil, where 2127 cases were reported for Brazilian spotted fever between 2000 and 2019, with São Paulo being the state with the highest number of cases and a case fatality rate of 48% (Galvão et al. 2003; Faccini-Martínez et al. 2021). Regarding the vectors, Ribeiro et al. (2021) showed that the countries in Latin America with the highest prevalence of *R. rickettsii* ticks were Colombia (17.0%), Mexico (9.9%), Panama (2.8%), and Brazil (2.0%). Moreover, the species mainly infected were *Rhipicephalus sanguineus* and *Amblyomma mixtum*. Brazil is the only

Neotropical country where evidence of rodent reservoirs of *R. rickettsii* has been reported (Pacheco et al. 2007; Milagres et al. 2013; Szabó et al. 2013; Krawczak et al. 2014; Binder et al. 2016; Luz et al. 2019; de Oliveira et al. 2020).

Plague is a disease caused by the gram-negative bacillus bacterium *Yersinia pestis*, the main reservoir of which is rodents, and is transmitted indirectly by fleas (Schneider et al. 2014; Barbieri et al. 2020). In America, the main vector is the flea *Xenopsylla cheopis*, although other species have been identified (Ruiz 2001). The infection has five clinical forms: bubonic, septicemic, pneumonic, meningeal, and pharyngeal plague, with the first three being the most frequent (Perry and Fetherston 1997; Bezerra and de Almeida 2022). Historically, the first evidence of the role of rodents in the transmission of this zoonosis occurred during the first plague pandemic. The literature describes three plague pandemics in history: the Justinian plague (541–750 AD), Black Death (1346–1353), and the nineteenth-century plague (1854; Gibert 2019). The most devastating period occurred during the Black Death, which killed one-third of the European human population (Barbieri et al. 2020; Bezerra and de Almeida 2022). Recently, the Neotropical countries that have recorded clustered cases of plague are Peru, Ecuador, Brazil, and Bolivia, with Peru having the highest percentage of cases at 68.80% (Schneider et al. 2014; Barbieri et al. 2020). In the Neotropics, 23 rodent reservoirs have been identified for *Y. pestis* in Argentina, Brazil, and Peru (de la Barrera 1953; Almeida et al. 1987, 1989; Martin-Alonso et al. 2014). In turn, Bezerra and de Almeida (2022) found that at least 50 species of wild rodents can be identified as hosts. It is important to consider that this disease continues to be emerging and seems to be endemic in South American countries. Given the high number of rodent reservoir species and the associated high mortality, surveillance, timely diagnosis, and pest control in synanthropic environments are critically important.

7.5 Vectorial Transmission: Rodents as a Source of Zoonotic Pathogens and Their Transmission Through Free-Living Vectors

Three zoonotic diseases transmitted by free-living arthropod vectors have been identified in the Neotropics (Table 7.1): Venezuelan equine encephalitis, caused by a virus; Chagas disease, produced by the protozoan *Trypanosoma cruzi*; and leishmaniasis, which is due to infection by protozoan organisms of the genus *Leishmania*. Together, these diseases have been detected in 95 species of Neotropical rodents.

Venezuelan equine encephalitis is caused by the Venezuelan equine encephalitis virus, an arbovirus of the genus *Alphavirus* of the family Togaviridae (Weaver et al. 2004; Lundberg et al. 2017). The main reservoirs are rodents, and transmission is through mosquitoes, mainly of the genus *Culex* (Weaver and Barrett 2004), as the most important vector. Female mosquitoes ingest the virus in the blood of the infected host and, when biting another animal, transfer the virus through their saliva to the new host (Weaver and Barrett 2004; Guzmán-Terán et al. 2020). Equines (horses, donkeys, and mules) function as amplification hosts, and humans living in close association with these animals can then become infected accidentally (Greene et al. 2005). The etiologic agent was first recognized in 1938 in Venezuela, where the virus was isolated from the brain of a horse that died of encephalitis (Weaver et al. 2004). In 1954, the virus was first isolated in human cases during an outbreak in Colombia, and associated neurological complications and mortality were described (Sanmartin-Barberi et al. 1954). General symptoms are fever, chills, myalgia in the legs and lower back, and general malaise, and there may be neurological signs such as convulsions, drowsiness, confusion, and photophobia. In lethal cases, there is diffuse congestion and edema in the brain, pulmonary hemorrhage, and sometimes meningoencephalitis (Guzmán-Terán et al. 2020). In human infections, neurological symptoms occur in 14% of cases, while 1% of cases result in lethal encephalitis (Guzmán-Terán et al. 2020). In the Neotropical region, human and equine cases have been described in at least 12 countries: Mexico, Belize, Guatemala, El Salvador, Costa Rica, Panama, Colombia, Venezuela, Ecuador, Peru, Brazil, and Argentina (Aguilar et al. 2011; Guzmán-Terán et al. 2020). One of the largest outbreaks occurred in 1962, beginning in Colombia and expanding to Venezuela, where 23,283 cases were reported, including 960 neurological cases and 156 deaths (Aguilar et al. 2011). At least six species of Neotropical rodents have been reported as reservoirs of Venezuelan equine encephalitis virus: in Mexico, *Handleyomys alfaroi*, *Oryzomys couesi*, and *Sigmodon hispidus* (Estrada-Franco et al. 2004; Deardorff et al. 2011); in Panama, *Proechimys semispinosus* and *S. hispidus* (Grayson and Galindo 1969); in Colombia, *Proechimys chrysaeolus* (Barrera et al. 2002); and in Venezuela, *Proechimys guairae* and *S. hispidus* (Navarro et al. 2005). Despite the evidence, studies are still required to determine the role of these hosts in zoonotic transmission.

Chagas disease or American trypanosomiasis is an infection caused by the protozoan agent *Trypanosoma cruzi*, a hemoflagellate intracellular parasite belonging to the family Trypanosomatidae (Echeverria and Morillo 2019; Lidani et al. 2019). This parasitosis is endemic to the Americas and is distributed from the southern United States to Argentina (Galaviz-Silva et al. 2017). Transmission to humans is mainly through a triatomine insect vector (Molina et al. 2016). Arthropod vectors belong to the subfamily Triatominae, and at least 157 species, represented by 18 genera, have been reported (Alevi et al. 2021). However, the main vectors are bedbugs of the genera *Triatoma*, *Panstrongylus*, and *Rhodnius* (Pérez-Molina and Molina 2018). It has been reported that *T. cruzi* can be transmitted to more than 180 species of mammals belonging to seven orders and 25 families (WHO 2002; Ibarra-Cerdeña et al., 2017). Transmission occurs when the vector defecates or urinates after sucking blood and the parasite comes into contact through a wound in the skin (the bite) or the eyes or mouth (Guhl and Ramírez 2021). Other routes of transmission are from mothers to infants, transfusions of blood products and tissue transplants, ingestion of contaminated food, or accidents in laboratories or hospitals (Molina et al. 2016; Guhl and Ramírez 2021). In humans, *T. cruzi* infection dates back to more than 7050 years BC. B.C. since it has been identified in naturally mummified human remains in northern Chile and southern Peru (Aufderheide et al. 2004). In 1909, Carlos Chagas first described *T. cruzi* protozoan disease in Lassance, Brazil (Kropf and Sá 2009). Human *T. cruzi* infection can be characterized by two successive phases: the acute phase, characterized by high parasitemia, is usually asymptomatic or may present symptoms such as a self-limited febrile illness (Rassi and Marin-Neto 2010; Lidani et al. 2019). Manifestations disappear spontaneously in 90% of cases, and 60–70% of infected individuals are asymptomatic (Lidani et al. 2019). In the case of vectorial transmission, symptoms may be observed through the skin (chagoma) or in the ocular mucous membranes (Rassi and Marin-Neto 2010). The symptomatic phase develops in 30–40% of patients and may include neurological, cardiac, digestive, or cardiodigestive clinical symptoms (Dias 1995; Vago et al. 2000). Currently, an estimated six to seven million people are infected with *T. cruzi,* and more than 65 million people live in areas of exposure at risk of infection (Lidani et al. 2019). Chagas disease is a major public health problem in Latin America, with an estimated annual incidence of 30,000 new cases and 14,000 deaths (González-Zambrano et al. 2021). Moreover, the disease causes damage to the economic sector; in Latin America, an estimated 752,000 workdays are lost per year due to premature deaths, along with USD 1.2 billion in lost productivity (Conteh et al. 2010). In Colombia in 2017, an annual loss of USD 13 million was estimated (Olivera and Buitrago 2020). Affections in the symptomatic phase cause high morbidity and mortality, with frequent hospitalizations and surgical procedures, incurring a significant economic cost (González-Zambrano et al. 2021). In the Neotropics, 78 rodent species have been identified as confirmed reservoirs of *T. cruzi*, and their distribution covers 12 countries in the region: Argentina, Bolivia, Brazil, Chile, Colombia, Costa Rica, Ecuador, El Salvador, Mexico, Panama, Peru, and Venezuela (Table 7.1 and Fig. 7.2). The diversity of rodent reservoirs of *T. cruzi* includes wild species found in conserved environments and species with affinity to agroecosystems and synanthropic areas. Synanthropic rodents are important hosts in domestic and peridomestic environments of the Chagas disease endemic area since they transport the parasite to human dwellings, where it can then be transmitted to humans and domestic animals via triatomines (Pinto et al. 2006; Ramsey et al. 2012; Lopez-Cancino et al. 2015; Yefi-Quinteros et al. 2018). Control of these reservoir populations in synanthropic settings is important to reduce the risk of vectorial transmission. Moreover, studies are required to assess the role of rodents in maintaining infections and the frequency of transmission to humans (Battersby 2015).

Leishmaniases are a group of diseases caused by the protozoan genus *Leishmania*, an obligate intracellular parasite of the family Trypanosomatidae (Roatt et al. 2020; Mann et al. 2021). Its distribution includes Europe, Asia, Africa, and the Americas, where at least 23 *Leishmania* species have the potential for transmission to humans and 18 of these are distributed in the Neotropical region of the Americas (Sasidharan and Saudagar 2021). In the Americas, at least 87 mammalian species can host *Leishmania* (Roque and Jansen 2014), and transmission takes place through phlebotomine dipterans, female sandflies of the genus *Lutzomyia* that inoculate parasites into the host through their bite (Burza et al. 2018; Sasidharan and Saudagar 2021; Serafim et al. 2021). The parasite was described in 1903 by Leishman and Donovan separately but in the same year. The species now known as *Leishmania donovani* was found in the spleen of patients in India diagnosed with visceral leishmaniasis (Herwaldt 1999). There are three clinical manifestations of the disease: visceral leishmaniasis, cutaneous leishmaniasis, and mucocutaneous leishmaniasis (Mann et al. 2021; Sasidharan and Saudagar 2021). Globally, leishmaniasis is endemic in 102 countries, and it is estimated that there are approximately 1.3 million cases annually, causing between 20,000 and 30,000 deaths (PAHO 2019). In Latin America, approximately 60,000 new cases are estimated each year (Torres-Guerrero et al. 2017). In the Neotropical region, at least 42 rodent species are confirmed hosts of *Leishmania*, and the presence of infected rodents has been reported in 14 countries: Argentina, Belize, Bolivia, Brazil, Colombia, Costa Rica, Ecuador, French Guyana, Honduras, Mexico, Panama, Peru, Trinidad and Tobago, and Venezuela (Table 7.1). The countries with the highest incidence of leishmaniasis are Brazil, Colombia, and Venezuela (Alvar et al. 2012), and the countries with the highest number of host species are Brazil (n = 25) and Mexico (n = 7). The presence of rodents susceptible to this infection in rural and urban environments, as well as in agroecosystems, reflects the potential risk of infection for both humans and domestic fauna since they persist in anthropogenic environments (De Lima et al. 2002; Marcelino et al. 2011; Brandão-Filho et al. 2003; Lima et al. 2013).

7.6 Rodents and Their Role as Propagators of Parasites in Fragmented Landscapes and Agricultural Areas

In the tropical regions of the planet, where mammal biodiversity is highest and land use change occurs, the risk of the emergence of zoonotic infectious disease is high (Allen et al. 2017). The main types of land-use change that are associated with infectious zoonotic disease transmission are deforestation and habitat fragmentation, agricultural development, and urbanization (Gottdenker et al. 2014; White and Razgour 2020), as has been demonstrated in the Amazon region of the American Neotropics (Ellwanger et al. 2020). Agricultural expansion influences the emergence of zoonoses through a consequent increased rate of contact with wildlife, such that humans and domestic animals that become sympatric with wild fauna come closer together, increasing the probability of contact with wild host species, vectors of zoonotic diseases, and the sylvatic cycles of potential zoonotic pathogens (Jones et al. 2013). On the one hand, land use change tends to cause a loss in biodiversity, with rare and range-limited species particularly experiencing reduced abundance, while generalist species tend to increase in abundance (Newbold et al. 2018). Indirectly, habitat fragmentation that leads to the decline in certain species of the animal community, particularly those that function as predators or competitors of rodents, triggers a predator- and competitor-release situation that promotes rodent abundance, including abundance of zoonotic species (Young et al. 2017; Glidden et al. 2021).

Land use change, particularly for agricultural development, can also influence rodent population dynamics. Agricultural interventions lead to an increase in certain food resources that, when found in greater quantities, promote an increase in the abundance of rodents that can exploit those resources (Gibb et al. 2020). In cases where this increase is substantial, exponential increases in rodent populations can occur, turning them into pests (Krebs 1999). Thus, high rodent density or abundance values may be related to a higher prevalence of parasites and pathogens (Begon et al. 2002), such that alterations to ecosystems can create favorable habitats that promote an increase in the host species of zoonotic pathogens (McMahon et al. 2018).

Rodents are one of the main groups in which the effect of land use change on their populations is observed (Gibb et al. 2020; Mendoza et al. 2020). However, studies focused on describing the relationship between the impacts of agricultural intensification and habitat fragmentation and the emergence of zoonoses is still scarce (White and Razgour 2020). It has been reported that a high percentage (73%) of rodent species that are reservoirs of zoonotic pathogens are found in anthropogenic habitats compared to natural habitats (Ecke et al. 2022). Agricultural expansion and intensification cause the loss of large consumer species, increasing rodent abundance and the displacement of animals to human-modified environments, thus increasing human exposure to zoonotic pathogens (Young et al. 2017; Glidden et al. 2021). The influence of human activity on pathogen infection has been observed in different studies. For example, McCauley et al. (2015) found that the prevalence of *Y. pestis* doubled in rodents captured in agricultural sites compared to conserved habitats; Morand et al. (2019) described a link between human-intervened habitats and the prevalence of pathogens harbored by rodents, including important pathogens that cause leptospirosis, bartonellosis, and Hantavirus infections; Shilereyo et al. (2022) found a higher average of ectoparasite load in small mammals on agricultural and pastoral land compared to those found in conserved environments.

Thus, the evidence indicates that human land use change activities can cause an increase in the abundance of generalist rodents that can amplify the transmission of zoonotic diseases to humans, their domestic fauna, and livestock as a consequence of the increased contact rate between infected and susceptible individuals (Mendoza et al. 2020).

7.7 Concluding Remarks

Zoonoses are undoubtedly among the most important causes of human morbidity and mortality. Mammals are the reservoirs of some of the most prominent emerging and re-emerging human diseases. Within mammals, rodents represent a group of concern because they are hosts for a remarkable diversity of pathogens, parasites, and ectoparasites. They can also tolerate the human modification of their habitats, and some species survive in extremely anthropogenic conditions (i.e., cities). These two conditions make rodents a host group with a high likelihood of being the source of human zoonotic diseases going forward. This chapter constitutes the most updated and comprehensive source of information on the role of Neotropical rodents in zoonotic transmission in this critical region of the globe. Given the increasingly omnipresent anthropogenic impact on natural ecosystems, particularly deforestation, fragmentation, agricultural conversion, and, indirectly, differential defaunation (all of which favors the abundance of rodents and the group's zoonotic species), extended, interdisciplinary research on the ecology and health implications of this mammalian lineage represents a critical agenda to address the challenges of the Anthropocene.

Several rodent species are reported as zoonotic hosts that can tolerate anthropogenic landscapes, including agricultural and fragmented landscapes. Of these, three species are highly anthropophilic (*Mus musculus*, *Rattus rattus*, and *R. norvegicus*). They can be reservoirs for at least 31 zoonotic pathogens. However, several other wild-tolerant species have an essential role in transmitting important diseases, such as Chagas or Leishmaniasis, among others. Because there is a rampant trend in tropical landscape degradation, rodents' role as reservoirs of zoonoses is expected to become more critical. A predicted consequence of landscape degradation is the loss of functional biodiversity, for instance, the loss of medium-size predators, which are a group that control rodent populations. Predator release in degraded landscapes (i.e., agriculture) has an impact not only on zoonoses but also on crop damage because rodents also act as agricultural pests. Although rodents are not the only piece in the pathogen circulation system, they are an essential source of pathogen amplification and a relatively easy group for transmission monitoring. Establishing a surveillance system that quantifies the population trends of rodents and ectoparasites with a systematic program for pathogen detection for some sentinel pathogens or parasites will provide timely indicators of the risk of zoonoses.

Acknowledgments We are grateful for the funding provided by Deutsche Gesellschaft für Internationale Zusammenarbeit (GIZ) GmbH. through grant no. 81278705 awarded to CNIC. CRRL thanks Conacyt for the postdoctoral grant. EEPA and MPIL thank Conacyt for the grant for graduate studies in the Human Ecology program of Cinvestav.

References

Adams JR, Schmidtmann ET, Azad AF (1990) Infection of colonized cat fleas, *Ctenocephalides felis* (Bouché), with a rickettsia-like microorganism. Am J Trop Med Hygiene 43(4):400–409

Aguilar PV, Estrada-Franco JG, Navarro-Lopez R, Ferro C, Haddow AD, Weaver SC (2011) Endemic Venezuelan equine encephalitis in the Americas: hidden under the dengue umbrella. Futur Virol 6(6):721–740

Albery GF, Becker DJ (2021) Fast-lived hosts and zoonotic risk. Trends Parasitol 37(2):117–129

Albery GF, Eskew EA, Ross N, Olival KJ (2020) Predicting the global mammalian viral sharing network using phylogeography. Nat Commun 11(1):2260

Alevi KCC, de Oliveira J, da Silva RD, Galvão C (2021) Trends in taxonomy of Chagas disease vectors (Hemiptera, Reduviidae, Triatominae): from Linnaean to integrative taxonomy. Pathogens 10(12):1627

Ali AJ, Jones KR (2020) Nutritive value and physical properties of Neo-tropical rodent meat-with emphasis on the Capybara (*Hydrochoerus hydrochaeris*). Animals 10(11):2134

Allen T, Murray KA, Zambrana-Torrelio C, Morse SS, Rondinini C, Di Marco M, Breit N, Olival KJ, Daszak P (2017) Global hotspots and correlates of emerging zoonotic diseases. Nat Commun 8(1):1124

Almeida AMPD, Brasil DP, Carvalho FGD, Almeida CRD (1987) Pesquisa de *Yersinia pestis* em roedores e outros pequenos mamiferos nos focos pestosos do nordeste do Brasil no periodo 1966 a 1982. Rev Saude Publica 21:265–267

Almeida AMP, Brasil DP, Leal NC, Melo MEBD, Rêgo RVBD, Almeida CRD (1989) Estudos bacteriológicos e sorológicos de um surto de peste no Estado da Paraíba, Brasil. Mem Inst Oswaldo Cruz 84:249–256

Almeida A, Corrigan R, Sarno R (2013) The economic impact of commensal rodents on small businesses in Manhattan's Chinatown: trends and possible causes. Suburban Sustain 1(1):2

Alvar J, Vélez ID, Bern C, Herrero M, Desjeux P, Cano J, Jannin J, den Boer M, WHO Leishmaniasis Control Team (2012) Leishmaniasis worldwide and global estimates of its incidence. PLoS One 7(5):e35671

Angelakis E, Raoult D (2010) Q fever. Vet Microbiol 140(3–4):297–309

Armstrong C, Lillie RD (1934) Experimental lymphocytic choriomeningitis of monkeys and mice produced by a virus encountered in studies of the 1933 St. Louis encephalitis epidemic. Publ Health Rep (1896–1970) 49(35):1019–1027

Aufderheide AC, Salo W, Madden M, Streitz J, Buikstra J, Guhl F, Arriaza B, Renier C, Wittmers LE Jr, Fornaciari G, Allison M (2004) A 9,000-year record of Chagas' disease. Proc Natl Acad Sci 101(7):2034–2039

Avšic-Zupanc T, Saksida A, Korva M (2016) Hantavirus infections. Clin Microbiol Infect 30:1–14

Azad AF (1990) Epidemiology of murine typhus. Annu Rev Entomol 35(1):553–570

Barbieri R, Signoli M, Chevé D, Costedoat C, Tzortzis S, Aboudharam G, Raoult D, Drancourt M (2020) *Yersinia pestis*: the natural history of plague. Clin Microbiol Rev 34(1):e00044–e00019

Barrera R, Ferro C, Navarro JC, Freier J, Liria J, Salas R, Ahumada M, Clovis V, González M, Kang W, Boshell J, Weaver SC (2002) Contrasting sylvatic foci of Venezuelan equine encephalitis virus in northern South America. Am J Trop Med Hygiene 67(3):324–334

Bass JW, Vincent JM, Person DA (1997) The expanding spectrum of Bartonella infections: I. Bartonellosis and trench fever. Pediatr Infect Dis J 16(1):2–10

Bastiaan GM, Aize K (2007) Role of rodents in transmission of Salmonella and Campylobacter. J Food Sci Agric 87:2774–2781

Battersby AS (2015) Rodents as carriers of disease. In: Buckle AP, Smith RH (eds) Rodent pests and their control, 2nd edn. CABI, London, pp 81–100

Bedi JS, Vijay D, Dhaka P (2022) Textbook of zoonoses, 1st edn. Wiley Blackwell

Begon M, Bennett M, Bowers RG, French NP, Hazel SM, Turner J (2002) A clarification of transmission terms in host-microparasite models: numbers, densities and areas. Epidemiol Infect 129:147–153

Bermúdez CSE, Troyo A (2018) A review of the genus Rickettsia in Central America. Res Rep Trop Med 9:103–112

Bezerra MF, de Almeida AMP (2022) Important infectious diseases in Latin America and the Caribbean: Plague. In: Infectious tropical diseases and One Health in Latin America. Springer International Publishing, Cham, pp 45–70

Bigna JJ, Tochie JN, Tounouga DN, Bekolo AO, Ymele NS, Youda EL, Sime PS, Nansseu JR (2020) Global, regional, and country seroprevalence of Toxoplasma gondii in pregnant women: a systematic review. Modeling and Meta-Analysis. Sci Rep 10:12102

Billeter SA, Levy MG, Chomel BB, Breitschwerdt EB (2008) Vector transmission of Bartonella species with emphasis on the potential for tick transmission. Med Vet Entomol 22(1):1–15

Binder LDC, Krawczak FDS, Sponchiado J, Melo GL, Moraes-Filho J, Bastos FAN, Cáceres NC, Labruna MB (2016) Serosurvey of *Rickettsia* spp. in small mammals from Mato Grosso do Sul state, Brazil. Ciência Rural 47(1):1

Bogitsh BJ, Carter CE, and Oeltmann TN (2012) Human parasitology. Academic Press, New York

Bolaños M, Angel-Moreno A, Pérez-Arellano JL (2004) Tifus endémico (murino). Una enfermedad en la que pensar aquí y ahora. Med Clin 122(10):383–389

Brandão-Filho SP, Brito ME, Carvalho FG, Ishikaw EA, Cupolillo E, Floeter-Winter L, Shaw JJ (2003) Wild and synanthropic hosts of Leishmania (Viannia) braziliensis in the endemic cutaneous leishmaniasis locality of Amaraji, Pernambuco State, Brazil. Trans R Soc Trop Med Hyg 97(3):291–296

Brenière SF, Waleckx E, Barnabé C (2016) Over six thousand *Trypanosoma cruzi* strains classified into discrete typing units (DTUs): attempt at an inventory. PLoS Negl Trop Dis 10(8):e0004792

Brown LD, Macaluso KR (2016) *Rickettsia felis*, an emerging flea-borne rickettsiosis. Curr Trop Med Rep 3:27–39

Burgdorfer W, Barbour AG, Hayes SF, Benach JL, Grunwaldt E, Davis JP (1982) Lyme disease—a tick-borne spirochetosis? Science 216(4552):1317–1319

Burgin CJ, Colella JP, Kahn PL, Upham NS (2018) How many species of mammals are there? J Mammal 99(1):1–14

Burza S, Croft SL, Boelaert M (2018) Leishmaniasis. Lancet 392:951–970

Capizzi D, Bertolino S, Mortelliti A (2014) Rating the rat: global patterns and research priorities in impacts and management of rodent pests. Mammal Rev 44(2):148–162

Chomel BB, Kasten RW, Williams C, Wey AC, Henn JB, Maggi R, Carrasco S, Mazet J, Boulouis HJ, Maillard R, Breitschwerdt EB (2009) Bartonella endocarditis: a pathology shared by animal reservoirs and patients. Ann N Y Acad Sci 1166(1):120–126

Christen JR, Edouard S, Lamour T, Martinez E, Rousseau C, de Laval F, Catzeflis F, Djossou F, Raoult D, de Santi VP, Epelboin L (2020) Capybara and brush cutter involvement in Q fever outbreak in remote area of Amazon Rainforest, French Guiana, 2014. Emerg Infect Dis 26(5):993

Civen R, Ngo V (2008) Murine typhus: an unrecognized suburban vectorborne disease. Clin Infect Dis 46(6):913–918

Cleaveland S, Laurenson MK, Taylor LH (2001) Diseases of humans and their domestic mammals: pathogen characteristics, host range and the risk of emergence. Philos Trans R Soc Lond B Biol Sci 356(1411):991–999

Conover MR, Vail RM (2015) Human diseases from wildlife. CRC Press, Boca Raton, Florida

Colunga-Salas P, Sánchez-Montes S, Volkow P, Ruiz-Remigio A, Becker I (2020) Lyme disease and relapsing fever in Mexico:an overview of human and wildlife infections. PLoS One 15(9):e0238496.

Conteh L, Engels T, Molyneux DH (2010) Socioeconomic aspects of neglected tropical diseases. Lancet 375(9710):239–247

Córdoba-Aguilar A, Ibarra-Cerdeña CN, Castro-Arellano I, Suzan G (2021) Tackling zoonoses in a crowded world: lessons to be learned from the COVID-19 pandemic. Acta Trop 214:105780

Cosson JF, Picardeau M, Mielcarek M, Tatard C, Chaval Y, Suputtamongkol Y, Buchy P, Jittapalapong S, Herbreteau V, Morand S (2014) Epidemiology of Leptospira transmitted by rodents in Southeast Asia. PLoS Negl Trop Dis 8(6):e2902

Costa F, Hagan JE, Calcagno J, Kane M, Torgerson P, Martinez-Silveira MS, Stein C, Abela-Ridder B, Ko AI (2015) Global morbidity and mortality of Leptospirosis: a systematic review. PLoS Negl Trop Dis 9(9):e0003898

D'Souza MH, Patel TR (2020) Biodefense implications of new-world hantaviruses. Front Bioeng Biotechnol 8:925

Dahmana H, Granjon L, Diagne C, Davoust B, Fenollar F, Mediannikov O (2020) Rodents as hosts of pathogens and related zoonotic disease risk. Pathogens 9(3):202

Dantas-Torres F (2007) Rocky Mountain spotted fever. Lancet Infect Dis 7(11):724–732

Davies RH, Wray C (1995) Observations on disinfection regimens used on Salmonella enteritidis infected poultry units. Poultry Sci 74:638–647

de la Barrera J (1953) Rongeurs sauvages infectés par Pasteurella pestis en Argentine. Bull World Health Organ 9(5):701

de Lima H, De Guglielmo Z, Rodríguez A, Convit J, Rodriguez N (2002) Cotton rats (*Sigmodon hispidus*) and black rats (*Rattus rattus*) as possible reservoirs of *Leishmania* spp. in Lara State, Venezuela. Mem Inst Oswaldo Cruz 97:169–174

de Lima BG, de Almeida Vitor RW, dos Santos M-DE (2021) Toxoplasma gondii in South America: a differentiated pattern of spread, population structure and clinical manifestations. Parasitol Res 120(9):3065–3076

de Oliveira GMB, da Silva IWG, da Cruz Ferreira Evaristo AM, de Azevedo Serpa MC, Silva Campos AN, Dutra V, Nakazato L, de Aguiar DM, Bahia Labruna MB, Horta MC (2020) Tick-borne pathogens in dogs, wild small mammals and their ectoparasites in the semiarid Caatinga biome, northeastern Brazil. Ticks Tick-Borne Dis 11(4):101409

Deardorff ER, Estrada-Franco JG, Freier JE, Navarro-Lopez R, Da Rosa AT, Tesh RB, Weaver SC (2011) Candidate vectors and rodent hosts of Venezuelan equine encephalitis virus, Chiapas, 2006–2007. Am J Trop Med Hygiene 85(6):1146

Del Guercio VMF, Rocha MMM, Melles HH, de Lima VC, Pignatti MG (1997) Febre maculosa no município de Pedreira, SP, Brasil. Inquérito sorológico. Rev Soc Bras Med Trop 30:47–52

DGE (Dirección General de Epidemiología) (2021) Anuario de morbilidad, 1984–2021. Disponible en http://www.epidemiologia.salud.gob.mx/anuario/html/anuarios.html [Fecha de consulta: marzo 2023]

Dhaliwal BS, Juyal PD (2013) Parasitic zoonoses. Springer, New Delhi

Dias JP (1995) Natural history of Chagas disease. Arq Bras Cardiol 65(4):359–366

Drancourt M, Tran-Hung L, Courtin J, Lumley HD, Raoult D (2005) *Bartonella quintana* in a 4000-year-old human tooth. J Infect Dis 191(4):607–611

Dzul-Rosado KR, Lugo-Caballero C, Salcedo-Parra A, López-Soto RD, Faccini-Martínez ÁA (2019) Long term neurologic sequelae in a Mexican rocky mountain spotted fever case. Braz J Infect Dis 23:121–123

Echeverria LE, Morillo CA (2019) American trypanosomiasis (Chagas disease). Infect Dis Clin 33(1):119–134

Ecke F, Han BA, Hörnfeldt B, Khalil H, Magnusson M, Singh NJ, Ostfeld RS (2022) Population fluctuations and synanthropy explain transmission risk in rodent-borne zoonoses. Nat Commun 13(1):7532

Ellis WA (2015) Animal Leptospirosis. In: Adler B (ed) Leptospira and Leptospirosis. Current topics in microbiology and immunology, vol 387. Springer, Berlin/Heidelberg

Ellwanger JH, Kulmann-Leal B, Kaminski VL, Valverde-Villegas JM, Veiga ABGD, Spilki FR, Fearnside PM, Caesar L, Giatti LL, Wallau GL, Almeida SEM, Borba MR, Da Hora VP, Chies JAB (2020) Beyond diversity loss and climate change: impacts of Amazon deforestation on infectious diseases and public health. Anais da Academia Brasileira de Ciencias 92(1):e20191375

Enria DA, Briggiler AM, Sánchez Z (2008) Treatment of Argentine hemorrhagic fever. Antivir Res 78(1):132–139

Epelboin L, Eldin C, Thill P, de Santi VP, Abboud P, Walter G, Melzani A, Lettre-Gibert P, Pérez L, Demar M, Bourou M, Fernández J, Cermeño JR, Panizo MM, Vreden SGS, Djossou F, Beillard E, de Waard JH, de Lemos ERS (2021) Human Q fever on the Guiana shield and Brazil: recent findings and remaining questions. Curr Trop Med Rep 8:173–182

Estavillo C, Weyland F, Herrera L (2022) Zoonotic disease risk and life-history traits: are reservoirs fast life species? EcoHealth 19(3):390–401

Estrada-Franco JG, Navarro-Lopez R, Freier JE, Cordova D, Clements T, Moncayo A, Kang W, Gomez-Hernandez C, Rodriguez-Dominguez G, Ludwing GV, Weaver SC (2004) Venezuelan equine encephalitis virus, southern Mexico. Emerg Infect Dis 10(12):2113

Estripeaut D, Morales MS, Sánchez AR, González A, Hernández M, Zaldívar Y, Poveda A, Martínez-Monter M, Guenther E, Gundacker N, Suarez JA (2023) Clinical manifestations of *Rickettsia rickettsii* in a familial outbreak in Panama. Travel Med Infect Dis 52:102542

Faccini-Martínez ÁA, Krawczak FDS, Oliveira SVD, Labruna MB, Angerami RN (2021) Rickettsioses in Brazil: distinct diseases and new paradigms for epidemiological surveillance. Rev Soc Bras Med Trop 54:e07322020

Favacho AR, Roger I, Akemi AK, Pessoa AA Jr, Varon AG, Gomes R, Godoy DT, Pereira S, Lemos ER (2014) Molecular identification of Bartonella henselae in a seronegative cat scratch disease patient with AIDS in Rio de Janeiro, Brazil. Rev Inst Med Trop Sao Paulo 56:363–365
Figueiredo LTM, de Souza WM, Ferrés M, Enria DA (2014) Hantaviruses and cardiopulmonary syndrome in South America. Virus Res 187:43–54
Firth C, Tokarz R, Simith DB, Nunes MR, Bhat M, Rosa ES, Medeiros DBA, Palacios G, Vasconcelos PFC, Lipkin WI (2012) Diversity and distribution of hantaviruses in South America. J Virol 86(24):13756–13766
Frank MG, Beitscher A, Webb CM, Raabe V, Beitscher A, Bhadelia N et al (2021) South American Hemorrhagic fevers: a summary for clinicians. Int J Infect Dis 105:505–515
Galaviz-Silva L, Mercado-Hernández R, Zárate-Ramos JJ, Molina-Garza ZJ (2017) Prevalence of *Trypanosoma cruzi* infection in dogs and small mammals in Nuevo León, Mexico. Revista Argentina de Microbiologia 49(3):216–223
Galvão MAM, Calic SB, Chamone CB, Mafra SCL, Cesarino Filho G, Olano JP, Walker DH (2003) Spotted fever rickettsiosis in Coronel Fabriciano, minas gerais state. Rev Soc Bras Med Trop 36:479–481
Galvão MAM, Zavala-Velázquez JE, Zavala-Castro JE, Mafra CL, Calic SB, Walker DH (2006) *Rickettsia felis* in the Americas. Ann N Y Acad Sci 1078(1):156–158
Gardon J, Héraud JM, Laventure S, Ladam A, Capot P, Fouquet E, Favre J, Weber S, Hommel D, Hulin A, Couratte Y, Talarmin A (2001) Suburban transmission of Q fever in French Guiana: evidence of a wild reservoir. J Infect Dis 184(3):278–284
Garg SR (2014) Rabies in man and animals. Springer, New Delhi
Gibb R, Redding DW, Chin KQ, Donnelly CA, Blackburn TM, Newbold T, Jones KE (2020) Zoonotic host diversity increases in human-dominated ecosystems. Nature 584(7821):398–402
Gibert CR (2019) La peste a lo largo de la historia. Revista Enfermedades Emergentes 18(3):119–127
Glidden CK, Nova N, Kain MP, Lagerstrom KM, Skinner EB, Mandle L, Sokolow SH, Plowright RK, Dirzo R, De Leo GA, Mordecai EA (2021) Human-mediated impacts on biodiversity and the consequences for zoonotic disease spillover. Curr Biol 31(19):R1342–R1361
González-Zambrano H, Amaya-Tapia G, Franco-Ramos MC, León-Murguía OJL (2021) Prevalence of Chagas heart disease in dilated cardiomyopathy. Arch Cardiol Mex 91(1):50–57
Gottdenker NL, Streicker DG, Faust CL, Carroll CR (2014) Anthropogenic land use change and infectious diseases: a review of the evidence. EcoHealth 11:619–632
Graeff-Teixeira C, Ávila-Pires FDD, Machado RDCC, Camillo-Coura L, Lenzi HL (1990) Identificação de roedores silvestres como hospedeiros do *Angiostrongylus costaricensis* no sul do Brasil. Rev Inst Med Trop Sao Paulo 32:147–150
Grayson MA, Galindo P (1969) Ecology of Venezuelan equine encephalitis virus in Panama. J Am Vet Med Assoc 155(12):2141–2145
Greca H, Langoni H, Souza LCD (2008) Brazilian spotted fever: a reemergent zoonosis. J Venom Anim Toxins Includ Trop Dis 14:3–18
Greene IP, Paessler S, Austgen L, Anishchenko M, Brault AC, Bowen RA, Weaver SC (2005) Envelope glycoprotein mutations mediate equine amplification and virulence of epizootic Venezuelan equine encephalitis virus. J Virol 79(14):9128–9133
Guerrant RL, Walker DH, Weller PF (2006) Tropical infectious diseases: principles, pathogens and practice, 2nd edn. Elsevier, Philadelphia
Guhl F, Ramírez JD (2021) Poverty, migration, and chagas disease. Curr Trop Med Rep 8:52–58
Gutiérrez R, Krasnov B, Morick D, Gottlieb Y, Khokhlova IS, Harrus S (2015) Bartonella infection in rodents and their flea ectoparasites: an overview. Vector-borne Zoonotic Dis 15(1):27–39
Guzmán-Terán C, Calderón-Rangel A, Rodriguez-Morales A, Mattar S (2020) Venezuelan equine encephalitis virus: the problem is not over for tropical America. Ann Clin Microbiol Antimicrob 19(1):1–8

Haake DA, Levett PN (2015) Leptospirosis in humans. In: Adler B (ed) Leptospira and Leptospirosis. Current Topics in Microbiology and Immunology, vol 387. Springer, Berlin/Heidelberg

Hampson K, Coudeville L, Lembo T, Sambo M, Kieffer A, Attlan M, Barrat J, Blanton JD, Briggs DJ, Cleaveland S, Costa P, Freuling CM, Hiby E, Knopf L, Leanes F, Meslin FX, Metlin A, Miranda ME, Müller T, Nel LH, Recuenco S, Rupprecht CE, Schumacher C, Taylor L, Vigilato MAN, Zinsstag J, Dushoff J (2015) Estimating the global burden of endemic canine rabies. PLoS Negl Trop Dis 9(5):e0003786

Han BA, Schmidt JP, Bowden SE, Drake JM (2015) Rodent reservoirs of future zoonotic diseases. Proc Natl Acad Sci 112(22):7039–7044

Han BA, Kramer AM, Drake JM (2016) Global patterns of zoonotic disease in mammals. Trends Parasitol 32(7):565–577

Henzler DJ, Opitz HM (1992) The role of mice in the epizootiology of Salmonella enteritidis infection on chicken layer farms. Avian Dis 36:625–631

Herrera EA, Barreto GR (2013) Capybaras as a source of protein: utilization and management in Venezuela. In: Moreira JR, Ferraz KMPMB, Herrera EA, Madonald DW (eds) Capybara: biology, use and conservation of an exceptional neotropical species. Springer, New York, pp 305–319

Herwaldt BL (1999) Leishmaniasis. Lancet 354(9185):1191–1199

Heyman P, Thoma BR, Marié J, Cochez C, Essbauer SS (2012) In search for factors that drive Hantavirus epidemics. Front Physiol 3:237

Holland GN (2003) Ocular toxoplasmosis: a global reassessment. Part I: epidemiology and course of disease. Am J Ophthalmol 136(6):973–988

Huber C, Finelli L, Stevens W (2018) The economic and social burden of the 2014 Ebola outbreak in West Africa. J Infect Dis 218(Supplement 5):S698–S704

Ibarra-Cerdeña CN, Valiente-Banuet L, Sanchez-Cordero V, Stephens CR, Ramsey JM (2017) *Trypanosoma cruzi* reservoir—triatomine vector co-occurrence networks reveal meta-community effects by synanthropic mammals on geographic dispersal. PeerJ 5:e3152

Izquierdo-Rodríguez E, Fernández-Álvarez Á, Martín-Carrilo N, Feliu C, Marchand B, Quilichini Y, Foronda P (2019) Rodents as reservoirs of the zoonotic pathogens *Coxiella burnetii* and *Toxoplasma gondii* in Corsica (France). Vector-Borne Zoonotic Dis 19(12):879–883

Jansen AM, Xavier SCDC, Roque ALR (2018) *Trypanosoma cruzi* transmission in the wild and its most important reservoir hosts in Brazil. Parasites Vectors 11(1):1–25

Jones KE, Patel NG, Levy MA, Storeygard A, Balk D, Gittleman JL, Daszak P (2008) Global trends in emerging infectious diseases. Nature 451(7181):990–993

Jones BA, Grace D, Kock R, Alonso S, Rushton J, Said MY et al (2013) Zoonosis emergence linked to agricultural intensification and environmental change. Proc Natl Acad Sci 110(21):8399–8404

Jonsson CB, Figueiredo LTM, Vapalahti O (2010) A global perspective on hantavirus ecology, epidemiology, and disease. Clin Microbiol Rev 23(2):412–441

Joynson DHM (1992) Epidemiology of toxoplasmosis in the U.K. Scandinavian J Infect Dis Suppl 23(84):65–69

Karesh WB, Dobson A, Lloyd-Smith JO, Lubroth J, Dixon MA, Bennett M, Aldrich S, Harrington T, Formenty P, Loh EH, Machalaba CC, Thomas MJ, Heymann DL (2012) Ecology of zoonoses: natural and unnatural histories. Lancet 380(9857):1936

Klaren VNA, Kijlstra A (2002) Toxoplasmosis, an overview with emphasis on ocular involvement. Ocul Immunol Inflamm 10(1):1–26

Krawczak FS, Nieri-Bastos FA, Nunes FP, Soares JF, Moraes-Filho J, Labruna MB (2014) Rickettsial infection in *Amblyomma cajennense* ticks and capybaras (*Hydrochoerus hydrochaeris*) in a Brazilian spotted fever-endemic area. Parasites Vectors 7:1–7

Krebs CJ (1999) Ecological methodology, 2nd edn. Addison-Wesley Educational Publishers, Inc., Menlo Park

Krebs CJ, Gaines MS, Keller BL, Myers JH, Tamarin RH (1973) Population cycles in small rodents: demographic and genetic events are closely coupled in fluctuating populations of field mice. Science 179(4068):35–41

Kropf SP, Sá MR (2009) The discovery of *Trypanosoma cruzi* and Chagas disease (1908-1909): tropical medicine in Brazil. História, Ciências, Saúde-Manguinhos 16:13–34

Krügel M, Król N, Kempf VA, Pfeffer M, Obiegala A (2022) Emerging rodent-associated Bartonella: a threat for human health? Parasites Vectors 15(1):113

Kugeler KJ, Schwartz AM, Delorey MJ, Mead PS, Hinckley AF (2021) Estimating the frequency of Lyme disease diagnoses, United States, 2010–2018. Emerg Infect Dis 27(2):616

Labruna MB (2009) Ecology of rickettsia in South America. Ann N Y Acad Sci 1166(1):156–166

Labruna MB, Mattar S, Nava S, Bermudez S, Venzal JM, Dolz G, Abarca K, Romero L, de Sousa R, Oteo J, Zavala-Castro J (2011) Rickettsiosis en America Latina, el Caribe, Espana y Portugal. Revista MVZ (Medicina Veterinaria y Zootecnia) 16(2):2435–2458

Lee HW, Lee PW, Johnson KM (1978) Isolation of the etiologic agent of Korean hemorrhagic fever. J Infect Dis 137(3):298–308

Li HY, Zhu GJ, Zhang YZ, Zhang LB, Hagan EA, Martinez S, Chmur AA, Francisco L, Tai H, Miller M, Daszak P (2020) A qualitative study of zoonotic risk factors among rural communities in southern China. Int Health 12(2):77–85

Lidani KCF, Andrade FA, Bavia L, Damasceno FS, Beltrame MH, Messias-Reason IJ, Sandri TL (2019) Chagas disease: from discovery to a worldwide health problem. Front Public Health 7:166

Lima BS, Dantas-Torres F, de Carvalho MR, Marinho-Junior JF, de Almeida EL, Brito ME, Gomes F, Brandão-Filho SP (2013) Small mammals as hosts of *Leishmania* spp. in a highly endemic area for zoonotic leishmaniasis in northeastern Brazil. Trans R Soc Trop Med Hyg 107(9):592–597

Loayza-Mafayle R, Morales-Betoulle ME, Romero C, Cossaboom CM, Whitmer S, Alvarez Aguilera CE, Avila-Ardaya C, Cruz-Zambrana M, Dávalos-Anajia A, Mendoza-Loayza N, Montaño A-M et al (2022) Chapare hemorrhagic fever and virus detection in rodents in Bolivia in 2019. N Engl J Med 386(24):2283–2294

Lopez-Cancino SA, Tun-Ku E, De la Cruz-Felix HK, Ibarra-Cerdeña CN, Izeta Alberdi A, Pech-May A, Mazariegos-Hidalgo CJ, Valdez-Tah A, Ramsey JM (2015) Landscape ecology of *Trypanosoma cruzi* in the southern Yucatan Peninsula. Acta Trop 151:58–72

Luis AD, Hayman DTS, O'Shea TJ, Cryan PM, Gilbert AT, Pulliam JRC, Mills JN, Timonin ME, Willis CKR, Cunningham AA, Fooks AR, Rupprecht CE, Wood JLN, Webb CT (2013) A comparison of bats and rodents as reservoirs of zoonotic viruses: are bats special? Proc R Soc B Biol Sci 280:20122753

Lundberg L, Carey B, Kehn-Hall K (2017) Venezuelan equine encephalitis virus capsid—the clever caper. Viruses 9(10):279

Luz HR, Costa FB, Benatti HR, Ramos VN, de Serpa AMC, Martins TF, Acosta ICL, Ramirez DG, Muñoz-Leal S, Ramirez-Hernandez A, Binder LC, Carvalho MP, Rocha V, Dias TC, Simeoni CL, Brites-Neto J, Brasil J, Nievas AM, Monticelli PF, Moro MEG, Lopez B, Aguiar DM, Pacheco RC, Sousa CE, Piovezan U, Juliano R, Ferras KMPMB, Zsabó MPJ, Labruna MB (2019) Epidemiology of capybara-associated Brazilian spotted fever. PLoS Negl Trop Dis 13(9):e0007734

Lytras S, Xia W, Hughes J, Jiang X, Robertson DL (2021) The animal origin of SARS-CoV-2. Science 373(6558):968–970

Ma J, Zhang S, Zhang X (2021) Structure of Machupo virus polymerase in complex with matrix protein Z. Nat Commun 12(1):6163

Macdonald DW, Fenn MGP, Gelling M (2015) The natural history of rodents: Preadaptations to pestilence. In: Buckle AP, Smith RH (eds) Rodent pests and their control, 2nd edn. CAB International, Croydon

Maco V, Maguiña C, Tirado A, Maco CV, Vidal JE (2004) Carrion's disease (*Bartonellosis bacilliformis*) confirmed by histopathology in the High Forest of Peru. Rev Inst Med Trop Sao Paulo 46:171–174

Majowicz SE, Musto J, Scallan E, Angulo FJ, Kirk M, O'Brien SJ, Jones TF, Fazil A, Hoekstra RM (2010) The global burden of nontyphoidal Salmonella gastroenteritis. Clin Infect Dis 50(6):882–889

Mann S, Frasca K, Scherrer S, Henao-Martínez AF, Newman S, Ramanan P, Suarez JA (2021) A review of leishmaniasis: current knowledge and future directions. Curr Trop Med Rep 8(2):121–132

Marcelino AP, Ferreira EC, Avendanha JS, Costa CF, Chiarelli D, Almeida G, Moreira EC, Leite RC, dos Reis JKP, Gontijo CM (2011) Molecular detection of *Leishmania braziliensis* in *Rattus norvegicus* in an area endemic for cutaneous leishmaniasis in Brazil. Vet Parasitol 183(1–2):54–58

Marquardt W (ed) (2004) Biology of disease vectors, 2nd edn. Elsevier Academic Press, Burlington

Martin-Alonso A, Soto M, Foronda P, Aguilar E, Bonnet G, Pacheco R, . Valladares B, and Quispe-Ricalde MA. 2014. *Bartonella* spp. and *Yersinia pestis* reservoirs, Cusco, Peru. Emerg Infect Dis 20(6):1069

Maurin M, Birtles R, Raoult D (1997) Current knowledge of Bartonella species. Eur J Clin Microbiol Infect Dis 16:487–506

McCauley DJ, Salkeld DJ, Young HS, Makundi R, Dirzo R, Eckerlin RP, Lambin EF, Gaffikin L, Barry M, Helgen KM (2015) Effects of land use on plague (*Yersinia pestis*) activity in rodents in Tanzania. Am J Trop Med Hygiene 92(4):776–783

McMahon BJ, Morand S, Gray JS (2018) Ecosystem change and zoonoses in the Anthropocene. Zoonoses Public Health 65(7):755–765

Meerburg BG, Kijlstra A (2007) Role of rodents in transmission of Salmonella and Campylobacter. J Sci Food Agric 87(15):2774–2781

Meerburg BG, Singleton GR, Kijlstra A (2009) Rodent-borne diseases and their risks for public health. Crit Rev Microbiol 35(3):221–270

Melis M, Espinoza-Carniglia M, Savchenko E, Nava S, Lareschi M (2020) Molecular detection and identification of *Rickettsia felis* in Polygenis (Siphonaptera, Rhopalopsyllidae, Rhopalopsyllinae) associated with cricetid rodents in a rural area from central Argentina. Veterin Parasitol Reg Stud Rep 21:100445

Mendoza H, Rubio AV, García-Peña GE, Suzán G, Simonetti JA (2020) Does land-use change increase the abundance of zoonotic reservoirs? Rodents say yes. Eur J Wildl Res 66(1):6

Milagres BS, Padilha AF, Montandon CE, Freitas RN, Pacheco R, Walker DH, Labruna MB, Mafra CL, Galvão MAM (2013) Spotted fever group rickettsia in small rodents from areas of low endemicity for Brazilian spotted fever in the eastern region of Minas Gerais State, Brazil. Am J Trop Med Hygiene 88(5):937

Mills JN (2006) Biodiversity loss and emerging infectious disease: an example from the rodent-borne hemorrhagic fevers. Biodiversity 7(1):9–17

Mills JN, Childs JE (1998) Ecologic studies of rodent reservoirs: their relevance for human health. Emerg Infect Dis 4(4):529–537

Moher D, Liberati A, Tetzlaff J, Altman DG (2010) Preferred reporting items for systematic reviews and meta-analyses: the PRISMA statement. Int J Surg 8(5):336–341

Molina I, Salvador F, Sánchez-Montalvá A (2016) Actualización en enfermedad de Chagas. Enferm Infecc Microbiol Clin 34(2):132–138

Mollentze N, Streicker DG (2020) Viral zoonotic risk is homogenous among taxonomic orders of mammalian and avian reservoir hosts. Proc Natl Acad Sci 117(17):9423–9430

Mones A, Ojasti J (1986) *Hydrochoerus hydrochaeris* Mammalian species 264:1–7

Morens DM, Folkers GK, Fauci AS (2004) The challenge of emerging and re-emerging infectious diseases. Nature 430(6996):242–249

Morrone JJ (2014) Biogeographical regionalization of the Neotropical region. Zootaxa 3782(1):1–110

Morand S, Blasdell K, Bordes F, Buchy P, Carcy B, Chaisiri K, Chaval Y, Claude J, Cosson JF, Desquesnes M, Jittapalapong S, Jiyipong T, Karnchanabanthoen A, Pornpan P, Rolain JM, and Tran A (2019) Changing landscapes of Southeast Asia and rodent-borne diseases: decreased diversity but increased transmission risks. Ecological Applications 29(4):e01886

Morrone JJ, Escalante T, Rodríguez-Tapia G, Carmona A, Arana M, Mercado-Gómez JD (2022) Biogeographic regionalization of the Neotropical region: new map and shapefile. An Acad Bras Cienc 94(1):e20211167

Morse SS, Mazet JA, Woolhouse M, Parrish CR, Carroll D, Karesh WB, Zambrana-Torrelio C, Lipkin WI, Daszak P (2012) Prediction and prevention of the next pandemic zoonosis. Lancet 380(9857):1956–1965

Navarro JC, Medina G, Vasquez C, Coffey LL, Wang E, Suárez A, Biord H, Salas M, Weaver SC (2005) Postepizootic persistence of Venezuelan equine encephalitis virus, Venezuela. Emerg Infect Dis 11(12):1907–1915

Nelson CA, Moore AR, Perea AE, Mead PS (2018) Cat scratch disease: US clinicians' experience and knowledge. Zoonoses Public Health 65(1):67–73

Newbold T, Hudson LN, Contu S, Hill SL, Beck J, Liu Y, Meyer C, Phillips HRP, Scharlemann JPW, Purvis A (2018) Widespread winners and narrow-ranged losers: land use homogenizes biodiversity in local assemblages worldwide. PLoS Biol 16(12):e2006841

Olivera MJ, Buitrago G (2020) Economic costs of Chagas disease in Colombia in 2017: a social perspective. Int J Infect Dis 91:196–201

Organización Panamericana de la Salud (2003) Zoonosis y enfermedades transmisibles comunes al hombre y a los animales: parasitosis, v.3 (3rd ed)–Publicación Científica y Técnica No. 580. Organización Panamericana de la Salud, Washington, DC

Organización Panamericana de la Salud (OPS) (2019) Manual de procedimientos para la vigilancia y control de las leishmaniasis en las Américas. OPS, Washington, DC

Pan American Health Organization (PAHO) (2019). Manual de procedimientos para la vigilancia y control de las leishmaniasis en las Américas. OPS. Washington, DC

Pacheco RC, Horta MC, Moraes-Filho J, Ataliba AC, Pinter A, Labruna MB (2007) Rickettsial infection in capybaras (*Hydrochoerus hydrochaeris*) from São Paulo, Brazil: serological evidence for infection by *Rickettsia bellii* and *Rickettsia parkeri*. Biomedica 27(3):364–371

Paddock CD, Sumner JW, Comer JA, Zaki SR, Goldsmith CS, Goddard J, McLellan SLF, Cl T, Ohl CA (2004) *Rickettsia parkeri*: a newly recognized cause of spotted fever rickettsiosis in the United States. Clin Infect Dis 38(6):805–811

Panti-May JA, Duarte-Jiménez A, Hernández-Betancourt SF, Rodríguez-Vivas RI (2021) A checklist of the Helminth parasites of invasive murid rodents in Mexico. Therya 12(1):85–93

Parker RR, Kohls GM, Cox GW, Davis GE (1939) Observations on an infectious agent from *Amblyomma maculatum*. Public Health Rep 54(1896–1970):1482–1484

Parker NR, Barralet JH, Bell AM (2006) Q fever. Lancet 367(9511):679–688

Parola P (2011) *Rickettsia felis*: from a rare disease in the USA to a common cause of fever in sub-Saharan Africa. Clin Microbiol Infect 17(7):996–1000

Patterson M, Grant A, Paessler S (2014) Epidemiology and pathogenesis of Bolivian hemorrhagic fever. Curr Opin Virol 5:82–90

Peters CJ (2002) Human infection with Arenaviruses in the Americas. Pp. 65–74. In M. B. Oldstone (ed.), Arenaviruses I, vol. 262. Springer-Verlag, Berlin, Germany

Peniche-Lara G, Dzul-Rosado KR, Zavala Velázquez JE, Zavala-Castro J (2012) Murine typhus: clinical and epidemiological aspects. Colomb Med 43(2):175–180

Peniche-Lara G, Dzul-Rosado KR, Perez-Osorio C, Zavala-Castro J (2015) *Rickettsia typhi* in rodents and *R. felis* in fleas in Yucatán as a possible causal agent of undefined febrile cases. Rev Inst Med Trop Sao Paulo 57:129–132

Pérez-Molina JA, Molina I (2018) Chagas disease. Lancet 391(10115):82–94

Pérez-Osorio CE, Zavala-Velázquez JE, León JJA, Zavala-Castro JE (2008) *Rickettsia felis* as emergent global threat for humans. Emerg Infect Dis 14(7):1019

Perry RD, Fetherston JD (1997) *Yersinia pestis*-etiologic agent of plague. Clin Microbiol Rev 10(1):35–66

Peters CJ (2002) Human infection with Arenaviruses in the Americas. In: Oldstone MB (ed) Arenaviruses I, vol 262. Springer-Verlag, Berlin, pp 65–74

Pinto CM, Ocaña-Mayorga S, Lascano MS, Grijalva MJ (2006) Infection by trypanosomes in marsupials and rodents associated with human dwellings in Ecuador. J Parasitol 92(6):1251–1255

Pinto-Junior VL, Hamidad AM, Albuquerque Filho DO, dos Santos VM (2014) Twenty years of hantavirus pulmonary syndrome in Brazil: a review of epidemiological and clinical aspects. J Infect Dev Ctries 8(2):137–142

Pires MM, Galett M (2022) Beyond the "empty forest": The defaunation syndromes of Neotropical forests in the Anthropocene. Global Ecol Conserv 41:e02362

Quaresma PF, Rêgo FD, Botelho HA, da Silva SR, Moura AJ, Neto RGT, Maderira FM, Carvalho MB, Paglia AP, Melo MN, Gontijo CMF (2011) Wild, synanthropic and domestic hosts of Leishmania in an endemic area of cutaneous leishmaniasis in Minas Gerais State, Brazil. Trans R Soc Trop Med Hyg 105(10):579–585

Quintal D (1996) Historical aspects of the rickettsioses. Clin Dermatol 14(3):237–242

Rabiee MH, Mahmoudi A, Siahsarvie R, Kryštufek B, Mostafavi E (2018) Rodent-borne diseases and their public health importance in Iran. PLoS Negl Trop Dis 12(4):e0006256

Rahman MT, Sobur MA, Islam MS, Ievy S, Hossain MJ, El Zowalaty ME, Rahman AT, Ashour HM (2020) Zoonotic diseases: etiology, impact, and control. Microorganisms 8(9):1405

Ramsey JM, Gutierrez-Cabrera AE, Salgado L, Peterson AT, Sanchez-Cordero V, Ibarra-Cerdeña CN (2012) Ecological connectivity of *Trypanosoma cruzi* reservoirs and *Triatoma pallidipennis* hosts in an anthropogenic landscape with endemic Chagas disease. PLoS One 7(9):e46013

Rassi A, Marin-Neto JA (2010) Chagas disease. Lancet 375(9723):1388–1402

Raufu IA, Ahmed OA, Aremu A, Odetokun IA, Raji MA (2019) Salmonella transmission in poultry farms: the roles of rodents, lizards and formites. Savannah Veterin J 2(1):1–4

Reif KE, Macaluso KR (2009) Ecology of *Rickettsia felis*: a review. J Med Entomol 46(4):723–736

Rewar S, Mirdha D, Rewar P (2015) Treatment and prevention of pandemic H1N1 influenza. Ann Glob Health 81(5):645–653

Ribeiro CM, Carvalho JLBD, Bastos PADS, Katagiri S, Batalha EY, Okano W, Costa VMD (2021) Prevalence of *Rickettsia rickettsii* in ticks: systematic review and meta-analysis. Vector-Borne Zoonotic Dis 21(8):557–565

Roatt BM, de Oliveira Cardoso JM, De Brito RCF, Coura-Vital W, de Oliveira Aguiar-Soares RD, Reis AB (2020) Recent advances and new strategies on leishmaniasis treatment. Appl Microbiol Biotechnol 104(21):8965–8977

Roberts T (1988) Salmonellosis control: estimated economic costs. Poult Sci 67(6):936–943

Robles A, Fong J, Cervantes J (2018) Borrelia infection in Latin America. Rev Investig Clin 70(4):158–163

Rodríguez-Rojas JJ, Rodríguez-Moreno Á, Sánchez-Casas RM, Hernández-Escareño JJ (2020) Molecular detection of *Leptospira interrogans* and *Borrelia burgdorferi* in wild rodents from Mexico. Vector-Borne Zoonotic Dis 20(11):860–863

Roiz D, Pontifes P, Jourdain F, Diagne C, Leroy B, Vaissière AC, Tolsá MJ, Salles JM, Simard F, Courchamp F (2023) The rising global economic costs of Aedes and Aedes-borne diseases. https://doi.org/10.21203/rs.3.rs-2679030/v1

Roque ALR, Jansen AM (2014) Wild and synanthropic reservoirs of Leishmania species in the Americas. Int J Parasitol Parasites Wildlife 3(3):251–262

Rosa EST, Mills JN, Padula PJ, Elkhoury MR, Ksiazek TG, Mendes WS, Santos ED, Araujo GCB, Martinez VP, Sosa JFST, Edelstein A, Vasconcelos PF (2005) Newly recognized hantaviruses associated with hantavirus pulmonary syndrome in northern Brazil: partial genetic characterization of viruses and serologic implication of likely reservoirs. Vector-Borne Zoonotic Dis 5(1):11–19

Rozental T, Ferreira MS, Guterres A, Mares-Guia MA, Teixeira BR, Gonçalves J, Bonvicino CR, D'Andrea PS, de Lemos ERS (2017) Zoonotic pathogens in Atlantic Forest wild rodents in Brazil: Bartonella and Coxiella infections. Acta Trop 168:64–73
Ruiz A (2001) Plague in the Americas. Emerg Infect Dis 7(3):539
Runge M, von Keyserlingk M, Braune S, Becker D, Plenge-Bönig A, Freise JF, Pelz HJ, Esther A (2013) Distribution of rodenticide resistance and zoonotic pathogens in Norway rats in Lower Saxony and Hamburg, Germany. Pest Manag Sci 69(3):403–408
Sabour S, Azimi T, Nasser A, Hadi N, Mohsenzadeh A, Shariati A (2022) A global overview of the most important zoonotic bacteria pathogens transmitted from *Rattus norvegicus* to humans in urban environments. In: Infectious medicine, vol 1, Issue 3. Elsevier B. V, pp 192–207
Sanmartin-Barberi C, Groot H, Osorno-Mesa E (1954) Human epidemio in Colombia caused by the Venezuelan equine encephalomyelitis virus. Am J Trop Med Hyg 3(2):283–293
Sasidharan S, Saudagar P (2021) Leishmaniasis: where are we and where are we heading? Parasitol Res 120:1541–1554
Schmaljohn C (2009) Vaccines for hantaviruses. Vaccine 27(4):61–64
Schmid GP (1985) The global distribution of Lyme disease. Rev Infect Dis 7(1):41–50
Schneider MC, Najera P, Aldighieri S, Galan DI, Bertherat E, Ruiz A, Dumit E, Gabastou JM, Espinal MA (2014) Where does human plague still persist in Latin America? PLoS Negl Trop Dis 8(2):e2680
Schriefer ME, Sacci JB Jr, Dumler JS, Bullen MG, Azad AF (1994) Identification of a novel rickettsial infection in a patient diagnosed with murine typhus. J Clin Microbiol 32(4):949–954
Schwartzman JD, Maguire JH (2011) Toxoplasmosis. In: Tropical infectious diseases: principles, pathogens and practice. W.B. Saunders, pp 722–728
Scott AT, Vondrak CJ, Sanderlin AG, Lamason RL (2022) Rickettsia parkeri. Trends Microbiol 30(5):511–512
Serafim TD, Coutinho-Abreu IV, Dey R, Kissinger R, Valenzuela JG, Oliveira F, Kamhawi S (2021) Leishmaniasis: the act of transmission. Trends Parasitol 37(11):976–987
Serra-Medeiros S, Ortega Z, Antunes PC, Herrera HM, Oliveira-Santos LGR (2021) Space use and activity of capybaras in an urban area. J Mammal 102(3):814–825
Shapiro ED (2014) *Borrelia burgdorferi* (Lyme disease). Pediatr Rev 35(12):500–509
Sharma U, Singh S (2008) Insect vectors of Leishmania: distribution, physiology and their control. J Vector Borne Dis 45(4):255–272
Shilereyo M, Magige F, Ranke PS, Ogutu JO, Røskaft E (2022) Ectoparasite load of small mammals in the Serengeti Ecosystem: effects of land use, season, host species, age, sex and breeding status. Parasitol Res 121(3):823–838
Silva-Ramos CR, Hidalgo M, Faccini-Martínez ÁA (2021) Clinical, epidemiological, and laboratory features of *Rickettsia parkeri* rickettsiosis: a systematic review. Ticks Tick-Borne Dis 12(4):101734
Sing A (ed) (2015) Zoonoses—Infections affecting humans and animals: focus on public health aspects. Springer, New York
Singleton GR, Leirs H, Hinds L, Zhang Z (1999) Ecologically based management of rodent pests: reevaluating our approach to an old problem. In: Singleton GR, Hinds L, Leirs H, Zhang Z (eds) Ecologically based management of Rodent Pests. ACIAR, Canberra, pp 17–29
Smadel JE (1953) Epidemic hemorrhagic fever. Am J Publ Health Nations Health 43(10):1327–1330
Solís-Henández A, Rodríguez-Vivas RI, Esteve-Gassent MD, Villegas-Pérez SL (2016) Prevalencia de *Borrelia burgdorferi sensu lato* en roedores sinantrópicos de dos comunidades rurales de Yucatán, México. Biomedica 36:109–117
Stanek G, Wormser GP, Gray J, Strle F (2012) Lyme borreliosis. Lancet 379(9814):461–473
Steere AC, Malawista SE, Snydman DR, Shope RE, Andiman WA, Ross MR, Steele FM (1977) An epidemic of oligoarticular arthritis in children and adults in three Connecticut communities. Arthritis Rheum 20(1):7–17
Stenseth NC, Leirs H, Skonhoft A, Davis SA, Pech RP, Andreassen HP et al (2003) Mice, rats, and people: the bio-economics of agricultural rodent pests. Front Ecol Environ 1(7):367–375

Szabó MPJ, Nieri-Bastos FA, Spolidorio MG, Martins TF, Barbieri AM, Labruna MB (2013) In vitro isolation from *Amblyomma ovale* (Acari: Ixodidae) and ecological aspects of the Atlantic rainforest Rickettsia, the causative agent of a novel spotted fever rickettsiosis in Brazil. Parasitology 140(6):719–728

Taylor LH, Latham SM, Woolhouse ME (2001) Risk factors for human disease emergence. Philos Trans R Soc Lond B Biol Sci 356(1411):983–989

Thorner AR, Walker DH, Petri WA Jr (1998) Rocky Mountain spotted fever. Clin Infect Dis 27:1353–1359

Tilly K, Rosa PA, Stewart PE (2008) Biology of infection with *Borrelia burgdorferi*. Infect Dis Clin N Am 22(2):217–234

Torres-Castro M, Martínez-Ortiz D, Panti-May A, Koyoc-Cardeña E, López-Ávila K, Dzul-Rosado K, Zavala-Castro J, Chablé-Santos J, Manrique-Saide P (2018) *Rickettsia typhi* in rodents from a community with history of murine typhus from Yucatan, Mexico. Revista MVZ Córdoba 23:6974–6980

Torres-Castro M, Sánchez-Montes S, Colunga-Salas P, Noh-Pech H, Reyes-Novelo E, Rodríguez-Vivas RI (2022) Molecular confirmation of *Rickettsia parkeri* in humans from Southern Mexico. Zoonoses Public Health 69(4):382–386

Torres-Guerrero E, Quintanilla-Cedillo MR, Ruiz-Esmenjaud J, Arenas R (2017) Leishmaniasis: a review. F1000Research 6:750

Tsioutis C, Zafeiri M, Avramopoulos A, Prousali E, Miligkos M, Karageorgos SA (2017) Clinical and laboratory characteristics, epidemiology, and outcomes of murine typhus: a systematic review. Acta Trop 166:16–24

Vago AR, Andrade LO, Leite AA, Reis DDÁ, Macedo AM, Adad SJ, Toster S Jr, Moreira MCV, Filho GB, Pena SD (2000) Genetic characterization of *Trypanosoma cruzi* directly from tissues of patients with chronic Chagas disease: differential distribution of genetic types into diverse organs. Am J Pathol 156(5):1805–1809

Verdade LM, Ferraz KMPMB (2006) Capybaras in an anthropogenic habitat in southeastern Brazil. Braz J Biol 66:371–378

Vial PA, Ferrés M, Vial C, Klingström J, Ahlm C, López R, Le Corre N, Mertz GJ (2023) Hantavirus in humans: a review of clinical aspects and management. Lancet Infect Dis. In press. https://doi.org/10.1016/S1473-3099(23)00128-7

Vilibic-Cavlek T, Savic V, Ferenc T, Mrzljak A, Barbic L, Bogdanic M, Stevanovic V, Tabain I, Ferencak I, Zidovec-Lepej S (2021) Lymphocytic Choriomeningitis—emerging trends of a neglected Virus: a narrative review. Trop Med Infect Dis 6(2):88

Walton BC, de Arjona I, Benchoff BM (1966) Relationship of Toxoplasma antibodies to altitude. Am J Trop Med Hygiene 15(4):492–495

Weaver SC, Barrett AD (2004) Transmission cycles, host range, evolution and emergence of arboviral disease. Nat Rev Microbiol 2(10):789–801

Weaver SC, Ferro C, Barrera R, Boshell J, Navarro JC (2004) Venezuelan equine encephalitis. Ann Rev Entomol 49(1):141–174

Weil A (1886) Ueber einer eigenhuemliche, mit Milztumor, Icterus un Nephritis einhergehende, acute Infektionskrankheit. Dtsch Arch Klin Med 39:109

White RJ, Razgour O (2020) Emerging zoonotic diseases originating in mammals: a systematic review of effects of anthropogenic land-use change. Mammal Rev 50(4):336–352

Wilson DE, Reeder DM (eds) (2005) Mammal species of the world: a taxonomic and geographic reference, 2nd edn. Smithsonian Institution Press, Washington, DC

Wolcott KA, Margos G, Fingerle V, Becker NS (2021) Host association of *Borrelia burgdorferi sensu lato*: a review. Ticks Tick-Borne Dis 12(5):101766

Woolhouse MEJ (2006) Where do emerging pathogens come from? Microbe Magaz 1(11):511–515

World Health Organization (WHO) (2002) Control of Chagas disease: second report of the WHO Expert Committee, vol 2. World Health Organization, Geneva

World Health Organization (WHO) (2012) Zoonoses. World Health Organization, Geneva. Available at: https://www.who.int/es/news-room/fact-sheets/detail/zoonoses. Accessed 20 March 2023

Yefi-Quinteros E, Muñoz-San Martín C, Bacigalupo A, Correa JP, Cattan PE (2018) *Trypanosoma cruzi* load in synanthropic rodents from rural areas in Chile. Parasites Vectors 11:1–7

Yeyati EL, Filippini F (2021) Social and economic impact of COVID-19. Brookings Global Working Paper #158, Global Economy and Development program at Brookings, Washington, DC

Young HS, McCauley DJ, Dirzo R, Nunn CL, Campana MG, Agwanda B, Otarola-Castillo ER, Castillo ER, Pringle RM, Veblen KE, Salkeld DJ, Stewardson K, Fleischer R, Lambin EF, Palmer TM, Helgen KM (2017) Interacting effects of land use and climate on rodent-borne pathogens in central Kenya. Phil Trans R Soc B Biol Sci 372(1722):20160116

Zavala-Velázquez JE, Ruiz-Sosa JA, Sánchez-Elias RA, Becerra-Carmona G, Walker DH (2000) *Rickettsia felis* rickettsiosis in Yucatan. Lancet 356(9235):1079–1080

Zavala-Velázquez J, Laviada-Molina H, Zavala-Castro J, Perez-Osorio C, Becerra-Carmona G, Ruiz-Sosa JA, Bouyer DH, Walker DH (2006) *Rickettsia felis*, the agent of an emerging infectious disease: report of a new case in Mexico. Arch Med Res 37(3):419–422

Zeppelini CG, Callefe JLR, Coelho R, Silveira MSM, Khalil H, Belmain SR, Bertherat EGG, Begon M, Costa F (2022) A systematic review of rodent control as part of infectious disease control programs. Int J Pest Manag:1–8

Zhang YY, Sun YQ, Chen JJ, Teng AY, Wang T, Li H, Hay SI, Fang L-Q, Yang Y, Liu W (2022) Mapping the global distribution of spotted fever group rickettsiae: a systematic review with modeling analysis. Lancet Digit Health 5(1):e5–e15

Chapter 8
Bats, Pathogen Diversity and Rabies in a Changing Neotropic Landscape

Oscar Rico-Chávez, Nuri Flores-Pérez, Karen Uxue Martínez-Pérez, María del Carmen Villalobos-Segura, and Rafael Ávila-Flores

8.1 Introduction

There are misconceptions that have led to the underestimation of bats' ecological roles, even though bats are essential to the health of natural and modified ecosystems (Kalka et al. 2008; de Carvalho-Ricardo et al. 2014; Medellin et al. 2017; Kemp et al. 2019). More than one-third of all bat species assessed by the International Union for Conservation of Nature (IUCN) are classified as threatened or data deficient, and more than half have unknown population trends (Fenton et al. 2020). According to the most recent IUCN Red List, 77 species are endangered. In addition, at least 988 of the approximately 1400 known bat species require immediate conservation measures and additional research (Frick et al. 2020). Due primarily to their nocturnal and cryptic behaviors, we are unable to accurately evaluate the ecosystem services bats provide. The primary hazards to bats are habitat degradation, roost damage, and hunting (Frick et al. 2020). In addition, bats are erroneously stigmatized in many regions of the world, and the current acute respiratory syndrome COVID-19 has generated fear to the point where it is endangering bat populations (Fenton et al. 2020; Rocha et al. 2021).

Most of the research on pathogens in bats has focused on the diversity and distribution of medically significant emergent pathogens, such as Ebola, Hendra, Nipah, Marburg, and multiple coronaviruses (Epstein et al. 2006; Woo et al. 2006; Smith

O. Rico-Chávez (✉) · N. Flores-Pérez · K. U. Martínez-Pérez
M. del Carmen Villalobos-Segura
Departamento de Etología, Fauna Silvestre y Animales de Laboratorio, Facultad de Medicina Veterinaria y Zootecnia, Universidad Nacional Autónoma de México, Mexico City, Mexico
e-mail: orich@unam.mx

R. Ávila-Flores
División Académica de Ciencias Biológicas, Universidad Juárez Autónoma de Tabasco, Villahermosa, Tabasco, Mexico

G. Acosta-Jamett, A. Chaves (eds.), *Ecology of Wildlife Diseases in the Neotropics*, https://doi.org/10.1007/978-3-031-50531-7_8

et al. 2011; Hayman 2016). Bats are significant reservoirs of numerous viruses (Streicker and Gilbert 2020), but their unique function as hosts of zoonotic viruses relative to other mammalian taxa is questionable. While some studies assert that bats host more zoonotic viruses per species than other mammalian orders, such as rodents (Luis et al. 2013; Olival et al. 2017), more recent research has demonstrated that the number of zoonotic pathogens found in bats is largely a function of host species diversity and that the proportion of zoonotic pathogens varies little among mammalian orders (Mollentze and Streicker 2020).

Rabies is the most significant zoonosis transmitted by bats in Latin America due to its global distribution, human cases, a wide range of potential reservoirs, and veterinary and economic cost implications, representing a public and animal health concern for which local governments allocate substantial funds for control and eradication programs. In the case of dog-transmitted rabies, these programs have been successful, notably in urban areas, but programs focused on bat eradication have not been effective (Rupprecht et al. 2002; Viana et al. 2023). Bat-mediated transmission of rabies virus only occurs in North, Central, and South America. It is believed that rabies has existed in tropical America since pre-Columbian times and is transmitted primarily by hematophagous vampire bats. However, recent phylogenetic reconstructions indicate that the rabies virus in the Americas is unlikely to have originated from vampire bats. Now that diagnostic techniques have advanced, it is possible to comprehend the dynamics of bat-rabies; nevertheless, a transdisciplinary approach that considers the ecology and habitat use of the main reservoir, the common vampire (*Desmodus rotundus*), is needed.

The risk of zoonoses of wild origin has led to several programs that sample wildlife to discover new viruses and assess their zoonotic potential (Olival et al. 2017). These programs include the PREDICT project (Kelly et al. 2017), the Global Virome Project (Carroll et al. 2018), and SpillOver: Virus Risk Ranking (Grange et al. 2021). The first two programs focused on conducting targeted surveillance to detect emerging viral pathogens with zoonotic potential, with the goal of preventing emerging diseases and possible pandemics. On the other hand, SpillOver is an open-source tool that evaluates the potential of wild-origin viruses for zoonotic transmission.

It has been observed that there is a significant bias in the investigation of emerging infectious disease events, with greater infectious disease surveillance efforts in wealthy and developed countries in Europe, North America, Australia, and parts of Asia compared to developing countries located in tropical regions. These areas are described as hotspots for the emergence of infectious diseases due to their high human population density and greater mammalian diversity. In other words, economic and scientific resources for pathogen detection and surveillance are located in nations that have a low probability of pathogen emergence (Jones et al. 2008). This is also the case for pathogen research in the order Chiroptera, where regions in Latin America and parts of Asia have comparatively few characterized viruses despite their bat abundance (Olival et al. 2012, 2017; Szentivanyi et al. 2023).

The economic losses resulting from zoonotic outbreaks are significant. The World Bank estimates that six zoonotic outbreaks (Nipah virus, West Nile virus, severe acute respiratory syndrome, highly pathogenic avian influenza virus, bovine spongiform encephalopathy, and Rift Valley fever) from 1997 to 2009 caused $80 billion in losses, equivalent to $6.7 billion per year. In 2016, Fan et al. estimated that a pandemic with the potential to harm 720,000 people could result in $570 billion in annual economic costs. In 2017, Daszak et al. estimated that the total damage caused by emerging infectious diseases of zoonotic origin over the next 50 years could reach $3.6 trillion, considering the global frequency of pandemics, mortality rates from emerging infectious diseases, and gross domestic product. According to Dobson et al. (2020), the COVID-19 pandemic was expected to reduce the global gross domestic product by $5.6 trillion by 2020.

Opposed to the mentioned economic effects, the World Bank estimated that upgrading public health and animal health systems in developing countries to meet World Health Organization (WHO) and Organization for Animal Health (OIE) standards would cost between $1.9 and $3.4 billion per year. In contrast, the PREDICT Project, which began in 2009 and lasted for 10 years, spent $200 million per year to enhance global capabilities to detect viruses with pandemic potential. The Global Virome Project, a 10-year initiative to discover, characterize, and evaluate 70% of viral diversity, is estimated to cost $1.2 billion annually (Dobson et al. 2020). This initiative focuses on collecting samples from key zoonotic reservoirs, such as bats, in highly biodiverse tropical developing countries with limited capacity to respond to public health crises (Carroll et al. 2018). Preventive measures to avert pandemics, including wildlife trade monitoring, reduction of spillovers, control and early detection, halving the rate of deforestation, and terminating the bushmeat trade in China, are estimated to cost between 22 and 31.2 billion annually (Dobson et al. 2020).

As the demand for natural resources rises worldwide, land-use change in tropical and subtropical ecosystems is occurring at an unprecedented rate (Song et al. 2018). Several recent zoonotic infections have been linked to land-use change; however, little is known about the connection between changing land use and the rise of diseases (Gibb et al. 2020). Several authors have suggested that land-use change may increase the risk of disease emergence by promoting human–animal interactions or by influencing pathogen diversity, either directly by changing the prevalence and/or diversity of pathogens or indirectly through impacts on host assemblages (Bradley et al. 2008; Vittor et al. 2009; Murray and Daszak 2013; Rulli et al. 2017). However, empirical research examining the impact of land-use change on pathogen assemblages in Neotropical bats remains limited.

8.2 Methodology

A comprehensive search of the Web of Science Core Collection for current information on parasite ecology in Neotropical bats from 1990 to 2021 was conducted. The initial search focused on pathogen groups (viruses, bacteria, and parasites), bats (bats, Chiroptera), and the Neotropical region (Neotropics, South America, and Latin America). Review articles and laboratory/experimental investigations reporting pathogens were excluded from the study. The second search was conducted to identify studies that demonstrate how anthropogenic (fragmentation, land use, landscape, urban, habitat perturbation) and ecological (temperature, latitude, altitude, precipitation, humidity, climate change, functional traits) factors are associated with the prevalence and diversity of each pathogen group. A database was constructed where a single observation consisted of a pathogen reported in a single bat species. Bat family, bat species, year, pathogen type (virus, bacteria, or parasite), pathogen family, and country where the report was made were collected.

The bias in the surveillance of pathogens with zoonotic potential may be due to the high costs associated with identifying and sampling wild species. Therefore, an analysis was conducted to investigate the relationship between gross domestic expenditure on research and development as a percentage of gross domestic product and detection and sampling effort to assess pathogen richness in Neotropic bat species. The goal was to determine whether these factors could affect the surveillance of viral pathogens in bats and how this could impact research and viral surveillance efforts. Information was collected on gross domestic expenditure on research and development (R&D), bat-associated viral richness, and bat richness in Latin America and the Caribbean.

The indicator of gross domestic expenditure allocated to research and development (R&D), expressed as a percentage of gross domestic product (GDP), was utilized to assess the participation of various nations in epidemiological surveillance. This indicator can be used to evaluate and analyze scientific research and experimental development activities (United Nations 2017). It was obtained from the World Bank's databank (World-Bank 2022), analyzing data from 1996 to 2018, as the countries with submissions have records for these years. Since not all countries have entries for all years, the average annual expenditure for each country was calculated. A list of the number of bat species present in each country was prepared using data from the International Union for Conservation of Nature (IUCN 2022). Linear regression analyses were performed between total reported bat viruses per country and gross domestic expenditure on research and development as a percentage of gross domestic product and between bat richness and total reported bat viruses per country. Finally, a literature review was conducted to describe rabies control strategies with emphasis on the ecology of *Desmodus rotundus*, the main species involved in the maintenance and dispersal of the rabies virus.

8.3 Bat-Associated Pathogen Diversity in the Neotropics

A total of 403 studies were selected. Of these, 24 of 33 existing Latin American and Caribbean countries were represented (Brazil, Peru, Argentina, Colombia, Chile, Ecuador, Mexico, Venezuela, Costa Rica, Dominican Republic, French Guiana, Panama, Guatemala, Bolivia, Uruguay, Jamaica, Trinidad and Tobago, Martinique, Paraguay, Grenada, St. Kitts, Cuba, Nicaragua, and Surinam). Brazil was the country with the highest number of publications (123), followed by Mexico (85),

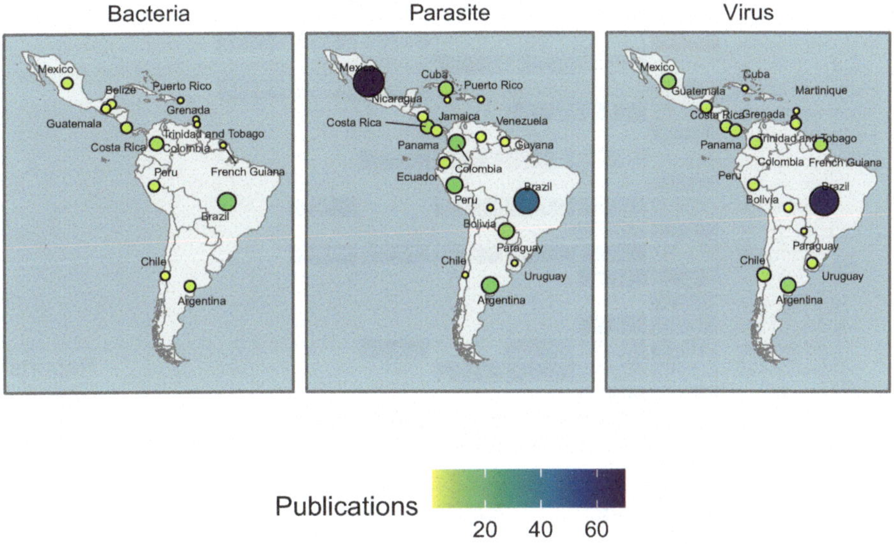

Fig. 8.1 Distribution map of the number of publications on pathogens associated with bats

Argentina (29), Colombia (28), Peru (22), Costa Rica (17), Paraguay (13), Chile (11), and Cuba (10) (Fig. 8.1). In terms of groups of pathogens, Brazil, Mexico, Argentina, Costa Rica, Chile, and Peru were the countries that contained publications belonging to all three groups of pathogens on which we aimed to focus in this chapter: Virus, Bacteria, and Parasites (Hemosporidians, *Babesia, Hepatozoon, Trypanosoma*, and Helminths; Nematoda and Platyhelminthes). Brazil mainly contained publications of viruses such as Rhabdoviridae, Coronaviridae, and Paramyxoviridae but also contained a significant number of publications of pathogenic bacteria (Bartonella, Mycoplasma, and Rickettsia). Regarding parasite publications, Brazil, together with Mexico, has the highest number of publications reporting Nematoda, Platyhelminthes, and Trypanosoma parasites.

Apart from containing the highest number of publications, Brazil represented the country with the highest pathogen richness (840), followed by Argentina (220), Mexico (203), Chile (180), Costa Rica (168), and Peru (126). These six countries contained most of the studies that we were able to retrieve. Additionally, our first search revealed that almost all pathogens included in our study (over 1380 parasite observations out of 2214 total pathogen observations) are being recorded and studied in bat species of the family Phyllostomidae (1383), followed by bats of the family Molossidae (675), Vespertilionidae (414), Mormoopidae (247), Emballonuridae (56), Natalidae (38), Noctilionidae (10), and Thyropteridae (1). Most records observed in the family Phyllostomidae belong to Nematoda parasites, followed by the viral families Rhabdoviridae and Coronaviridae (Fig. 8.2). Several pathogens (e.g., *Anaplasma,* Hepeviridae, Hepadnaviridae, Arenaviridae, and Adenoviridae) were found to be uniquely associated with the Phyllostomidae family. Other

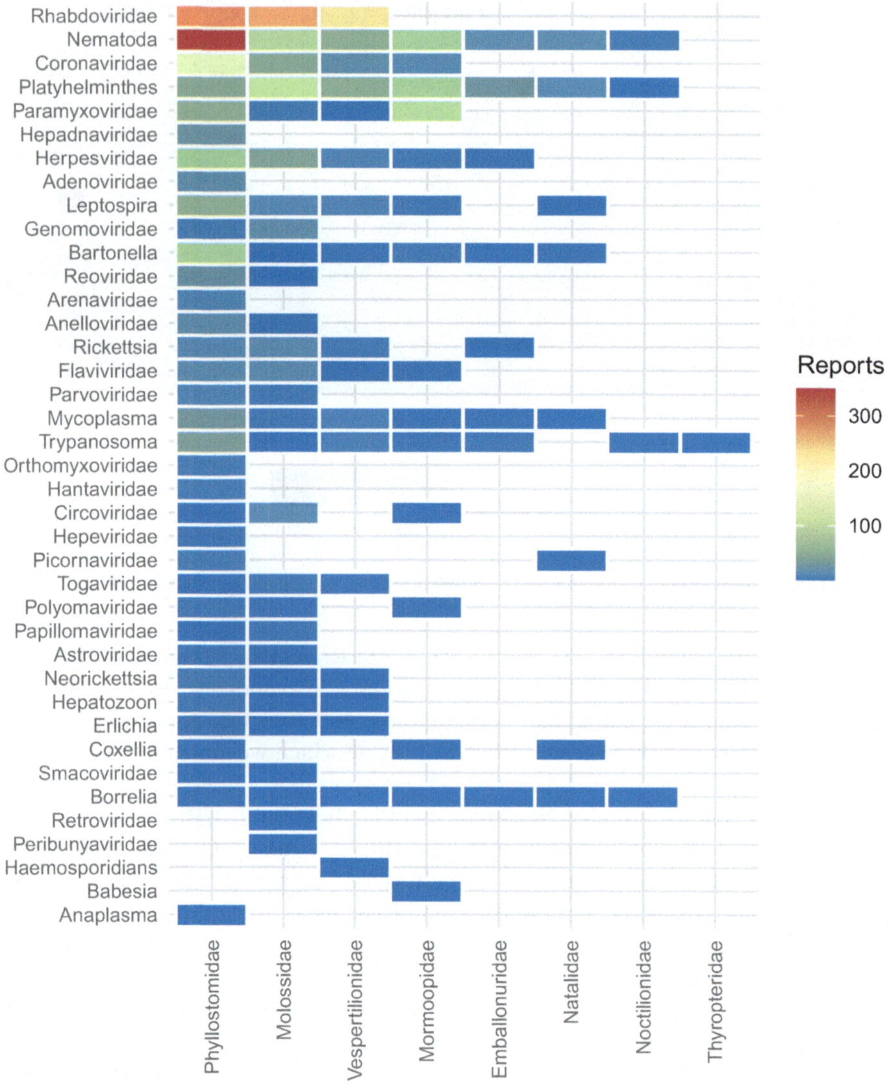

Fig. 8.2 Heatmap of the pathogen diversity associated with bat families

families with exclusive associations were Molossidae with retroviruses and peribunyaviruses, Vespertilonidae with haemosporidians, and Mormoopidae with *Babesia* (Fig. 8.2). The bat species with the highest number of publications were *Desmodus rotundus* (75), *Carollia perspicillata* (58), *Tadarida brasiliensis* (56), *Molossus molossus* (41) and *Artibeus lituratus* (41) (Fig. 8.3). The high number of pathogens associated with these species can be attributed to their widespread distribution; as they are found throughout the majority of the Neotropics, they are considered habitat generalists that are able to tolerate the alteration of their habitat. In particular,

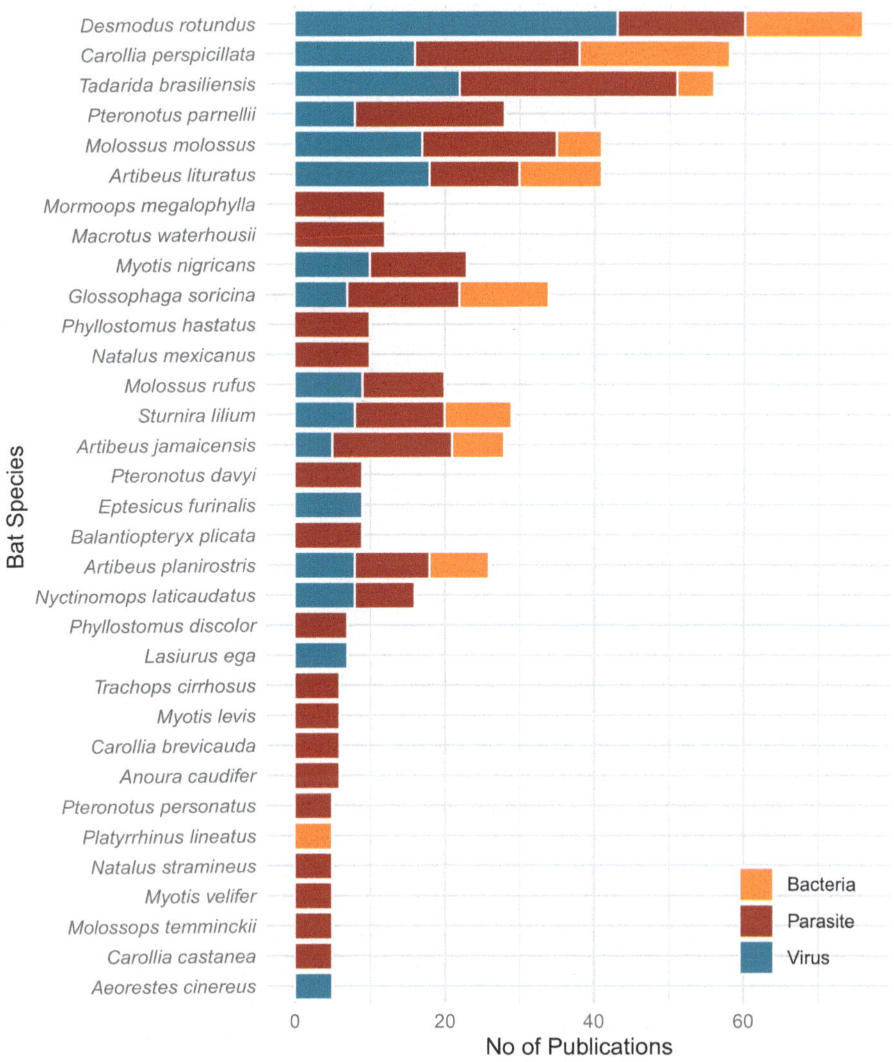

Fig. 8.3 Number of publications by bat species. Species with more than four publications are shown

D. rotundus and *C. perspicillata* are considered bioindicators of altered ecosystems (Medellín et al. 2000). When analyzing the relationship between species richness and the richness of reported pathogens, a strong positive correlation was found between the richness of viruses ($R^2_{adj} = 0.89$, $p < 0.001$) and parasites ($R^2_{adj} = 0.85$, $p < 0.001$) and the richness of bats, whereas the positive correlation between bacteria ($R^2_{adj} = 0.34$, $p = 0.015$) and bats was not as strong (Fig. 8.4).

In the second search, 20 studies were selected where anthropogenic or ecological factors (fragmentation, habitat loss, habitat type, or ecological factors) were

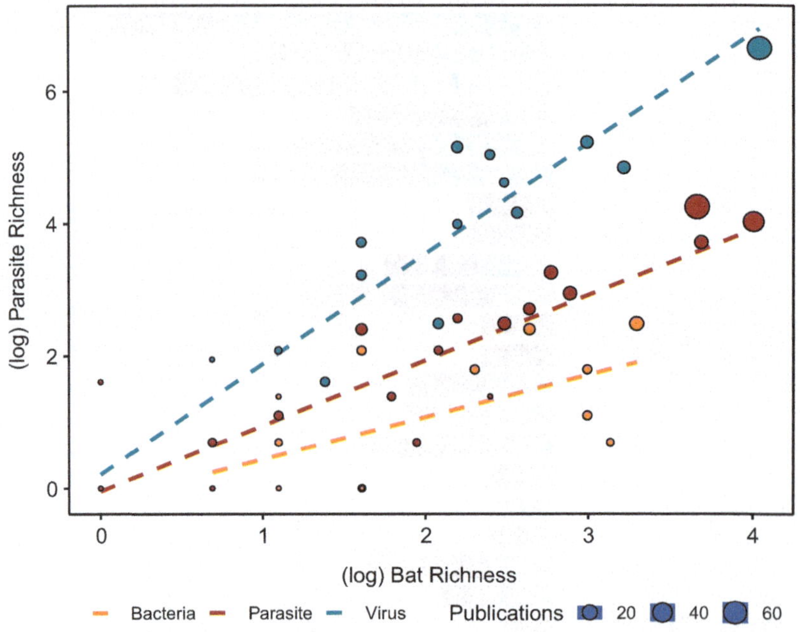

Fig. 8.4 Relationship between bat richness per country and the number of pathogen reports

examined in relation to the presence, prevalence, or diversity of pathogens. Only six studies correlated anthropogenic variables such as fragmentation, habitat loss, or habitat type with the presence, diversity, and presence of pathogens. Fourteen studies analyzed ecological variables such as host specificity, vegetation cover, functional characteristics, climate, temperature, trophic guilds, population properties, and elevational gradient (Table 8.1). In these 20 studies, the following pathogen types were evaluated: bacteria (*Mycoplasma*), hemoparasites (*Trypanosoma*), helminths (Trematoda, Cestoda, and Nematoda), and viruses (primarily Rabies, Coronaviridae, Herpesviridae, and Hepadnaviridae).

Clearly, there is a bias in the study of viruses associated with bats, while most studies focus on detecting, sequencing, and reporting viruses of medical significance. There are studies that attempt to comprehend the patterns of association and distribution between viral diversity and host diversity (Luis et al. 2013, 2015; Nieto-Rabiela et al. 2018, 2019). It is predicted that viral richness will be higher in species with a trophic frugivorous guild (Luis et al. 2015; Willoughby et al. 2017), that are long-lived (Luis et al. 2013; Guy et al. 2020) and have a larger body size (Maganga et al. 2014), that form large colonies (Turmelle and Olival 2009; Guy et al. 2020) and that have extensive geographical ranges that overlap with those of multiple species (Luis et al. 2013; Maganga et al. 2014; Olival et al. 2017).

There is evidence to suggest that deforestation can increase the prevalence and diversity of viruses in bat hosts. A study conducted in the Atlantic Brazilian Forest provides additional proof. The study identified 13 distinct viral species belonging to

Table 8.1 Summary of the selected studies that explore the association between anthropogenic/ecological factors and pathogen prevalence/diversity in Neotropical bats

Factor	Region	Parasite Studied	Bat Family	Main findings	References
Anthropogenic	Brazil	Bacteria: *Mycoplasma spp*	Phyllostomidae	Hemotropic *Mycoplasma* spp. were identified in one (*Sturnira parvidens*) out of 15 bats tested from forest fragments	Collere et al. (2022)
Ecological	Belize	Bacteria: *Mycoplasma spp*	Phyllostomidae	Hemoplasmas displayed high genetic diversity and strong host specificity.	Becker et al. (2020)
Ecological	Peru and Belize	Bacteria: *Mycoplasma spp*	Phyllostomidae	Prevalence was highest in nonreproductive and young individuals, did not differ by country, and was relatively stable over time.	Volokhov et al. (2017)
Anthropogenic	Panama	Hemoparasite: *Trypanosome sp*	Phyllostomidae	Trypanosome prevalence was significantly higher in bats from forest fragments than in bats captured in continuous forest.	Cottontail et al. (2009)
Ecological	Mexico	Parasites: Trematoda, Cestoda and Nematoda	Phyllostomidae, Noctilionidae, Molossidae, Vespertilionidae, Emballonuridae	Prevalence and mean intensity of infection values ranged from 7.1% to 100% and from 1 to 56, respectively.	Moguel-Chin et al. (2023)
Anthropogenic	Trinidad and Tobago	Virus: Rabies	Phyllostomidae	Juvenile bats were more likely to be seropositive than adults, suggesting early exposure to the RABV. RVNA was found in bats from both rural and nonrural areas, as well as in both hematophagous and nonhematophagous bat species.	Seetahal et al. (2020)

(continued)

Table 8.1 (continued)

Factor	Region	Parasite Studied	Bat Family	Main findings	References
Anthropogenic	Mexico	Virus: Coronaviridae, Herpesviridae, Adenoviridae and Astroviridae	Phyllostomidae, Mormoopidae, Vespertilonidae	A positive relationship was found between the phylogenetic diversity of hosts and viral diversity and with viral richness. Statistical models categorized host species and habitat types that have a higher risk of being associated with a greater viral richness.	Rico-Chávez et al. (2015)
Anthropogenic	Brazil	Virus: Coronaviridae, Herpesviridae, Adenoviridae, Hantaviridae, Paramyxoviridae and Astroviridae	Phyllostomidae, Molossidae, and Vespertilionidae	Deforested sites had a less diverse bat community and higher viral prevalence than forested sites Viral detection was more likely in juvenile males located in deforested sites. Coronaviridae family were detected more frequently in generalist species.	Loh et al. (2022)
Ecological	Panama	Virus: Hepadnavirus	Phyllostomidae	Distribution of TBHBV in bats was significantly affected by forest cover, with higher infection rates in areas with lower forest cover.	Hiller et al. (2019)
Ecological	Uruguay	Virus: Herpesviridae, Rhabdovirus and Pneumovirus	Phyllostomidae, Molossidae, and Vespertilionidae	The viral community in the bat species is more strongly associated with ecological traits of the hosts, rather than their taxonomy. Herpesviruses may be used to track the population dynamics of their hosts, including movement and demographics.	Moreira Marrero et al. (2021)
Ecological	French Guiana	Virus: Rabies	Phyllostomidae	Species' seroprevalence ranged from 0% to 20%, and the risk of seropositivity was higher in bats with a hematophagous diet, living in monospecific colonies and in dense forests. Temporal fluctuations in seroconversions and waning immunity were observed.	de Thoisy et al. (2016)

Ecological	French Guiana	Virus: Nairoviridae, Circoviridae, Retroviridae, Herpesviridae, Papillomaviridae	Phyllostomidae and Molossidae	Viral diversity was shaped by the diet of the bat species, with a high proportion of plant and insect-related viral families for *M. molossus* and a high proportion of vertebrate-related viral families for *D. rotundus*.	Salmier et al. (2017)
Ecological/ Anthropogenic	Peru	Virus: Rabies	Phyllostomidae	Seroprevalence of rabies virus (RV) in vampire bats ranged from 3% to 28%, with highest prevalence in juvenile and subadult bats. RV exposure was independent of bat colony size, suggesting an absence of population density thresholds for viral invasion and extinction. Culling campaigns implemented during the study failed to reduce seroprevalence and may have been counterproductive for disease control.	Streicker et al. (2012b)
Ecological	Puerto Rico	Virus: Herpesviridae	Phyllostomidae, Molossidae and Mormoopidae	Reproductively active bats, especially females, have significantly higher viral richness. Host sex significantly affects viral turnover across host populations.	Sjodin et al. (2020)
Ecological	Costa Rica, Panama, Ecuador and Brazil	Virus: Coronaviridae	Phyllostomidae, Noctilionidae, Molossidae, Vespertilionidae and Emballonuridae	The 50 viruses reported represented five different alphacoronavirus RGUs and two betacoronavirus RGUs, and closely related alphacoronaviruses were detected in *Carollia perspicillata* and *C. brevicauda* across a geographical distance exceeding 5600 km.	Corman et al. (2013)
Ecological	Argentina	Virus: Rabies	Phyllostomidae	Phylogenetic analysis of N gene sequences from vampire bat-related RABV variants showed a geographical and temporal structure. Multiple introductions of the infection were detected.	Torres et al. (2014)

(continued)

Table 8.1 (continued)

Factor	Region	Parasite Studied	Bat Family	Main findings	References
Ecological	Neotropic	Virus: Rabies	Phyllostomidae	Viral evolutionary rates were labile following historical jumps between bat species and nearly four times faster in tropical and subtropical bats compared to temperate species. The association between geography and viral evolution could not be explained by host metabolism, phylogeny, or variable selection pressures, and instead appeared to be a consequence of reduced seasonality in bat activity and virus transmission associated with climate.	Streicker et al. (2012a)
Ecological	Peru	Virus: Rabies	Phyllostomidae	Spatially compartmentalized transmission cycles of rabies virus were observed in vampire bats, but no effects of bat demography or environmental context on seroprevalence were found. Seroprevalence in bats was only weakly correlated with outbreaks in livestock, indicating the challenge of predicting spillover.	Meza et al. (2022)
Ecological	Peru	Virus: Rabies	Phyllostomidae	Gradual spread of RABVs and involvement of different vampire bat subpopulations. New RABV lineages were found in 3 non–vampire bat species, suggesting new virus reservoirs.	Condori-Condori et al. (2013)
Ecological	Peru	Viroma	Phyllostomidae	Viral communities in vampire bats vary significantly and are not consistently predicted by spatial or genetic distances. Viral diversity follows an elevational gradient, is enriched by juvenile-biased age structure, and declines with local anthropogenic food resources.	Bergner et al. (2020)

four viral families known to infect humans with Coronaviridae viruses, the majority of which were generalists. The likelihood of viral detection was higher in young male bats, who were more likely to interact and share food sources with unfamiliar host species, increasing their exposure to potential viruses (Loh et al. 2022). The study demonstrated that despite having fewer bat species, deforested regions maintain higher virus richness. Although the risk of disease was not directly assessed, people living near disturbed forest edges may be particularly vulnerable to zoonotic infections due to increased exposure to disease reservoirs and because deforested areas have higher viral richness. These findings are consistent with a published study that examined the association between bat diversity and the variety of four medically important virus families across an environmental gradient in human-dominated settings in southern Mexico (Rico-Chávez et al. 2015). The generalist species were linked with more viruses than predicted when examining the kind of host response to habitat change, and a positive correlation between the beta diversity of both viral and bat communities and the habitat gradient was found.

The habitat heterogeneity hypothesis (Lawton 1983) postulates a strong relationship between environmental diversity; in this case, phylogenetic host diversity and biological diversity (pathogen diversity) were supported by the significant positive correlation of phylogenetic diversity with viral richness and viral diversity. The findings imply that variables such as fragmentation and habitat loss drive species assemblages, resulting in regions of increased risk for the establishment of zoonotic species (Rubio et al. 2014; Kamiya et al. 2014; Gay et al. 2014). Viral community composition changes throughout the assessed anthropogenic settings revealed a significant dependency on host species change. The study discovered significant rates of turnover in viral populations, validating the perturbation theory, which states that changes in land use alter parasite dynamics in multihost systems by modifying the mode of transmission across species (Murray and Daszak 2013). According to several studies (Lloyd-Smith et al. 2009; Murray and Daszak 2013; Brearley et al. 2013), habitat change causes the exposure of new hosts to a diverse pool of parasites, particularly in locations with high levels of biodiversity.

Future research is needed to determine which characteristics of the host determine microparasite community assemblages, but greater microparasite diversity does not necessarily translate into higher health risks. In fact, the relationship between parasites and host diversity suggests that a pathogen's chances of posing a health risk are only increased in disturbed, deforested environments. Using machine learning, it has been possible to accurately identify reservoir hosts (Plowright et al. 2019; Becker et al. 2022). In addition, data-driven identification of vector species of a wide range of zoonotic diseases has been carried out, evaluating the vectorial capacity of various mosquito and tick species (Yang and Han 2018). Utilizing these predictive tools could assist in prioritizing sampling of bat species and pathogen surveillance. Moreover, combining refined molecular tools (e.g., metagenomic sequencing) with noninvasive collection methods (e.g., eDNA) could not only optimize pathogen surveillance but also contribute to the conservation of the targeted species (Giles et al. 2021).

8.4 Investment in Research and Development and Surveillance of Viral Diversity in Bats

From 1996 to 2018, the World Bank database contains information on gross domestic expenditure on research and development for 15 Latin American countries. The average expenditure varies across the Neotropical region, ranging from 0.042% in Guatemala to 1.12% in Brazil. Among the countries with the greatest bat richness, Brazil (bacteria = 103, parasite = 160, virus = 620) and Mexico (bacteria = 15, parasite = 334, virus = 115) are the only countries with a significant number of pathogen registers, which coincides with their higher R&D expenditure (1.12%). In contrast, Colombia (bacteria = 40, parasite = 131, virus = 11) and Peru (bacteria = 33, parasite = 38, virus = 62), which also have a high diversity of bat species, have very few reports due to minimal pathogen surveillance efforts (Turmelle and Olival 2009; Olival et al. 2017).

The linear regression between total reported bat pathogens per country and gross domestic expenditure on research and development as a percentage of the gross domestic product showed a positive trend (bacteria: $r = 0.74$, $p < 0.001$, parasite: $r = 0.55$, $p < 0.001$, virus: $r = 0.88$, $p < 0.001$). On the other hand, the relationship between bat richness and total reported bat viruses per country did not show a statistically significant correlation (Fig. 8.5).

Most countries in the region spend less than 0.5% of their GDP on research and development, which translates to a low national investment in the detection and surveillance of bat-associated viruses. Despite the importance of the relationship between bats and viruses, the pathogen diversity described in Latin America depends largely on foreign investment. This highlights the need to prioritize the allocation of resources to the most vulnerable regions, especially those experiencing high rates of land use change, and to direct epidemiological surveillance to areas with a higher likelihood of infectious disease emergence. Encouraging transdisciplinary approaches aimed at both conserving bats and their ecosystems and improving human health should be a priority.

8.5 A Multidisciplinary Approach to Bat Rabies Virus in Latin America

Rabies, an acute progressive viral encephalitis (family Rhabdoviridae, genus Lyssavirus), is one of the oldest known zoonoses associated with bats, but it continues to pose a threat to public and animal health systems on most continents. The natural history and evolution of lyssaviruses are largely attributed to bats (Rupprecht et al. 2002). Rabies transmitted by *Desmodus rotundus*, commonly known as the common vampire bat, has become an increasingly concerning and previously underestimated threat. It currently represents the leading cause of rabies-related deaths in both humans and livestock in Latin American countries, where canine rabies is on

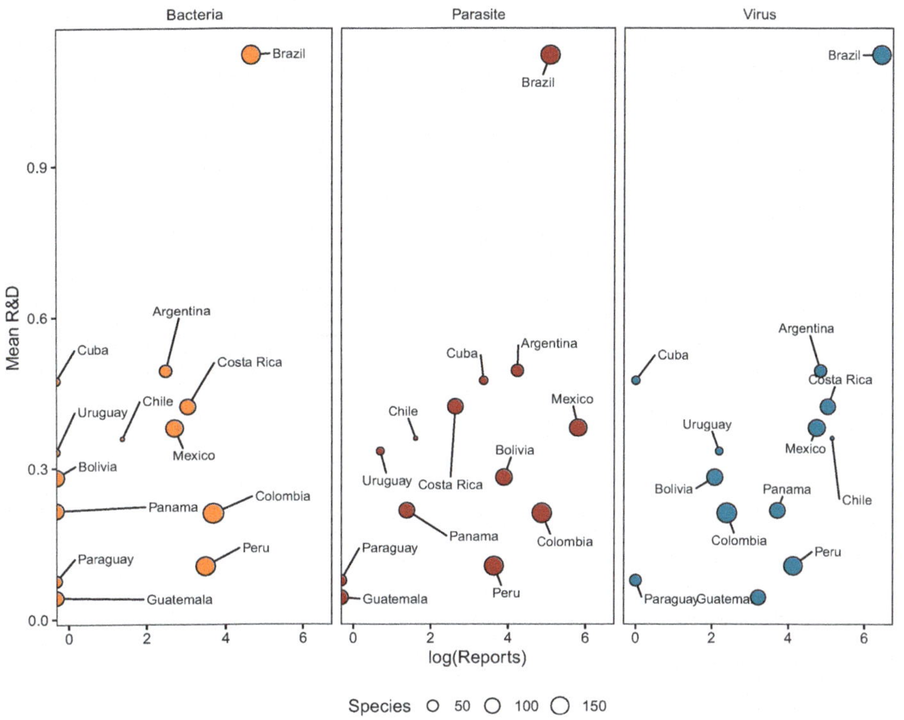

Fig. 8.5 Relationship between research and development spending and pathogen richness reported per country

the verge of eradication (Vigilato et al. 2013; Velasco-Villa et al. 2017; Benavides et al. 2017). Among more than a dozen lyssaviruses associated with bats, rabies virus (RABV) is the most significant member and species of the genus from a public health perspective. It is unknown how RABV spreads to new hosts and why spill-over events typically result in one-time infections as opposed to sustained infections. Due to the frequency of interaction between conspecifics, it is hypothesized that host taxonomy and social behavior play a role in RABV persistence (Hughes et al. 2005). Species barriers also limit the host range of the rabies virus (Streicker et al. 2010).

Although measuring the human health burden of RABV is difficult due to the geographic isolation of infected communities and underreporting, the rates of bat depredation on humans and the associated rabies mortality can be alarming. Surveys indicate that in high-risk areas, 23–88% of residents are bitten by *D. rotundus*, causing the death of 1–7% of the local human population during epidemics. Agricultural losses are also substantial, with tens of thousands of livestock being killed by RABV annually, resulting in a cost of more than $30 million USD before accounting for underreporting and recurring investments in surveillance, diagnostics, and prevention (Benavides et al. 2017).

In Latin America, efforts to control and prevent RABV have led to the establishment of various programs, including surveillance, livestock vaccination, pre- and postexposure human prophylaxis, and bat eradication (Johnson et al. 2014). The only method currently employed to reduce RABV transmission within the reservoir is the eradication of bats to reduce their population density. Theoretically, eradication of disease reservoirs is most effective when pathogen transmission rates are dependent on host density, with a critical density threshold below which the pathogen cannot be maintained by the population (Lloyd-Smith et al. 2005). However, the elimination of vampire bat colonies has not had a significant effect on reducing the circulation of the virus; in contrast, there is evidence that this type of program could favor the persistence of RAVB in local populations (Choisy and Rohani 2006; Viana et al. 2023).

Several countries, such as Belize, Costa Rica, Ecuador, Guatemala, Mexico, Paraguay, and Peru, have reported an increase in the rate of RABV cases over the past decade, which requires a review of control and prevention strategies and the incorporation of multidisciplinary strategies, such as landscape management, new forms of livestock management, and the development of bat vaccines (Benavides et al. 2020).

8.5.1 *Landscape Features and the Presence of* Desmodus rotundus

There is substantial evidence that human activities influence the dynamics and distribution of animal populations (Lewis et al. 2021). Although these effects are typically detrimental to wild animal species (Chase et al. 2020), livestock production has benefited vampire bat populations in Latin America, the primary reservoir of RABV (Fenton et al. 1992; García-Morales et al. 2013). The common vampire bat (*Desmodus rotundus*) is one of three hematophagous bats found in the Neotropics. It mainly feeds on livestock, with a preference for cattle, frequently choosing juveniles, weak individuals, or those dozing at the herd's perimeter (Johnson et al. 2014; Anderson et al. 2014), while occasionally feeding on humans. The other two species, the hairy-legged vampire bat (*Diphylla ecaudata*) and the white-winged vampire bat (*Diaemus youngi),* primarily feed on birds and other wildlife (Fenton and Kunz 2003; Mayen 2003).

Understanding the factors that influence the geographic distribution and regional movement of *D. rotundus* is crucial to comprehending the RABV risk. *D. rotundus* is distributed throughout tropical and subtropical Latin America, from northern Mexico to northern Argentina and Chile (Lee et al. 2012). It inhabits diverse ecosystems, including tropical rainforests, coastal deserts, xeric shrublands, and montane regions as high as 3600 m (Lee et al. 2012). At large spatial scales, temperature, altitude, and precipitation are important factors influencing bat presence and rabies risk, whereas, at local scales, factors such as livestock density, human-induced forest fragmentation, and density/proximity of highways and rivers appear to be important (Lee et al. 2012; de Andrade et al. 2016). In the past decade, efforts have been made to investigate and better understand the population ecology and the influence of local landscape characteristics on the abundance and bite rates of *D. rotundus* on livestock.

The highest levels of forest loss and fragmentation in the Neotropics are currently associated with forest conversion into extensive pastures used for cattle raising, which appears to be the most important factor in explaining the abundance of *D. rotundus* in sites dominated by a pasture matrix (Sanchez-Gomez et al. 2022). It has been found that *D. rotundus* tends to be more abundant at sites that are embedded within pasture matrices with high levels of forest fragmentation. This pattern was characterized by a large number of agricultural areas and high landscape heterogeneity (Bolívar-Cimé et al. 2019; Sanchez-Gomez et al. 2022). In contrast, the abundance of *D. rotundus* tends to be considerably lower at sites with extensive fragments of continuous forest, nonfragmented forest or in areas with very low levels of fragmentation, where there are no easily accessible food sources for this hematophagous bat, such as cattle and other types of domestic livestock (Medellín et al. 2000).

This pattern suggests that *D. rotundus* is highly opportunistic when searching for food, as it frequently approaches rural areas with human settlements, where corrals, chicken coops, and pig farms, which typically contain confined animals, represent valuable and easily accessible food resources (Bobrowiec 2012). Thus, vampire bat populations are favored by their adaptability to human-altered environments, and they thrive in landscapes with high concentrations of cattle and other domestic livestock raised for human consumption (Bobrowiec et al. 2015). Therefore, the vampire bat is an excellent indicator species of human habitat disturbance or land-use change resulting from cattle husbandry in the Neotropics (Medellín et al. 2000).

It is known that *D. rotundus* prefers to fly using linear elements of the landscape, although no preference has been found for one type of element. However, it has been suggested that the use of linear elements of the landscape as a reference for flight is influenced by the level of contrast between the linear feature and the open areas. Moreover, food abundance is a factor promoting the use of linear features by insectivorous bats (Ávila-Flores and Sánchez Gómez 2019).

Studies indicate that the detection range of objects, including experimental "prey," is limited for *D. rotundus* through the use of echolocation, scent, thermoreception, and passive hearing of sounds emitted by prey. The role of vision in *D. rotundus* prey detection is uncertain, but it is suggested that their visual detection of prey decreases as the distance from their flight paths increases. This implies that *D. rotundus* is more likely to commute near wooded areas because they have greater prey availability. Domestic animals kept near wooded areas and *D. rotundus* roosting sites are the most vulnerable to hematophagous bat attacks. These findings have strong implications for livestock management in tropical regions, as production animals resting near linear landscape features are at higher risk of attacks by *D. rotundus* (Ávila-Flores and Sánchez Gómez 2019).

8.5.2 Effectiveness of Bat Culling on Rabies Prevalence

Reducing wildlife populations by lethal methods is a common strategy for preventing the spread of pathogens into human or domestic animal populations (Velasco-Villa et al. 2017). Lower densities of susceptible organisms are believed to reduce the incidence of infection in the reservoir and, consequently, the risk of transmission to other species. In extreme cases, reservoir populations may fall below a critical threshold that triggers pathogen extinction (Benavides et al. 2017). While culling has been effective in managing some host–pathogen systems, its effectiveness in other systems has been hampered by the underestimation of ecological complexity (Vigilato et al. 2013). Therefore, understanding the factors that contribute to the success or failure of culls in reducing disease transmission is essential for improving their efficacy, promoting investments in alternatives for disease control, such as vaccination or reproductive suppression, or increasing the attractiveness of such investments.

Few studies have modeled the effect of bat culling on the prevalence and circulation of the rabies virus in bat populations, and even fewer empirical studies have tested the effect of this type of management on the prevalence (Benavides et al. 2016, 2017; Viana et al. 2023). One of these studies conducted in Peru found that a large, geographically synchronized cull was associated with widespread reductions in vampire bat populations but had negligible effects on the occurrence or intensity of spillover to local livestock (Viana et al. 2023). Although preventive culls halted the virus's spread, culls in regions with active viral circulation had the opposite impact, accelerating the spread of rabies across the landscape. These findings support that the long-term maintenance of rabies through spatial processes limits the efficacy of culling as currently practiced. They also provide a rare empirical example of perturbation effects, in which changes in reservoir host behavior following incomplete eradication compromise spillover management by promoting pathogen spatial spread. The authors observed a reduction in the population density of the common vampire and a decrease in the rate of bites to livestock, but there was no reduction in the transmission of rabies to livestock. Rabies is only ephemerally present in individual bat colonies and can spread at low bat densities due to virus-induced behavioral changes that facilitate biting (Benavides et al. 2020). The social disruption caused by bat control may facilitate the spread of rabies by increasing bat dispersal (Blackwood et al. 2013). If culls reduce population immunity by preferentially removing adult immune individuals or if vacated niche space increases juvenile survival or immigration of naive individuals (Choisy and Rohani 2006), additional complications may arise. While empirical data remain limited, a comparison of seroprevalence between *D. rotundus* colonies with varying histories of culling suggests that culling is associated with higher rather than lower rabies seroprevalence in bats (Streicker et al. 2012b).

Additionally, population size is unrelated to rabies incidence, and there are no population thresholds below which rabies cannot invade. This emphasizes the importance of analyzing epidemiological and genomic data alongside real-world culls to provide fundamental insights into the determinants of viral maintenance within bat reservoirs and to assess how ecological and behavioral responses of bats to culling translate into pathogen spillover risk and spatial spread (Benavides et al. 2020).

8.5.3 Vaccination in Bats

The vaccination of reservoirs is a fundamental component of human rabies prevention for all nonbat reservoirs (Hampson et al. 2015; Benavides et al. 2019). Vaccines have been effectively applied to both canines and wild carnivores. One effective approach to combat wildlife rabies in some European regions and the dog/coyote rabies virus variant in the United States has been the use of oral rabies vaccine baits containing an attenuated recombinant vaccinia virus vector vaccine expressing the rabies virus glycoprotein gene (VRG) (Maki et al. 2017). A long-term oral rabies vaccine barrier program in the United States is preventing the spread of the raccoon rabies virus variant (Maki et al. 2017). However, to date, there is no official vaccination program against rabies in free-living bats, largely due to the difficulties caused by their ecology and immunology, especially in the common vampire bat. In laboratory experiments, recombinant viral vaccines using vaccinia (Almeida et al. 2005) and raccoon pox (Stading et al. 2016, 2017) vectors have been shown to be immunogenic and protective in hematophagous and insectivorous bats. Since both vectors are already used in large-scale campaigns targeting wildlife, their safety and lack of virulence reversion in nontarget species have been extensively evaluated (Brochier et al. 1991; Rocke et al. 2017). Alternatively, transmissible vaccines are being developed, which have a greater potential for spread, ensuring that the potential negative effects of a genetically modified microorganism in wild populations are minimized (Nuismer et al. 2016).

A recent experiment used spatiotemporally replicated deep sequencing to parameterize competing epidemiological mechanistic models of *D. rotundus* betaherpesvirus (DrBHV), a candidate vector for a vaccine against vampire bat-transmitted rabies (Griffiths et al. 2023). The study determined that lifelong infections with cycles of latency and reactivation, along with high transmission rates, are required to explain patterns of DrBHV infection observed in feral bats. Simulations demonstrated that inoculating a single bat with a DrBHV-vectored rabies vaccine could immunize greater than 80% of a bat population, thereby reducing the magnitude, frequency, and duration of rabies outbreaks by 50–95% (Griffiths et al. 2023).

Important next steps include investigating the dynamics of vaccine spread in the environment through field trials. However, even if vaccination of bats reduced rabies transmission, the deleterious effects of bat bites on humans and livestock would necessitate enhanced bat population management strategies (Benavides et al. 2020).

8.6 Impact of Bat-Virus Relationships on the Conservation of Latin American Bats

The natural link of bats with the night and darkness, in combination with their peculiar external appearance, has historically produced fear, aversion, or even repugnance in many people around the world. In many cultures, associating bats with evil, death, and magic has reinforced negative emotions and feelings toward bats. In the last century, the identification of bats as reservoirs of the rabies virus and the association of bat guano with the growth of *Histoplasma* fungus contributed to creating the popular notion that bats are synonyms of disease and danger. Not surprisingly, direct attacks on bat colonies and destruction of bat roosting sites were the most important causes of human-induced bat mortality during the nineteenth and twentieth centuries (O'Shea et al. 2016).

In tropical regions of Latin America, the public image of bats has been further deteriorated by the association of common vampire bats with the transmission of the rabies virus to cattle. The inefficient culling campaigns promoted by Latin American governments to control bovine paralytic rabies have contributed to reinforcing the idea that killing wild animals is the best option to reduce zoonotic disease transmission. In addition to the annual killing of dozens of thousands of common vampire bats under official culling campaigns, individuals of other species of bats have been erroneously poisoned or attacked by untrained people to control bovine paralytic rabies (Saldaña-Vázquez et al. 2013). In fact, direct kills and roost destruction are, at present, the main causes of multiple-mortality events of bats in Latin America (O'Shea et al. 2016).

In 2019, the emergence of COVID-19 raised the concern of bat biologists and conservationists (Fenton et al. 2020) because the causative pathogen (SARS-CoV-2) was closely related (96.2% genome sequence identity) to the RaTG13 coronavirus found in *Rhinolophus affinis* from Yunnan Province in China (Zhou et al. 2020). The only suggestion that this coronavirus could have a bat origin triggered a cascade of news in traditional media and social networks that put bats in the global spotlight (Cerri et al. 2022). By then, there was growing evidence that bats harbor a great diversity of viruses, including many SARS-like coronaviruses, and more research was generated to identify new bat-virus associations (Fig. 8.6). The worry of being infected by SARS-CoV-2 renewed the fear, aversion, and repugnance of bats in many people around the world (Lu et al. 2021). In Latin America, at the beginning of the pandemic, there were sound cases of attacks to bat colonies and roosts that resulted in the death of many individuals (e.g., https://phys.org/news/2020-03-peru-blamed-coronavirus.html). Further studies demonstrated that SARS-CoV-2 was sufficiently different from any coronavirus present in bats (or any other mammal) so that it should be considered a human-specific virus (Boni et al. 2020). To date, available information does not allow us to identify the precise origin of SARS. CoV-2 (Fenton et al. 2020). However, most scientists agree that the activities of modern human societies set the stage for the emergence of COVID-19 and other zoonotic

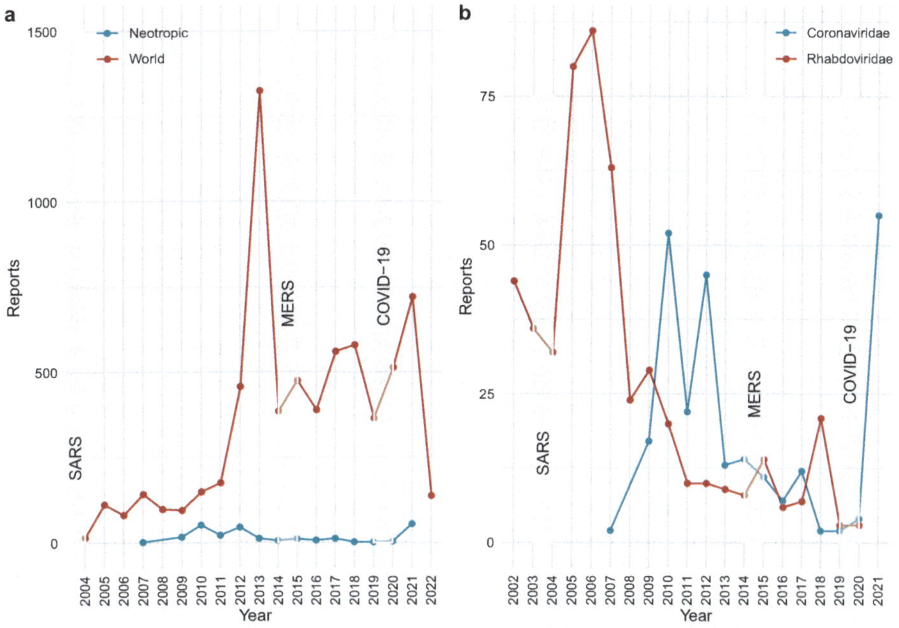

Fig. 8.6 Temporal trend of coronavirus reports in bats. The years of the beginning of the outbreaks of severe acute respiratory syndrome (SARS), Middle East respiratory syndrome (MERS), and COVID-19 are highlighted. (**a**) Trend of reports in the Neotropics and the rest of the world; (**b**) trend of reports of rabies virus and coronaviruses in the Neotropical region

diseases. In particular, it has been suggested that the risk of pathogen spillover has increased in recent times by (at least), (1) the reduction in the disease's dilution effect due to biodiversity loss, (2) the increase in contacts between people, domestic animals, and free-living wildlife due to habitat loss and land-use change, and (3) the accelerated exchange of native and exotic pathogens (e.g., in local wet markets) due to illegal trade, the appearance of new and distant markets that demand wildlife, and the movement of sick individuals (Fenton et al. 2020; Platto et al. 2021; Córdoba-Aguilar et al. 2021). Both the history of paralytic rabies control and the experience of the COVID-19 pandemic have left important lessons to manage enzootic and emergent zoonotic diseases in Latin America. First, the interaction between bats, domestic animals, and people is strongly mediated by human activities, so preventive or reactive strategies must consider the human dimension. Second, scientists from different disciplines should improve communication to design more integrative strategies. In this sense, the one-health approach might provide a suitable framework for guiding research efforts. Third, official protocols to manage diseases must be based on the best available evidence, so local authorities should maintain open communication channels with scientists.

8.7 Conclusions

In this chapter, we identify research opportunities to improve our knowledge of the complex relationship between humans, bats, and pathogens from a one-health perspective. Our review revealed that viruses and parasites are the most studied pathogen groups in bats throughout the Neotropical region. Currently, there is a significant bias in favor of viral research compared to bacterial research in bats. Despite the fact that viruses pose a significant risk to public health and host conservation due to their potential to spread to susceptible species, our understanding of infectious bacterial pathogens in bats is limited. Bats and their ectoparasites harbor potentially zoonotic bacterial pathogens (Mühldorfer 2013; Hornok et al. 2019). It was observed that epidemics caused by viruses genetically related to specific bat viruses have a direct impact on virus detection efforts (Fig. 8.6), reflecting a public health policy-driven bias.

The collected data revealed that Brazil, Mexico, Argentina, Costa Rica, Chile, and Peru had a greater diversity of parasite species than the other analyzed Neotropical countries. This was expected due to the wide diversity of bat species in these countries, but it should also be correlated with the large number of studies we were able to retrieve for these countries. Reduced funding for research in Latin America seems to limit pathogen surveillance based on molecular tools, although international collaboration might facilitate such efforts. In contrast, the use of inexpensive microbiological techniques appears to be an alternative to learning about different kinds of parasites. We encourage the science, agricultural, and health authorities of Latin American countries to increase investment in pathogen surveillance and zoonotic disease diagnostics in bat populations to identify conditions that increase the risk of pathogen or parasite spillover. We identified a variety of anthropogenic and ecological factors that influence the prevalence of pathogens throughout the Neotropics. However, few studies have analyzed anthropogenic factors, with most research focusing on contrasting conserved and altered sites. There is a lack of knowledge regarding the temporal and spatial diversity of pathogens associated with bats.

Due to its significance in public health and its economic effects on livestock production, the rabies virus is the Neotropical region's most studied bat-associated virus. To implement multidisciplinary strategies to prevent and monitor rabies, particularly in regions with high rates of habitat loss, it is necessary to integrate knowledge on population ecology and response to landscape characteristics by bats. There is evidence that the strategy based on vampire bat extermination has not been entirely effective and may, in some cases, contribute to the spread of rabies. Undoubtedly, the integration of multidisciplinary, multi-institutional, and even multinational teams could provide more fruitful results and help to design more effective strategies for disease prevention and control.

References

Almeida MF, Martorelli LFA, Aires CC et al (2005) Indirect oral immunization of captive vampires, Desmodus rotundus. Virus Res 111:77–82. https://doi.org/10.1016/j.virusres.2005.03.013

Anderson A, Shwiff S, Gebhardt K et al (2014) Economic evaluation of vampire bat (Desmodus rotundus) rabies prevention in Mexico. Transbound Emerg Dis 61:140–146. https://doi.org/10.1111/tbed.12007

Ávila-Flores R, Sánchez Gómez WS (2019) Use of linear features by the common vampire bat (Desmodus rotundus) in a tropical cattle-ranching landscape. Theyra 10:229–234

Becker DJ, Speer KA, Brown AM et al (2020) Ecological and evolutionary drivers of haemoplasma infection and bacterial genotype sharing in a Neotropical bat community. Mol Ecol 29:1534–1549. https://doi.org/10.1111/mec.15422

Becker DJ, Albery GF, Sjodin AR et al (2022) Optimising predictive models to prioritise viral discovery in zoonotic reservoirs. Lancet Microbe 3:e625–e637. https://doi.org/10.1016/S2666-5247(21)00245-7

Benavides JA, Valderrama W, Streicker DG (2016) Spatial expansions and travelling waves of rabies in vampire bats. Proc R Soc B Biol Sci 283:20160328

Benavides JA, Rojas Paniagua E, Hampson K et al (2017) Quantifying the burden of vampire bat rabies in Peruvian livestock. PLoS Negl Trop Dis 11:e0006105

Benavides JA, Megid J, Campos A et al (2019) An evaluation of Brazil's surveillance and prophylaxis of canine rabies between 2008 and 2017. PLoS Negl Trop Dis 13:e0007564

Benavides JA, Valderrama W, Recuenco S et al (2020) Defining new pathways to manage the ongoing emergence of bat rabies in Latin America. Viruses 12:1002

Bergner LM, Orton RJ, Benavides JA et al (2020) Demographic and environmental drivers of metagenomic viral diversity in vampire bats. Mol Ecol 29:26–39. https://doi.org/10.1111/mec.15250

Blackwood JC, Streicker DG, Altizer S, Rohani P (2013) Resolving the roles of immunity, pathogenesis, and immigration for rabies persistence in vampire bats. Proc Natl Acad Sci U S A 110:20837–20842. https://doi.org/10.1073/pnas.1308817110

Bobrowiec PED (2012) A Chiroptera preliminary survey in the middle Madeira River region of Central Amazonia, Brazil. 76:277–283. https://doi.org/10.1515/mammalia-2011-0065

Bobrowiec PED, Lemes MR, Gribel R (2015) Prey preference of the common vampire bat (Desmodus rotundus, Chiroptera) using molecular analysis. J Mammal 96:54–63. https://doi.org/10.1093/jmammal/gyu002

Bolívar-Cimé B, Flores-Peredo R, García-Ortíz SA et al (2019) Influence of landscape structure on the abundance of Desmodus rotundus (Geoffroy 1810) in northeastern Yucatan, Mexico. Ecosistemas y recur Agropecu 16, pp 263–271

Boni MF, Lemey P, Jiang X et al (2020) Evolutionary origins of the SARS-CoV-2 sarbecovirus lineage responsible for the COVID-19 pandemic. Nat Microbiol 5:1408–1417. https://doi.org/10.1038/s41564-020-0771-4

Bradley CA, Gibbs SEJ, Altizer S (2008) Urban land use predicts West Nile virus exposure in songbirds. Ecol Appl 18:1083–1092. https://doi.org/10.1890/07-0822.1

Brearley G, Rhodes J, Bradley A et al (2013) Wildlife disease prevalence in human-modified landscapes. Biol Rev Camb Philos Soc 88:427–442. https://doi.org/10.1111/brv.12009

Brochier B, Kieny MP, Costy F et al (1991) Large-scale eradication of rabies using recombinant vaccinia-rabies vaccine. Nature 354:520–522. https://doi.org/10.1038/354520a0

Carroll D, Daszak P, Wolfe ND et al (2018) The global virome project. Science 359:872–874. https://doi.org/10.1126/science.aap7463

Cerri J, Mori E, Ancillotto L et al (2022) COVID-19, media coverage of bats and related Web searches: a turning point for bat conservation? Mamm Rev 52:16–25. https://doi.org/10.1111/mam.12261

Chase JM, Blowes SA, Knight TM et al (2020) Ecosystem decay exacerbates biodiversity loss with habitat loss. Nature 584:238–243. https://doi.org/10.1038/s41586-020-2531-2

Choisy M, Rohani P (2006) Harvesting can increase severity of wildlife disease epidemics. Proc Biol Sci 273:2025–2034. https://doi.org/10.1098/rspb.2006.3554

Collere FCM, Ferrari LDR, Drozino RN et al (2022) Detecção e caracterização de micoplasmas hemotrópicos em morcegos de fragmentos florestais, Paraná, Sul do Brasil. Semin Ciênc Agrár 43:431–440. https://doi.org/10.5433/1679-0359.2022v43n1p431

Condori-Condori RE, Streicker D, Cabezas-Sanchez C, Velasco-Villa A (2013) Enzootic and epizootic rabies associated with vampire bats, Peru. Emerg Infect Dis J 19:1463. https://doi.org/10.3201/eid1909.130083

Córdoba-Aguilar A, Ibarra-Cerdeña CN, Castro-Arellano I, Suzan G (2021) Tackling zoonoses in a crowded world: lessons to be learned from the COVID-19 pandemic. Acta Trop 214:105780. https://doi.org/10.1016/j.actatropica.2020.105780

Corman VM, Rasche A, Diallo TD et al (2013) Highly diversified coronaviruses in neotropical bats. J Gen Virol 94:1984–1994. https://doi.org/10.1099/vir.0.054841-0

Cottontail VM, Wellinghausen N, Kalko EKV (2009) Habitat fragmentation and haemoparasites in the common fruit bat, Artibeus jamaicensis (Phyllostomidae) in a tropical lowland forest in Panamá. Parasitology 136:1133–1145. https://doi.org/10.1017/S0031182009990485

de Andrade FAG, Gomes MN, Uieda W et al (2016) Geographical analysis for detecting high-risk areas for bovine/human rabies transmitted by the common hematophagous bat in the Amazon region, Brazil. PLoS One 11:e0157332. https://doi.org/10.1371/journal.pone.0157332

de Carvalho-Ricardo MC, Uieda W, Fonseca RCB, Rossi MN (2014) Frugivory and the effects of ingestion by bats on the seed germination of three pioneering plants. Acta Oecol 55:51–57. https://doi.org/10.1016/j.actao.2013.11.008

de Thoisy B, Bourhy H, Delaval M et al (2016) Bioecological drivers of rabies virus circulation in a neotropical bat community. PLoS Negl Trop Dis 10:e0004378

Dobson AP, Pimm SL, Hannah L et al (2020) Ecology and economics for pandemic prevention. Science 369:379–381. https://doi.org/10.1126/science.abc3189

Epstein JH, Field HE, Luby S et al (2006) Nipah virus: impact, origins, and causes of emergence. Curr Infect Dis Rep 8:59–65

Fenton M, Kunz TH (2003) Bat ecology. University of Chicago Press, Chicago

Fenton MB, Acharya L, Audet D et al (1992) Phyllostomid bats (Chiroptera: Phyllostomidae) as indicators of habitat disruption in the neotropics. Biotropica 24:440–446. https://doi.org/10.2307/2388615

Fenton MB, Mubareka S, Tsang SM et al (2020) COVID-19 and threats to bats. FACETS 5:349–352. https://doi.org/10.1139/facets-2020-0028

Frick WF, Kingston T, Flanders J (2020) A review of the major threats and challenges to global bat conservation. Ann N Y Acad Sci 1469:5–25. https://doi.org/10.1111/nyas.14045

García-Morales R, Badano EI, Moreno CE (2013) Response of neotropical bat assemblages to human land use. Conserv Biol 27:1096–1106. https://doi.org/10.1111/cobi.12099

Gay N, Olival KJ, Bumrungsri S et al (2014) Parasite and viral species richness of southeast Asian bats: fragmentation of area distribution matters. Int J Parasitol Parasites Wildl 3:161–170. https://doi.org/10.1016/j.ijppaw.2014.06.003

Gibb R, Franklinos LHV, Redding DW, Jones KE (2020) Ecosystem perspectives are needed to manage zoonotic risks in a changing climate. BMJ 371:m3389. https://doi.org/10.1136/bmj.m3389

Giles JR, Peel AJ, Wells K et al (2021) Optimizing noninvasive sampling of a zoonotic bat virus. Ecol Evol 11:12307–12321. https://doi.org/10.1002/ece3.7830

Grange ZL, Goldstein T, Johnson CK et al (2021) Ranking the risk of animal-to-human spillover for newly discovered viruses. Proc Natl Acad Sci U S A 118. https://doi.org/10.1073/pnas.2002324118

Griffiths ME, Meza DK, Haydon DT, Streicker DG (2023) Inferring the disruption of rabies circulation in vampire bat populations using a betaherpesvirus-vectored transmissible vaccine. Proc Natl Acad Sci U S A 120:e2216667120. https://doi.org/10.1073/pnas.2216667120

Guy C, Ratcliffe JM, Mideo N (2020) The influence of bat ecology on viral diversity and reservoir status. Ecol Evol 10:5748–5758. https://doi.org/10.1002/ece3.6315

Hampson K, Coudeville L, Lembo T et al (2015) Estimating the global burden of endemic canine rabies. PLoS Negl Trop Dis 9:e0003709. https://doi.org/10.1371/journal.pntd.0003709

Hayman DTS (2016) Bats as viral reservoirs. Annu Rev Virol 3:77–99. https://doi.org/10.1146/annurev-virology-110615-042203

Hiller T, Rasche A, Brändel SD et al (2019) Host biology and anthropogenic factors affect hepadnavirus infection in a neotropical bat. EcoHealth 16:82–94. https://doi.org/10.1007/s10393-018-1387-5

Hornok S, Szőke K, Meli ML et al (2019) Molecular detection of vector-borne bacteria in bat ticks (Acari: Ixodidae, Argasidae) from eight countries of the Old and New Worlds. Parasit Vectors 12:50. https://doi.org/10.1186/s13071-019-3303-4

Hughes GJ, Orciari LA, Rupprecht CE (2005) Evolutionary timescale of rabies virus adaptation to North American bats inferred from the substitution rate of the nucleoprotein gene. J Gen Virol 86:1467–1474. https://doi.org/10.1099/vir.0.80710-0

IUCN (2022) The IUCN red list of threatened species. Bersion 2022–2. https://www.iucnredlist.org. Accessed 10 May 2023

Johnson N, Aréchiga-Ceballos N, Aguilar-Setien A (2014) Vampire bat rabies: ecology, epidemiology and control. Viruses 6:1911–1928. https://doi.org/10.3390/v6051911

Jones KE, Patel NG, Levy MA et al (2008) Global trends in emerging infectious diseases. Nature 451:990–993. https://doi.org/10.1038/nature06536

Kalka MB, Smith AR, Kalko EKV (2008) Bats limit arthropods and herbivory in a tropical forest. Science (80-) 320:71. https://doi.org/10.1126/science.1153352

Kamiya T, O'Dwyer K, Nakagawa S, Poulin R (2014) Host diversity drives parasite diversity: meta-analytical insights into patterns and causal mechanisms. Ecography 37:689–697. https://doi.org/10.1111/j.1600-0587.2013.00571.x

Kelly TR, Karesh WB, Johnson CK et al (2017) One Health proof of concept: bringing a transdisciplinary approach to surveillance for zoonotic viruses at the human-wild animal interface. Prev Vet Med 137:112–118. https://doi.org/10.1016/j.prevetmed.2016.11.023

Kemp J, López-Baucells A, Rocha R et al (2019) Bats as potential suppressors of multiple agricultural pests: a case study from Madagascar. Agric Ecosyst Environ 269:88–96. https://doi.org/10.1016/j.agee.2018.09.027

Lawton JH (1983) Plant architecture and the diversity of phytophagous insects. Annu Rev Entomol 28:23–39. https://doi.org/10.1146/annurev.en.28.010183.000323

Lee DN, Papeş M, Van Den Bussche RA (2012) Present and potential future distribution of common vampire bats in the Americas and the associated risk to cattle. PLoS One 7:e42466

Lewis JS, Spaulding S, Swanson H et al (2021) Human activity influences wildlife populations and activity patterns: implications for spatial and temporal refuges. Ecosphere 12:e03487. https://doi.org/10.1002/ecs2.3487

Lloyd-Smith JO, Cross PC, Briggs CJ et al (2005) Should we expect population thresholds for wildlife disease? Trends Ecol Evol 20:511–519

Lloyd-Smith JO, George D, Pepin KM et al (2009) Epidemic dynamics at the human-animal interface. Science 326:1362–1367. https://doi.org/10.1126/science.1177345

Loh EH, Nava A, Murray KA et al (2022) Prevalence of bat viruses associated with land-use change in the Atlantic Forest, Brazil. Front Cell Infect Microbiol 12:921950

Lu M, Wang X, Ye H et al (2021) Does public fear that bats spread COVID-19 jeopardize bat conservation? Biol Conserv 254:108952. https://doi.org/10.1016/j.biocon.2021.108952

Luis AD, Hayman DTS, Shea TJO et al (2013) A comparison of bats and rodents as reservoirs of zoonotic viruses: are bats special?. Proceedings of the Royal Society B, and the title breviation is Proc Biol Sci. https://doi.org/10.1098/rspb.2012.2753

Luis AD, O'Shea TJ, Hayman DTS et al (2015) Network analysis of host-virus communities in bats and rodents reveals determinants of cross-species transmission. Ecol Lett 18:1153–1162

Maganga GD, Bourgarel M, Vallo P et al (2014) Bat distribution size or shape as determinant of viral richness in african bats. PLoS One 9:e100172. https://doi.org/10.1371/journal.pone.0100172

Maki J, Guiot A-L, Aubert M et al (2017) Oral vaccination of wildlife using a vaccinia-rabies-glycoprotein recombinant virus vaccine (RABORAL V-RG(®)): a global review. Vet Res 48:57. https://doi.org/10.1186/s13567-017-0459-9

Mayen F (2003) Haematophagous bats in Brazil, their role in rabies transmission, impact on public health, livestock industry and alternatives to an indiscriminate reduction of bat population. J Veterinary Med Ser B 50:469–472. https://doi.org/10.1046/j.1439-0450.2003.00713.x

Medellín R, Equihua M, Amin M (2000) Bat diversity and abundance as indicators of disturbance in neotropical rainforests. Conserv Biol 14:1666–1675. https://doi.org/10.1111/j.1523-1739.2000.99068.x

Medellin RA, Wiederholt R, Lopez-Hoffman L (2017) Conservation relevance of bat caves for biodiversity and ecosystem services. Biol Conserv 211:45–50. https://doi.org/10.1016/j.biocon.2017.01.012

Meza DK, Mollentze N, Broos A et al (2022) Ecological determinants of rabies virus dynamics in vampire bats and spillover to livestock. Proc Biol Sci 289:20220860. https://doi.org/10.1098/rspb.2022.0860

Moguel-Chin WI, Hernández-Mena DI, Torres-Castro M et al (2023) Survey on helminths of bats in the Yucatan Peninsula: infection levels, molecular information and host-parasite networks. Parasitology 150:172–183. https://doi.org/10.1017/S0031182022001627

Mollentze N, Streicker DG (2020) Viral zoonotic risk is homogenous among taxonomic orders of mammalian and avian reservoir hosts. Proc Natl Acad Sci 117:9423–9430. https://doi.org/10.1073/pnas.1919176117

Moreira Marrero L, Botto Nuñez G, Malta L et al (2021) Ecological and conservation significance of herpesvirus infection in neotropical bats. EcoHealth 18:123–133. https://doi.org/10.1007/s10393-021-01530-2

Mühldorfer K (2013) Bats and bacterial pathogens: a review. Zoonoses Public Health 60:93–103. https://doi.org/10.1111/j.1863-2378.2012.01536.x

Murray KA, Daszak P (2013) Human ecology in pathogenic landscapes: two hypotheses on how land use change drives viral emergence. Curr Opin Virol 3:79–83. https://doi.org/10.1016/j.coviro.2013.01.006

Nieto-Rabiela F, Suzán G, Wiratsudakul A, Rico-Chávez O (2018) Viral metacommunities associated to bats and rodents at different spatial scales. Community Ecol 19:168–175. https://doi.org/10.1556/168.2018.19.2.9

Nieto-Rabiela F, Wiratsudakul A, Suzán G, Rico-Chávez O (2019) Viral networks and detection of potential zoonotic viruses in bats and rodents: a worldwide analysis. Zoonoses Public Health 66. https://doi.org/10.1111/zph.12618

Nuismer SL, Althouse BM, May R et al (2016) Eradicating infectious disease using weakly transmissible vaccines. Proc R Soc B Biol Sci 283:20161903. https://doi.org/10.1098/rspb.2016.1903

O'Shea TJ, Cryan PM, Hayman DTS et al (2016) Multiple mortality events in bats: a global review. Mamm Rev 46:175–190. https://doi.org/10.1111/mam.12064

Olival KJ, Epstein JH, Wang L-F et al (2012) Are bats exceptional viral reservoirs? In: Aguirre AA, Ostfeld RS, Daszak P (eds) New directions in conservation medicine. Applied cases of ecological health. Oxford University Press, New York, pp 195–212

Olival KJ, Hosseini PR, Zambrana-Torrelio C et al (2017) Host and viral traits predict zoonotic spillover from mammals. Nature 546:646–650. https://doi.org/10.1038/nature22975

Platto S, Zhou J, Wang Y et al (2021) Biodiversity loss and COVID-19 pandemic: the role of bats in the origin and the spreading of the disease. Biochem Biophys Res Commun 538:2–13. https://doi.org/10.1016/j.bbrc.2020.10.028

Plowright RK, Becker DJ, Crowley DE et al (2019) Prioritizing surveillance of Nipah virus in India. PLoS Negl Trop Dis 13:e0007393

Rico-Chávez O, Ojeda-Flores R, Zambrana-Torrelio C et al (2015) Viral diversity of bat communities in human-dominated landscapes in Mexico. Vet Mex 2:1–22
Rocha R, Aziz SA, Brook CE et al (2021) Bat conservation and zoonotic disease risk: a research agenda to prevent misguided persecution in the aftermath of COVID-19. Anim Conserv 24:303–307. https://doi.org/10.1111/acv.12636
Rocke TE, Tripp DW, Russell RE et al (2017) Sylvatic plague vaccine partially protects prairie dogs (Cynomys spp.) in field trials. EcoHealth 14:438–450. https://doi.org/10.1007/s10393-017-1253-x
Rubio AV, Avila-Flores R, Suzán G (2014) Responses of small mammals to habitat fragmentation: epidemiological considerations for rodent-borne hantaviruses in the Americas. EcoHealth. https://doi.org/10.1007/s10393-014-0944-9
Rulli MC, Santini M, Hayman DTS, D'Odorico P (2017) The nexus between forest fragmentation in Africa and Ebola virus disease outbreaks. Sci Rep 7:41613. https://doi.org/10.1038/srep41613
Rupprecht CE, Hanlon CA, Hemachudha T (2002) Rabies re-examined. Lancet Infect Dis 2:327–343. https://doi.org/10.1016/s1473-3099(02)00287-6
Saldaña-Vázquez RA, Sosa VJ, Iñiguez-Dávalos LI, Schondube JE (2013) The role of extrinsic and intrinsic factors in Neotropical fruit bat–plant interactions. J Mammal 94:632–639. https://doi.org/10.1644/11-MAMM-A-370.1
Salmier A, Tirera S, de Thoisy B et al (2017) Virome analysis of two sympatric bat species (Desmodus rotundus and Molossus molossus) in French Guiana. PLoS One 12:e0186943
Sanchez-Gomez WS, Selem-Salas CI, Cordova-Aldana DI, Erales-Villamil JA (2022) Common vampire bat (Desmodus rotundus) abundance and frequency of attacks to cattle in landscapes of Yucatan, Mexico. Trop Anim Health Prod 54:130. https://doi.org/10.1007/s11250-022-03122-w
Seetahal JFR, Greenberg L, Satheshkumar PS et al (2020) The serological prevalence of rabies virus-neutralizing antibodies in the bat population on the Caribbean Island of Trinidad. Viruses 12:178
Sjodin AR, Anthony SJ, Willig MR, Tingley MW (2020) Accounting for imperfect detection reveals role of host traits in structuring viral diversity of a wild bat community. bioRxiv 2020.06.29.178798. https://doi.org/10.1101/2020.06.29.178798
Smith I, Broos A, de Jong C et al (2011) Identifying Hendra virus diversity in pteropid bats. PLoS One 6:e25275
Song X-P, Hansen MC, Stehman SV et al (2018) Global land change from 1982 to 2016. Nature 560:639–643. https://doi.org/10.1038/s41586-018-0411-9
Stading BR, Osorio JE, Velasco-Villa A et al (2016) Infectivity of attenuated poxvirus vaccine vectors and immunogenicity of a raccoonpox vectored rabies vaccine in the Brazilian Free-tailed bat (Tadarida brasiliensis). Vaccine 34:5352–5358. https://doi.org/10.1016/j.vaccine.2016.08.088
Stading B, Ellison JA, Carson WC et al (2017) Protection of bats (Eptesicus fuscus) against rabies following topical or oronasal exposure to a recombinant raccoon poxvirus vaccine. PLoS Negl Trop Dis 11:e0005958. https://doi.org/10.1371/journal.pntd.0005958
Streicker DG, Gilbert AT (2020) Contextualizing bats as viral reservoirs. Science (80-) 370:172–173. https://doi.org/10.1126/science.abd4559
Streicker DG, Turmelle AS, Vonhof MJ et al (2010) Host phylogeny constrains cross-species emergence and establishment of rabies virus in bats. Science 329:676–679. https://doi.org/10.1126/science.1188836
Streicker DG, Altizer SM, Velasco-Villa A, Rupprecht CE (2012a) Variable evolutionary routes to host establishment across repeated rabies virus host shifts among bats. Proc Natl Acad Sci U S A 109:19715–19720. https://doi.org/10.1073/pnas.1203456109
Streicker DG, Recuenco S, Valderrama W et al (2012b) Ecological and anthropogenic drivers of rabies exposure in vampire bats: implications for transmission and control. Proc R Soc B Biol Sci 279:3384–3392. https://doi.org/10.1098/rspb.2012.0538
Szentivanyi T, McKee C, Jones G, Foster JT (2023) Trends in bacterial pathogens of bats: global distribution and knowledge gaps. Transbound Emerg Dis 2023:9285855. https://doi.org/10.1155/2023/9285855

Torres C, Lema C, Dohmen FG et al (2014) Phylodynamics of vampire bat-transmitted rabies in Argentina. Mol Ecol 23:2340–2352. https://doi.org/10.1111/mec.12728

Turmelle AS, Olival KJ (2009) Correlates of viral richness in bats (order Chiroptera). EcoHealth 6:522–539. https://doi.org/10.1007/s10393-009-0263-8

United Nations (2017) Research and development expenditure as a proportion of GDP. https://unstats.un.org/sdgs/metadata/files/Metadata-09-05-01.pdf

Velasco-Villa A, Escobar LE, Sanchez A et al (2017) Successful strategies implemented towards the elimination of canine rabies in the Western Hemisphere. Antivir Res 143:1–12. https://doi.org/10.1016/j.antiviral.2017.03.023

Viana M, Benavides JA, Broos A et al (2023) Effects of culling vampire bats on the spatial spread and spillover of rabies virus. Sci Adv 9:eadd7437. https://doi.org/10.1126/sciadv.add7437

Vigilato MAN, Clavijo A, Knobl T et al (2013) Progress towards eliminating canine rabies: policies and perspectives from Latin America and the Caribbean. Philos Trans R Soc Lond B, Biol Sci 368:20120143. https://doi.org/10.1098/rstb.2012.0143

Vittor AY, Pan W, Gilman RH et al (2009) Linking deforestation to malaria in the Amazon: characterization of the breeding habitat of the principal malaria vector, Anopheles darlingi. Am J Trop Med Hyg 81:5–12

Volokhov DV, Becker DJ, Bergner LM et al (2017) Novel hemotropic mycoplasmas are widespread and genetically diverse in vampire bats. Epidemiol Infect 145:3154–3167. https://doi.org/10.1017/S095026881700231X

Willoughby AR, Phelps KL, Olival KJ (2017) A comparative analysis of viral richness and viral sharing in cave-roosting bats. Diversity 9:35

Woo PCY, Lau SKP, Li KSM et al (2006) Molecular diversity of coronaviruses in bats. Virology 351:180–187. https://doi.org/10.1016/j.virol.2006.02.041

World-Bank (2022) DataBank. https://databank.worldbank.org

Yang LH, Han BA (2018) Data-driven predictions and novel hypotheses about zoonotic tick vectors from the genus Ixodes. BMC Ecol 18:7. https://doi.org/10.1186/s12898-018-0163-2

Zhou P, Yang X-L, Wang X-G et al (2020) A pneumonia outbreak associated with a new coronavirus of probable bat origin. Nature 579:270–273. https://doi.org/10.1038/s41586-020-2012-7

Chapter 9
Neotropical Primates and Humans: Risk of Bidirectional Parasite Transmission and Disease Sharing in Fragmented and Pristine Landscapes

Andrea Chaves, María del Carmen Villalobos-Segura, Juan Ricardo Sánchez Ayala, Óscar M. Chaves, Júlio César Bicca-Marques, Brenda Solórzano-García, Marco Antônio Barreto de Almeida, and Gustavo A. Gutiérrez-Espeleta

The original version of the chapter has been revised. A correction to this chapter can be found at https://doi.org/10.1007/978-3-031-50531-7_15

Supplementary Information The online version contains supplementary material available at https://doi.org/10.1007/978-3-031-50531-7_9.

A. Chaves (✉) · G. A. Gutiérrez-Espeleta
Escuela de Biología, Universidad de Costa Rica, San José, Costa Rica
e-mail: andrea.chavesramirez@ucr.ac.cr

M. del Carmen Villalobos-Segura · J. R. S. Ayala
Facultad de Medicina Veterinaria y Zootecnia, Universidad Nacional Autónoma de México, Mexico City, México

Ó. M. Chaves
Escuela de Biología, Universidad de Costa Rica, San José, Costa Rica

Laboratorio de Ensayos Biológicos, Universidad de Costa Rica, San José, Costa Rica

J. C. Bicca-Marques
Laboratório de Primatologia, Escola de Ciências da Saúde e da Vida, Pontifícia Universidade Católica do Rio Grande do Sul, Porto Alegre, RS, Brazil

B. Solórzano-García
Laboratorio de Parasitología y Medicina de la Conservación, ENES-Mérida, Universidad Nacional Autónoma de México, Mérida, Yucatán, Mexico

M. A. B. de Almeida
Núcleo de Vigilância dos Riscos e Agravos Ambientais Biológicos, Divisão de Vigilância Ambiental em Saúde, Centro Estadual de Vigilância em Saúde, Secretaria Estadual da Saúde do Rio Grande do Sul, Porto Alegre, RS, Brazil

G. Acosta-Jamett, A. Chaves (eds.), *Ecology of Wildlife Diseases in the Neotropics*, https://doi.org/10.1007/978-3-031-50531-7_9

9.1 Introduction: Neotropical Non-human Primate Diversity and Its Relationships with the Risk of Bidirectional Parasite Spillover and Disease Sharing with Humans

Primates are the fourth most diverse mammal order (523 species + 278 subspecies and varieties), after Artiodactyla (551 species), Chiroptera (1386 species), and Rodentia (2552 species, Burgin et al. 2018). The 179 Neotropical nonhuman primate (NHP) species represent 34% of the species diversity of the order primates (IUCN 2022). The 22 Neotropical primate genera are distributed into five families of the Platyrrhini infraorder: Callitrichidae, Cebidae, Aotidae, Pithecidae, and Atelidae (Rylands and Mittermeier 2009). Neotropical NHP occurs from ca. 24°N in southern Mexico (*Ateles geoffroyi*; Cortés-Ortiz et al. 2021) to 31°S in southern Brazil (*Alouatta guariba clamitans*; Printes et al. 2001) and possibly in northwestern Uruguay (*Alouatta caraya*; Jardim et al. 2020). Neotropical NHPs are arboreal, forest-living monkeys, with a few species having adapted to exploit more open habitats (e.g., *Callithrix jacchus*: Garber et al. 2019; *Sapajus libidinosus*, Wright et al. 2019; *Sapajus flavius*, Ferreira et al. 2009; *Saimiri oerstedii*, Solano Rojas 2018).

The smallest Platyrrhini monkeys from Callitrichidae are distributed into eight genera: *Callithrix*, *Cebuella*, *Callibella*, *Mico*, *Saguinus*, *Leontocebus*, *Leontopithecus*, and *Callimico*. According to the IUCN Red List of Threatened Species (IUCN 2022), *Callithrix* (marmosets) and *Leontopithecus* (lion tamarins) are restricted to the Atlantic Forest, *Cebuella* (pigmy marmosets), *Callibella* (dwarf marmosets), *Leontocebus* (tamarins), and *Callimico* (Goeldi's monkeys) are restricted to the Amazon, and *Saguinus* (tamarins) and *Mico* (marmosets) are mostly Amazonian. Callitrichid adult weight ranges from 130 g (*Cebuella*) to 700 g (*Leontopithecus*; Bicca-Marques et al. 2011). Callitrichids live in groups ranging from 2 to 20 individuals who use home ranges of 0.1 to ca. 400 ha (Digby et al. 2011). They are omnivorous and feed on a mix of fruit, seeds, flowers, nectar, exudate, invertebrates, and small vertebrate prey (Digby et al. 2011). Overall, callitrichids are well adapted to living in forest fragments, where they are more likely to use the ground (JC Bicca-Marques, personal communications; Santos et al. 1995). Most, if not all, genera have species used as pets (Costa-Araújo et al. 2022; de la Torre et al. 2021; Freire Filho et al. 2021; Link et al. 2021; Oliveira et al. 2021; Savage et al. 2022; Shanee et al. 2017). Although rare, callitrichids can also be hunted (Pereira et al. 2019; Roncal et al. 2018; Shaffer et al. 2017).

Only three genera belong to Cebidae: *Cebus*, *Sapajus*, and *Saimiri*. *Cebus* (gracile capuchin monkeys) occurs from the Amazon into the Guyanan Shield and the Andes of Colombia and Ecuador up to Guatemala in Central America. *Sapajus* (robust capuchin monkeys) is found throughout the Atlantic Forest, Caatinga, Cerrado, Pantanal, Chaco, and Amazon, and *Saimiri* (squirrel monkeys) occurs in most of the Amazon and in a narrow region of the Pacific coast of Panama and Costa Rica in Central America (IUCN 2022). Adult weight ranges from 620 g (*Saimiri*) to 4800 g (*Sapajus*; Jack 2011). Cebids live in groups ranging from 6 to 75 individuals who use home, ranging from 12 to almost 300 ha (Bicca-Marques et al. 2011).

All cebids exploit a highly omnivorous diet composed mostly of fruit and invertebrate and small vertebrate prey (Jack 2011). Capuchin and squirrel monkeys are predominantly canopy dwellers. Whereas squirrel monkeys rarely descend to the ground, gracile capuchins are considered the most terrestrial Neotropical NHP (Jack 2011). Populations of robust capuchins that inhabit dry Caatinga environments, such as *S. libidinosus* and *S. flavius*, also spend considerable amounts of time on the ground (Bezerra et al. 2014; Ferreira et al. 2009; Wright et al. 2019). Capuchins are more frequently hunted than squirrel monkeys, and all cebids are found in the pet trade (Pereira et al. 2019; Shaffer et al. 2017; Shanee et al. 2017).

Aotidae contains a single genus of nocturnal monkeys—*Aotus* (night or owl monkeys)—found throughout most of the Amazon, the Chaco, the foothills of the Andes in northwestern South America, and the forests of Panama in Central America (Fernandez-Duque 2011). Adult night monkeys weigh from ca. 600 to 1600 g and live in small groups of two to six individuals who use home ranges between 3 and 18 ha. Night monkeys are omnivorous, feeding on fruit, nectar, flowers, leaves, and insects (Fernandez-Duque 2011). Given their importance for malaria research, large numbers of night monkeys have been harvested to supply biomedical facilities (Maldonado and Peck 2014; Svensson et al. 2016). Despite their small body size, occasional hunting is also a threat (Pereira et al. 2019; Roncal et al. 2018).

Pithecidae comprises six genera: *Callicebus*, *Plecturocebus*, *Cheracebus*, *Pithecia*, *Chiropotes*, and *Cacajao*. Except for the Atlantic Forest *Callicebus* (titi monkeys), including *C. barbarabrownae* that also ranges in the Caatinga (Printes et al. 2011), and one *Plecturocebus* titi monkey that ranges in the Chaco and Pantanal (Byrne et al. 2016), the other titi monkeys (*Plecturocebus* and *Cheracebus*) and the pithecine *Pithecia* (sakis), *Chiropotes* (cuxiús), and *Cacajao* (uakaris) are restricted to the Amazon (Norconk 2011). Adult pithecids weigh from 850 g (*Plecturocebus*) to 3700 g (*Chiropotes*; Norconk 2011). Whereas titi monkeys and sakis live in small groups composed of as few as two individuals, cuxiús live in groups of up to a few dozen, and uakaris can be found in groups of over 100 individuals (Norconk 2011). The home ranges of pithecid groups also vary widely, from 1.4 to 550 ha (Norconk 2011).

The larger *Cacajao*, *Chiropotes*, and *Pithecia* are specialized seed predators that complement their diets with fleshy fruit, flowers, leaves, and a small proportion of insects (Norconk 2011). Titi monkeys rely more on fruit with leaves and invertebrate prey complementing their diets (Bicca-Marques and Heymann 2013). The use of the ground is rare among the larger, canopy dweller *Cacajao*, *Chiropotes*, and some species of *Pithecia*. However, this behavior is more frequent in smaller *Pithecia* species that spend longer times vertically clinging and leaping in the understory (Barnett et al. 2012), especially in *Callicebus* and *Plecturocebus* (Souza-Alves et al. 2019). Data available on *Cheracebus* are less reliable to evaluate whether they follow the pattern of the other titi monkeys or that of the larger Pithecids (Souza-Alves et al. 2019). Despite their less frequent use of the lower forest strata and ground, larger Pithecids are more vulnerable to hunting than titi monkeys (Pereira et al. 2019; Shaffer et al. 2017).

The largest Platyrrhini monkeys belong to four Atelidae genera: *Alouatta*, *Ateles*, *Brachyteles*, and *Lagothrix*. *Alouatta* (howler monkeys) presents the largest distribution of all primate genera in the Americas, occurring from 21°N in Mexico (Cortés-Ortiz et al. 2020) to 31°S in Brazil (Printes et al. 2001), *Ateles* (spider monkeys) ranges from the Amazon to the northern limit of primate distribution in the Americas (24°N; Cortés-Ortiz et al. 2021), *Lagothrix* (woolly monkeys) occurs only in the Amazon, and *Brachyteles* (muriquis) is restricted to a narrow portion of the Atlantic Forest of southeastern Brazil (Di Fiore et al. 2011). Adult weight ranges from 3100 g (*Alouatta palliata*) to 13,800 g (*Brachyteles hypoxanthus*; Di Fiore et al. 2011). Atelids live in groups ranging from 2 to ca. 80 individuals who use home ranges varying from <1 ha (Fortes et al. 2015) to 460 ha (García-Toro et al. 2019). Whereas *Alouatta* and *Brachyteles* are characterized by a folivorous-frugivorous diet, *Ateles* and *Lagothrix* are predominantly frugivorous, with *Lagothrix* often complementing the diet with animal prey (Di Fiore et al. 2011). Although all atelids are canopy dwellers, howler monkeys, well known for coping well with disturbed and small habitat patches immersed in fragmented landscapes (Bicca-Marques 2003; Bicca-Marques et al. 2020), often descend to the forest floor to cross canopy gaps, move between habitat patches, drink water, or access supplementary resources in the matrix (Bicca-Marques et al. 2020; Chaves et al. 2021a, b; Pozo-Montuy and Serio-Silva 2007; Pozo-Montuy et al. 2013). Additionally, one population of muriquis has been reported drinking, resting, feeding, playing, and traveling on the ground (Mourthé et al. 2007; Tabacow et al. 2009). As the largest Neotropical NHP, atelids are important targets of subsistence and commercial hunting irrespective of their legality (Freire Filho et al. 2021; Pereira et al. 2019; Peres 1990; Shaffer et al. 2017; Shanee 2011; Shanee et al. 2017). Both *Brachyteles* species, currently critically endangered because of habitat loss and high hunting pressure in the past (Melo et al. 2021; Talebi et al. 2021), continue to be vulnerable to illegal hunters (ICMBio 2011).

All characteristics of the platyrrhine genera described above influence the risk of parasite spillover between nonhuman primates and humans. Given marked regional differences in human population density and the resulting level of conservation of forest ecosystems, Neotropical NHP living in pristine forests of western Amazonia, for example, have less contact with humans than those inhabiting the highly fragmented Atlantic Forest.

On the one hand, interactions occur in the former context when humans hunt or keep monkeys as pets. Whereas these intimate contacts impose risks of bidirectional spillover, their consequences for monkeys are limited to those individuals kept as pets. For humans, on the contrary, the spillover from monkey carcasses or pets may spread through the population. On the other hand, the higher proximity between humans and wild monkeys in fragmented landscapes increases the risk that spillovers in any direction cause population-level negative consequences for both species. High contact between humans and NHPs is also common in some national parks and forest fragments of Nicaragua, Costa Rica, and Panama, where at least three species (*Alouatta palliata*, *Cebus imitator*, and *Saimiri oerstedii*) interact closely with tourists, farmers, and local inhabitants in subsistence orchards or gardens on the forest edge (e.g., Mansell and McKinney 2021; McKinney 2014; Ó. M. Chaves, personal observations). Monkeys inhabiting fragmented landscapes are more likely to use the ground, where they are exposed to parasites with fecal-oral transmission and those whose infective stages penetrate the skin (Nunn and Altizer 2006). Arboreal animals traveling on the ground are also more vulnerable to hunting and trapping, whose consequences for parasite sharing are described below. Additionally, monkeys inhabiting periurban or urban landscapes that forage in garbage cans are more likely to be exposed to human disease agents (Nunn and Altizer 2006).

Although the extent of the distribution of a given taxon is predicted to have a direct influence on its overall parasite diversity (Nunn and Altizer 2006), this relationship is unlikely to translate into higher risks of spillover at the local level. As mentioned above, differences in geographic distribution affect the likelihood of contact with humans and, consequently, the likelihood and modes of parasite sharing. They also contribute to marked differences in our knowledge on the parasitic fauna of the platyrrhine genera, as taxa living near developed urban centers are more frequent targets of research. This trend is clear in the comparison of the number of parasitological studies on wild platyrrhine populations. Whereas *Alouatta* (39 studies), *Callithrix* (13), and *Sapajus* (11) were targets of over 10 studies, strictly Amazonian genera were investigated in three or fewer studies (*Cebuella*, *Mico*, *Cheracebus*, *Cacajao*, *Chiropotes*, and *Pithecia*) or have not been studied at all (*Callimico* and *Callibella*, Rondón et al. 2021).

Combining the likely positive relationship between body mass and disease risk in primates in general (Nunn and Altizer 2006) with the stronger hunting pressure on larger NHPs in the Americas leads us to expect that there are higher risks of spillover from taxa used as bushmeat, except at tourist sites and orchards near the forest edge, as mentioned above. At the other extreme of the body mass gradient, the risks come mostly from individuals kept as pets. Moreover, all platyrrhine infants are kept as pets when their adult carriers are killed for food or trapped for the biomedical trade (Roncal et al. 2018).

Group size and aggregation influence the spread of infectious agents with direct transmission among individuals (Nunn and Altizer 2006). Group size also influences exposure to parasites via changes in ranging behavior. On the one hand, large home ranges and long day ranges expose hosts to a greater diversity of parasites. On the other hand, the overuse of small home ranges increases the risk of reinfection, with consequences on parasite prevalence and intensity of infection (Nunn and Altizer 2006; see also Bicca-Marques and Calegaro-Marques 2016). Neotropical NHPs that thrive in human-modified landscapes may reach high population densities in habitat fragments, increasing both the risk of within-group disease spread and the contact and sharing of parasites with humans and domestic animals.

Finally, host diet has direct and indirect implications for platyrrhine-parasite interactions. In addition to parasites with direct cycles and being vector-borne, monkeys feeding on animal matter are exposed to parasites whose life cycles involve intermediate invertebrate hosts. Diet composition is related to primate body size and influences survival in forest fragments, social organization, population density, and other variables that may modulate the likelihood of interactions with humans and bidirectional spillover.

9.1.1 Conservation of Neotropical Primates

Human activities, such as extensive agriculture, unplanned urbanization, deforestation, and mining, severely impact the natural habitats of free-ranging primates in the tropics (Estrada et al. 2017, 2019). For instance, ca. 60 Mha of tropical primary forest were lost from 2002 to 2019. This massive habitat destruction was headed by Brazil, which was responsible for almost half of it (24.5 Mha; Weisse and Gladman 2020). Therefore, Neotropical NHP are probably the most affected primates by this land cover change given their arboreality and dependence on resources - including plant and animal food, water, and refuge - provided by forest trees (Chapman et al. 2013; Peres 1994).

Deforestation, forest fragmentation, hunting, illegal pet trade, and urbanization, among other human disturbances, have caused ca. 42% of the 216 recognized platyrrhine taxa (i.e., 179 species + 37 subspecies) to be threatened with extinction and to be considered within the IUCN categories of vulnerable, endangered, and critically endangered (IUCN 2022). Nine of the 22 critically endangered Neotropical NHPs (Fig. 9.1) live in Brazil, four in Colombia, and the remaining are distributed

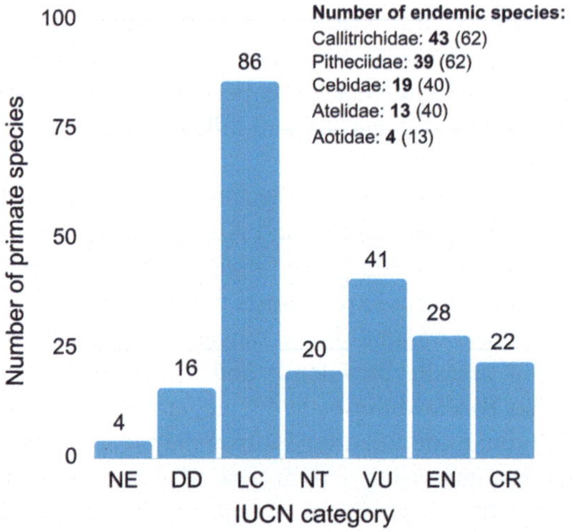

Fig. 9.1 Number of Neotropical NHP taxa ($N = 217$) in each IUCN category in 2022. Categories: *NE* not evaluated, *DD* data deficient, *LC* least concern, *NT* near threatened, *VU* vulnerable, *EN* endangered, *CR* critically endangered. The number of species in each category is shown above the bars. The box shows the number of endemic species (in bold) out of the total number of species (in parentheses) in each Platyrrhini family

among Peru, Ecuador, Bolivia, Venezuela, Panama, Costa Rica, and Nicaragua (Supplementary Table 9.1).

Given the impact of these habitat disturbances on population persistence, endemic Neotropical NHP may be more vulnerable to disappear or to experience sharp population declines than taxa ranging over wider regions, as reported for many terrestrial vertebrates (Chichorro et al. 2019, Manes et al. 2021). Overall, taxa endemic to a smaller distribution are 3 to 10 times more affected by climate change and other environmental disturbances than wide-ranging native taxa (Manes et al. 2021). For instance, most critically endangered Neotropical NHPs (i.e., 18 out of 22) are also endemic species to a single country (Fig. 9.1 and Supplementary Table 9.1).

Therefore, efforts to conserve Neotropical NHP must focus on preventing, controlling, and mitigating habitat loss and degradation, forest fragmentation, hunting, fire, and the growing threats of climate change, infectious diseases, and habitat pollution (Chapman and Peres 2021; IUCN 2022). Given the financial limitations of most Latin American countries, Neotropical NHPs that are endemic and critically endangered and those occurring in biomes severely impacted by human activities (e.g., Tropical Dry Forests: Hasnat and Hossain 2020; Atlantic Forest: de Lima et al. 2020) should be priority targets of conservation efforts. Ideally, long-term (multi-year) conservation initiatives should integrate national, regional, and local governments, academia, nongovernmental environmental organizations, and, certainly, local communities to be more effective in promoting both biodiversity conservation and the improvement of the well-being of the peoples living inside or near the remaining platyrrhine habitats. However, the viability and success of such initiatives depend on the often low and between-country variable access to public and private financial resources and qualified personnel.

We collected data on conservation initiatives focused on Neotropical NHPs and their habitats via a Google Forms short interview sent to the community of Latin American primatologists. We received information on 30+ long-term projects aiming to conserve 14+ taxa (Supplementary Table 9.2). Most (63%) of these projects are concentrated in Costa Rica, Mexico, and Brazil and focus on *Alouatta* (especially *A. palliata*, *A. guariba clamitans*, *A. caraya*, and *A. belzebul*) and, to a lesser degree, on *Ateles*, *Cebus*, *Saimiri*, *Brachyteles*, and *Leontopithecus*. Projects in Colombia, Paraguay, and Venezuela have focused on endemic and threatened species, such *as Plecturocebus caquetensis*, *Saguinus leucopus*, and *Cebus malitiosus* (Supplementary Table 9.2). Despite wide variation in the major objectives of these projects, we identified six general conservation goals: (i) improve our understanding of primate behavior and ecology, particularly in human-disturbed and/or small habitat patches, (ii) assess the distribution and demography of endemic or poorly studied species, (iii) develop management strategies to improve habitat connectivity and gene flow between isolated populations, (iv) prevent or minimize negative interactions with humans, including spillover events of zoonoses and zooanthropozoonosis in urban and periurban regions, (v) inform and sensitize the lay public and the media about the important role of free-ranging primates as sentinels of the circulation of agents of infectious diseases also affecting humans, and (vi) keep

Platyrrhini taxonomy and the main threats faced by each taxon constantly updated (Supplementary Table 9.2). Furthermore, most initiatives emphasize the importance of citizen science, volunteering, and the participation of local human communities in conservation efforts.

Although these efforts can improve the conservation of the target taxa and their habitats in the medium or long term, they cannot address all threats that Neotropical NHPs face. In fact, until today, most conservation efforts in Latin America that we present in Supplementary Table 9.2 have been focused on <10% of the 219 recognized primate taxa or are restricted to very small geographical areas, and their efficiency is rarely evaluated. This latter point has been stressed in recent reviews (e.g., Chapman and Peres 2021, Junker et al. 2020). For instance, <1% of 13,000 NHP studies published between 1971 and 2015 evaluated conservation efficiency, and those that addressed it did not include quantitative data (Junker et al. 2020). Therefore, there are many challenges in addition to funding limitations to qualify primate conservation efforts in Latin America. Many of them have been the focus of recent attention (Chapman and Peres 2021; Junker et al. 2020; Oxley et al. 2022; Waters et al. 2021; Webber et al. 2022). We hope that research on Neotropical NHP continues flourishing and contributing to increasing our knowledge and improving the health of NHPs, ecosystems, and humans in Latin America, ultimately promoting the effective conservation of Neotropical biodiversity.

9.2 Drivers of Neotropical Primate Infectious Agent Transmission

Among the main threats for Neotropical NHP are human activities such as agriculture, logging, wood harvesting, hunting, expansion of roads and railways, livestock farming, and ranching (Estrada et al. 2017). These activities change the landscape configuration and environmental conditions, which have a great influence on parasite diversity, distribution, abundance, and prevalence on their hosts (Bregnard et al. 2020; Fecchio et al. 2019, 2021; Solórzano-García et al. 2023). The ongoing expansion of the human frontier into the wild and urbanization of the landscape generates interfaces where Neotropical NHPs are forced to live in closer proximity to humans and domestic animals, altering NHP-parasite interactions and dynamics, which could boost opportunities for parasite spreading between humans and NHPs and have an impact on human and animal health and conservation (Cândido et al. 2022; dos Santos et al. 2022; Dietz et al. 2019; Ehlers et al. 2022; Lorenz et al. 2017; Ribeiro Prist et al. 2022).

Most parasites are embedded in interactions with multiple hosts; in this regard, shared traits among hosts could facilitate cross-transmission. Parasite cross-transmission is expected to be more feasible among evolutionarily related host species; therefore, parasite-host switches will be more common between closely related host species than between distantly related ones (Engelstädter and Fortuna 2019; Huang et al. 2014). Shared physiological and immunological similarities between humans and NHPs are expected to facilitate bidirectional transmission of infectious agents (i.e., adenovirus, retrovirus, respirovirus) (Han et al. 2016; Mollentze and Streicker 2020; Negrey et al. 2019).

Furthermore, in their environment, Neotropical NHPs present biological traits and face anthropogenic and ecological variables that could increase opportunities for contact and pathogen exchange. Landscape disturbances either naturally and manmade can generate smaller and spatially isolated populations going through nutritional deficiencies and stress, which makes them vulnerable to pathogen infections (Chapman et al. 2007; Deb et al. 2021) but can also favor the selection of generalist parasites and the expansion of vectors, leading to the emergence of infectious diseases (Dharmarajan et al. 2021; Failloux 2019). The incorporation of these variables in NHP infectious disease studies could provide a broader understanding of NHP-pathogen dynamics and consequently generate strategies for the mitigation of infectious disease effects in NHP health and conservation.

We conducted a systematic search of the literature on the infectious agents detected in Neotropical NHP throughout Latin America. We conducted a search following the recommendations of the PRISMA guide (Moher et al. 2010). We use the Web of Science search engine Scopus for the period 1950–2023. The search strategy included the following words ((virus OR bacteria OR helminth) AND (monkey OR primate) AND (Trinidad and Tobago, Costa Rica OR Peru OR Panama OR Mexico OR Guatemala OR Honduras OR Nicaragua OR Salvador OR Belize OR Colombia OR Bolivia OR Argentina OR Brazil OR Guiana OR Venezuela OR Ecuador OR Paraguay OR Suriname)). All articles published in English, Spanish, and Portuguese were included. The selection criteria for the articles were as follows: (1) full text; (2) developed in the American region; and (3) reported the presence of infectious agents mentioned in the search that were detected in Neotropical NHPs. For data extraction, location variables (i.e., country), Neotropical NHP-specific data (i.e., genus, species) and infectious agent-specific data (i.e., etiologic agent) were selected. After eliminating duplicates, a total of 13,779 studies were identified, and studies were reviewed through titles and abstracts; thereafter, in cases where more information was needed, the full text was used. The search yielded a total of 1473 data points (Supplementary Table 9.3), grouped into acanthocephalans (59), cestodes (76), protozoa (513), nematodes (446), trematodes (54), bacteria (54) and viruses (271) (Fig. 9.2 and Supplementary Table 9.3). Additionally, the infectious agents were grouped according to transmission routes, with the aim of making visible the modes of transmission that can potentially be more sensitive to external factors that may favor contact between the primate and the infectious agent (Fig. 9.3).

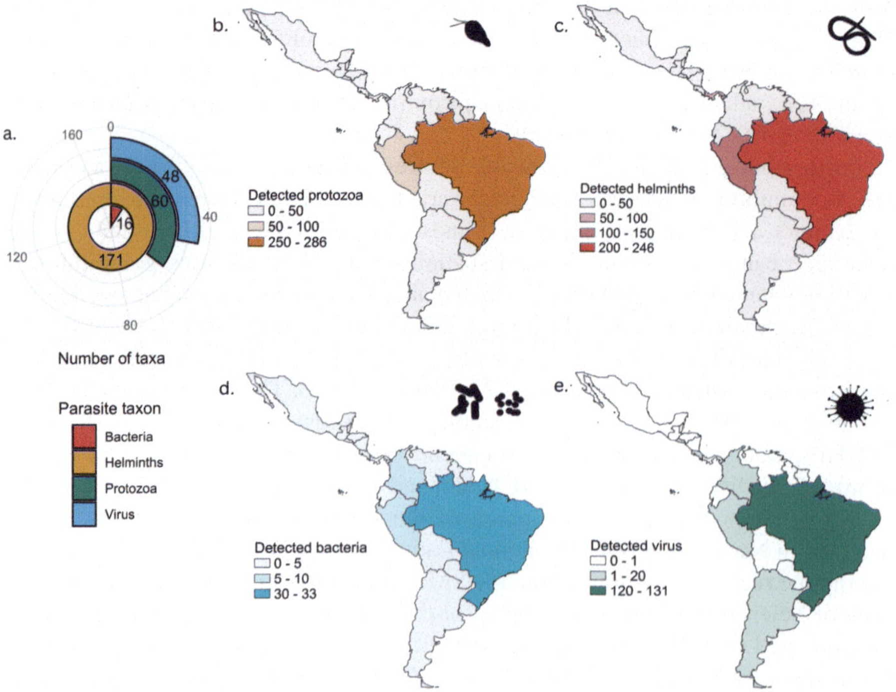

Fig. 9.2 Infectious agents found in Neotropical NHP. (**a**) Number of parasite species in each taxon; (**b**) Frequency of protozoan records per country; (**c**) Frequency of helminth records per country; (**d**) Frequency of bacteria records per country; (**e**) Frequency of viral records per country

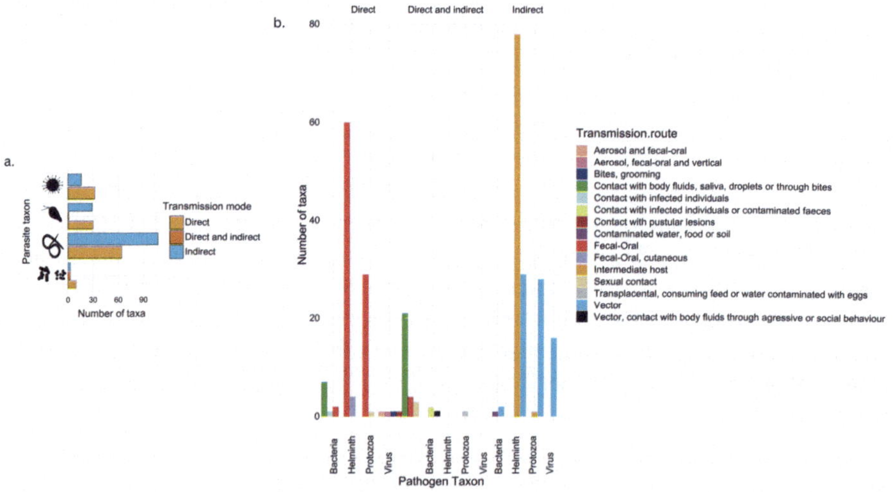

Fig. 9.3 Infectious agents found in Neotropical NHP throughout Latin America. (**a**) Number of parasite taxa with direct, direct and indirect, or indirect transmission modes in each major taxonomic group; (**b**) Number of parasite taxa with specific modes of direct, direct and indirect, or indirect transmission in each taxonomic group

9.2.1 Anthropogenic Variables

9.2.1.1 Habitat Loss and Fragmentation

Wildlife living in tropical forests degraded or fragmented by human activities (e.g., agriculture, livestock ranching, logging) face not only alterations in their dispersal activities, availability, and quality of resources but also in their biological interactions with their habitat and other species (including parasites) (Benítez-Malvido and Arroyo-Rodríguez 2008; Schwitzer et al. 2011). This last factor in turn affects the incidence and transmission of infectious diseases. Overall, deforestation and habitat fragmentation generate a mixture of land covers (remnants of forests, urban areas, agricultural land, and pastures) that favor interfaces where frequent contact between wild animals, domestic animals, and people occurs, facilitating parasite transmission (Guégan et al. 2020). For example, howler monkeys *A. palliata* and *A. pigra* have a higher prevalence of *Controrchis biliophilus* in fragmented habitats than in protected or continuous forests, probably because of the proliferation of intermediate hosts. Furthermore, *A. palliata* presented a higher prevalence of *Trypanoxyuris minutus,* likely due to the large groups concentrated in forest fragments (Trejo-Macias and Estrada 2012).

Moreover, the removal and logging of vegetation create habitat patches and edges. Some edges are permeable to host movement, allowing them to cross to other patches. In this sense, these types of edges, known as "soft edges," act as corridors where opportunities for host–parasite interactions can occur (Warburton and Blanar 2021). For instance, the presence of fruit trees along patches and edges attracts not only NHP but also invertebrate intermediate hosts, resulting in a higher prevalence of parasites (Valdespino et al. 2010). Moreover, favorable microclimatic conditions and host availability in edges can also promote the abundance of vector species of pathogens such as Oropuche or Mayaro virus (da Silva Pessoa Vieira et al. 2021).

9.2.1.2 Land-Use Changes

The land conversion to satisfy human necessity for resources and spaces has created mixed landscapes where different human activities are embedded. This could increase the frequency of contact among wild animals, domestic animals, vectors, and people, which could induce the emergence of infectious diseases (Guégan et al. 2020).

For instance, the presence of *Blastocystis hominis* in *Alouatta seniculus* suggests environmental contamination with human and domestic animal feces containing parasite cysts coming from grassland surrounding forest fragments (Rondon et al. 2017)). Similarly, *A. caraya* and *Callithrix penicillata* living in fragmented forests surrounded by agricultural lands were positive for the presence of *Toxoplasma gondii*, probably due to contact with contaminated feline feces (García et al. 2005; Molina et al. 2014). Conversely, Zika virus (ZIKV) genome sequences obtained from *Callithrix jacchus* in Ceará State in Northeast Brazil were 100% similar to ZIKV circulating in humans in South America, and yellow fever cases in humans have been reported simultaneously with findings of viral infection in NHP, suggesting that NHP could act as reservoirs for these viruses in landscapes shared with humans and domestic animals (Favoretto et al. 2016, 2019; Mares-Guia et al. 2020).

9.2.1.3 Hunting

In poorer countries, especially in rural areas, the extraction of wild meat or "bushmeat" is a solution for food scarcity generated by poverty, civil unrest, and wars. Furthermore, the consumption of bushmeat is perceived as healthy, tasty, and, in some instances, part of traditional culture or even as a luxury commodity in urban areas. This expanding luxury market has made commercial hunting more important than subsistence hunting. This promotes the illegal traffic of hunting products to the US and European countries, posing a considerable risk for zoonotic disease transmission (Muehlenbein 2017).

In the American continent, the presence of parasites such as *Capillaria hepatica* and *T. gondii* has been reported in the bushmeat of NHP species (Van Vliet et al. 2017). However, the risk this practice represents for the transmission of more hazardous agents should not be discarded. For instance, in countries of West and Central Africa, where the bushmeat extraction rates are highest, the consumption of NHP meat and the exposure to fluids during hunting, butchering, or transportation to markets has been a source of several infectious agents, such as Ebola virus (which can remain viable in carcasses for up to 3–4 days) or simian retroviruses (Cawthorn and Hoffman 2015; Kazanji et al. 2015; Maurice et al. 2017).

9.2.1.4 Tourism

Having an encounter with an NHP represents an experience with great touristic appeal for foreign and local visitors, despite the warnings on not providing food, keeping their distance, avoiding touching the NHP, and fines punishing these activities. Biting and scratches are a common result of visiting temples and shrines in Asian countries. Similarly, other attractions where interactions with wildlife are possible (e.g., theme parks, resort restaurants, petting zoos) provide opportunities for pathogen exchange among humans and other animal species.

Durán (2014) reported a higher intensity of *Cryptosporidium* infection in NHP living in the surrounding forest of a Mexican archaeological site, probably as a result of mishandling of human feces in that tourist attraction. Likewise, *Leptospira* infections and seroprevalence found in Neotropical NHP kept in zoological collections were likely the result of the lack of rodent control and proper sanitization of the enclosures (Ferreira et al. 2001).

Furthermore, although ecotourism holds potential as a tool for primate conservation, tourists often travel with gastrointestinal and respiratory infections or without the recommended vaccines, posing a risk for NHP health (Muehlenbein et al. 2010). Thus, providing information on the risk of disease transmission from tourists to NHPs and vice versa, as well as implementing safety measures such as wearing facemasks and providing evidence of vaccinations, could help to reduce exposure to infectious diseases for both humans and NHPs (Setchell et al. 2017).

9.2.1.5 Exotic Pets

Charismatic animals such as howler monkeys (genus *Alouatta*), capuchin monkeys (genera *Cebus* and *Sapajus*), squirrel monkeys (genus *Saimiri*), lion tamarin (genus *Leontopithecus*), and marmosets (genus *Callithrix*) are often kept as pets in different places around the world, either acquired by legal, illegal commerce or as young orphans. NHPs in these situations can show aggressive behaviors toward children, unfamiliar people, and even owners when facing uncomfortable or frightening settings. This type of negative interaction between humans and NHP (e.g., *Alouatta* spp., *Ateles geoffroyi*, *Cebus imitator*, and *Sapajus nigritus*) is relatively frequent in urban and periurban areas of northern Costa Rica and southern Brazil (Ó.M. Chaves, personal observations). Moreover, the predation of insects and domestic animals by pet NHP has also been documented (Ceballos-Mago and Chivers 2010). These behaviors open an opportunity for pathogen acquisition and transmission (Tregle Jr. et al. 2011; Johnston et al. 2015). In addition, owners can also transmit pathogens to their pet NHP, some with fatal consequences (Huemer, et al. 2002; Imura et al. 2014; Quevedo and Lescano 2014).

9.2.2 *Ecological Variables*

9.2.2.1 Climate Change

Human activities have increased the emission of the greenhouse gases carbon dioxide (CO_2), methane (CH_4), nitrous oxide (N_2O), and aerosols (e.g., sulfate, nitrate, carbon) and, consequently, alterations in the levels of solar and infrared radiation leaving the Earth's surface. This imbalance in energy flux has generated increments in Earth's surface temperature and contributed to changes in air and water circulation around the globe, which has led to changes in the presentation and intensity of precipitation or extreme climate events (e.g., cyclones, hurricanes, droughts) (IPCC 2014).

High temperatures, heavy rainfall, and drought events resulting from climate change could influence vector-borne disease expansion and transmission (Hamrick et al. 2017). Models considering two future climate scenarios (i.e., increase by 1.0 °C or an increase 2.0 to 3.7 °C) show an expansion in the distribution of Oropuche, Saint Louis Encephalitis, and Mayaro virus in Brazil (Lorenz et al. 2017). These diseases may have putative sylvatic cycles involving NHP, as antibodies against the pathogens have been detected in them (Almeida et al. 2016, 2019; Batista et al. 2013; Valentine et al. 2019; Chaves et al. 2020a, b; Gibrail et al. 2016). Further monitoring of these agents is still needed to detect sylvatic cycle establishment and to assess risks to wildlife and human health.

Whereas heavy rainfall events can decrease mosquito abundance by flushing larvae, drought events can favor some mosquito species by the drying up of streams, creating pools, eutrophication in standing waters or increasing household water storage in human settlements, making them more suitable larval habitats (Brown et al. 2014). These fluctuations in mosquito populations could influence the transmission probability of vector-borne diseases (Beck-Johnson et al. 2017).

9.2.2.2 Natural Disasters

Hurricanes drastically modify the habitat of NHP due to the mortality of trees used as food or refuge sources. The removal of canopy after these events generates large forest gaps that are rapidly colonized by pioneer plant species, impacting NHP diets and disease exposure. The dominance of *Cecropia peltata* trees in forests of Belize after the Iris Hurricane resulted in an increased consumption of these plants by howler monkeys but also a higher *Controrchis* spp. prevalence and intensity of infection. *C. peltata* serves as a refuge and food source for ants (a probable intermediate host), which could increase the exposure to this parasite. In addition, due to higher monkey population density and contact, the prevalence of both *Trichuris* sp. and strongylid eggs was also increased (Behie and Pavelka 2013; Behie et al. 2014).

Climatic phenomena such as El Niño have generated severe droughts and fires in tropical forests of Central and South America. Like other wildlife, primates are affected by the loss of resources and space; these events can generate increased exposure to parasites and injuries due to violent encounters with other troops to obtain access to resources (Mittermeier et al. 2005; Stark et al. 2019).

9.2.3 *Biological Variables*

9.2.3.1 Behavioral Traits

Some Neotropical NHPs living in disturbed tropical forests exploit resources present in the surroundings of forest patches, living fences, secondary vegetation, agricultural lands, gardens, and semiurban environments and even come to the ground to access pastures, fruit plantations, and isolated trees. This implicates changes in diet composition (Chaves et al. 2012), activity patterns (Carretero-Pinzón et al. 2016; Bustamante-Manrique et al. 2021) and travel (Martínez-Mota et al. 2007).

Descending to the ground to move across forest fragments or to access resources exposes NHP to human-primate conflict, electrocution, canine attacks, and vehicle collisions, along with possible health issues due to dietary changes

(Cristobal-Azkarate and Arroyo-Rodríguez 2007; Pozo-Montuy et al. 2013; Azofeifa-Rojas et al. 2021; Chaves et al. 2022). This behavioral flexibility represents an opportunity for pathogen infection and transmission, since NHP can be exposed to pathogens not found in treetops (where they normally live). *Cebus imitator*, an omnivorous NHP, frequently forages on the ground, consuming fruits, insects, small vertebrates and drinking water from puddles; furthermore, they present crop-feeding behavior, all of which could expose them to infective stages of parasites such as *T. gondii* (Niehaus et al. 2020).

Neospora caninum, a protozoan that causes abortion in cattle, has also been detected in free-living *Alouatta caraya, Alouatta guariba clamitans, Aotus azarae, Aotus infulatus, Mico melanurus, Sapajus apella, and Sapajus nigritus cucullatus* in Brazil (Cândido et al. 2022; dos Santos et al. 2022). Likewise, these NHPs go down to the understory or the ground (Catão-Dias et al. 2013), consume plants and fruits also used by humans as well as feces from definitive hosts (i.e., dogs, cats) in peri-urban environments (Lins and Ferreira 2019; Bustamante-Manrique et al. 2021), and enter into contact with these parasitic oocysts, which could result in mortality infections (Costa et al. 2018).

Social systems in NHP also play a role in the susceptibility and transmission of infectious diseases. Interindividual contact-associated behaviors (e.g., allogrooming, huddling, aggression, and mating), position in the social network, and the directionality of social interactions can either lead to a greater probability of pathogen transmission due to increased contact with conspecifics or reduce host susceptibility to pathogen infection due to less stress and immunosuppressive effects (Balasubramaniam et al. 2016; Rimbach et al. 2015).

Likewise, social structures can influence social interactions among individuals and regulate pathogen or parasite transmission. Fission-fusion dynamics (i.e., group division into subgroups of different sizes, compositions, and spatial cohesion that vary over time) in the NHP result in individual variation in the use of the shared space and the amount of social contact with other conspecifics, which could influence the acquisition and transmission of pathogens (Deere et al. 2021). On the other hand, NHP cohesive social groups could serve as an abundant supply of blood for vector populations to thrive, influencing vector distribution, population dynamics, and abundance; hence, altering the transmission cycle of parasites, as has been described in *Leishmania* spp. in wild and peri-domestic environments (Martínez et al. 2019).

9.2.3.2 Physiological Stress

Living in disturbed environments represents a physiological challenge. Levels of cortisol in NHPs living in fragmented habitats are often higher than those living in continuous forests (Dunn et al. 2013; Martínez-Mota et al. 2007; Rangel-Negrín et al. 2009, but see Chaves et al. 2019). However, other factors, such as the type of human activities carried out in the fragments (e.g., logging, hunting), the presence of tourists, the protection status of the habitat, and multispecies parasite infections, can also increase cortisol levels (Behie and Pavelka 2013; Behie et al. 2014; Cantarelli et al. 2017; Durán 2014; Rangel-Negrín et al. 2014; Rimbach et al. 2013).

When these factors are constant, as occurs in captivity or anthropogenically disturbed environments, a chronic stress response is established in the individual; this response can influence animal fitness, affecting growth and decreasing fertility and immune functions and making it susceptible to infections (Beehner and Bergman 2017). A higher intensity of parasite infection has been found in NHPs with high levels of cortisol (Durán 2014). Moreover, high levels of cortisol also increase participation in agonistic interactions, which could facilitate contact-mediated transmission of pathogens (e.g., *Shigella flexneri*; Balasubramaniam et al. 2016).

9.3 Bidirectional Transmission Between Humans and Neotropical NHP

According to the most recent bibliographic reviews (Carrillo-Bilbao et al. 2021; Rondón et al. 2021; Solórzano-García and Pérez-Ponce de León 2018), in the American continent, humans and NHP share approximately 89 taxa of parasites, making a total of 472 records; more than 70% constitute microparasites such as viruses, bacteria, and protozoa, and the rest are macroparasites such as helminths and arthropods (Fig. 9.4). Almost half of these parasitic taxa can also infect domestic animals, such as dogs, cats, and farm animals. *Alouatta* is the genus that shares the highest number of parasite taxa with humans, almost half of them protozoa, followed by *Callithrix*, *Sapajus*, *Saguinus*, *Ateles*, *Saimiri*, and *Aotus*, all among the most trafficked and laboratory-preserved NHP species (De Souza Fialho et al. 2016; Shanee et al. 2017). The number of parasite species that a particular NHP genus shares with humans is related to the diversity of NHP parasitic fauna; thus, primates parasitized by a great variety of parasite species tend to share more parasites with humans than primates parasitized by fewer species of parasites (Fig. 9.5). Zoonotic potential seems to be higher in protozoa; particularly, hemoparasites such as *Plasmodium* and *Trypanosoma* have the largest number of records and infect the broadest range of NHP species. Among viruses, the families Flaviviridae and Retroviridae are among the most important zoonotic agents, at least in terms of the number of records. Additionally, several species of tapeworms and roundworms are commonly shared between humans and NHP, while ectoparasites seem rarely transmitted between humans and NHP (Fig. 9.6).

Studies have shown that great apes and Old World NHP having a closer phylogenetic proximity to humans carry a higher risk of sharing pathogens compared to other NHP groups (Gomez et al. 2013). However, Neotropical NHPs are distributed in environments with different degrees of disturbance (Batista et al. 2012). Since they are arboreal, changes in the landscape force them to use other altitudinal strata that cause them to face unknown risks, including increased contact with pathogens transmitted indirectly (e.g., oro-fecal) or by vectors (mosquitoes) that are not distributed in the tree canopy (Chaves et al. 2019).

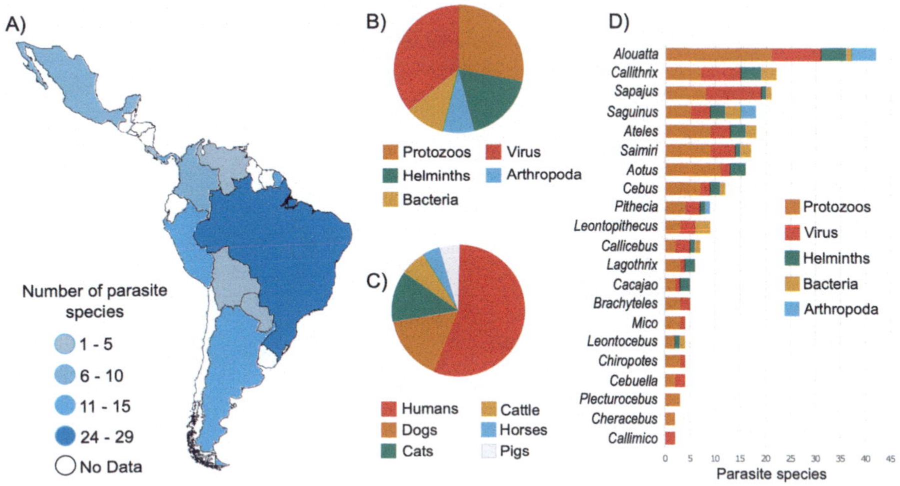

Fig. 9.4 Parasites shared between humans and Neotropical NHP. (**a**) Number of shared parasite species reported per country. (**b**) Shared parasite species by taxonomic group. (**c**) Proportion of parasite species shared between Neotropical NHP, humans and domestic animals. (**d**) Number of parasite species shared with humans and each genus of Neotropical NHP

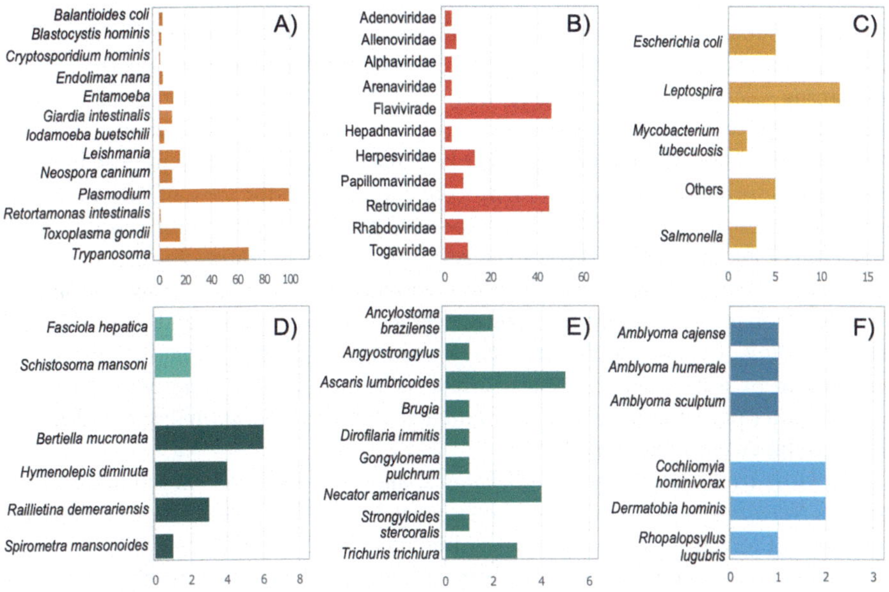

Fig. 9.5 Number of records of zoonotic parasites of Neotropical NHP. (**a**) protozoa, (**b**) Viruses, (**c**) Bacteria, (**d**) Trematodes and Cestodes, (**e**) Nematodes, (**e**) Arachnida and Insecta

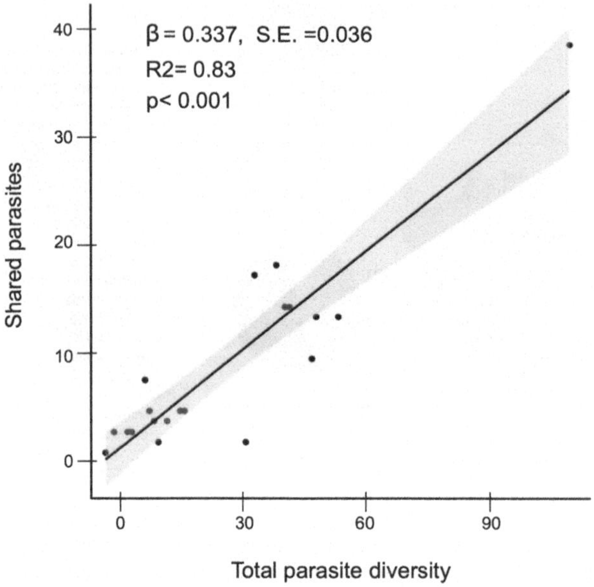

Fig. 9.6 Relation between the number of shared parasite species and the total diversity of parasites reported for each genus of Neotropical NHP

Close contact between humans and wildlife is key for the exchange of infection agents. Land use and land cover changes, especially urbanization processes, constitute strong selective forces pushing wild organisms to develop strategies that enable them to cope with these pressures to survive in anthropic environments (CITA). This includes parasites and pathogens, which must adapt to host demographic fluctuations, as well as to potential new hosts and vectors. In the case of the tapeworm *Bertiella,* humans are accidental hosts since parasite transmission requires the ingestion of an intermediate host, usually a mite. Levels of *Bertiella* infection have been related to habitat anthropogenization, with urban NHP troops showing a higher prevalence than those living in more conserved habitats (Kane and Smith 2020). The crowding effect due to the reduced amount of remaining habitat in urban areas surrounded by a harsh matrix that restricts NHP movements, along with the presence of other potential hosts (such as humans and domestic animals), could facilitate high tapeworm abundance.

Trichuris trichiura, *Ascaris lumbricoides,* and other soil-transmitted nematodes, such as *Necator americanus,* have been reported in NHPs held *ex situ* or inhabiting remnants of forests frequently visited by humans. Since these are common parasites of humans, these reports suggest anthropozoonotic transmission of these parasites caused by environmental contamination with human feces (Agostini et al. 2018; Phillips et al. 2004; Stuart et al. 1990). Similarly, the protozoan *Giardia intestinalis* was found in howler monkey individuals living close to human settlements in Belize (Vitazkova and Wade 2006). These examples show that human-induced habitat perturbation disrupts natural host–parasite dynamics, as well as parasite community composition, by facilitating the spread of some agents, restricting the transmission and survival of others, and introducing new ones. Such conditions increase exposure to zoonotic and human-borne pathogens, affecting both human and NHP health.

9.3.1 Spillover and Spillback

For spillover events to occur, two conditions must be fulfilled: first, the parasite needs to be in contact with a novel host species, and second, it has to be able to successfully establish itself in the novel host. These two conditions are triggered by ecological and evolutionary opportunities that, in the case of zoonosis, promote contact between humans and wildlife and facilitate the success of the infectious agent (Fig. 9.7). However, the effects of these anthropogenic disturbances are not ubiquitous, and parasitosis tends to either increase or decrease depending on the infectious agent species, host species, and type of disturbance. Studies with howler monkeys have reported higher parasite richness in *Alouatta guariba* troops living in highly disturbed areas (Klain et al. 2021), but the opposite trend was observed for *A. palliata*, *A. pigra*, and *Saguinus leucopus*, with higher parasite richness in more conserved areas (Cristobal-Azkarate et al. 2010; Martínez-Mota et al. 2018; Soto-Calderón et al. 2016). No significant differences were observed in the parasite diversity between *A. g. clamitans* groups inhabiting small and large Atlantic Forest fragments in southern Brazil (Lopes et al. 2022).

These contrasting patterns could be explained by host group size and density, type of parasite, and mode of transmission. A meta-analysis on the patterns of parasitic infections in South American howler monkeys showed that the prevalence of infectious agents such as bacteria and protozoans was higher in rural areas near

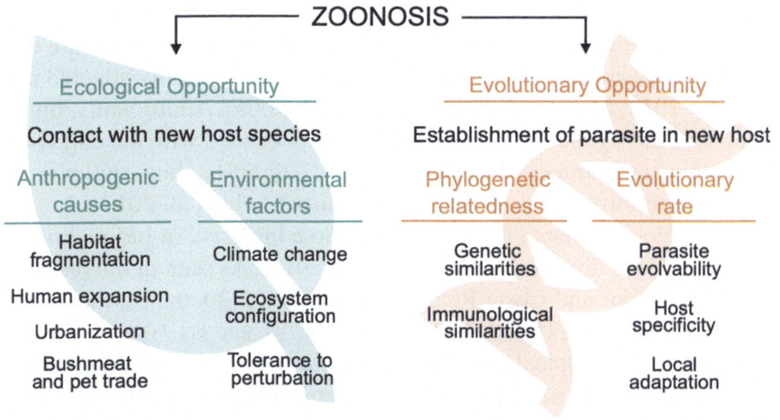

Fig. 9.7 Elements that promote ecological and evolutionary opportunities for zoonosis to occur

human settlements, while the average prevalence of helminths was slightly higher in remote undisturbed areas (Kowalewski and Gillespie 2009). In addition, habitat perturbation could affect the abundance of intermediate hosts and vectors. For example, a high prevalence of infection with the acanthocephalan *Prosthenorchis elegans* was reported in tamarin monkeys living near human facilities and garbage dumps where the parasite's intermediate hosts (cockroaches) are also highly abundant (Wenz et al. 2010). Likewise, the rate of *Plasmodium* infection in mosquito vectors has been reported to increase with the degree of forest fragmentation, linking environmental perturbation with the risk of contracting malaria (Rondón et al. 2019).

The impact of the spread of pathogens from humans to wildlife species is an emerging threat that is not well understood, but such spreads are likely to increase in the future, establishing new reservoirs of disease propagation with important implications for public health and economy, as well as for wildlife conservation (Thompson 2013). Environmental changes and land-use transformation are globally affecting the dynamics of disease transmission between wild populations of NHP and humans (Rondón et al. 2017), with pathogens of human importance being found more frequently in NHP living in degraded environments, potentially serving as reservoirs and transmitters of these pathogens to humans. Pathogen spillover requires a permeability (the degree to which an organism is able or willing to cross a given habitat) for at least one of the actors involved in spillover: source host, intermediate host/vector, recipient host, or pathogen (Borremans et al. 2019). Each of these actors is influenced by several factors and nonlinear interactions, as well as by the processes of contact with the source host, the intensity and duration of exposure, the potential for pathogen invasion into the recipient host (release, transport, and survival), and subsequent transmission and adaptation within host populations (Plowright et al. 2017).

Some species of parasites are more prone to host switching than others, and certain host species are more likely to share parasites. Parasites' evolutionary rate along with phylogenetic relatedness among host species will create evolutionary opportunities facilitating parasite cross-transmission. Those parasite species capable of infecting a wide variety of hosts—multihost parasites—either because they have a generalist strategy or given the complexity of their life cycles, are expected to show greater genetic variability than highly specialized locally adapted parasites (Cooper et al. 2012; Matthee 2020; Pedersen et al. 2005). Additionally, high mutational rates and short generation times have been associated with the ability of parasites and pathogens to infect a broader range of NHP hosts; thus, spillover tends to be greater in generalist versus specialized parasites (Auld et al. 2017).

NHP being closely related to humans leads to a high risk of bidirectional transmission of infectious diseases (Sharp and Hahn 2011). As seen in the research conducted in El Salvador and Costa Rica by Chaves et al. (2020a, b), the presence of human parvovirus species (PARV4 14%, HBoV 3.7%, and B19 0.67%) in stool and blood samples of Neotropical NHP, determined by molecular evidence, suggested possible transmission between humans and Neotropical. This could have an impact on NHP conservation, since an organism that is relatively nonvirulent in one host can be extremely virulent in a different host species (Hope et al. 2004; Smith et al.

2009; Ye et al. 2012). Parvoviruses have characteristics that make them particularly risky in natural environments: they are highly resistant to environmental conditions and remain viable outside the host for months (Suzan and Ceballos 2005), and their main route of transmission is the oro-fecal route (López-Pérez et al. 2019). Thus, the possibility of infection should be greater in fragmented environments, where NHPs are expected to descend to the ground more frequently to move across remnants of forest (Valdespino et al. 2010), increasing the possibility of contact with the infectious agent not only excreted by individuals of the same species but also by other primates, including humans (Chaves et al. 2020a, b).

Occasionally, recipient hosts can play a crucial epidemiological role by transmitting the pathogen back to a potential source host (spillback) (Nugent 2011). This host will not automatically transmit the pathogen; for this to happen, it must be a competent host for the pathogen, possess demographic characteristics that can facilitate the spread of the pathogen and be able to disseminate the infective stages of the pathogen, sometimes being a more competent host than the source host. If this is not the case, the recipient host would be a diluent of pathogen transmission in an ecosystem (Kelly et al. 2009). Pathogen spillback could affect source hosts when interspecific transmission equals or exceeds intraspecific transmission and when pathogen virulence in a source host is greater than that in the recipient host (Hatcher et al. 2006). In addition to directly amplifying pathogen infection by acting as a definitive host, they could also potentially increase infection in source hosts by fulfilling other roles in the pathogen life cycle, such as an intermediate host or as a vector (Kelly et al. 2009).

Multiple vector-borne zoonotic diseases are caused by the spillover of pathogens from NHP to humans and other animals at the boundary between rural and natural ecosystems (Borremans et al. 2019). Several arboviruses of medical importance have caused outbreaks worldwide in recent decades, including the Zika virus and yellow fever virus, both Flaviviruses originating in Africa and transmitted by vectors, with NHP as a reservoir host (spillover). Like the other arboviruses of the American continent, yellow fever is established in an urban cycle, which has been identified in the American continent since the seventeenth century (Bryant et al. 2007). Until 1930, only the urban transmission cycle had been described. However, the establishment of a sylvatic cycle was documented after the confirmation of human cases in a rural area free of *Aedes aegypti* in southeastern Brazil (Soper et al. 1933). Subsequently, it was shown that this virus had adapted to Neotropical NHP (spillback) and to sylvatic mosquitoes such as *Hemagogus* spp. In the case of other arboviruses, although Zika virus has never been isolated from any other vertebrate besides humans, nor in sylvatic mosquitoes that feed on NHP in the Americas, findings of RNA fragments and antibodies compatible against this virus were found in marmosets and capuchin monkeys captured in peri-urban areas of Brazil (Terzian et al. 2018). It is known that the dengue virus (flavivirus) transmitted by mosquitoes of the genus *Aedes* infects Old World NHP, some serving as amplifying hosts for the enzootic transmission of dengue. However, the role of Neotropical NHP in the transmission or maintenance of this virus is unknown, although there are several reports of dengue virus (I, II, III, and IV) or antibodies against this virus in Neotropical NHP from various regions of Latin America (Chaves et al. 2020a, b and Box 9.1).

Box 9.1: Arboviruses–Sylvatic Cycles

Yellow fever virus, a member of the genus *Flavivirus* (family Flaviviridae), is the etiologic agent of yellow fever. The natural transmission cycle of yellow fever involves NHP (monkeys and great apes) and mosquitoes (Monath and Vasconcelos 2015). The virus was introduced into the Americas from Africa in the seventeenth century through ships that transported the anthropophilic mosquito species *Ae. aegypti*, establishing itself in coastal cities, with the subsequent establishment of an enzootic jungle cycle in tropical forests (Monath and Vasconcelos 2015).

Although antibodies against yellow fever have been reported in several species of mammals, humans, and NHPs are the main hosts. New World monkeys (Platyrrhini infraorder) are more susceptible to the disease than Old World monkeys (Catarrhini infraorder) (Brasil et al. 2017). In South America, sylvatic yellow fever occurs sporadically in regions where people are vaccinated; however, epidemics can also occur, usually preceded by epizootics in the NHP (Almeida et al. 2014). In NHP, the disease may have severe results if enough time has elapsed for new populations of susceptible NHP to emerge after each epizootic wave. Platyrrhine species have different levels of susceptibility, and howler monkeys (*Alouatta* spp.) are the most sensitive, much more so than humans (Araújo et al. 2011). Howler monkeys experience high mortality rates, often dying three to seven days after infection. In Rio Grande do Sul, approximately 2000 black and golden howler monkeys (*Alouatta caraya*) and brown howler monkeys (*Alouatta guariba clamitans*) were lost after an epizootic yellow fever event in 2008 and 2009 (Almeida et al. 2012). If howler monkeys survive the disease, they develop permanent immunity and consequently act as virus amplifiers during this short time (Georgiev 2009). Howler monkeys are among the most widely distributed monkeys in the New World, inhabiting Neotropical regions from central Mexico to northeastern Argentina (Groves et al. 2005). Due to its wide distribution area and high mortality levels when facing a yellow fever virus spread, *Alouatta* is an ideal species to monitor new cases of yellow fever (Almeida et al. 2012). Recovery of howler monkey populations becomes increasingly compromised after yellow fever outbreaks.

In Brazil from 1999, cases of yellow fever were detected in northern and midwestern regions outside the natural foci of the disease (North and Midwest) (Torres et al. 2003; Brasil et al. 2017). From 2014 to 2022, Brazil witnessed a wide yellow fever virus circulation, which killed thousands of NHPs and generated over 700 human deaths (i.e., Brazil 2021, Bicca-Marques et al. 2017).

(continued)

Box 9.1 (continued)

Zika viruses (Flavivirus) invaded the Americas in 2013–14; both are transmitted to humans by the blood-feeding *Aedes aegypti* vector. To date, there is an unconfirmed report of Zika virus RNA in a capuchin monkey in Brazil (Favoretto et al. 2016), and an experimental infection in squirrel monkeys (*Saimiri* spp.) demonstrated viremia (Vanchiere et al. 2018).

Dengue virus belongs to the *Flavivirus* genus (Flaviviridae family) and circulates in anthroponotic cycles transmitted by mosquitoes, mainly *Aedes aegypti* species. In the Americas, its possible maintenance through sylvatic cycles involving different species of Neotropical NHP, such as *Alouatta* spp., *Cebus* spp., and *Ateles* spp., has been suggested by serological and molecular detection (Morales et al. 2017; Moreira-Soto et al. 2018a, b; Chaves et al. 2021a, b).

Saint Louis encephalitis virus belongs to the genus *Flavivirus*, and this virus is maintained in nature in an enzootic cycle involving mosquitoes, mainly of the genus *Culex*, and birds of the orders Passeriformes and Columbiformes (Beltrán et al. 2015). In the Neotropics, NHP has been reported to be positive for this virus, suggesting possible alternative transmission cycles involving NHP instead of birds (Chaves et al. 2020a, b). A high prevalence of antibodies against this virus has been reported in black howlers (*Alouatta caray*) in Argentina (Morales et al. 2017), and serological findings have also occurred in southern Brazil (Almeida et al. 2016; Almeida et al. 2019). Chaves et al. (2020a, b) evaluated 86 free-living and captive Neotropical NHPs and found that 39.5% (34/86) had evidence of prior infection with a St. Louis encephalitis virus. However, the role that Neotropical NHP might play in maintaining this virus in the wild is still unknown (Morales et al. 2017).

Oropouche virus, a member of the genus *Orthobunyavirus* (family Peribunyaviridae), is the etiological agent of Oropouche fever, a zoonotic disease transmitted mainly by mosquitoes of the species *Culicoides paraensis*. Mammals and wild birds act as natural reservoirs of this virus in its sylvatic cycle (Romero-Alvarez and Escobar 2018). Immunity to the Oropouche virus was found in *A. caraya* in southern Brazil (Almeida et al. 2016).

Mayaro virus, a member of the genus *Alphavirus* (Family Togaviridae), causes Mayaro fever, often confused with dengue. Its sylvatic cycle is similar to that of yellow fever, involving mosquitoes of the genus *Hemagogus* and NHP as reservoirs (Muñoz and Navarro 2012). The virus has been isolated in the marmoset monkey (*Callithrix penicillata*) and the weeping capuchin (*Cebus olivaceus*) (Navarro et al. 2016). Likewise, two studies performed in Brazil by Batista and collaborators (2012, 2013) examining 35 and 16 NHPs, respectively, reported the presence of the Mayaro, Oropouche, and Cacipacoré viruses.

9.3.2 Transmission Risk: Direct and Indirect Contact

The anthropogenic influence on ecological systems dictates the level of risk in the transmission of zoonotic diseases. Ecosystem edge sites (rural and natural ecosystems) are recognized as potential foci of pathogen transmission, as they form the boundaries of occurrence of many species. This implies a higher rate of contact between species occupying adjacent ecosystems (Borremans et al. 2019). Among Neotropical NHPs, the genus *Alouatta* has shown great ecological flexibility, allowing them to survive and even make use of new resources from the surrounding matrix (Argüello-Sánchez 2012), documenting that they frequent habitat edges as well as buffer zones (Conatser 2016). Another species resilient to these changes is the white-faced capuchin monkey (*Cebus imitator*), which has adapted very well to edge sites in Costa Rica and Panama (Mckinney 2010, Mansell and McKinney 2021). Additionally, callitrichid monkeys (tamarins and marmosets) exhibit a particular ability to adapt to disturbed and urbanized environments, an example being white-footed tamarin (*Saguinus leucopus*), an endangered species endemic to the central Andes in northwestern Colombia (Soto-Calderón et al. 2016).

An important component of transmission is contact, which can occur directly between hosts or indirectly through vectors, intermediary hosts, fomites, or environmental sources (Yang et al. 2021). Diseases that are spread indirectly by pathogens outside the host and their survival are very important for their epidemiological dynamics (Fig. 9.3). This mechanism contrasts with direct transmission, which postulates that the pathogen is acquired through contact with an infected individual or contaminated source (Li et al. 2017). NHPs are particularly vulnerable to the effects of pathogens because they live in close social groups, facilitating contact and thus pathogen transmission (Friant et al. 2016). In addition, anthropogenic changes have favored contact between humans and domestic animals with NHP, resulting in an increased opportunity for the transmission of pathogens directly and indirectly between these groups.

The mechanisms and behaviors that facilitate host–host contact are varied, i.e., direct contact often requires spatiotemporal colocation between hosts, whereas indirect contact requires spatial colocation within a given time window (Godfrey 2013). This distinction in host contact mechanisms can result in variation in contact rates within and between host species and across the landscape (Yang et al. 2021). Understanding the magnitude of variation between different types of contact (direct/indirect) between host populations is crucial for optimizing disease control strategies in multihost systems.

Arboviruses are a worldwide public health problem, representing a threat to both humans and animals. They are transmitted and maintained in nature through sylvatic cycles, where hematophagous arthropod species act as vectors and wild vertebrates can act as reservoir hosts (Batista et al. 2013). They belong to several viral families, such as Togaviridae (genus *Alphavirus*), Flaviviridae (genus *Flavivirus*),

Peribunyaviridae (genus *Orthobunyavirus*), Phenuiviridae (genus *Phlebovirus*) and Sedoreoviridae (https://ictv.global/taxonomy). Many of these pathogens circulate permanently in nature in an enzootic sylvatic cycle involving wild animals that inhabit nearby environments where anthropic activities are developed (rural and peri-urban areas) (Weaver 2006). In Latin America, the role of Neotropical NHP in the maintenance of the sylvatic cycles of arboviruses is inconclusive, except for yellow fever virus. However, Neotropical NHPs are the only mammals classified as high risk for being part of the maintenance of arbovirus sylvatic cycles (Pandit et al. 2018). In the Neotropics, NHPs have been reported to be infected with different species of arboviruses, such as yellow fever virus, Mayaro virus, Chikungunya virus, Zika virus, dengue virus, Cacipacore virus, St. Louis encephalitis virus and Oropouche virus (Carrillo-Bilbao et al. 2021).

9.4 Challenges and Future Directions

The accelerated rates of landscape transformation in the Neotropics (Antonelli 2022), the continuous human encroachment into the wild, and the fact that nearly 40% of NHP species are at risk by human-induced threats, such as habitat loss and fragmentation, hunting and pet trade (Estrada et al. 2020), make the monitoring of NHP health, the diagnosis of NHP parasites and pathogens, as well as their transmission potential, research priorities. Nonetheless, conducting studies of diseases in NHP entails several logistical, methodological, and ethical challenges. To understand how anthropogenic pressures affect primate-parasite interactions, we first need to know how these associations function in undisturbed conditions. Describing primatological disease dynamics in natural pristine environments is not simple since most of the remaining forested areas in the Neotropics are of difficult access given their orography, remoteness, and, in most recent times, security concerns. Carrying out such fieldwork requires specialized transportation, equipment, and, most importantly, the support of local human communities.

In biologically healthy wild populations, disease outbreaks are natural phenomena and usually have minimal impact; however, in small populations with low genetic diversity living in high-stress environments, they can have catastrophic effects (Junge and Louis 2002); thus, infectious diseases are a major factor in the decline of free-living NHP populations (Nunn and Altizer 2006). For example, in South America, yellow fever has caused reductions in NHP wild populations as a result of different virus circulations, such as those occurring between 2007 and 2009 and from 2014 to 2021, which killed thousands of howler monkeys and other species, including highly endangered ones (Moreno et al. 2017; Almeida et al. 2012, 2014; Romano et al. 2014; Strier et al. 2019; Bicca-Marques et al. 2017; da Cardoso et al. 2010; Dietz et al. 2019; Andrade et al. 2021).

When designing a study on primatological diseases, we need to address several questions, such as which is the best strategy to monitor the health of a given NHP species? What type of samples are needed? How do we get those samples? Depending on the collection strategy, *in situ* challenges may include capture, manipulation, and treatment of animals, as well as handling and transportation of samples, especially when working in remote areas. On the other hand, *ex situ* challenges deal with the best protocol to process and analyze samples to obtain as much accurate and useful data as possible to assess health and disease risks. There are many microscopic methods, immunological tests, and DNA sequencing techniques to detect parasites and pathogens; however, their application is not universal, and the best choice will be based on the type of sample and the etiological agent studied.

In many cases, invasive sampling is required to detect the presence of certain parasites and pathogens. Capture and manipulation of any wildlife animal has its risks, and it is our obligation to pursue safe practices and to follow ethical protocols that ensure the protection of primates and handlers. Since many NHP populations are threatened, deciding the appropriate sampling size is not trivial; we need to find a balance between gathering enough data and the risk to the target populations. Additionally, it is very important to communicate the aims and the social benefits of the study to the local people to avoid confusion with activities of wildlife extraction and trafficking.

The application of noninvasive sampling methods in combination with molecular diagnostic techniques has proven to be a good option for the study of diseases and health monitoring in NHPs without disturbing their natural populations (e.g., Chaves et al. 2020a, b; Rondón et al. 2019; Solórzano-García and Pérez-Ponce de León 2017; Wenz et al. 2010). In the case of NHP, the use of this type of sampling has been crucial and very efficient since most species live in dense forests and tend to be elusive, making observations difficult (Argüello-Sánchez and García-Feria 2014). Nonetheless, they also pose important methodological challenges. The advancement of molecular technology allows us to obtain relevant information regarding the health of an organism from different sources. For example, through fecal samples, we can monitor hormones, immunology, genetics, parasites and some pathogens, toxins, diet, and microbiome, among other parameters (Wasser et al. 2002). These techniques are constantly refined and standardized, but they should be tested for their diagnostic accuracy and utility, since their application might not be appropriate in all cases, being more suitable for studying certain types of hosts, parasites, or disease dynamics. For example, not all disease agents are excreted in the feces, or the genetic material could not be enough to detect their presence. Additionally, distinguishing some parasite species could be impossible only through coproscopic examinations, hindering the diagnostic potential of these tools. Moreover, ecological estimates such as parasite abundance and intensity of infection are hard to obtain when using noninvasive sampling. We need to keep working on developing tools that will help us overcome these challenges and enhance the applicability of noninvasive sampling methods, especially when studying endangered species in which capture is not desired or even feasible, but the assessment of their vulnerability to diseases is fundamental.

9.4.1 Research Needs

Although the study of parasites and diseases in NHP has been growing in recent years, there are still many NHP species and locations across their range that lack information (Solórzano-García and Pérez-Ponce de León 2018). Greater efforts are needed to monitor understudied endangered species to assess the role of diseases in population declines and identify risk zones. Several Neotropical NHP species tolerate a certain amount of habitat disturbance and use anthropic land covers for different purposes, making human-transformed environments important components of their home ranges and daily activities (Galán-Acedo et al. 2019). It is a priority to develop studies assessing the diversity of parasites and pathogens and the disease dynamics in these anthropic land cover types and across transformation gradients to identify those scenarios that can facilitate the occurrence of spillover and spillback events between humans and the NHP.

Habitat disturbance and forest fragmentation are expected to constitute a risk for infectious diseases in NHP, facilitating pathogen cross-species transmission, hence triggering zoonotic diseases. It is critical to assess the ecological aspects that could alter disease morbidity and mortality, along with the challenges faced by both host and parasites in these modified ecosystems. Additionally, it is critical to include evaluations on the parasites and diseases in humans and domestic animals living in proximity to NHP, along with the corresponding molecular characterization. In this way, we will continue increasing the genetic library on parasites and pathogens, which is crucial for the identification of variants and evolutionary lineages to corroborate the origin and zoonotic potential of a disease and to determine the role of NHP in human disease and vice versa (Hasegawa et al. 2014; Helenbrook et al. 2015; Van Lieshout et al. 2005; Garcia et al. 2005). The embracement of approaches such as One health will facilitate the construction of interdisciplinary research programs to assess primate–parasite dynamics in the human–NHP interface and to develop conservation actions that consider the interactions among human health–primate health–environmental health (Deem 2016).

Finally, the study of primatological diseases should focus not only on ecological associations such as parasites–host–environment but also on the intrinsic factors that contribute to the host's susceptibility or resistance. Assessments from the holobiont perspective (Lederberg and McCray 2001) will aid in uncovering the interactions among the organism and the different microorganisms that live within it and how they relate to the incidence and intensity of a disease, as well as how these internal interactions react to ongoing environmental changes.

References

Agostini I, Vanderhoeven E, Beldomenico PM, Pfoh R, Notarnicola J (2018) First coprological survey of helminths in a wild population of black capuchin monkeys (*Sapajus nigritus*) in northeastern Argentina. Mastozoologia Neotrop 25:269–281

Almeida MABD, Santos ED, Cardoso JDC, Fonseca DFD, Noll CA, Silveira VR, Maeda AY, Pereira de Souza R, Kanamura C, Brasil RA (2012) Yellow fever outbreak affecting *Alouatta* populations in southern Brazil (Rio Grande do Sul State), 2008–2009. Am J Primatol 74:68–76

Almeida MA, Cardoso J da C, Dos Santos E, da Fonseca DF, Cruz LL, Faraco FJ, Bercini MA, Vettorello KC, Porto MA, Mohrdieck R, Ranieri TM, Schermann MT, Sperb AF, Paz FZ, Nunes ZM, Romano AP, Costa ZG, Gomes SL, Flannery B (2014) Surveillance for yellow Fever virus in non-human primates in southern Brazil, 2001–2011: a tool for prioritizing human populations for vaccination. PLoS Negl Trop Dis 8(3):e2741. https://doi.org/10.1371/journal.pntd.0002741

Almeida MAB, Cardoso JC, Santos E, Romano APM, Chiang JO, Martins LC, Vasconcelos PFC, Bicca-Marques JC (2016) Immunity to Yellow Fever, Oropouche and Saint Louis viruses in a wild howler monkey. Neotrop Primates 23:19–21

Almeida MAB, Santos E, Cardoso JC, Noll CA, Lima MM, Silva FA, Ferreira MS, Martins LC, Vasconcelos PFC, Bicca-Marques JC (2019) Detection of antibodies against Icoaraci, Ilhéus, and Saint Louis Encephalitis arboviruses during yellow fever monitoring surveillance in non-human primates (*Alouatta caraya*) in southern Brazil. J Med Primatol 48(4):211–217. https://doi.org/10.1111/jmp.12417

Andrade MS, Campos FS, Campos AAS, Abreu FVS, Melo FL, Sevá ADP, Cardoso JDC, Dos Santos E, Born LC, Silva CMDD, Müller NFD, Oliveira CH, Silva AJJD, Simonini-Teixeira D, Bernal-Valle S, Mares-Guia MAMM, Albuquerque GR, Romano APM, Franco AC, Ribeiro BM, Roehe PM, Almeida MAB (2021) Real-time genomic surveillance during the 2021 re-emergence of the Yellow Fever Virus in Rio Grande do Sul State, Brazil. Viruses 13(10):1976. https://doi.org/10.3390/v13101976

Antonelli A (2022) The rise and fall of neotropical biodiversity. Bot J Linn Soc 199:8–24

Araújo FAA, Ramos DG, Santos AL, Passos PHO, Elkhoury ANSM, Costa ZGA, Leal SG, Romano APM (2011) Epizootics in nonhuman primates during reemergence of yellow fever virus in Brazil, 2007 to 2009. Epidemiol Serv Saúde 20:527–536

Argüello-Sánchez LE (2012) Genética de la conservación en *Alouatta palliata mexicana*: evaluación del efecto de la fragmentación de hábitat y sus poblaciones en Veracruz. Master thesis, Instituto de Ecología, Xalapa, Veracruz, México

Argüello-Sánchez LE, García-Feria LM (2014) La genética como herramienta para el estudio y conservación del género *Alouatta* en México. Acta Zool Mex 30:387–394

Auld SKJR, Searle CL, Duffy MA (2017) Parasite transmission in a natural multihost-multiparasite community. Philos Trans R Soc B Biol Sci 372:1–10

Azofeifa-Rojas I, Sánchez-Porras R, Daniele S (2021) Mortalidad por Electrocución de Monos Congo (Alouatta palliata) Debido a Líneas Eléctricas en Guanacaste, Costa Rica. Mesoamericana 25:15–21

Balasubramaniam K, Beisner B, Vandeleest J, Atwill E, McCowan B (2016) Social buffering and contact transmission: network connections have beneficial and detrimental effects on Shigella infection risk among captive *Rhesus* macaques. PeerJ 4:e2630

Barnett AA, Boyle SA, Norconk MM, Palminteri S, Santos RR, Veiga LM et al (2012) Terrestrial activity in pitheciins (*Cacajao*, *Chiropotes*, and *Pithecia*). Am J Primatol 74:1106–1127

Batista PM, Andreotti R, Chiang JO, Ferreira MS, Vasconcelos PFDC (2012) Seroepidemiological monitoring in sentinel animals and vectors as part of arbovirus surveillance in the state of Mato Grosso do Sul, Brazil. Rev Soc Bras Med Trop 45:168–173

Batista PM, Andreotti R, de Almeida PS, Marques AC, Rodrigues SG, Chiang JO, da Vasconcelos CPF (2013) Detection of arboviruses of public health interest in free-living New World primates (*Sapajus* spp.; *Alouatta caraya*) captured in Mato Grosso do Sul, Brazil. Rev Soc Bras Med Trop:46684–46690

Beck-Johnson LM, Nelson WA, Paaijmans KP, Read AF, Thomas MB, Bjørnstad ON (2017) The importance of temperature fluctuations in understanding mosquito population dynamics and malaria risk. RSoc Pen Sci 4:160969

Beehner JC, Bergman TJ (2017) The next step for stress research in primates: to identify relationships between glucocorticoid secretion and fitness. Horm Behav 91:68–83

Behie AM, Pavelka MS (2013) Interacting roles of diet, cortisol levels, and parasites in determining population density of Belizean howler monkeys in a hurricane damaged forest fragment. In: Behie AM, Pavelka MS (eds) Primates in fragments. Springer, New York, pp 447–456

Behie AM, Kutz S, Pavelka MS (2014) Cascading effects of climate change: do hurricane-damaged forests increase risk of exposure to parasites? Biotropica 46(1):25–31

Beltrán FJ, Díaz LA, Konigheim B, Molina J, Beaudoin JB, Contigiani M, Spinsanti LI (2015) Evidencia serológica de circulación del virus de la encefalitis de San Luis en aves de la Ciudad Autónoma de Buenos Aires, Argentina. Rev Argent Microbiol 47:312–316

Benítez-Malvido J, Arroyo-Rodríguez V (eds) (2008) Habitat fragmentation, edge effects and biological corridors in tropical ecosystems. Eolss Publishers, Oxford

Bezerra BM, Bastos M, Souto A, Keasey MP, Eason P, Schiel N, Jones G (2014) Camera trap observations on nonhabituated Critically Endangered wild blonde capuchins, *Sapajus flavius* (formerly *Cebus flavius*). Int J Primatol 35:895–907

Bicca-Marques JC (2003) How do howler monkeys cope with habitat fragmentation? In: Marsh LK (ed) Primates in fragments: ecology and conservation. Kluwer Academic/Plenum Publishers, New York, pp 283–303

Bicca-Marques JC, Calegaro-Marques C (2016) Ranging behavior drives parasite richness: a more parsimonious hypothesis. Am J Primatol 78:923–927

Bicca-Marques JC, Heymann EW (2013) Ecology and behaviour of titi monkeys (genus Callicebus). In: Veiga LM, Barnett AA, Ferrari SF, Norconk MA (eds) Evolutionary biology and conservation of titis, sakis, and uacaris. Cambridge University Press, Cambridge, pp 196–207

Bicca-Marques JC, Silva VM, Gomes DF (2011) Orden primates. In: Reis NR, Peracchi AL, Pedro WA, Lima IP (eds) Mamíferos do Brasil, 2nd edn, Londrina, pp 107–150

Bicca-Marques JC, Chaves ÓM, Hass GP (2020) Howler monkey tolerance to habitat shrinking: lifetime warranty or death sentence? Am J Primatol 82:e23089

Bicca Marques JC, Calegaro Marques C, Rylands A, Strier KB, Mittermeier R, De Almeida MA, De Castro PH, Chaves OM, Ferraz LP, Fortes VB, Hirano Z (2017) Yellow fever threatens Atlantic Forest primates. American Association for the Advancement of Science

Borremans B, Faust C, Manlove KR, Sokolow SH, Lloyd-Smith JO (2019) Cross-species pathogen spillover across ecosystem boundaries: mechanisms and theory. Philos Trans R Soc B Biol Sci 374:1782

Brasil P, Zalis MG, de Pina-Costa A, Siqueira AM, Júnior CB, Silva S et al (2017) Outbreak of human malaria caused by *Plasmodium simium* in the Atlantic Forest in Rio de Janeiro: a molecular epidemiological investigation. Lancet Glob Health 5:e1038–e1046

Bregnard C, Rais O, Voordouw MJ (2020) Climate and tree seed production predict the abundance of the European Lyme disease vector over a 15-year period. Parasit Vectors 13(1):1–12

Brown L, Medlock J, Murray V (2014) Impact of drought on vector-borne diseases – how does one manage the risk? Public Health 128:29–37

Bryant JE, Holmes EC, Barrett ADT (2007) Out of Africa: a molecular Perspective on the introduction of Yellow Fever Viruses into the Americas. PLoS Pathog 3:e75

Burgin CJ, Colella JP, Kahn PL, Upham NS (2018) How many species of mammals are there? J Mammal 99:1–14

Bustamante-Manrique S, Botero-Henao N, Castaño JH, Link A (2021) Activity budget, home range and diet of the Colombian night monkey (*Aotus lemurinus*) in peri-urban forest fragments. Primates 62:529–536

Byrne H, Rylands AB, Carneiro JC, Lynch AJW, Bertuol F, Silva MNF et al (2016) Phylogenetic relationships of the New World titi monkeys (Callicebus): first appraisal of taxonomy based on molecular evidence. Front Zool 13:1–26

Cândido SL, Fonseca MJDOR, Pacheco RDC, Campos CGD, Morgado TO, Colodel EM et al (2022) Molecular detection of *Toxoplasma gondii* and *Neospora caninum* in neotropical primates. J Med Primatol 51:149–154

Cantarelli VI, Perez-Rueda MA, Kowalewski MM, Mastromonaco GF, Ponzio MF (2017) Validation of an enzyme immunoassay and comparison of fecal cortisol metabolite levels in black and gold howler monkeys (*Alouatta caraya*) inhabiting fragmented and continuous areas of the humid Chaco region, Argentina. Am J Primatol 79:e22625

Carretero-Pinzón X, Defler TR, Ruiz-Garcia M (2016) Chapter 15: How does the Colombian squirrel monkey cope with habitat fragmentation? Strategies to survive in small fragments. In: Ruiz-Garcia M, Shostell JM (eds) Evolutionary biology and conservation of the neotropical primates. Nova Publishers, New York, pp 491–506

Carrillo-Bilbao G, Martin-Solano S, Saegerman C (2021) Zoonotic blood-borne pathogens in non-human primates in the Neotropical Region: a systematic review. Pathogens 10:1009

Catão-Dias JL, Epiphanio S, Kierulff MCM (2013) Neotropical primates and their susceptibility to *Toxoplasma gondii*: new insights for an old problem. In: Brinkworth J, Pechenkina K (eds) Primates, pathogens, and evolution. Springer, New York, pp 253–289

Cawthorn DM, Hoffman LC (2015) The bushmeat and food security nexus: a global account of the contributions, conundrums and ethical collisions. Food Res Int 76:906–925

Ceballos-Mago N, Chivers DJ (2010) Local knowledge and perceptions of pet primates and wild Margarita capuchins on Isla de Margarita and Isla de Coche in Venezuela. Endanger Species Res 13:63–72

Chapman CA, Peres CA (2021) Primate conservation: lessons learned in the last 20 years can guide future efforts. Evol Anthropol Issues News Rev 30:345–361

Chapman CA, Saj TL, Snaith TV (2007) Temporal dynamics of nutrition, parasitism, and stress in colobus monkeys: implications for population regulation and conservation. Am J Phys Anthropol 134(2):240–250

Chapman CA, Bonnell TR, Gogarten JF, Lamber JE, Omeja PA, Twinomugisha D et al (2013) Are primates ecosystem engineers? Int J Primatol 34:1–14

Chaves ÓM, Stoner KE, Arroyo-Rodríguez V (2012) Differences in diet between spider monkey groups living in forest fragments and continuous forest in Mexico. Biotropica 44:105–113

Chaves ÓM, Amorim Fernandes F, Turcato Oliveira G, Bicca-Marques JC (2019) Assessing the influence of biotic, abiotic, and social factors on the physiological stress of a large Neotropical Primate in Atlantic Forest fragments. Sci Total Environ 690:705–716

Chaves A, Ibarra-Cerdeña CN, López-Pérez AM, Monge O, Avendaño R, Ureña-Saborio H, Chavarría M, Zaldaña K, Sánchez L, Ortíz-Malavassi E, Suzan G, Foley J, Gutiérrez-Espeleta GA (2020a) Bocaparvovirus, erythroparvovirus and tetraparvovirus in new world primates from Central America. Transbound Emerg Dis 67:377–387

Chaves A, Piche-Ovares M, Corrales E, Suzán G, Moreira-Soto A, Gutiérrez-Espeleta GA (2020b) Flaviviruses infections in neotropical primates suggest long-term circulation of Saint Louis Encephalitis and Dengue virus spillback in socioeconomic regions with high numbers of Dengue human cases in Costa Rica. Preprints 2020:2020110222

Chaves OM, Fortes VB, Hass GP, Azevedo RB, Stoner KE, Bicca-Marques JC (2021a) Flower consumption, ambient temperature and rainfall modulate drinking behavior in folivorous-frugivorous arboreal mammal. PLoS One 16:e0236974

Chaves A, Piche-Ovares M, Ibarra-Cerdeña CN, Corrales-Aguilar E, Suzán G, Moreira-Soto A, Gutiérrez-Espeleta GA (2021b) Serosurvey of nonhuman primates in Costa Rica at the human–wildlife interface reveals high exposure to flaviviruses. Insects 12:554

Chaves Ó, Júnior JS, Buss G, Hirano ZM, Jardim MMA, Amaral ELS, Godoy JC, Peruchi AR, Michel T, Bicca-Marques JC (2022) Wildlife is imperiled in peri-urban landscapes: threats to arboreal mammals. Sci Total Environ 821:152883

Chichorro F, Juslén A, Cardoso P (2019) A review of the relation between species traits and extinction risk. Biol Conserv 237:220–229

Conatser JM (2016) A coprological survey of Golden Mantled Howler monkeys (*Alouatta palliata palliata*) in the Osa Peninsula, Costa Rica. M.Sc. theses, Dissertations and Capstones, p 1045

Cooper N, Griffin R, Franz M, Omotayo M, Nunn CL (2012) Phylogenetic host specificity and understanding parasite sharing in primates. Ecol Lett 15:1370–1377

Cortés-Ortiz L, Rosales-Meda M, Marsh LK, Mittermeier RA (2020) Alouatta pigra. The IUCN Red List of Threatened Species 2020

Cortés-Ortiz L, Solano-Rojas D, Rosales-Meda M, Williams-Guillén K, Méndez-Carvajal PG, Marsh LK, Canales-Espinosa D, Mittermeier RA (2021) *Ateles geoffroyi* (amended version of 2020 assessment). The IUCN Red List of Threatened Species 2021

Costa TLC, Iglesias GA, Rosa JMA, Bento HJ, Rondelli LAS, Furlan F, Morgado TO, Dutra V, Corrêa SHR (2018) Molecular detection of *Neospora caninum* in Azara's night monkey (*Aotus azarae*) of wildlife in the state of Mato Grosso: case report. Arquivo Brasileiro de Medicina Veterinária e Zootecnia:701227–701232

Costa-Araújo R, Silva Júnior JS, Canale GR, Melo R, Boubli JP, Farias I, Hrbek T (2022) Mico schneideri. The IUCN Red List of Threatened Species 2022:e.T210363264A210363416. https://doi.org/10.2305/IUCN.UK.2022-1.RLTS.T210363264A210363416.en

Cristobal-Azkarate J, Arroyo-Rodríguez V (2007) Diet and activity patterns of howler monkeys (*Alouatta palliata*) in Los Tuxtlas, Mexico: Effects of habitat fragmentation and implications for conservation. Am J Primatol 69:1013–1029

Cristobal-Azkarate J, Hervier B, Vegas-Carrillo S, Osorio-Sarabia D, Rodriguez-Luna E, Vea JJ (2010) Parasitic infections of three mexican howler monkey groups (*Alouatta palliata mexicana*) exposed to differing degrees of habitat disturbance in fragmented forests. Primates 51:231–239

da Cardoso CJ, de Almeida MA, dos Santos E, da Fonseca DF, Sallum MA, Noll CA, Monteiro HA, Cruz AC, Carvalho VL, Pinto EV, Castro FC, Nunes Neto JP, Segura MN, Vasconcelos PF (2010) Yellow fever virus in *Haemagogus leucocelaenus* and *Aedes serratus* mosquitoes, Southern Brazil, 2008. Emerg Infect Dis 16(12):1918–1924. https://doi.org/10.3201/eid1612.100608

da Silva Pessoa Vieira CJ, Steiner São Bernardo C, Ferreira da Silva DJ, Rigotti Kubiszeski J, Serpa Barreto E, de Oliveira Monteiro HA, Rodrigues Canale G, Peres CA, Leigh Massey A, Levi T, Vieira de Morais Bronzoni R (2021) Land-use effects on mosquito biodiversity and potential arbovirus emergence in the Southern Amazon, Brazil. Transbound Emerg Dis 00:1–12

de la Torre S, Shanee S, Palacios E, Calouro AM, Messias MR, Valença-Montenegro MM (2021) *Cebuella pygmaea* (amended version of 2021 assessment). The IUCN Red List of Threatened Species 2021:e.T136926A200203263. https://doi.org/10.2305/IUCN.UK.2021-2.RLTS.T136926A200203263.en

de Lima RA, Oliveira AA, Pitta GR, de Gasper AL, Vibrans AC, Chave J et al (2020) The erosion of biodiversity and biomass in the Atlantic Forest biodiversity hotspot. Nat Commun 11(1):1–16. https://doi.org/10.1038/s41467-020-20217-w

De Souza Fialho M, Ludwig G, Valença-Montenegro MM (2016) Legal international trade in live neotropical primates originating from South America. Primate Conserv 30:1–6

Deb M, Choudhury S, Sharma I, Bhattacharjee PC, Formicki G, Bulla J, Roychoudhury S (2021) Nutritional stress in western hoolock gibbon due to adverse changes in diet pattern. J Microbiol Biotechnol Food Sci 2021:152–156

Deem SL (2016) Conservation medicine: a solution-based approach for saving nonhuman primates. In: Waller MT (ed) Ethnoprimatology, developments in primatology: progress and prospects. Springer, pp 63–76

Deere JR, Schaber KL, Foerster S, Gilby IC, Feldblum JT, VanderWaal K et al (2021) Gregariousness is associated with parasite species richness in a community of wild chimpanzees. Behav Ecol Sociobiol *75*(5):1–11

Dharmarajan G, Gupta P, Vishnudas CK, Robin VV (2021) Anthropogenic disturbance favours generalist over specialist parasites in bird communities: implications for risk of disease emergence. Ecol Lett 24(9):1859–1868

Di Fiore A, Link A, Campbell CJ (2011) The atelines: behavioral and socioecological diversity in a New World monkey radiation. In: Campbell CJ, Fuentes A, MacKinnon KC, Bearder SK, Stumpf RM (eds) Primates in perspective, 2nd edn. Oxford University Press, Oxford, pp 155–188

Dietz JM, Hankerson SJ, Alexandre BR, Henry MD, Martins AF, Ferraz LP, Ruiz-Miranda CR (2019) Yellow fever in Brazil threatens successful recovery of endangered golden lion tamarins. Sci Rep 9(1):1–13

Digby LJ, Ferrari SF, Saltzman W (2011) Callitrichines: the role of competition in cooperatively breeding species. In: Campbell CJ, Fuentes A, MacKinnon KC, Bearder SK, Stumpf RM (eds) Primates in perspective, 2nd edn. Oxford University Press, Oxford, pp 91–107

Dos Santos EO, Klain VF, Manrique SB, Roman IJ, Dos Santos HF, Sangioni LA et al (2022) The influence of landscape structure on the occurrence of Neospora caninum, Toxoplasma gondii, and Sarcocystis spp. Free-Living Neotrop Primates Acta Parasitologica 67(4):1680–1696

Dunn JC, Cristóbal-Azkarate J, Schulte-Herbrüggen B, Chavira R, Veà JJ (2013) Travel time predicts fecal glucocorticoid levels in free-ranging howlers (Alouatta palliata). Int J Primatol 34:246–259

Durán ND (2014) Cortisol y endoparásitos de mono araña *(Ateles geoffroyi yucatanensis)* bajo diferente grado de presencia humana. Doctoral dissertation, El Colegio de la Frontera Sur.

Ehlers LP, Slaviero M, Bianchi MV, de Mello LS, De Lorenzo C, Surita LE et al (2022) Causes of death in neotropical primates in Rio Grande do Sul State, Southern Brazil. J Med Primatol 51(2):85–92

Engelstädter J, Fortuna NZ (2019) The dynamics of preferential host switching: host phylogeny as a key predictor of parasite distribution*. Evolution (NY) 73:1330–1340. https://doi.org/10.1111/evo.13716

Estrada A, Garber PA, Rylands AB, Roos C, Fernandez-Duque E, Di Fiore A, Nekaris KAI, Nijman V, Heymann EW, Lambert JE, Rovero F, Barelli C, Setchell JM, Gillespie TR, Mittermeier RA, Arregoitia LV, Guinea M, Gouveia S, Dobrovolski R, Shanee S, Shanee N, Boyle SA, Fuentes A, MacKinnon KC, Amato KR, Meyer ALS, Wich S, Sussman RW, Pan R, Kone I, Li B (2017) Impending extinction crisis of the world's primates: why primates matter. Sci Adv 3:e1600946

Estrada A, Garber PA, Chaudhary A (2019) Expanding global commodities trade and consumption place the world's primates at risk of extinction. PeerJ 7:e7068

Estrada A, Garber PA, Chaudhary A (2020) Current and future trends in socio-economic, demographic and governance factors affecting global primate conservation. PeerJ 8:1–35. https://doi.org/10.7717/peerj.9816

Failloux AB (2019) Human activities and climate change in the emergence of vector-borne diseases. Comptes Rendus. Biologies 342(7–8):269–270

Favoretto S, Araújo D, Oliveira D, Duarte N, Mesquita F, Zanotto P, Durigon E (2016) First detection of Zika virus in neotropical primates in Brazil: a possible new reservoir. BioRxiv 10:049395

Favoretto SR, Araujo DB, Duarte NF, Oliveira DB, da Crus NG, Mesquita F, Leal F, Machado RGR, Gaio F, Oliviera WF, Zanotto PMA, Durigon EL (2019) Zika Virus in peridomestic neotropical primates, Northeast Brazil. Ecohealth 16(1):61–69

Fecchio A, Wells K, Bell JA, Tkach VV, Lutz HL, Weckstein JD et al (2019) Climate variation influences host specificity in avian malaria parasites. Ecol Lett 22(3):547–557

Fecchio A, Lima MR, Bell JA, Schunck F, Corrêa AH, Beco R et al (2021) Loss of forest cover and host functional diversity increases prevalence of avian malaria parasites in the Atlantic Forest. Int J Parasitol 51(9):719–728

Fernandez-Duque E (2011) Aotidae: social monogamy in the only nocturnal anthropoid. In: Campbell CJ, Fuentes A, MacKinnon KC, Bearder SK, Stumpf RM (eds) Primates in perspective, 2nd edn. Oxford University Press, Oxford, pp 140–154

Ferreira DRA, Laroque PO, Wagner PGC, Higino SS, Azevedo SS, Rego EW, Mota RA (2001) Ocorrência de anticorpos e fatores de risco associados à infecção por *Leptospira* spp. em *Cebus* spp. mantidos em cativeiro no Nordeste do Brasil. Pesqui Vet Bras 31:1019–1023

Ferreira RG, Jerusalinsky L, Silva TCF, Fialho MS, Roque AA, Fernandes A, Arruda F (2009) On the occurrence of *Cebus flavius* (Schreber 1774) in the Caatinga, and the use of semi-arid environments by *Cebus* species in the Brazilian state of Rio Grande do Norte. Primates 50:357–362. https://doi.org/10.1007/s10329-009-0156-z

Fortes VB, Bicca-Marques JC, Urbani B, Fernández VA, Pereira TS (2015) Ranging behavior and spatial cognition of howler monkeys. In: Kowalewski MM, Garber PA, Cortés-Ortiz L, Urbani B, Youlatos D (eds) Howler monkeys: behavior, ecology, and conservation. Springer, New York, pp 219–255. https://doi.org/10.1007/978-1-4939-1960-4_9)

Freire Filho R, Castro CSS, Casanova C, Bezerra BM (2021) Uses of nonhuman primates by humans in northeastern Brazil. Primates 62:777–788. https://doi.org/10.1007/s10329-021-00919-5

Friant S, Ziegler TE, Golberg TL (2016) Primate reinfection with gastrointestinal parasites: behavioral and physiological predictors of parasite acquisition. Anim Behav 117:105–113

Galán-Acedo C, Arroyo-Rodríguez V, Andresen E, Verde Arregoitia L, Vega E, Peres CA, Ewers RM (2019) The conservation value of human-modified landscapes for the world's primates. Nat Commun 10:1–8. https://doi.org/10.1038/s41467-018-08139-0

Garber PA, Caselli CB, McKenney AC, Abreu F, De la Fuente MF, Araújo A, Arruda MF, Souto A, Schiel N, Bicca-Marques JC (2019) Trait variation and trait stability in common marmosets (*Callithrix jacchus*) inhabiting ecologically distinct habitats in northeastern Brazil. Am J Primatol 81(7):e23018. https://doi.org/10.1002/ajp.23018

Garcia JL, Svoboda WK, Chryssafidis AL, de Souza Malanski L, Shiozawa MM, de Moraes Aguiar L, Teixeira GM, Ludwig G, da Silva LR, Hilst C, Navarro IT (2005) Sero-epidemiological survey for toxoplasmosis in wild New World monkeys (Cebus spp.; Alouatta caraya) at the Paraná river basin, Paraná State, Brazil. Vet Parasitol 133:307–311

García-Toro LC, Link A, Páez-Crespo EJ, Stevenson PR (2019) Home range and daily traveled distances of highland Colombian woolly monkeys (*Lagothrix lagothricha lugens*): comparing spatial data from GPS collars and direct follows. In: Reyna-Hurtado R, Chapman C (eds) Movement ecology of neotropical forest mammals. Springer, Cham, pp 173–193

Georgiev VST (2009) National Institute of Allergy and Infectious Diseases, NIH, Volume 2: Impact on global health. Humana Press, New York

Gibrail MM, Fiaccadori FS, Souza M, Almeida TNV, Chiang JO, Martins LC, Ferreira MS, Cardoso DDDP (2016) Detection of antibodies to Oropouche virus in non-human primates in Goiânia City, Goiás. Rev Soc Bras Med Trop 49:357–360

Godfrey SS (2013) Networks and the ecology of parasite transmission: a framework for wildlife parasitology. Int J Parasitol Parasites Wildl 2:235–245. https://doi.org/10.1016/j.ijppaw.2013.09.0

Gomez JM, Nunn CL, Verdu M (2013) Centrality in primate-parasite networks reveals the potential for the transmission of emerging infectious diseases to humans. Proc Natl Acad Sci 110(19):7738–7741. https://doi.org/10.1073/pnas.1220716110

Groves CP, Wilson DE, Reeder DM (eds) (2005) Mammal species of theworld: a taxonomic and geographic reference, 3rd edn. Johns Hopkins University Press, Baltimore, pp 148–152

Guégan JF, Ayouba A, Cappelle J, De Thoisy B (2020) Forests and emerging infectious diseases: unleashing the beast within. Environ Res Lett 15(8):083007

Hamrick PN, Aldighieri S, Machado G, Leonel DG, Vilca LM, Uriona S, Schneider MC (2017) Geographic patterns and environmental factors associated with human yellow fever presence in the Americas. PLoS Negl Trop Dis 11(9):e0005897

Han BA, Kramer AM, Drake JM (2016) Global patterns of zoonotic disease in mammals. Trends Parasitol 32:565–577

Hasegawa H, Modrý D, Kitagawa M, Shutt KA, Todd A, Kalousová B, Profousová I, Petrželková KJ (2014) Humans and Great Apes cohabiting the forest ecosystem in Central African Republic Harbour the same Hookworms. PLoS Negl Trop Dis 8:1–10. https://doi.org/10.1371/journal.pntd.0002715

Hasnat GT, Hossain MK (2020) Global overview of tropical dry forests. In: Bhadouria R, Tripathi S, Srivastava P, Singh P (eds) Handbook of research on the conservation and restoration of tropical dry forests. IGI Global, Hershey, pp 1–23

Hatcher MJ, Dick JTA, Dunn AM (2006) How parasites affect interactions between competitors and predators. Ecol Lett 9:1253–1271

Helenbrook WD, Shields WM, Whipps CM (2015) Characterization of *Blastocystis* species infection in humans and mantled howler monkeys, *Alouatta palliata aequatorialis*, living in close proximity to one another. Parasitol Res 114:2517–2525. https://doi.org/10.1007/s00436-015-4451-x

Hope K, Goldsmith ML, Graczyk T (2004) Parasitic health of olive baboons in Bwindi Impenetrable National Park, Uganda. Vet Parasitol 122(2):165–170

Huang S, Bininda-Emonds OR, Stephens PR, Gittleman JL, Altizer S (2014) Phylogenetically related and ecologically similar carnivores harbour similar parasite assemblages. J Anim Ecol 83(3):671–680

Huemer HP, Larcher C, Czedik-Eysenberg T, Nowotny N, Reifinger M (2002) Fatal infection of a pet monkey with human herpesvirus 1. Emerg Infect Dis 8(6):639

ICMBio (2011) In: Jerusalinsky L, Talebi M, Melo R (eds) Plano de Ação Nacional para a Conservação dos Muriquis. Instituto Chico Mendes de Conservação da Biodiversidade, Brasília

Imura K, Chambers JK, Uchida K, Nomura S, Suzuki S, Nakayama H, Miwa Y (2014) Herpes simplex virus type 1 infection in two pet marmosets in Japan. J Vet Med Sci 14:0374

International Committee on Taxonomy of Viruses (ICTV) (n.d.). https://ictv.global/taxonomy/

IPCC (2014) Summary for policymakers. In: Field CB, Barros VR, Dokken DJ, Mach KJ, Mastrandrea MD, Bilir TE, Chatterjee M, Ebi KL, Estrada YO, Genova RC, Girma B, Kissel ES, Levy AN, MacCracken S, Mastrandrea PR, White LL (eds) Climate change 2014: impacts, adaptation, and vulnerability. Part A: Global and sectoral aspects. Contribution of Working Group II to the fifth assessment report of the Intergovernmental Panel on Climate Change. Cambridge University Press, Cambridge/New York, pp 1–32

IUCN (2022) The IUCN Red List of Threatened Species. Version 2022–1. https://www.iucnredlist.org

Jack KM (2011) The cebines: toward an explanation of variable social structure. In: Campbell CJ, Fuentes A, MacKinnon KC, Bearder SK, Stumpf RM (eds) Primates in perspective, 2nd edn. Oxford University Press, Oxford, pp 108–122

Jardim MMA, Queirolo D, Peters FB, Mazim FD, Favarini M, Tirelli FP, Trindade RA, Bonatto SL, Bicca-Marques JC, Mourthe Í (2020) Southern extension of the geographic range of black-and-gold howler monkeys (*Alouatta caraya*). Mammalia 84(1):102–106. https://doi.org/10.1515/mammalia-2018-0127

Johnston WF, Yeh J, Nierenberg R, Procopio G (2015) Exposure to macaque monkey bite. J Emerg Med 49(5):634–637

Junge RE, Louis EE (2002) Medical evaluation of free-ranging primates in Betampona Reserve, Madagascar. Lemur News 7:23–25

Junker J, Petrovan SO, Arroyo-Rodríguez V, Boonratana R, Byler D, Chapman CA et al (2020) A severe lack of evidence limits effective conservation of the world's primates. Bioscience 70(9):794–803. https://doi.org/10.1093/biosci/biaa082

Kane J, Smith RL (2020) *Bertiella* sp. (Meyner, 1895) infection of *Alouatta caraya* (Humboldt, 1812) in urban and natural environments in Ñeembucú, Southwest Paraguay. Am J Primatol 82. https://doi.org/10.1002/ajp.23166

Kazanji M, Mouinga-Ondémé A, Lekana-Douki-Etenna S, Caron M, Makuwa M, Mahieux R, Gessain A (2015) Origin of HTLV-1 in hunters of nonhuman primates in Central Africa. J Infect Dis 211(3):361–365

Kelly DW, Paterson RA, Townsend CR, Poulin R, Tompkins DM (2009) Parasite spill-back: a neglected concept in invasion ecology? Ecology 90(8):2047–2056. https://doi.org/10.1890/08-1085.1

Klain VF, Mentz MB, Bustamante-Manrique S, Bicca-Marques JC (2021) Habitat patch size and landscape structure influence, although weakly, the parasite richness of an arboreal folivorous-frugivorous primate in anthropogenic landscapes. Res Sq:1–25. https://doi.org/10.21203/rs.3.rs-509824/v1

Kowalewski MM, Gillespie TR (2009) Ecological and anthropogenic influences on patterns of parasitism in free-ranging primates: a meta-analysis of the genus Alouatta. In: Garber PA (ed) South American Primates, developments in primatology: progress and prospects. Springer, pp 433–461. https://doi.org/10.1007/978-0-387-78705-3

Lederberg J, McCray A (2001) 'Ome Sweet' Omics – a genealogical treasury of words. Scientist 15

Li MT, Jin Z, Sun GQ, Zhang J (2017) Modeling direct and indirect disease transmission using multi-group model. J Math Anal Appl 446(2):1292–1309. https://doi.org/10.1016/j.jmaa.2016.09.043

Link A, Guzmán-Caro DC, Roncancio N, Mittermeier RA, Rodríguez V (2021) *Saguinus leucopus* (amended version of 2020 assessment). The IUCN Red List of Threatened Species 2021:e.T19819A192550769. https://doi.org/10.2305/IUCN.UK.2021-1.RLTS.T19819A192550769.en

Lins PGADS, Ferreira RG (2019) Competition during sugarcane crop raiding by blond capuchin monkeys (*Sapajus flavius*). Primates 60(1):81–91

Lopes S, Calegaro-Marques C, Klain V, Chaves ÓM, Bicca-Marques JC (2022) Necropsies disclose a low helminth parasite diversity in Periurban Howler monkeys. Am J Primatol 84. https://doi.org/10.1002/ajp.23346

López-Pérez AM, Moreno K, Chaves A, Ibarra-Cerdeña CN, Rubio A, Foley J, List R, Suzán G, Sarmiento RE (2019) Carnivore protoparvovirus 1 at the wild-domestic carnivore interface in Northwestern Mexico. EcoHealth 16(3):502–511. https://doi.org/10.1007/s10393-019-01436-0

Lorenz C, Azevedo TS, Virginio F, Aguiar BS, Chiaravalloti-Neto F, Suesdek L (2017) Impact of environmental factors on neglected emerging arboviral diseases. PLoS Negl Trop Dis 11(9):e0005959

Maldonado AM, Peck MR (2014) Research and *in situ* conservation of owl monkeys enhances environmental law enforcement at the Colombian-Peruvian border. Am J Primatol 76:658–669. https://doi.org/10.1002/ajp.22260

Manes S, Costello MJ, Beckett H, Debnath A, Devenish-Nelson E, Grey KA et al (2021) Endemism increases species' climate change risk in areas of global biodiversity importance. Biol Conserv 257:109070. https://doi.org/10.1016/j.biocon.2021.109070

Mansell NL, McKinney T (2021) Interactions between humans and Panamanian White-Faced Capuchin Monkeys (*Cebus imitator*). Int J Primatol 42:548–562

Mares-Guia MAMDM, Horta MA, Romano A, Rodrigues CD, Mendonça MC, Dos Santos CC, Torres MC, Araujo ESM, Fabri A, de Souza ER, Ribeiro ROR, Lucena FP, Junior LCA, da Cunha RV, Nogueira RMR, Sequeira PC, de Filippis AMB (2020) Yellow fever epizootics in non-human primates, Southeast and Northeast Brazil (2017 and 2018). Parasit Vectors 13(1):1–8

Martínez MF, Santini MS, Kowalewski MM, Salomón OD (2019) Phlebotominae in peri-domestic and forest environments inhabited by *Alouatta caraya* in northeastern Argentina. Med Vet Entomol 33(3):367–374

Martínez-Mota R, Valdespino C, Sánchez-Ramos MA, Serio-Silva JC (2007) Effects of forest fragmentation on the physiological stress response of black howler monkeys. Anim Conserv 10(3):374–379

Martínez-Mota R, Pozo-Montuy G, Bonilla-Sánchez YM, Gillespie TR (2018) Effects of anthropogenic stress on the presence of parasites in a threatened populations of black howler monkeys (*Alouatta pigra*) in Southern Mexico. 2Therya 9:161–169. https://doi.org/10.12933/therya-18-572. ISSN:2007-3364

Matthee CA (2020) The influence of host dispersal on the gene flow and genetic diversity of generalist and specialist ectoparasites. Afr Zool 55:119–126. https://doi.org/10.1080/15627020.2020.1762512

Maurice ME, Manyi ES, Munge EL (2017) Primate bushmeat consumption: a source of zoonotic disease risk in Tombel Area, Southwest Region, Cameroon. Int J Environ Sci Nat Resour 5(2)

McKinney T (2010) The effects of provisioning and crop-raiding on the diet and foraging activities of human-commensal white-faced capuchins (*Cebus capucinus*). Am J Primatol 73:*439–448*

McKinney T (2014) Species-specific responses to tourist interactions by white-faced capuchins (*Cebus imitator*) and mantled howlers (*Alouatta palliata*) in a Costa Rican wildlife refuge. Int J Primatol 35:573–589

Melo FR, Boubli JP, Mittermeier RA, Jerusalinsky L, Tabacow FP, Ferraz DS, Talebi M (2021) *Brachyteles hypoxanthus* (amended version of 2019 assessment). The IUCN Red List of Threatened Species 2021:e.T2994A191693399. https://doi.org/10.2305/IUCN.UK.2021-1.RLTS.T2994A191693399.en

Molina CV, Catão-Dias JL, Ferreira Neto JS, Vasconcellos SA, Gennari SM, do Valle RDR, de Souza GO, de Morais ZM, Vitaliano SN, Strefezzi Rde F, Bueno MG (2014) Sero-epidemiological survey for brucellosis, leptospirosis, and toxoplasmosis in free-ranging *Alouatta caraya* and *Callithrix penicillata* from São Paulo State, Brazil. J Med Primatol 43(3):197–201

Mollentze N, Streicker DG (2020) Viral zoonotic risk is homogenous among taxonomic orders of mammalian and avian reservoir hosts. Proc Natl Acad Sci U S A 117:9423–9430. https://doi.org/10.1073/pnas.1919176117

Monath TP, Vasconcelos PF (2015) Yellow fever. J Clin Virol 64:160–173. https://doi.org/10.1016/j.jcv.2014.08.030

Morales MA, Fabbri CM, Zunino GE, Kowalewski MM, Luppo VC, Enría DA, Levis SC, Calderón GE (2017) Detection of the mosquito-borne flaviviruses, West Nile, Dengue, Saint Louis Encephalitis, Ilheus, Bussuquara, and Yellow Fever in free-ranging black howlers (*Alouatta caraya*) of Northeastern Argentina. PLoS Negl Trop Dis 11(2):e0005351. https://doi.org/10.1371/journal.pntd.00053

Moreira-Soto A, Carneiro IDO, Fischer C, Feldmann M, Kümmerer BM, Silva NS, Santos UG, Souza BFCD, Liborio FA, Valença-Montenegro MM, Laroque PO, da Fontoura FR, Oliveira AVD, Drosten C, de Lamballerie X, Franke CR, Drexler JF (2018a) Limited evidence for infection of urban and peri-urban nonhuman primates with Zika and chikungunya viruses in Brazil. Msphere 3(1):e00523–e00517

Moreira-Soto A, Torres MC, de Mendonça ML, Mares-Guia MA, dos Santos Rodrigues CD, Fabri AA, Dos Santos CC, Machado Araújo ES, Fischer C, Ribeiro Nogueira RM, Drosten C, Sequeira PC, Drexler JF, de Filippis AB (2018b) Evidence for multiple sylvatic transmission cycles during the 2016–2017 yellow fever virus outbreak, Brazil. Clin Microbiol Infect 24(9):1019–10e1

Moreno ES, Agostini I, Holzmann I, Di Bitetti MS, Oklander LI, Kowalewski MM, Beldomenico PM, Goenaga S, Martínez M, Lestani E, Desbiez AL, Miller P (2017) Yellow fever impact on brown howler monkeys (*Alouatta guariba clamitans*) in Argentina: a metamodelling approach based on population viability analysis and epidemiological dynamics. Mem Inst Oswaldo Cruz 110(7):865–876. https://doi.org/10.1590/0074-02760150075

Mourthé IMC, Guedes D, Fidelis J, Boubli JP, Mendes SL, Strier KB (2007) Ground use by northern muriquis (*Brachyteles hypoxanthus*). Am J Primatol 69(2):706–712. https://doi.org/10.1002/ajp.20405

Muehlenbein MP (2017) Primates on display: potential disease consequences beyond bushmeat. Am J Phys Anthropol 162:32–43

Muehlenbein MP, Martinez LA, Lemke AA, Ambu L, Nathan S, Alsisto S, Sakong R (2010) Unhealthy travelers present challenges to sustainable primate ecotourism. Travel Med Infect Dis 8(3):169–175

Muñoz M, Navarro JC (2012) Virus Mayaro: un arbovirus reemergente en Venezuela y Latinoamérica. Biomedica 32:286–302

Mittermeier RA, Da Fonseca GA, Rylands AB, Brandon K (2005) A brief history of biodiversity conservation in Brazil. Conserv Biol 601–607

Moher D, Liberati A, Tetzlaff J, Altman DG, Prisma Group (2010) Preferred reporting items for systematic reviews and meta-analyses: the PRISMA statement. Int J Surg 8(5):336–341

Navarro J-C, Vianez JLSG, Hernandez R, Giambalvo D, Weaver SC, Tesh RB, Weaver SC, Montañez H, Liria J, Lima A, Travassos da Rosa JF, da Silva SP, Vasconcelos JM, Oliveira R, Vianez JL Jr, Nunes MRT (2016) Isolation of Madre de Dios Virus (Orthobunyavirus; Bunyaviridae), an Oropouche Virus Species Reassortant, from a Monkey in Venezuela. Am J Trop Med Hyg 95(2):328–338. https://doi.org/10.4269/ajtmh.15-0679

Negrey JD, Reddy RB, Scully EJ, Phillips-Garcia S, Owens LA, Langergraber KE et al (2019) Simultaneous outbreaks of respiratory disease in wild chimpanzees caused by distinct viruses of human origin. Emerg Microbes Infect 8(1):139–149

Niehaus C, Spínola M, Su C, Rojas N, Rico-Chávez O, Ibarra-Cerdeña CN, Foley J, Suzán G, Gutiérrez-Espeleta GA, Chaves A (2020) Environmental factors associated with *Toxoplasma gondii* exposure in Neotropical Primates of Costa Rica. Front Vet Sci 7:583032

Norconk MA (2011) Sakis, uakaris, and titi monkeys: behavioral diversity in a radiation of primate seed predators. In: Campbell CJ, Fuentes A, MacKinnon KC, Bearder SK, Stumpf RM (eds) Primates in perspective, 2nd edn. Oxford University Press, Oxford, pp 122–139

Nugent G (2011) Maintenance, spillover and spillback transmission of bovine tuberculosis in multi-host wildlife complexes: a New Zealand case study. Vet Microbiol 151(1-2):34–42. https://doi.org/10.1016/j.vetmic.2011.02.0

Nunn CL, Altizer S (2006) Infectious diseases in primates. Oxford University Press, Oxford

Oliveira LC, Neves LG, Kierulff MCM, Jerusalinsky L, Mittermeier RA, Rylands AB (2021) *Leontopithecus chrysomelas* (amended version of 2020 assessment). The IUCN Red List of Threatened Species 2021:e.T40643A192327573. https://doi.org/10.2305/IUCN.UK.2021-1.RLTS.T40643A192327573.en

Oxley AS, Donati G, Hill CM (2022) What works and what doesn't work? The challenges of doing effective applied conservation research in human-modified habitats. Int J Primatol. https://doi.org/10.1007/s10764-022-00336-5

Pandit PS, Doyle MM, Smart KM, Young CC, Drape GW, Johnson CK (2018) Predicting wildlife reservoirs and global vulnerability to zoonotic Flaviviruses. Nat Commun 9(1):1–10

Pedersen AB, Altizer S, Poss M, Cunningham AA, Nunn CL (2005) Patterns of host specificity and transmission among parasites of wild primates. Int J Parasitol 35:647–657. https://doi.org/10.1016/j.ijpara.2005.01.005

Pereira PM, Valsecchi J, Queiroz H (2019) Spatial patterns of primate hunting in riverine communities in Central Amazonia. Oryx 53(1):165–173. https://doi.org/10.1017/S0030605317000199

Peres CA (1990) Effects of hunting on western Amazonian Primate communities. Biol Conserv 54:47–59

Peres CA (1994) Primate responses to phenological changes in an Amazonian Terra Firme Forest. Biotropica 26:98–112. https://doi.org/10.2307/2389114

Phillips KA, Haas ME, Grafton BW, Yrivarren M (2004) Survey of the gastrointestinal parasites of the primate community at Tambopata National Reserve, Peru. J Zool 264:149–151. https://doi.org/10.1017/S0952836904005680

Plowright RK, Parrish CR, McCallum H, Hudson PJ, Ko AI, Graham AL, Lloyd-Smith JO (2017) Pathways to zoonotic spillover. Nat Rev Microbiol 15(8):502–510

Pozo-Montuy G, Serio-Silva JC (2007) Movement and resource use by a group of *Alouatta pigra* in a forest fragment in Balancán, México. Primates 48:102–107. https://doi.org/10.1007/s10329-006-0026-x

Pozo-Montuy G, Serio-Silva JC, Chapman CA, Bonilla-Sánchez YM (2013) Resource use in a landscape matrix by an arboreal primate: evidence of supplementation in black howlers (*Alouatta pigra*). Int J Primatol 34:714–731. https://doi.org/10.1007/s10764-013-9691-y

Printes RC, Liesenfeld MVA, Jerusalinsky L (2001) *Alouatta guariba clamitans* Cabrera, 1940: a new southern limit for the species and for Neotropical primates. Neotrop Primates 9(3):118–121

Printes RC, Rylands AB, Bicca-Marques JC (2011) Distribution and status of the critically endangered blond titi monkey *Callicebus barbarabrownae* of North-East Brazil. Oryx 45(3):439–443. https://doi.org/10.1017/S0030605311000111

Quevedo M, Lescano J (2014) Infección por *Klebsiella pneumoniae* en un mono choro (*Lagothrix lagotricha*) criado como mascota en Lima, Perú. Revista de Investigaciones Veterinarias del Perú 25(2):317–323

Rangel-Negrín A, Alfaro JL, Valdez RA, Romano MC, Serio-Silva JC (2009) Stress in Yucatan spider monkeys: effects of environmental conditions on fecal cortisol levels in wild and captive populations. Anim Conserv 12(5):496–502

Rangel-Negrín A, Coyohua-Fuentes A, Chavira R, Canales-Espinosa D, Dias PAD (2014) Primates living outside protected habitats are more stressed: the case of black howler monkeys in the Yucatán Peninsula. PLoS One 9(11):e112329

Ribeiro Prist P, Reverberi Tambosi L, Filipe Mucci L, Pinter A, Pereira de Souza R, de Lara Muylaert R et al (2022) Roads and forest edges facilitate yellow fever virus dispersion. J Appl Ecol 59(1):4–17

Rimbach R, Link A, Heistermann M, Gómez-Posada C, Galvis N, Heymann EW (2013) Effects of logging, hunting, and forest fragment size on physiological stress levels of two sympatric ateline primates in Colombia. Conserv Physiol 1(1):cot031

Rimbach R, Bisanzio D, Galvis N, Link A, Di Fiore A, Gillespie TR (2015) Brown spider monkeys (*Ateles hybridus*): a model for differentiating the role of social networks and physical contact on parasite transmission dynamics. Philos Trans R Soc B Biol Sci 370(1669):20140110

Romano AP, Costa ZG, Ramos DG, Andrade MA, de Jayme VS, Almeida MA, Vettorello KC, Mascheretti M, Flannery B (2014) Yellow Fever outbreaks in unvaccinated populations, Brazil, 2008–2009. PLoS Negl Trop Dis 8(3):e2740. https://doi.org/10.1371/journal.pntd.0002740

Romero-Alvarez D, Escobar LE (2018) Oropouche fever, an emergent disease from the Americas. Microbes Infect 20(3):135–146. https://doi.org/10.1016/j.micinf.2017.11.0

Roncal CM, Bowler M, Gilmore MP (2018) The ethnoprimatology of the Maijuna of the Peruvian Amazon and implications for primate conservation. J Ethnobiol Ethnomed 14:19. https://doi.org/10.1186/s13002-018-0207-x

Rondón S, Ortiz M, León C, Galvis N, Link A, González C (2017) Seasonality, richness and prevalence of intestinal parasites of three neotropical primates (*Alouatta seniculus, Ateles hybridus and Cebus versicolor*) in a fragmented forest in Colombia. Int J Parasitol Parasites Wildl 6(3):202–208. https://doi.org/10.1016/j.ijppaw.2017.07.006

Rondón S, León C, Link A, González C (2019) Prevalence of Plasmodium parasites in non-human primates and mosquitoes in areas with different degrees of fragmentation in Colombia. Malar J 18:1–10. https://doi.org/10.1186/s12936-019-2910-z

Rondón S, Cavallero S, Renzi E, Link A, González C, D'Amelio S (2021) Parasites of free-ranging and captive American primates: a systematic review. Microorganisms 9:2546. https://doi.org/10.3390/microorganisms9122546

Rylands AB, Mittermeier RA (2009) The diversity of the New World primates (Platyrrhini): an annotated taxonomy. In: Garber PA, Estrada A, Bicca-Marques JC, Heymann EW, Strier KB (eds) South American Primates: comparative perspectives in the study of behavior, ecology, and conservation. Springer, New York, pp 23–54. https://doi.org/10.1007/978-0-387-78705-32

Santos FGA, Bicca-Marques JC, Calegaro-Marques C, Farias EMP, Azevedo MAO (1995) On the occurrence of parasites in free-ranging callitrichids. Neotrop Primates 3(2):46–47

Savage A, Díaz L, Pasion J, Torregroza K, Guillen R (2022) Proyecto Titi: teaching children that cotton-top tamarins (*Saguinus oedipus*) are not appropriate pets. Am J Primatol e23431. https://doi.org/10.1002/ajp.23431

Schwitzer C, Glatt L, Nekaris KAI, Ganzhorn JU (2011) Responses of animals to habitat alteration: an overview focussing on primates. Endanger Species Res 14(1):31–38

Setchell JM, Fairet E, Shutt K, Waters S, Bell S (2017) Biosocial conservation: Integrating biological and ethnographic methods to study human–primate interactions. Int J Primatol 38(2):401–426

Shaffer CA, Milstein MS, Yukuma C, Marawanaru E, Suse P (2017) Sustainability and comanagementof subsistence hunting in an indigenous reserve in Guyana. Conserv Biol 31(5):1119–1131. https://doi.org/10.1111/cobi.12891

Shanee S (2011) Distribution survey and threat assessment of the yellow-tailed woolly monkey (*Oreonax flavicauda*; Humboldt 1812), northeastern Peru. Int J Primatol 32:691–707. https://doi.org/10.1007/s10764-011-9495-x

Shanee N, Mendoza AP, Shanee S (2017) Diagnostic overview of the illegal trade in primates and law enforcement in Peru. Am J Primatol 79:e22516. https://doi.org/10.1002/ajp.22516

Sharp PM, Hahn BH (2011) Origins of HIV and the AIDS pandemic. Cold Spring Harbor Perspect Med 1(1):a006841. https://doi.org/10.1101/cshperspect.a006841

Smith KF, Acevedo-Whitehouse K, Pedersen AB (2009) The role of infectious disease in biological conservation. Anim Conserv 12:1–12

Solano Rojas D (2018) Importancia de los bosques secundarios para el mono tití centroamericano (*Saimiri oerstedii oerstedii*) en la península de Osa, Costa Rica. In: Urbani B, Kowalewski MM, Cunha RGT, de la Torre S, Cortéz-Ortíz L (eds) La Primatología en Latinoamérica. Instituto Venezolano de Investigaciones Científicas, Caracas, pp 385–395

Solórzano-García B, Pérez-Ponce de León G (2017) Helminth parasites of howler and spider monkeys in Mexico: insights into molecular diagnostic methods and their importance for zoonotic diseases and host conservation. Int J Parasitol Parasites Wildl 6:76–84

Solórzano-García B, Pérez-Ponce de León G (2018) Parasites of Neotropical primates: a review. Int J Primatol 39:155–182. https://doi.org/10.1007/s10764-018-0031-0

Solórzano-García B, White JM, Shedden A (2023) Parasitism in heterogeneous landscapes: association between conserved habitats and gastrointestinal parasites in populations of wild mammals. Acta Trop 237:106751

Soper FL, Penna H, Cardos E, Serafim J, Frobisher JRM, Pinheiro JRJ (1933) Yellow Fever without *Aedes aegypty* study of a real epidemic in the Valle do Chanaan, Espirito Santo, 1932. Am J Epidemiol 18(3):555–587

Soto-Calderón ID, Acevedo-Garcés YA, Álvarez-Cardona J, Hernández-Castro C, García-Montoya GM (2016) Physiological and parasitological implications of living in a city: the case of the white-footed tamarin (*Saguinus leucopus*). *Am J Primatol* 78(12):1272–1281. https://doi.org/10.1002/ajp.22581

Souza-Alves JP, Mourthé Í, Hilário R, Bicca-Marques JC, Rehg J, Gestich CC, Acero-Murcia AC, Adret P, Aquino R, Berthet M, Bowler MT, Calouro AM, Canale GR, Cardoso NA, Caselli CB, Cäsar C, Chagas RRD, Clyvia A, Corsini CF, Defler TR, DeLuycker AM, Di Fiore A, Dingess KA, Erkenswick GA, Ferreira MA, Fernández-Duque E, Ferrari SF, Fontes IP, Gomes JD, Gonçalves FPR, Guerra M, Haugaasen T, Heiduck S, Heymann EW, Hodges S, Huashuayo-Llamocca R, Jerusalinsky L, Kasper CB, Lawrence J, Lueffe TM, Lopes KGD, Martínez J, de Melo FR, Messias MR, Nagy-Reis MB, Nole I, Paciência FMD, Palacios E, Poirier A, de Oliveira Porfírio GE, Porter AM, Price EC, Printes RC, Quintino EP, Reis EA, Rocha A, Rodríguez A, Röhe F, Rumíz DI, Shanee S, Santana MM, Setz EZF, Souza FSC, de Spironello WR, Tirado Herrera ER, Vinhas L, Vulinec K, Wallace RB, Watsa M, Wright PC, Young RJ, Barnett AA (2019) Terrestrial behavior in titi monkeys (Callicebus, Cheracebus, and Plecturocebus): potential drivers, patterns and differences among genera. Int J Primatol 40(4–5):553–572. https://doi.org/10.1007/s10764-019-00105-x

Stark DJ, Fornace KM, Brock PM et al (2019) Long-tailed Macaque response to deforestation in a Plasmodium knowlesi-Endemic Area. EcoHealth 16:638–646. https://doi.org/10.1007/s10393-019-01403-9

Strier KB, Tabacow FP, de Possamai CB, Ferreira AIG, Nery MS, de Melo FR, Mendes SL (2019) Status of the northern muriqui (*Brachyteles hypoxanthus*) in the time of yellow fever. *Primates* 60(1):21–28. https://doi.org/10.1007/s10329-018-0701-8

Stuart MD, Greenspan LL, Glander KE, Clarke MR (1990) A coprological survey of parasites of wild mantled howling monkeys, *Alouatta palliata palliata*. J Wildl Dis 26:547–549

Suzán G, Ceballos G (2005) The role of feral mammals on wildlife infectious disease prevalence in two nature reserves within Mexico city limits. J Zoo Wildlife Med 36(3):479–484. https://doi.org/10.1638/04-078.1

Svensson MS, Shanee S, Shanee N, Bannister FB, Cervera L, Donati G, Huck M, Jerusalinsky L, Juarez CP, Maldonado AM, Mollinedo JM, Méndez-Carvajal PG, Molina Argandoña MA, Mollo Vino AD, Nekaris KAI, Pecko M, Rey-Goyeneche J, Spaan D, Nijman V (2016) Disappearing in the night: an overview on trade and legislation of night monkeys in South and Central America. Folia Primatol 87:332–348. https://doi.org/10.1159/000454803

Tabacow FP, Mendes SL, Strier KB (2009) Spread of a terrestrial tradition in an arboreal primate. Am Anthropol 11(2):238–249. https://doi.org/10.1111/j.1548-1433.2009.01116.x

Talebi M, Jerusalinsky L, Martins M, Mittermeier RA, Ingberman B, Ferraz DS, Melo FR, Boubli JP (2021) *Brachyteles arachnoides* (amended version of 2019 assessment). The IUCN Red List of Threatened Species 2021:e.T2993A191692658. https://doi.org/10.2305/IUCN.UK.2021-1.RLTS.T2993A191692658.en

Terzian ACB, Zini N, Saccetto L, Rocha RF, Parra MCP, Del Sarto JL, Dias ACS, Coutinho F, Rayna J, da Silva RA, Vasconcelos Costa V, Coelho Couto De Azevedo Fernandes N, Réssio R, Díaz-Delgado J, Guerra J, Cunha MS, Catão-Dias JL, Bittar C, Negri Reis AF, Penha dos Santos IN, Marascalchi Ferreira AC, Arão Antônio Cruz LE, Rahal P, Ullmann L, Malossi C, Pessoa de Araújo J, Widen S, de Rezende IM, Mello E, Pacca CC, Kroon EG, Trindade G, Drumond B, Chiaravalloti-Neto F, Vasilakis N, Teixeira MM, Lacerda Nogueira M (2018) Evidence of natural Zika virus infection in neotropical non-human primates in Brazil. Sci Rep 8:16034

Thompson RC (2013) Parasite zoonoses and wildlife: one health, spillover and human activity. Int J Parasitol 43(12–13):1079–1088. https://doi.org/10.1016/j.ijpara.2013.06.007

Torres MAN, Santos E, Almeida MAB, Cruz LL, Sperb AF (2003) Vigilância da febre amarela silvestre no Rio Grande do Sul. Boletim Epidemiológico 5(4):1–7

Tregle RW Jr, Loe CL, Earhart RH III, d'Autremont SB (2011) Cercopithecine herpesvirus 1 risk in a child bitten by a bonnet macaque monkey. J Emerg Med 41(4):e89–e90

Trejo-Macias G, Estrada A (2012) Risk factors connected to gastrointestinal parasites in mantled *Alouatta palliata mexicana* and black howler monkeys *Alouatta pigra* living in continuous and in fragmented rainforests in Mexico. Curr Zool 58(3):375–383

Valdespino C, Rico-Hernández G, Mandujano S (2010) Gastrointestinal parasites of howler monkeys (*Alouatta palliata*) inhabiting the fragmented landscape of the Santa Marta mountain range, Veracruz, Mexico. Am J Primatol 72(6):539–548

Valentine MJ, Murdock CC, Kelly PJ (2019) Sylvatic cycles of arboviruses in non-human primates. Parasit Vectors 12(1):1–18

Van Lieshout L, De Gruijter JM, Adu-Nsiah M, Haizel M, Verweij JJ, Brienen EAT, Gasser RB, Polderman AM (2005) *Oesophagostomum bifurcum* in non-human primates is not a potential reservoir for human infection in Ghana. Trop Med Int Health 10:1315–1320. https://doi.org/10.1111/j.1365-3156.2005.01527.x

Van Vliet N, Moreno Calderón JL, Gomez J, Zhou W, Fa JE, Golden C, Nobrega A, Romulo R, Nasi R (2017) Bushmeat and human health: assessing the evidence in tropical and sub-tropical forests. Ethnobiol Conserv 6. https://doi.org/10.15451/ec2017-04-6.3-1-45

Vanchiere J, Ruiz JC, Vasilakis N, Kuehl TJ, Weaver SC, Abee CR (2018) Experimental Zika virus infection in a neotropical Primate model. Open Forum Infect Dis 3(1):605

Vitazkova SK, Wade SE (2006) Parasites of free-ranging black howler monekys (*Alouatta pigra*) from Belize and Mexico. Am J Primatol 68:1089–1097. https://doi.org/10.1002/ajp.20309

Warburton EM, Blanar CA (2021) Life in the margins: host-parasite relationships in ecological edges. Parasitol Res 120(12):3965–3977

Wasser SK, Hunt KE, Clarke C (2002) Assessing stress and population genetics through non-invasive sampling means. In: Aguirre AA, Ostfeld RS, Tabor GM, House C, Pearl MC (eds) Conservation medicine: ecological health in practice. Oxford University Press, New York, pp 130–145

Waters S, El Harrad A, Bell S, Setchell JM (2021) Decolonizing primate conservation practice: a case study from North Morocco. Int J Primatol. https://doi.org/10.1007/s10764-021-00228-0

Weaver SC (2006) Evolutionary influences in arboviral disease. Springer, Berlin/Heidelberg, pp 285–314

Webber AD, Cotton S, McCabe GM (2022) Failure is the greatest teacher: embracing the positives of failure in primate conservation. Int J Primatol. https://doi.org/10.1007/s10764-022-00296-w

Weisse M, Gladman ED (2020) We lost a football pitch of primary rainforest every 6 seconds in 2019. World Resource Institute. Available in: https://www.wri.org/insights/we-lost-football-pitch-primary-rainforest-every-6-seconds-2019

Wenz A, Heymann EW, Petney TN, Taraschewski HF (2010) The influence of human settlements on the parasite community in two species of Peruvian tamarin. Parasitology 137:675. https://doi.org/10.1017/S0031182009991570

Wright KA, Biondi L, Visalberghi E, Ma Z, Izar P, Fragaszy D (2019) Positional behavior and substrate use in wild adult bearded capuchin monkeys (*Sapajus libidinosus*). Am J Primatol 81:e23067. https://doi.org/10.1002/ajp.23067

Yang A, Boughton RK, Miller RS, Wight B, Anderson WM, Beasley JC, VerCauteren KC, Pepin KM, Wittemyer G (2021) Spatial variation in direct and indirect contact rates at the wildlife-livestock interface for informing disease management. Prevent Vet Med 194:105423. https://doi.org/10.1016/j.prevetmed.2021.105423

Ye J, Xiao L, Ma J, Guo M, Liu L, Feng Y (2012) Anthroponotic enteric parasites in monkeys in public park, China. Emerg Infect Dis 18(10):1640–1643

Chapter 10
Pathogen Transmission and the Risk of Spillover for Wild Carnivores in the Neotropics

Gerardo Acosta-Jamett, Constanza Napolitano, Andrés M. López-Pérez, and Felipe A. Hernández

10.1 Infectious Diseases Affecting Wild Carnivores in Human-Dominated Landscapes

After thousands of years of humans impacting natural ecosystems, the scale and extent of human-ecosystem interactions have expanded considerably in recent decades (McMichael 2004), leading to a human-driven trend of high extinction rates and global loss of biodiversity caused by climate change and habitat loss and fragmentation (Ceballos et al. 2015; Dirzo et al. 2014). The global biodiversity crisis has extensive effects on human and animal welfare, including changes in the ecoepidemiology of infectious diseases. Studies have hypothesized that land use change affects pathogen transmission dynamics at different spatial scales by

Supplementary Information The online version contains supplementary material available at https://doi.org/10.1007/978-3-031-50531-7_10.

G. Acosta-Jamett (✉) · F. A. Hernández
Instituto de Medicina Preventiva Veterinaria y Center for Disease Surveillance and Evolution of Infectious Diseases, Facultad de Ciencias Veterinarias, Universidad Austral de Chile, Valdivia, Chile
e-mail: gerardo.acosta@uach.cl

C. Napolitano
Departamento de Ciencias Biológicas y Biodiversidad, Universidad de Los Lagos, Osorno, Chile

Institute of Ecology and Biodiversity (IEB), Concepción, Chile

Cape Horn International Center (CHIC), Puerto Williams, Chile

A. M. López-Pérez
Red de Biología y Conservación de Vertebrados, Instituto de Ecología, A.C., Xalapa, Mexico

G. Acosta-Jamett, A. Chaves (eds.), *Ecology of Wildlife Diseases in the Neotropics*, https://doi.org/10.1007/978-3-031-50531-7_10

changing the niches of vectors, hosts, and pathogens; the structure of host and vector communities; the behavior of vectors and hosts; and increasing human-domestic-wildlife interfaces as well as the probability of exchange of ectoparasites and pathogens (Bradley and Altizer 2007; Gottdenker et al. 2014). Human-induced landscape changes (e.g., human settlements, agriculture encroachment, and forestry) are among the most important drivers of current global emerging and re-emerging infectious diseases (Jones et al. 2013; Patz et al. 2004). In this regard, some pathogens have expanded their prevalence, geographical distribution, or host range, increasing concern for human and veterinary medicine worldwide (Kilpatrick and Randolph 2012).

While much of the attention on infectious diseases in wildlife is placed on the threat they pose to public health, understanding these infections in the context of wildlife conservation is important (Burnard and Polkinghorne 2016). Outbreaks of infectious diseases can have significant impacts on the population health of free-ranging wildlife, are of heightened importance in species of conservation concern, and therefore represent a serious threat to global biodiversity (e.g., O'Hanlon et al. 2018; Zipkin et al. 2020). Pathogens can interact with other driving factors, such as habitat loss, climate change, overexploitation, invasive species, and environmental pollution, to contribute to local and global extinctions, having negative consequences for threatened wildlife populations already affected by other risks (Pedersen et al. 2007; Smith et al. 2009).

Wild carnivores are particularly vulnerable to human disturbances due to their large body sizes, wide home ranges, low population densities, low recruitment, and specialized diet (Crooks 2002; Gittleman 2001). Urbanization, agriculture expansion, and habitat fragmentation can directly or indirectly modify their ecology and behavior by affecting habitat use, home range and activity patterns, reducing habitat and prey availability, decreasing reproductive success, impeding gene flow among populations isolated, and increasing alien species invasion (Quesnelle et al. 2014; Tuomainen and Candolin 2011).

By the end of the 1990s, the role of infectious diseases in mass mortality events, population declines, and/or species extinctions was often considered controversial or secondary to other factors. However, a series of well-reported deaths in wild carnivores identified infectious diseases as their most likely cause (e.g., Roelke-Parker et al. 1996). Pathogens were also implicated for the first time in species extinctions or near extinctions in several cases. For instance, canine distemper in the 1970s drove the black-footed ferret (*Mustela nigripes*) of North America to the brink of extinction (Daszak and Cunningham 1999; Williams et al. 1988). The devastating outbreak of canine distemper virus (CDV) in Serengeti lions (*Panthera leo*) was estimated to have killed over 1000 lions, a third of the Serengeti population, with a neurologic syndrome. Additionally, uncounted hyenas, bat-eared foxes, and leopards were also affected (Roelke-Parker et al. 1996). These events highlighted the importance of diseases in the conservation of wild carnivores worldwide.

10.2 Domestic Carnivores as a Source of Infection to Wild Counterparts

Biological invasions are one of the most important factors contributing to the loss of biodiversity, degradation of ecosystems, and decline in ecosystem services (Chapin et al. 1997; Pysek and Richardson 2010; Sala et al. 2000). The introduction of invasive species has contributed to the global spread of pathogens into naïve ecosystems (human-mediated pathogen invasion or "pathogen pollution") (Daszak et al. 2000; Strauss et al. 2012; Vitousek et al. 1996). Domestic animals usually act as invasive species that can play a role in infectious diseases, and the threat of spillover from them to wildlife is relevant to animal health globally (Murray and Daszak 2013). As domestic animals are globally distributed and maintained at high densities (often outnumbering the wildlife hosts of shared pathogens), they can easily act as reservoirs for pathogens shared with wildlife (Cleaveland and Dye 1995; Daszak et al. 2001; Lafferty and Gerber 2002; Pedersen et al. 2007).

Domestic dogs (*Canis lupus familiaris*) are among the most numerous carnivores in the world and have been identified as reservoirs for infectious agents that have led to numerous epidemics in different wild carnivore species (Gompper 2014b). For instance, they are known to be the source of CDV and canine parvovirus (CPV) (Behdenna et al. 2019; Cleaveland et al. 2002, 2007). In fact, domestic dogs were implicated as the source of the 1994 canine distemper (CD) epidemic that nearly killed 30% of the lion population in the Serengeti, which is believed to have originated from unvaccinated domestic dog populations near the park (Cleaveland et al. 2000; Roelke-Parker et al. 1996; Viana et al. 2015). The outbreak was probably initiated through other reservoir hosts, different from village dogs (which rarely get sufficiently close to a lion to pass on the virus), such as spotted hyenas (*Crocuta crocuta*), which mix with lions at the kill (Nikolin et al. 2017). Additionally, domestic dogs were also regarded as the source of rabies epidemics that have affected the Ethiopian wolf and are threatening the conservation of this endangered carnivore (Haydon et al. 2006; Sillero-Zubiri et al. 1996). They were also believed to be partly responsible for the extinction of the African wild dog (*Lycaon pictus*) in areas of the Serengeti ecosystem in 1991, which is thought to have occurred through transmission of rabies (Woodroffe 1997). Even in the absence of direct contact between domestic dogs and wild carnivores, the ability of some pathogens, such as CPV, to remain viable in the environment for extended periods means that domestic and wild carnivore sympatry may be sufficient for disease transmission (Gordon and Angrick 1986).

On the other hand, domestic cats (*Felis catus*) can also harbor many infectious diseases (Greene 1998). For instance, domestic cats as sources of feline immunodeficiency virus (FIV) and feline leukemia virus (FeLV) infections in wild felids have been widely recorded for different species and contexts (Cunningham et al. 2008; Meli et al. 2010; Nishimura et al. 1999; O'Brien et al. 2012). Pathology associated with FIV and FeLV infections in nondomestic felids has been recorded as clinical signs, hematologic abnormalities and even mortality with postmortem lesions in free-ranging African lions, Florida panthers (*Puma concolor coryi*), and Iberian lynx (*Lynx pardinus*) (Cunningham et al. 2008; Meli et al. 2010; O'Brien et al. 2012). In addition, domestic cats were thought to be the source of infection of an outbreak of FeLV that affected the endangered Iberian lynx in Doñana, Spain (Lopez et al. 2009). Additionally, FeLV caused a deadly outbreak in Florida panthers in 2002–2004 (Cunningham et al. 2008), spilling over from domestic cats with subsequent direct transmission among panthers (Brown et al. 2008), and ongoing FeLV spillover to and transmission among panthers (Chiu et al. 2019).

Free-roaming domestic dogs and cats have started to attract wide scientific attention in the conservation arena due to their considerable impacts on native wildlife through predation, competition, disturbance, hybridization, and disease transmission (Bischof et al. 2022; Gompper 2014b; Loss et al. 2022; Moseby et al. 2015). In particular, domestic dogs can act as pivotal "bridge hosts" for diseases of conservation (to wildlife) and zoonotic (to humans) concern (Caron et al. 2015; Sepulveda et al. 2014). They have been implicated as the source of infection for several disease outbreaks affecting wild carnivores of conservation concern (e.g., lions, African wild dogs, Ethiopian wolves, and foxes), involving several multihost pathogens, such as canine distemper virus (CDV), canine parvovirus (CPV), or rabies (Behdenna et al. 2019; Calatayud et al. 2019; Lembo et al. 2007, 2008; Viana et al. 2015).

Domestic dogs and cats are particularly abundant in urban areas of some developing countries, such as in the Americas (Gompper 2014a), where they can act as excellent reservoirs for pathogens, since they usually live in large populations, are not vaccinated, and are regularly allowed to roam freely, facilitating contact between infected and susceptible hosts (Acosta-Jamett et al. 2010, 2015b; Flores-Ibarra and Estrella-Valenzuela 2004; Knobel et al. 2014). Consequently, and similar to what has been found for measles in humans, domestic dogs inhabiting urban areas have been proposed as probable maintenance populations of directly transmitted pathogens such as CDV (Acosta-Jamett et al. 2011; Cleaveland et al. 2002; Lembo et al. 2007, 2008; Viana et al. 2015). In these studies, it has been suggested that the size of wild carnivore and rural dog populations is often not sufficient to maintain directly transmitted pathogens; therefore, these pathogens could be transmitted from spillover from more abundant hosts, such as domestic carnivores inhabiting urban sites, to less abundant populations, such as those in rural areas and/or wild species (Acosta-Jamett et al. 2011; Haydon et al. 2002; Viana et al. 2014). The transmission of highly virulent pathogens from domestic to wild carnivores requires close contact between individuals of different species (Dobson and Hudson 1995; Grenfell and Dobson 1995). In many developing countries, domestic dogs are frequently left to roam freely in rural areas (Hernandez et al. 2021). They may roam over livestock areas and/or wildlife areas searching for food and can live in sympatry with wild carnivores. This will increase the likelihood of disease transmission from infectious dogs to susceptible wild carnivores (Butler et al. 2004; Hernandez et al. 2021).

10.3 Interaction Among Domestic and Wild Carnivores as a Proxy of Pathogen Transmission

Interspecific interactions constitute behavioral patterns that modulate population and community dynamics at multiple ecological levels (Farris et al. 2020; Karanth et al. 2017). Regarding disease occurrence, interspecific interactions can facilitate the transmission of infectious agents through direct and indirect pathways. The first route relates to spatial and temporal co-occurrence and often a specific behavioral interaction (i.e., direct physical contact or very close proximity between individuals), which tends to be rare for between-species transmission (Godfrey 2013; Viana et al. 2014). The second route is based on the spatial co-occurrence of reservoirs and susceptible hosts within a certain time window at shared spaces, likely driven by movement or feeding/drinking behaviors that may determine indirect cross-species transmission of environmentally resistant pathogens, such as viruses, bacteria, prions, and macroparasites (Drewe et al. 2013; Lange et al. 2016).

Recently, Suzán et al. (2015) hypothesized that the combined occurrence of generalist (reservoir) and alternative host species at the wildlife-domestic interface would increase the prevalence of pathogen infections. This hypothesis strongly aligns with previous studies that have proposed the wildlife-domestic interface as an "edge between two worlds" (e.g., Murcia 1995), representing critical habitats where humans, vectors, and reservoir hosts (wildlife or domestic animals) overlap, thus increasing contact rates and the risk of cross-species transmission and the emergence of pathogens (Hahn et al. 2014; Hassell et al. 2017). Thus, studying the dynamics of interspecific interactions as potential promoters of pathogen exposure and the patterns of pathogen coinfection is warranted to shed light on the role of the interface as a zone of risk for disease transmission between sympatric domestic and wild species.

During the last few decades, several hypotheses about how wildlife-domestic interactions can contribute to infectious disease emergence have been tested within interface systems. Relatively frequent opportunities for indirect contact between domestic and wild mammals at shared resource locations are believed to play a relevant role in the transmission of several pathogens of economic, public health, and conservation concern, such as bovine tuberculosis (bTB) and canine distemper virus (Campbell et al. 2019; Hernandez et al. 2021). For example, the mechanism of interspecies *Mycobacterium bovis* (bacterial agent of bovine tuberculosis) transmission to wild mammals is relatively unknown but has been attributed to indirect interactions through contamination of water, pastures, feed, or browsing with bacteria shed by infected hosts. The interaction between cattle and carnivores as promoters of *M. bovis* transmission has also received significant research attention, particularly in reference to the cattle-badger model. Along these lines, it is believed that direct contact between both species via nose-to-nose is rare, with previous studies suggesting that badgers actively avoid farmyards when cattle are present, with close physical proximity almost never being recorded (Böhm et al. 2009; Campbell et al. 2019). Thus, direct contact between live badgers and cattle seems to be an unlikely route to maintain bTB through interspecific transmission, while indirect contact by causing space at segregated times may boost the potential transfer of infectious material deposited in soils contaminated with feces or urine (surrounding badger setts and latrines), water, or feedstuffs (Campbell et al. 2019).

10.4 Studies in Wild Carnivores in the Neotropics

The neotropics have suffered extensive and intense habitat fragmentation and deforestation, are the main causes of wild carnivore decline (e.g., Paviolo et al. 2016), and are projected to worsen in the coming decades (Oakleaf et al. 2015). While pathogens have also emerged as important factors affecting wild carnivores globally, research and surveillance efforts in the neotropics have increased in recent years. However, there remains a notable scarcity of studies examining how landscape changes impact the dynamics of pathogens in both wild and domestic carnivores, and this aspect remains poorly understood. This chapter aims to address this knowledge gap by reviewing and summarizing recent literature on selected pathogens that affect wild carnivores in the neotropics. This effort is crucial for a comprehensive understanding of the complex interplay between habitat change and pathogens affecting carnivores in one of the most biodiverse areas of the world.

We conducted a literature search of peer-reviewed literature published between 2000 and 2022 of selected pathogens affecting wild carnivore populations in Web of Science and Google Scholar using the following search query: "wild carnivores" OR "free-ranging carnivores" AND "adenovirus" OR "aleutian disease" OR "alphaherpervirus" OR "anaplasma" OR "babesia" OR "bartonella" OR "borrelia" OR "brucella" OR "calicivirus" OR "coronavirus" OR "coxiella" OR "cytauxzoon" OR "distemper" OR "ehrlichia" OR "Feline immunodeficiency virus" OR "feline leukemia virus" OR "feline morbillivirus" OR "feline panleukopenia virus" OR "FELV" OR "FIV" OR "gammaherpes virus" OR "hemoparasite" OR "hemoplasma" OR "hepatozoon" OR "herpesvirus" OR "influenza" OR "leishmania" OR "leptospira" OR "mink enteritis virus" OR "mycobacterium bovis" OR "mycoplasma" OR "neospora caninum" OR "parainfluenza" OR "parvovirus" OR "piroplasmids" OR "protoparvovirus" OR "puma lentivirus" OR "rabies" OR "rangelia" OR "rickettsia" OR "rotavirus" OR "sarcoptic mange" OR "sarcoptes scabiei" OR "theileria" OR "toxoplasma gondii" AND "Argentina" OR "Belize" OR "Belice" OR "Bolivia" OR "Brazil" OR "Chile" OR "Colombia" OR "Costa Rica" OR "Ecuador" OR "El Salvador" OR "French Guiana" OR "Guatemala" OR "Guyana" OR "Honduras" OR "Mexico" OR "Nicaragua" OR "Panama" OR "Paraguay" OR "Perú" OR "Suriname" OR "Uruguay" OR "Venezuela" OR "South America" OR "Central America" OR "Caribbean." We then looked at each of the found articles to filter only those conducted in the neotropics. Other pathogens, such as *Trypanosoma cruzi,* which were reported in some articles, were also included.

After reviewing the records and eliminating duplicates, they were grouped into the following sections: (1) direct transmission, (2) vector-borne diseases, and (3)

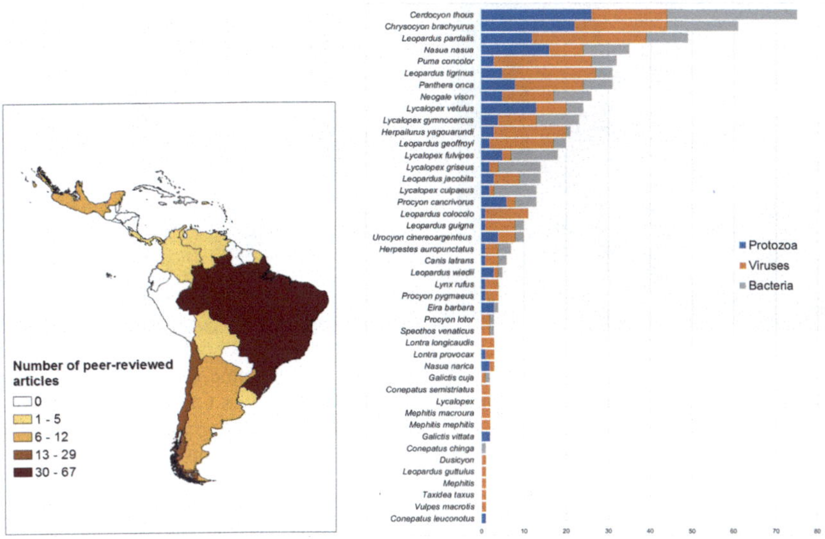

Fig. 10.1 Summary of articles included in this chapter. *Left:* Number of peer-reviewed articles per country about selected pathogens affecting wild carnivore populations in the neotropics; and *right:* number of references of protozoa, viruses, and bacteria reported by carnivore hosts across the reviewed literature in the neotropics

environmental transmission. It is worth noting that the pathogens included and discussed in these sections may not fully align with their respective categories. This categorization was performed to enhance the understanding of the studies conducted in the neotropics thus far.

After conducting our search, we observed a fairly unbalanced publication report by country and carnivore hosts carrying the main pathogens of interest (protozoa, viruses, and bacteria). Regarding the country of origin, approximately 47% ($n = 67$) of publications were developed in Brazil, followed by 37% of publications related to research from Chile ($n = 29$), Argentina ($n = 12$), and Mexico ($n = 11$) (Fig. 10.1 left). In terms of host species, the reviewed publications included a total of 45 wild carnivores, such as kodkod (*Leopardus guigna*) (Fig. 10.2 left) and culpeo fox (*Lycalopex culpaeus*) (Fig. 10.2 right). The crab-eating fox (*Cerdocyon thous*), the most cited carnivore, was included in 41 articles, which as a whole provided 75 references of pathogens (26 protozoa, 18 viruses, and 31 bacteria) that may be carried by this neotropical canid. The second most represented species were the maned wolf (*Chrysocyon brachyurus*) and the ocelot (*Leopardus pardalis*), being mentioned by 19 and 21 articles, respectively, which provided 61 references of pathogens (22 protozoa, 22 viruses, and 17 bacteria) in maned wolves and 49 references of pathogens (12 protozoa, 27 viruses, and 10 bacteria) in ocelots. The ring-tailed coati (*Nasua nasua*) was the most studied procyonid, being included by 18 articles that focused on 35 references of pathogens (16 protozoa, 8 viruses, and 11 bacteria) (Fig. 10.1 right). We also reviewed three articles that were focused on detecting the

Fig. 10.2 Examples of wild carnivore hosts included in the bibliographic review. *Left*: Kodkod (*Leopardus guigna*) photographed in the coastal area of Pucatrihue, Los Lagos region, Chile—credit to Eduardo Minte; and *right*: Culpeo fox (*Lycalopex culpaeus*) photographed in the Bosque Fray Jorge National Park. Coquimbo region, Chile—credit to Bernardita Julio

mite *Sarcoptes scabiei* in individuals of kinkajou (*Potos flavus*), crab-eating fox, and wild foxes of the genus *Lycalopex*; however, these data were not included in Fig. 10.1 right.

In the following sections, a representative selection of the reviewed studies is provided. The complete list of articles is included in Supplementary Material Table 10.1.

10.5 Direct Transmission

Canine distemper virus (CDV) is one of the most studied pathogens in the neotropics, with 31 studies carried out in seven countries covering 27 species at 61 different times. These studies have mostly relied on serological methods, and only seven have included molecular tools, suggesting the lack of understanding of this important virus capable of decimating populations worldwide (Roelke-Parker et al. 1996). Pioneers studies carried out by Gonzalez-Acuña et al. (2003) in Chile, Deems and Emmons (2005), Deems et al. (2004), and Fiorello et al. (2004, 2007) in Bolivia and the reports by Nava et al. (2008) and Megid et al. (2009) in Brazil showed the importance of studying domestic dog spillover from domestic dogs to wild carnivores. This is consistent with what was found by Acosta-Jamett et al. (2011) in two South American foxes (*Lycalopex griseus* and *L. culpaeus*) in Chile, where higher probabilities of being seropositive to canine CDV were found in foxes sampled closer to urban areas. The same pattern was also suggested for parvovirus in the same species and same area in Chile (Acosta-Jamett et al. 2015a), for protoparvovirus 1 infections in wild carnivores in northern Mexico (Lopez-Perez et al. 2019), and in wild guignas in Chile (Sacristan et al. 2021b), which were more likely in areas closer to human settlements. In addition, a recent study confirmed CDV exposure and CPV exposure/infection in free-ranging minks in southern Chile (Barros et al. 2022), supporting the previous hypothesis of Sepúlveda et al. (2014) regarding minks acting as potential bridge hosts between domestic dogs and river otters for these viral pathogens. The neotropics is one of the areas with the highest populations of domestic dogs in the world (Gompper 2014a), with most of them allowed to roam freely and usually not properly vaccinated, which can have disastrously negative effects on wild carnivore populations. Whether pathogen transmission from domestic to wild carnivores is influenced by urbanization in a fragmented landscape is worth studying to predict the effect that pathogens can have on wild carnivore populations in an area that is facing important habitat deterioration (Oakleaf et al. 2015; Paviolo et al. 2016).

In the neotropics, there are other examples of negative impacts associated with pathogen infections affecting wild carnivores. For instance, acute hemorrhagic diarrhea and mortality were documented in a kodkod infected with carnivore protoparvovirus in Chile (Ortega et al. 2021). In Mexico, a small group of white-nosed coatis (*Nasua narica*) from the Yucatan Peninsula (Kabah National Park in Cancun, Quintana Roo) were found dead and were confirmed to be infected from rabies, despite the low frequency of rabies in this species (Arechiga-Ceballos et al. 2010). Molecular detection of viral agents in 109 free-ranging and 68 captive neotropical felids in Brazil revealed felid alphaherpesvirus 1 (FHV-1) in 1% free-ranging and 3% captive, protoparvovirus 1 in 3% free-ranging and 2% captive, and FIV in 1% free-ranging (Furtado et al. 2017a). Other studies have found novel pathogens, such as gammaherpesvirus, affecting the critical endangered Darwin's fox (Cabello et al. 2013b) or a paramyxovirus in the guignas (Sieg et al. 2020) in southern Chile, emphasizing the need for further studies in the carnivores in this region.

Studies carried out in Brazil, Bolivia, Argentina, Mexico, and others found that canine adenovirus affects different species of terrestrial carnivores, such as *C. thous* or *Lycalopex gymnocercus*, as well as aquatic carnivores, such as *Lontra longicaudis* (de Almeida Curi et al. 2010; Deem and Emmons 2005; Michelazzo et al. 2020, 2022; Orozco et al. 2014; Uhart et al. 2012). These infections probably originate in domestic dog populations nearby (Deem and Emmons 2005) or even in wild carnivore populations, as suggested in other regions worldwide (Balboni et al. 2019; Walker et al. 2016).

In Bolivia, maned wolves have been found to be susceptible to and die from common infectious diseases of domestic dogs, including CDV, CPV, rabies virus, and canine adenovirus (CAV). In the remote Noel Kempff Mercado National Park in northeastern Bolivia, domestic dogs in villages bordering the park were found to have high levels of exposure to these pathogens. Free-ranging wolves in the park have been exposed to multiple infectious and parasitic agents of domestic carnivores and may be at increased disease risk due to these pathogens originating in the domestic dog population (Deem and Emmons 2005). Contact between wild carnivores and dogs has been documented in the sampled villages; therefore, dogs likely pose a substantial risk to the carnivores within and near the park and have been described as an additional threat to the survival of free-ranging maned wolves (Bronson et al. 2008).

The disease exposure of domestic carnivores (dogs and cats) living near Madidi National Park, Bolivia, was also studied. High levels of exposure to canine distemper virus, canine parvovirus, *Sarcoptes scabiei*, and *Toxoplasma gondii* were found among domestic dogs, with similarly high levels of exposure to feline parvovirus, feline calicivirus, and *T. gondii* being found among domestic cats. If contact occurs between domestic and wild carnivores, disease spillover may represent an important risk for the persistence of wild carnivores in the region. Additional research is therefore necessary to determine whether wild carnivores living in proximity to these domestic carnivore populations are exposed to these pathogens (Fiorello et al. 2004).

FeLV and FIV were studied in wild felids in the neotropics (11 studies), with observed prevalences ranging from 0% to 20% and 0% to 50%, respectively, in nine different species from five countries (Supplementary Material Table 10.1). Sacristán et al. (2021a) found a low frequency of clinical signs of disease in kodkods infected with FeLV and FIV. However, proviral DNA detected by the study's analysis is integrated into the host genome and could potentially reactivate and/or recombine with other viral subtypes, leading to emerging diseases and posing future threats for kodkods, including potential population extinctions and impacting the species' long-term viability. Two additional studies in Chile found that the infection of feline leukemia virus (FeLV) in kodkod was significantly associated with fragmented landscapes with resident domestic cats (Mora et al. 2015; Sacristan et al. 2021a), while parvovirus was more likely to be present in fragmented landscapes with domestic carnivore presence compared with continuous forest (Sacristan et al. 2021b).

Toxoplasma gondii is the protozoa that causes toxoplasmosis, a widely distributed disease whose transmission could occur by the consumption of food and water contaminated with oocysts of this protozoan. Different studies have detected the presence of this parasite on wild carnivores thriving on heterogeneous landscapes, ranging from pristine to highly disturbed land uses. By using multiple serological/molecular assays, studies in protected areas and natural reserves of southeastern Brazil have described the exposure to *T. gondii* in different wild canid species, such as crab-eating foxes (70%; 7/10) and maned wolves (86%; 6/7), but not in hoary foxes (*Lycalopex vetulus*) (de Almeida Curi et al. 2010). Afterwards, the same authors estimated a similar seroprevalence of *T. gondii* in maned wolves (75%; 6/8) (de Almeida Curi et al. 2012). Nascimento et al. (2015) detected an infection prevalence of 14% (7/49) derived from roadkill samples of hoary foxes collected from highways. Regarding felid hosts, jaguars (*Panthera onca*) of the Pantanal have exhibited up to 91% (10/11) seropositivity for *T. gondii*, suggesting a widespread sylvatic cycle of toxoplasmosis in large felids of Brazil (Onuma et al. 2014). Overall, wild carnivores might be potential sentinels of toxoplasmosis, especially in the presence of infected domestic and wild felids, maintaining environmental contamination across different Brazilian regions.

In other neotropical countries, such as Mexico, Chile, and Argentina, studies have revealed high-medium levels of exposure to *T. gondii* among wild carnivores. For instance, in cattle ranches and a biosphere reserve of northeastern Mexico, the prevalence of anti-*Toxoplasma* antibodies in free-ranging ocelots was 69% (18/26), where adult males were more exposed than subadult males due to longer exposure times (Rendon-Franco et al. 2012). On the other hand, Sepúlveda et al. (2011) reported relatively high exposure/infection (70%; 21/30) by the protozoan parasite *T. gondii* on invasive North American minks (*Neogale vision*), where adult individuals nearby cities were more likely to be exposed/infected than minks from rural sites in southern Chile. Similarly, Barros et al. (2018) detected high seroprevalences in minks (59%; 43/73) inhabiting an urban-rural gradient, including other native carnivores such as southern river otters (*Lontra provocax*) (77%; 10/13) and two seropositive kodkods, which were found to be related to medium/high presence of domestic cats and higher levels of vegetation loss. In contrast, Martino et al. (2017) found a *T. gondii* exposure prevalence of only 26% (23/87) in minks sampled in both Buenos Aires and Patagonian grasslands. As a whole, research findings have suggested that diverse carnivore groups may be naturally infected across contaminated environments; in particular, mustelids' semiaquatic behavior and closeness to human and domestic cat populations may have a relevant role in maintaining *T. gondii* in southern cone ecosystems.

Sarcoptic mange (caused by *S. scabiei*), commonly transmitted by infected domesticated animals, infects 104 species and can cause high-mortality epidemics in wildlife (Pence and Ueckermann 2002). Hosts that are threatened by other factors or are already limited to small population sizes can be pushed to extinction by mange outbreaks (i.e., red fox population on Borrnholm Island in Denmark) (Pence and Ueckermann 2002; Smith et al. 2009). *S. scabiei* was assessed in only three wild carnivore studies in the neotropics, in three different species from three countries, with observed prevalences ranging from 21% to 100% (Supplementary Material Table 10.1). Given the low number of wild carnivore studies for many pathogens in the region, more research is urgently needed, including both continuous surveillance and specific studies to gain a more in-depth understanding of pathogen dynamics in human-dominated landscapes. This is especially relevant given the taxonomic closeness and pathogen sharing among wild and domestic carnivores and the unsolved regional problem of high population numbers of unvaccinated domestic dogs and cats, where in many countries, a better management and public policy strategy is still pending, therefore constituting a conservation and public health problem in the neotropics.

10.6 Vector-Borne Diseases

Vector-borne diseases (VBDs) are illnesses caused by several pathogens (e.g., protozoa, viruses, and bacteria) that are transmitted by hematophagous arthropod species, including fleas, mites, mosquitoes, ticks, triatomine bugs, and sandflies (Müller et al. 2019). Many of these diseases have increased in prevalence and become more widespread worldwide, increasing concern for human and animal health. The emergence of VBDs has been mainly driven by anthropogenic land use changes and climate change (Guegan et al. 2020; Swei et al. 2020). Numerous studies have found that wild carnivores may have different roles in the ecoepidemiology of vector-borne pathogens. Several wild carnivore species have been suggested to be host reservoirs of the *Bartonella* species transmitted by fleas, flea-borne bacteria of human health concern, while raccoons (*Procyon lotor*) are reservoir-competent for *Anaplasma phagocytophilum*, a tick-borne bacterium that causes human granulocytic anaplasmosis (Stuen et al. 2013). In addition, several VDBs may also cause mortalities in wild carnivores. *Yersinia pestis* and *Cytauxzoon felis*, flea- and tick-vectored pathogens, may cause fatal infections in free-ranging wild felids (Elbroch et al. 2020; Nietfeld and Pollock 2002). Thus, wild carnivores could play roles as alternate or carrier hosts for ectoparasites (Dobler and Pfeffer 2011). For instance, swift and kit foxes (*Vulpes macrotis* and *V. velox*) have been proposed as carriers of plague-infective fleas (Salkeld and Stapp 2006).

Ecological processes between wild carnivores and vector-borne pathogens are poorly understood in the neotropics, with only a few studies focused on surveying the domestic and wildlife interface. Although most of these pathogens are well known to be of human health concern, little is known about their impact on wild carnivore populations. There is also little evidence from isolated reports, mainly in captivity, of wild carnivore individuals exhibiting clinical signs associated with fatal cases or illness caused by vector-borne infections. In the neotropics, a total of 50 studies have focused on VBD in 21 native and one invasive wild carnivore belonging to four families (Canidae, Felidae, Mustelidae, and Procyonidae). Overall, they were conducted using serological and molecular approaches to test for 23 species of vector-borne pathogens of 13 genera in ten countries from 2006 to 2022. Out of these studies, 68% were conducted in Brazil, followed by Argentina (10%), Chile (8%), Uruguay (6%), Bolivia (2%), and Mexico (2%), Panama (2%), Colombia (2%), and Venezuela (2%). The studied pathogens were mainly transmitted by ticks, followed by flies, kissing bugs, mosquitos, and fleas. The most studied genus of vector-borne pathogens was *Hepatozoon* sp. ($n = 19$), followed by *Leishmania* sp. ($n = 13$), *Babesia* sp. ($n = 13$), *Ehrlichia* sp. ($n = 9$), *Anaplasma* sp. ($n = 8$), *Rangelia* sp. ($n = 7$), *Rickettsia* sp. ($n = 7$), *Bartonella* sp. ($n = 5$), *Cytauxzoon* sp. ($n = 5$), *Borrelia* sp. ($n = 2$), and *Coxiella* ($n = 2$).

Hepatozoonosis constitutes a vector-borne parasitic infection caused by the apicomplexan protozoan of the genus *Hepatozoon*, and it is widely distributed across a range of wild and domestic vertebrate hosts (including mammals). The principal transmission route is by the consumption of a hematophagous arthropod definitive host containing oocysts or by preying on infected vertebrates containing *Hepatozoon* cysts in their tissues. In South America, the pathogen has been studied across diverse wild carnivores, especially in canids and felids from Brazil. For instance, Criado-Fornelio et al. (2006) reported a relatively high infection prevalence by *Hepatozoon* spp. in free-ranging canids such as crab-eating foxes (83%; 5/6) and pampas foxes (*Lycalopex gymnocercus*) (71%; 5/7) hunted in the state of Rio Grande do Sul. All foxes harbored the species *Hepatozoon canis*—a relatively widespread protozoan of domestic cats, dogs, and foxes—excluding one crab-eating fox that harbored an *H. americanum*-related organism (i.e., *H. americanum* is highly pathogenic for domestic dogs in the United States). Almeida et al. (2013) found a moderate infection prevalence in road-killed crab-eating foxes (50%; 29/58) collected from a highway crossing and Atlantic rainforest reserve, with isolates mostly related to *Hepatozoon* spp./*H. americanum*, except by one fox infected by an organism closely related to reptile-associated *Hepatozoon* agents. A similar *Hepatozoon* infection prevalence was reported for maned wolves (57%; 21/37) in a protected Cerrado area surrounded by small farms, revealing haplotypes genetically similar to *H. canis* and *H. americanum* (Arrais et al. 2021). Afterwards, relatively higher *Hepatozoon* infection positivity rates were reported for wild canids by de Sousa et al. (2017a), who detected the presence of the parasite in crab-eating foxes (91%; 71/78) caught in the central region of the Pantanal (isolated sequence phylogenetically related to *H. americanum*).

On the other hand, Metzger et al. (2008) provided the first molecular detection/characterization of *Hepatozoon* spp. in neotropical felids captured in Brazilian northeastern states, resulting in infection prevalences that ranged from 0% in jaguarundi (*Herpailurus yagouaroundi*) and margay (*Leopardus wiedii*) to 50% in ocelot (see Supplementary Material Table 10.1 for details). The majority of infected wild cats exhibited *H. canis* isolates, while Furtado et al. (2017b) demonstrated that 97% (29/30) of jaguars sampled within protected areas and adjacent rural properties in three different Brazilian biomes (i.e., Pantanal, Amazon, and Cerrado) were infected by *Hepatozoon* related to other *Helicobacter felis*. They found that jaguars (including sampled dogs and cats) from the Pantanal and Amazon were significantly more exposed to *Hepatozoon* spp. than animals from the Cerrado, and jaguars were more exposed than the other domestic species. These findings are likely explained by the higher opportunities for protozoan transmission through predation on infected prey by jaguars occupying contiguous habitats in the Pantanal/Amazon area, suggesting that jaguars may play an important role in the maintenance of *Hepatozoon* spp. in nature. Among other carnivore families in Brazil, coatis have been reported as *Hepatozoon* hosts in forest fragments inserted into urban areas of São Paulo, exhibiting an infection prevalence of 25% (21/83) caused by *H. procyonis*; thus, the authors suggested that coatis freely moving between urban and wild environments would facilitate exposure to protozoan vectors and consequently increase the likelihood of infection (da Silva et al. 2018).

Among studies conducted in other South American countries, Millán et al. (2019) reported chilla foxes with a *Hepatozoon* infection prevalence of 50% (24/48) in two remote protected areas of Santa Cruz Province in Argentina, principally associated with *H. felis* and *H. americanum* (and other related sequences). Interestingly, the authors found a higher *Hepatozoon* prevalence in chillas inhabiting the northern protected area Bosques Petrificados National Park compared to the colder Monte León National Park (located 300 km southward), likely associated with closeness to the limit for the suitability of potential tick vectors by the second location and demonstrating natural circulation of the protozoan among foxes thriving in remote areas without contact with dogs. In Uruguayan wild canids, Carvalho et al. (2021) found lower *Hepatozoon* detection rates of 13% (6/45) in crab-eating foxes and 6% (2/32) in pampas foxes, which corresponded to road-killed specimens retrieved from highways. Animals were infected with an *H. americanum*-like protozoan (closely related to *H. americanum*), whose presence may represent a probable threat to sympatric domestic canids.

Leishmania spp. encompass a group of bacteria transmitted by phlebotomid flies that have been one of the most studied in carnivores in the neotropics and almost exclusively in Brazil, the country reporting the highest number of human visceral leishmaniasis cases in the region (Azami-Conesa et al. 2021; de Sousa-Paula et al. 2020). Most studies in the neotropics have reported results from small sample sizes, with prevalence ranging from 0% to 48% by molecular approaches and 0% to 100% by serological tests. *Leishmania* studies have been mainly focused on wild carnivores because they are considered reservoirs of these pathogens. However, one clinical case of leishmaniasis has been described in captive wild carnivores (Roque and Jansen 2014; Souza et al. 2014). Among 15 captive canids from a zoo in Belo Horizonte, Minas Gerais, Brazil, two animals, a bush dog (*Spheotos venaticos*) and a hoary fox (*Lycalopex vetulus*), were serologically positive and developed clinical signs of canine visceral leishmaniasis, whereas three other canids, including a crab-eating fox, a maned wolf, and a hoary fox, had positive serological results without clinical signs (Luppi et al. 2008). Crab-eating fox and maned wolf are recognized as reservoirs of *Leishmania* spp. for humans (Azami-Conesa et al. 2021). Among free-ranging carnivores, several studies have been located in forested, rural, and human settings across Brazil. For instance, de Almeida Curi et al. (2006) reported variable exposure rates across carnivores, such as 4% (19/21) in hoary foxes, 17% (2/12) in crab-eating foxes, and 29% (2/7) in maned wolves near a protected area, while Richini-Pereira et al. (2014) reported a higher infection prevalence of 42% (5/12) in road-killed crab-eating foxes collected in highways of the central western of the state of São Paulo. In particular, the crab-eating fox is believed to be involved in an enzootic transmission cycle independent of dogs in peri-urban environments (e.g., Araujo Soares et al. 2017); however, based on the evidence of either exposure or infection to *Leishmania* spp. (50%; 31/62) of dog populations inhabiting villages or rural settlements in proximity to natural areas, it is not possible to rule out this domestic canid group as incidental hosts in the sylvatic cycle if the protozoan (e.g., through direct contact or feeding behavior on infected wild mammals) (Porfirio et al. 2018) and thus potentially affects coexisting wild canid populations at wildlife-domestic interfaces (Brandao et al. 2020). In other neotropical countries, the presence of *Leishmania* DNA has been tested in other canids, such as chilla foxes (38%; 12/32), in remote, nonendemic protected areas of the Argentinian Patagonia, where domestic dogs are scarce and sand flies are not known to be present, thus extending the geographic distribution of this infection in South America (Millan et al. 2016).

Bartonella species are vector-borne and gram-negative bacteria that infect the erythrocytes and endothelial cells of mammalian hosts (Boulouis et al. 2005; Chomel et al. 2009). In the neotropics, *Bartonella* has been poorly studied, with only four descriptive and one cross-sectional study conducted in three countries. In Brazil, two studies surveyed five species of felids, including *Leopardus geoffroyi*, *L. wiedii*, *L. pardalis*, *L. tigrinus*, and *Puma concolor*, and they found that seroprevalence ranged from 18% to 89%, while polymerase chain reaction (PCR) prevalence was low, ranging from 0% to 5% (Filoni et al. 2006; Souza et al. 2021).

Studies on wild canids are also scarce, with only two studies that did not detect any *Bartonella-PCR*-positive samples in *Lycalopex fulvipes*, *L. culpaeus*, and *L. griseus* in Argentina and Chile (Cabello et al. 2013a; Millan et al. 2019). Finally, a molecular survey in 246 invasive American minks from southern Chile found 9% infection with *Bartonella* spp. (Sepulveda-Garcia et al. 2021). *Bartonella* bacteria are mainly associated with asymptomatic persistence in host reservoirs, with only two documented cases of pathological manifestations in wild carnivores. The first one was three Florida pumas that were reintroduced into the wild after being in captivity exhibiting well-documented pathological signs of bartonellosis (Elsmo et al. 2018). The second was the finding of *Bartonella koehlerae/henselae*-like in a wild-captured and captive margay (*Leopardus wiedii*) in Brazil (Filoni et al. 2012).

Anaplasmataceae is a family of obligate intracellular bacteria that encompass the genera *Ehrlichia*, *Anaplasma*, *Neorickettsia*, and *Wolbachia* (Dixon et al. 2021). In the neotropics, Anaplasmataceae group studies have been mainly conducted using molecular approaches in wild canids, with only one serological study in three wild felids (Supplementary Material Table 10.1). In the latter serosurvey study, Filoni et al. (2006) did not find antibodies against *Anaplasma phagocytophilum* in *L. tigrinus*, *L. pardalis*, and *P. concolor* in Brazil. The molecular prevalence in three of seven studies ranged from 1% to 23%. One study conducted in a Pantanal region of Brazil found molecular evidence of *Anaplasma and Ehrlichia* in *Nasua nasua*, *L. pardalis*, and *C. thous*, as well as in dogs (de Sousa et al. 2017b). Almeida et al. (2013) found that 3% of 58 *C. thous* surveyed were PCR-positive for *Anaplasma* sp., while Di Cataldo et al. (2021a) found *Anaplasma platys* in three canids (*L. culpaeus*, *L. griseus*, and dogs) in Chile.

Of the two *Ehrlichia* species (*E. canis* and *E. chaffeensis*) that have been reported in wild carnivores worldwide, *E. canis* is the only one that has been reported in the neotropics. Antibodies have been found in three studies conducted in Brazil in wild carnivores, including *Chrysocyon brachyurus* (36%), *C. thous* (18%), *N. nasua* (3%), and *P. concolor* (11%) (Arrais et al. 2021; de Sousa et al. 2017b; Filoni et al. 2006). On the other hand, wild carnivores in the neotropics have a relatively low prevalence below 10%. In Brazil, de Sousa et al. (2017b) reported a prevalence of infection of 3% in both *N. nasua* and *C. thous*, while in Brazil, Almeida et al. (2013) and Collere et al. (2021) found a higher prevalence in *C. thous* (10%) and *N. nasua* (6%), respectively.

Rickettsioses are caused by obligate intracellular bacteria of the genus *Rickettsia*, which currently comprises approximately 30 validly described species, including 20 documented pathogens (Parola et al. 2013; Shpynov et al. 2018). Out of five studies aiming to find *Rickettsia* DNA in neotropical wild carnivores, only one study conducted in Argentina found *Rickettsia* DNA in one of 48 *L. griseus* (2%) (Millan et al. 2019). This is consistent with the fact that finding active infection in canids is unusual because the duration of bacteremia is limited since *Rickettsia* spp. have a tropism for endothelial cells (Levin et al. 2014; Lundgren et al. 1963). On the other hand, three studies carried out in Brazil and Panama have reported antibodies of *Rickettsia* sp. in four wild canid species (*C. brachyurus*, *Canis latrans*, *C. thous*, and

L. gymnocercus), with a prevalence ranging from 25% to 95% (Arrais et al. 2021; Bermudez et al. 2017; Dall'Agnol et al. 2018).

Piroplasmids (*Babesia* spp., *Rangelia* spp., *Cytauxzoon* spp., and *Theileria* spp.) are intracellular protozoans transmitted by ticks and are one of the most prevalent tick-borne diseases affecting several wild and domestic vertebrates worldwide (Schnittger et al. 2022). The diseases have been widely characterized among domestic dogs and their ectoparasites, but recent studies have impacted the understanding of the ecology of piroplasmosis toward the inclusion of novel wild hosts. In the neotropics, the presence of antibodies or infection of *Babesia* sp. in wild carnivores is often rare, with only four studies of 13 showing a prevalence ranging from 1% to 54%. A study by de Sousa et al. (2018) reported that 54% (42/78) of crab-eating foxes of the central region of the Brazilian Pantanal were seroreactive to the piroplasmid *Babesia vogeli*; in parallel, the same authors reported a high *B. vogeli* seroprevalence in dogs (93%; 39/42) cohabiting the same studied area, which suggested the role of dogs as a potential source of infection to wild carnivores. Other studies based on procyonid sampling confirmed that up to 38% (5/13) of road-killed crab-eating raccoons (*Procyon cancrivorus*) were *Babesia* spp. PCR-positive in different Uruguayan localities (Thompson et al. 2018), while coatis collected from an urban park in southeastern Brazil exhibited a nearly 2% (3/151) infection prevalence of this protozoan (Estevam et al. 2020), thus expanding the potential carnivore hosts for this pathogen.

The piroplasmid *Rangelia vitalii* (etiologic agent of rangeliosis) is a tick-borne protozoan parasite that affects domestic dog populations in Brazil, and it is phylogenetically related to the genus *Babesia*. In the neotropics, *Rangelia* spp. studies were focused only on wild canids, and the bacteria were present in six of seven studies that tested for them, with prevalence ranging from 0% to 30% (Supplementary Material Table 10.1). Based on PCR and sequencing, Soares et al. (2014) reported an *R. vitalii* infection prevalence of 30% (6/20) in crab-eating foxes (either rescued or road-killed) that were collected from different areas of the state of Rio Grande do Sul and São Paulo. Similarly, de Souza et al. (2019) reported that 25% (7/27) of crab-eating foxes were infected in protected and rural areas of Rio Grande do Sul, while De Lorenzo et al. (2021a) confirmed that 26% (8/31) of crab-eating foxes were found to be positive for *R. vitalii* in the metropolitan area of the same Brazilian state (including road-killed crab-eating foxes (13%; 5/38) from northern Uruguay (de Lorenzo et al. 2021b)); thus, these findings support this neotropical wild canid as the possible natural reservoir host of this infectious agent.

All five studies that tested for *Theileria* spp. were conducted using molecular approaches, and only one found a prevalence of 10% in *N. nasua* in Brazil (de Sousa et al. 2018). Regarding *Cytauxzoon felis*, it has been tested in six wild felid and one wild canid species but has been found in only two species (*Panthera onca* and *Leopardus pardalis*) in Brazil, with prevalence ranging from 57% to 97% (de Sousa et al. 2018; Furtado et al. 2017c). Both studies aimed to study the wild-domestic interface by testing cats and dogs, but none of these species were positive for *Cytauxzoon* sp.

As with the other pathogens found in the neotropics, information about their clinical importance in wild carnivores is scarce; however, piroplasmid infections are often asymptomatic. Few studies have reported illness or fatalities caused by piroplasmid pathogens mainly in captive wild carnivores, such as two captive gray wolves, one brown bear, and one African wild dog that were claimed to die due to *Babesia canis* and *Babesia* sp., while only one free-ranging red fox with hemolytic anemia was claimed to be associated with *Babesia microti* in Canada (Alvarado-Rybak et al. 2016). *Cytauxzoon* causes fatal infections in free-ranging bobcats (*Lynx rufus*) (Nietfeld and Pollock 2002) and captive wild felids such as tigers (*Panthera tigris*) and lions (*Panthera leo*) (Garner et al. 1996; Peixoto et al. 2007).

Mycoplasma spp. was assessed in wild carnivore studies in the neotropics (14 studies), with varying observed prevalences ranging from 0% to 100% in 13 different wild carnivore species from four countries (Supplementary Material Table 10.1). A molecular survey in 246 invasive American minks from southern Chile found 2% infection with hemotropic *Mycoplasma* (hemoplasmas) (Sepulveda-Garcia et al. 2021), and in wild mammals from Brazil, de Sousa et al. (2017c) detected *Mycoplasma* spp. in 77% (24/31) of *Nasua nasua*, 4% (3/78) of *Cerdocyon thous*, 14% (1/7) of *Leopardus pardalis*, and 5% (2/42) of dogs. Di Cataldo et al. (2021b) molecularly detected *Mycoplasma haemocanis/haemofelis* in 24%, 20%, and 27% of 626 rural dogs, 140 Andean foxes (*Lycalopex culpaeus*), and 83 South American gray foxes (*L. griseus*), respectively, and 13%, 7%, and 8% *Candidatus* Mycoplasma hematoparvum in rural dogs, Andean foxes, and South American gray foxes, respectively. *Mycoplasma* spp. were molecularly assessed and found in 57% (47/82) of Darwin's foxes (*L. fulvipes*) and 24% (60/250) of free-ranging dogs (Di Cataldo et al. 2020). Higher infection of *Mycoplasma* spp. (57%) was also found on a Chilean island in *Lycalopex fulvipes* (Cabello et al. 2013a).

10.7 Environmental Transmission

Several long-lasting pathogenic viruses, bacteria, prions, and macroparasites can be indirectly transmitted, where pathogens are shed by infectious hosts to the environment (e.g., through respiratory secretions, urine, and/or feces), thus facilitating multihost disease spillover/spillback events among sympatric species (Lange et al. 2016). Examples include highly contagious diseases affecting wildlife, domestic species, and/or humans, such as leptospirosis, canine parvoviruses, bovine tuberculosis, brucellosis, and salmonellosis (e.g., Acosta-Jamett et al. 2015a; Barros et al. 2014; Dwyer et al. 2020; Miller et al. 2015). Among neotropical carnivores, the high level of sympatry between both domestic dogs and wild canids promotes indirect contact opportunities with a higher risk of interspecific pathogen spread. For example, free-ranging dogs may represent a potential CPV/CDV source for sympatric crab-eating foxes in Brazil (Courtenay et al. 2001). In northern Chile, higher indirect contact rates between dogs and wild foxes were recorded in proximity to urbanized centers, which may have relevant implications for canid-borne pathogen sharing (both macro- and microparasites) among highly abundant urban/peri-urban dog populations and susceptible wild canids coexisting at the wildlife-domestic interface (Hernandez et al. 2021). Expanding to other carnivore families, the interplay among the presence of dogs, invasive North American minks (as bridge hosts), and southern river otters would increase the risk of exposure to pathogens such as CDV and *Leptospira* spp. in proximity to rural villages and protected areas in southern Chile (Barros et al. 2014; Sepulveda et al. 2014).

In the neotropics, a significant amount of research has been focused on leptospirosis and the role of carnivores as potential sources of infection. Caused by *Leptospira* bacteria, this disease constitutes the most common zoonotic infection worldwide and can be environmentally transmitted through direct contact with urine or contaminated water. Several investigations have confirmed the exposure to multiple *Leptospira* spp. serovars in neotropical wild carnivores—mostly by using the microscopic agglutination test (MAT)—associated with a diversity of land uses, such as protected areas, cattle farms, croplands, rural, and peri-urban/urban districts. For instance, significant research has been conducted in Brazil within different biomes, such as the Pantanal and Cerrado, including more developed, human-altered areas. Jorge et al. (2011) reported variable proportions of *Leptospira* spp.-seroreactive specimens of crab-eating fox (40%; 17/43), maned wolf (38%; 3/8), crab-eating raccoon (50%; 6/12), puma (29%; 2/7), and ocelot (75%; 3/4), including one bush dog in the northern Pantanal. In federal conservation units in the Pantanal, Onuma et al. (2015) found that only 18% (2/11) of jaguars were seroreactive for the *Leptospira* spp. antigen, with a Brazilian isolate of serovar *canicola* being the most infective serotype found. On the other hand, Furtado et al. (2015) reported that jaguars were seropositive to different serovars of *Leptospira* spp. (overall seropositivity of 42%; 13/31), with the higher prevalences exhibited by serovars *grippotyphosa* and *Pomona* in the Brazilian Cerrado. Among free-ranging wild canids inhabiting this biome, Rodrigues et al. (2015) estimated diverse *Leptospira* spp. seroprevalence (mostly represented by serovars *copenhageni* and *hardjo*), which varied among 14% (1/7) in hoary fox, 36% (5/14) in maned wolf, and 42% (8/19) in crab-eating fox. In a more populated area within the state of São Paulo, exposure/infection positivity to *Leptospira* spp. was highly variable across a guild of free-ranging carnivores, ranging from 0% in puma to 75% in crab-eating fox (see Supplementary Material Table 10.1 for details), which exhibit a dominance of serovars *djasiman* and *australis*; however, neither demographic (i.e., age, sex) nor environmental (i.e., type of environment, season) factors significantly predicted *Leptospira* positivity in coatis (36%; 20/56), which were the most numerous carnivore species sampled (Fornazari et al. 2018).

In other South American countries, research by Fiorello et al. (2007) was focused on determining whether small carnivores were exposed to common domestic carnivore pathogens such as *Leptospira interrogans* in the Bolivian Chaco, finding that only one pampas fox (11%; 1/9) exhibited antibodies to the serovar *grippotyphosa*, while individuals of other species such as ocelot, Geoffroy's cat, jaguarundi, and crab-eating fox were all seronegative. In protected areas in central Argentina, Uhart et al. (2012) reported Geoffroy's cats exhibiting an *L. interrogans* seroprevalence of

40% (15/38), including both serovars icterohaemorrhagica/icterhaemorrhagica and *ballum*. One experience in Central America within a highly populated area of Costa Rica reported that 33% (31/95) of sampled raccoons presented evidence of antibodies against *Leptospira* spp., without dependency on host sex, age, or weight, but harboring up to 19 serovars implicated in morbidity of both humans and domestic animals (Baldi et al. 2019).

By using molecular methods such as PCR, other studies have detected the presence of pathogenic *Leptospira* species in diverse tissue samples of invasive carnivores thriving across diverse habitats in the neotropics. For instance, Barros et al. (2014) reported an infection prevalence of 54% (31/57), involving both *L. interrogans* and *L. borgpetersenii*, in the blood and kidneys of minks in southern Chile, which would be associated with predation on rodents and contact with garbage and contaminated water from cattle in locations where livestock is common. Overall, *Leptospira* spp. infection prevalence was slightly lower (34%; 22/64) in minks sampled at different continental (i.e., Maullín) and insular (i.e., Ancud and Navarino Island) Chilean localities, corresponding to several pathogenic serovars of *L. interrogans*, *L. kirschneri*, *L. noguchii*, and *L. borgpetersenii* (Ramírez-Pizarro et al. 2019). A recent study by Alfaro et al. (2021) relied on *Leptospira* spp. genotyping from kidney DNA of minks captured in the Los Ríos region, Chile; although none of the 45 sampled minks exhibited antibodies against the bacterium, an infection prevalence of 9% (4/45) was detected, which was related to *L. interrogans*, serovars *hardjo-prajitno* (associated with cattle-farming areas), and *copenhageni* and *icterohaemorrhagiae* (related to small rodents), but infection status was not related to the density of dairy farming or mink sex. In other neotropical regions, a study on the Caribbean Island of Saint Kitts showed evidence of exposure (8%; 12/148) and infection (6%; 9/146) by *Leptospira* spp. in the small Indian mongoose (*Herpestes auropunctatus*), revealing significantly higher bacterial exposure in a densely human-inhabited peridomestic area compared to an arid and less populated site. Here, *Leptospira* isolates were closely related to serovar *copenhageni*, which were potentially acquired from contaminated environments characterized by forested, wet zones in close proximity to rodents and peridomestic waste (Shiokawa et al. 2019). Despite the recognized health risks posed by invasive minks and mongooses across their nonnative neotropical ranges, further studies are needed to elucidate the epidemiological role of these (or other) alien carnivores in maintaining environmentally transmissible multihost pathogens infecting humans, domestic animals, and wildlife, including their impacts on sympatric native carnivores.

10.8 Future Remarks

Continued management is needed to prevent future epidemics of deadly pathogens (Gilbertson et al. 2022). For example, FeLV inoculation with a domestic cat vaccine has been used previously in panthers but with unknown efficacy (Cunningham et al. 2008). Furthermore, the proportion of the population that must be vaccinated to prevent future FeLV outbreaks most efficiently is unknown, as is how proactive vaccination might interact with other reactive interventions to interrupt an FeLV outbreak. Such uncertainties are common among free-ranging wildlife threatened by infectious disease and hamper efforts to effectively control pathogen transmission (Gilbertson et al. 2022).

Although there is a growing recognition of the impact of pathogens on wildlife, there is a relative lack of infectious disease focus and diagnostic capability (Smith et al. 2006). Collaboration among ecologists, conservation biologists, and veterinary pathologists is relatively recent, and increased pathological and epidemiological involvement in studies of the causes of wildlife declines is critically needed to identify and understand disease threats to wildlife and how to mitigate them (Cunningham et al. 2017). For the sake of public health and well-being, we need to work better to conserve nature and preserve the ecosystem services, including disease regulation, that biodiversity provides, while also understanding and mitigating activities that lead to disease emergence. Holistic, One Health approaches to the management and mitigation of the risks of emerging infectious diseases have the greatest chance of success (Cunningham et al. 2017).

Proving that a disease is a cause of population declines in wildlife requires longitudinal population and pathogen data, which are often very difficult to collect (Cunningham et al. 2017). However, long-term datasets have since been published that provide convincing evidence that pathogens alone can cause mass mortalities leading to population declines, local (population) extinctions, and global (species) extinctions (Daszak and Cunningham 1999).

References

Acosta-Jamett G, Cleaveland S, Cunningham AA, Bronsvoort BMD (2010) Demography of domestic dogs in rural and urban areas in Coquimbo region of Chile and its implication for diseases transmission. Prev Vet Med 94:272–281

Acosta-Jamett G, Chalmers WSK, Cunningham AA, Cleaveland S, Handel IG, Bronsvoort BMD (2011) Urban domestic dog populations as a source of canine distemper virus for wild carnivores in the Coquimbo region of Chile. Vet Microbiol 152:247–257

Acosta-Jamett G, Cunningham AA, Bronsvoort BMD, Cleaveland S (2015a) Serosurvey of canine distemper virus and canine parvovirus in wild canids and domestic dogs at the rural interface in the Coquimbo Region, Chile. Eur J Wildl Res 61:329–332

Acosta-Jamett G, Surot D, Cortes M, Marambio V, Valenzuela C, Vallverdu A, Ward MP (2015b) Epidemiology of canine distemper and canine parvovirus in domestic dogs in urban and rural areas of the Araucania region in Chile. Vet Microbiol 178:260–264

Alfaro MAS, Raffo E, Bustos MI, Tomckowiack C, Tejeda C, Collado L, Medina-Vogel G (2021) New insights on the infection of pathogenic Leptospira species in American mink (*Neovison vison*) in southern Chile. Trop Anim Health Prod 53:2

Almeida AP, Souza TD, Marcili A, Labruna MB (2013) Novel *Ehrlichia* and *Hepatozoon* agents infecting the crab-eating fox (*Cerdocyon thous*) in southeastern Brazil. J Med Entomol 50:640–646

Alvarado-Rybak M, Solano-Gallego L, Millan J (2016) A review of piroplasmid infections in wild carnivores worldwide: importance for domestic animal health and wildlife conservation. Parasite Vector 9:538

Araujo Soares MR, Lopes Antunes JE, de Mendonca IL, Lima RN, Nery Costa CH (2017) Occurrence of *Lutzomyia longipalpis* Lutz & Neiva 1912 and *Cerdocyon thous* Linnaeus 1977, in a visceral leishmaniasis endemic area in Brazil. Acta Trop 174:118–121

Arechiga-Ceballos N, Velasco-Villa A, Shi M, Flores-Chavez S, Barron B, Cuevas-Dominguez E, Gonzalez-Origel A, Aguilar-Setien A (2010) New rabies virus variant found during an epizootic in white-nosed coatis from the Yucatan Peninsula. Epidemiol Infect 138:1586–1589

Arrais RC, Paula RC, Martins TF, Nieri-Bastos FA, Marcili A, Labruna MB (2021) Survey of ticks and tick-borne agents in maned wolves (Chrysocyon *brachyurus*) from a natural landscape in Brazil. Ticks Tick Borne Dis 12:101639

Azami-Conesa I, Gomez-Munoz MT, Martinez-Diaz RA (2021) A systematic review (1990-2021) of wild animals infected with zoonotic *Leishmania*. Microorganisms 9:1101

Balboni A, Tryland M, Mork T, Killengreen ST, Fuglei E, Battilani M (2019) Unique genetic features of canine adenovirus type 1 (CAdV-1) infecting red foxes (*Vulpes vulpes*) in northern Norway and arctic foxes (Vulpes lagopus) in Svalbard. Vet Res Commun 43:67–76

Baldi M, Hernandez-Mora G, Jimenez C, Hutter SE, Alfaro A, Walzert C (2019) Leptospira seroprevalence detection and rabies virus absence in an urban raccoon (*Procyon lotor*) population in a highly populated area, Costa Rica. Vector-Borne Zoonotic Dis 19:889–895

Barros M, Saenz L, Lapierre L, Nunez C, Medina-Vogel G (2014) High prevalence of pathogenic Leptospira in alien American mink (*Neovison vison*) in Patagonia. Rev Chil Hist Nat 87:1–5

Barros M, Cabezon O, Dubey JP, Almeria S, Ribas MP, Escobar LE, Ramos B, Medina-Vogel G (2018) *Toxoplasma gondii* infection in wild mustelids and cats across an urban–rural gradient. PLoS One 13:e0199085

Barros M, Pons DJ, Moreno A, Vianna J, Ramos B, Dueñas F, Coccia C, Saavedra-Rodríguez R, Santibañez A, Medina-Vogel G (2022) Domestic dog and alien North American mink as reservoirs of infectious diseases in the endangered Southern river otter. Austral J Vet Sci 54:65–75

Behdenna A, Lembo T, Calatayud O, Cleaveland S, Halliday JEB, Packer C, Lankester F, Hampson K, Craft ME, Czupryna A, Dobson AP, Dubovi EJ, Ernest E, Fyumagwa R, Hopcraft JGC, Mentzel C, Mzimbiri I, Sutton D, Willett B, Haydon DT, Viana M (2019) Transmission ecology of canine parvovirus in a multi-host, multi-pathogen system. Proc Biol Sci 286:20182772

Bermudez SE, Gottdenker N, Krishnvajhala A, Fox A, Wilder HK, Gonzalez K, Smith D, Lopez M, Perea M, Rigg C, Montilla S, Calzada JE, Saldana A, Caballero CM, Lopez JE (2017) Synanthropic mammals as potential hosts of tick-borne pathogens in Panama. PLoS One 12:e0169047

Bischof R, Hansen NR, Nyheim OS, Kisen A, Prestmoen L, Haugaasen T (2022) Mapping the "catscape" formed by a population of pet cats with outdoor access. Sci Rep 12:5964

Böhm M, Hutchings MR, White PCL (2009) Contact networks in a wildlife-livestock host community: identifying high-risk individuals in the transmission of bovine TB among badgers and cattle. PLoS One 4(4):e5016. https://doi.org/10.1371/journal.pone.0005016

Boulouis HJ, Chang CC, Henn JB, Kasten RW, Chomel BB (2005) Factors associated with the rapid emergence of zoonotic *Bartonella* infections. Vet Res 36:383–410

Bradley CA, Altizer S (2007) Urbanization and the ecology of wildlife diseases. Trends Ecol Evol 22:95–102

Brandao EMV, Xavier SCC, Rocha FL, Lima CFM, Candeias IZ, Lemos FG, Azevedo FC, Jansen AM, Roque ALR (2020) Wild and domestic canids and their interactions in the transmission cycles of *Trypanosoma cruzi* and *Leishmania* spp. in an area of the Brazilian Cerrado. Pathogens 9:818

Bronson E, Emmons LH, Murray S, Dubovi EJ, Deem SL (2008) Serosurvey of pathogens in domestic dogs on the border of Noel Kempff Mercado National Park, Bolivia. J Zoo Wildl Med 39:28–36

Brown MA, Cunningham MW, Roca AL, Troyer JL, Johnson WE, O'Brien SJ (2008) Genetic characterization of feline leukemia virus from Florida panthers. Emerg Infect Dis 14:252–259

Burnard D, Polkinghorne A (2016) Chlamydial infections in wildlife-conservation threats and/or reservoirs of 'spill-over' infections? Vet Microbiol 196:78–84

Butler JRA, du Toit JT, Bingham J (2004) Free-ranging domestic dogs (Canis familiaris) as predators and prey in rural Zimbabwe: threats of competition and disease to large wild carnivores. Biol Conserv 115:369–378

Cabello J, Altet L, Napolitano C, Sastre N, Hidalgo E, Davila JA, Millan J (2013a) Survey of infectious agents in the endangered Darwin's fox (Lycalopex fulvipes): high prevalence and diversity of hemotrophic mycoplasmas. Vet Microbiol 167:448–454

Cabello J, Esperon F, Napolitano C, Hidalgo E, Davila JA, Millan J (2013b) Molecular identification of a novel gammaherpesvirus in the endangered Darwin's fox (*Lycalopex fulvipes*). J Gen Virol 94:2745–2749

Calatayud O, Esperon F, Cleaveland S, Biek R, Keyyu J, Eblate E, Neves E, Lembo T, Lankester F (2019) Carnivore parvovirus ecology in the Serengeti ecosystem: vaccine strains circulating and new host species identified. J Virol 93:e02220-18

Campbell EL, Byrne AW, Menzies FD, McBride KR, McCormick CM, Scantlebury M, Reid N (2019) Interspecific visitation of cattle and badgers to fomites: a transmission risk for bovine tuberculosis? Ecol Evol 9:8479–8489

Caron A, Cappelle J, Cumming GS, de Garine-Wichatitsky M, Gaidet N (2015) Bridge hosts, a missing link for disease ecology in multi-host systems. Vet Res 46:83

Carvalho L, Felix ML, Bazzano V, da Costa A, Armua-Fernandez MT, Munoz-Leal S, Venzal JM (2021) An *Hepatozoon americanum*-like protozoan in crab-eating (*Cerdocyon thous*) and grey pampean (*Lycalopex gymnocercus*) foxes from Uruguay. Parasitol Res 120:3587–3593

Ceballos G, Ehrlich PR, Barnosky AD, Garcia A, Pringle RM, Palmer TM (2015) Accelerated modern human-induced species losses: entering the sixth mass extinction. Sci Adv 1:e1400253

Chapin FS, Walker BH, Hobbs RJ, Hooper DU, Lawton JH, Sala OE, Tilman D (1997) Biotic control over the functioning of ecosystems. Science 277:500–504

Chiu ES, Kraberger S, Cunningham M, Cusack L, Roelke M, VandeWoude S (2019) Multiple introductions of domestic cat feline leukemia virus in endangered Florida panthers. Emerg Infect Dis 25:92–101

Chomel BB, Kasten RW, Williams C, Wey AC, Henn JB, Maggi R, Carrasco S, Mazet J, Boulouis HJ, Maillard R, Breitschwerdt EB (2009) Bartonella endocarditis a pathology shared by animal reservoirs and patients. Rickettsiol Rickettsial Dis 1166:120–126

Cleaveland S, Dye C (1995) Maintenance of a microparasite infecting several host species: rabies in the Serengeti. Parasitology 111:S33–S47

Cleaveland S, Appel MGJ, Chalmers WSK, Chillingworth C, Kaare M, Dye C (2000) Serological and demographic evidence for domestic dogs as a source of canine distemper virus infection for Serengeti wildlife. Vet Microbiol 72:217–227

Cleaveland S, Hess GR, Dobson AP, Laurenson MK, McCallum HI, Roberts MG, Woodroffe R (2002) The role of pathogens in biologial conservation. In: Hudson RJ, Rizzoli AP, Grenfell BT, Heesterbeek H, Dobson AP (eds) The ecology of wildlife diseases. Oxford University Press, Oxford, pp 139–150

Cleaveland S, Mlengeya T, Kaare M, Haydon D, Lembo T, Laurenson MK, Packer C (2007) The conservation relevance of epidemiological research into carnivore viral diseases in the Serengeti. Conserv Biol 21:612–622

Collere FCM, Delai RM, Ferrari LDR, da Silva LH, Fogaca PLC, Rodrigues AN, Goncalves DD, Baggio RA, Moraes MFD, Lux Hoppe EG, Andre MR, Vieira T, Vieira RFC (2021) '*Candidatus* Mycoplasma haematonasua' and tick-borne pathogens in ring-tailed coatis (*Nasua nasua* Linnaeus, 1976) from the Iguacu National Park, Parana State, southern Brazil. Transbound Emerg Dis 68:3222–3229

Courtenay O, Quinnell RJ, Chalmers WSK (2001) Contact rates between wild and domestic canids: no evidence of parvovirus or canine distemper virus in crab-eating foxes. Vet Microbiol 81:9–19

Criado-Fornelio A, Ruas JL, Casado N, Farias NA, Soares MP, Muller G, Brumt JG, Berne ME, Buling-Sarana A, Barba-Carretero JC (2006) New molecular data on mammalian *Hepatozoon* species (Apicomplexa: Adeleorina) from Brazil and Spain. J Parasitol 92:93–99

Crooks KR (2002) Relative sensitivities of mammalian carnivores to habitat fragmentation. Conserv Biol 16:488–502

Cunningham MW, Brown MA, Shindle DB, Terrell SP, Hayes KA, Ferree BC, McBride RT, Blankenship EL, Jansen D, Citino SB, Roelke ME, Kiltie RA, Troyer JL, O'Brien SJ (2008) Epizootiology and management of feline leukemia virus in the Florida puma. J Wildl Dis 44:537–552

Cunningham AA, Daszak P, Wood JLN (2017) One Health, emerging infectious diseases and wildlife: two decades of progress? Philos Trans R Soc Lond Ser B Biol Sci 372:20160167

da Silva MRL, Fornazari F, Martins TF, Hippolito AG, Rolim LS, Bisca JM, Teixeira CR, O'Dwyer LH (2018) A survey of hemoparasites and ectoparasites in *Nasua nasua* Linnaeus, 1766 with a redescription of *Hepatozoon procyonis* Richards, 1961 based on morphological and molecular data. Parasitol Res 117:2159–2169

Dall'Agnol B, Souza UA, Weck B, Trigo TC, Jardim MMA, Costa FB, Labruna MB, Peters FB, Favarini MO, Mazim FD, Ferreira CAS, Reck J (2018) *Rickettsia parkeri* in free-ranging wild canids from Brazilian Pampa. Transbound Emerg Dis 65:e224–e230

Daszak P, Cunningham AA (1999) Extinction by infection. Trends Ecol Evol 14:279

Daszak P, Cunningham AA, Hyatt AD (2000) Wildlife ecology—emerging infectious diseases of wildlife—threats to biodiversity and human health. Science 287:443–449

Daszak P, Cunningham AA, Hyatt AD (2001) Anthropogenic environmental change and the emergence of infectious diseases in wildlife. Acta Trop 78:103–116

de Almeida Curi NH, Miranda I, Talamoni SA (2006) Serologic evidence of Leishmania infection in free-ranging wild and domestic canids around a Brazilian National Park. Mem Inst Oswaldo Cruz 101:99–101

de Almeida Curi NH, Araujo AS, Campos FS, Lobato ZIP, Gennari SM, Marvulo MFV, Silva JCR, Talamoni SA (2010) Wild canids, domestic dogs and their pathogens in Southeast Brazil: disease threats for canid conservation. Biodivers Conserv 19:3513–3524

de Almeida Curi NH, Coelho CM, Malta MDC, Magni EMV, Sabato MAL, Araujo AS, Lobato ZIP, Santos JLC, Santos HA, Ragozo AAM, de Souza SLP (2012) Pathogens of wild maned wolves (*Chrysocyon brachyurus*) in Brazil. J Wildl Dis 48:1052–1056

de Lorenzo C, Bianchi MV, Ehlers LP, Vielmo A, Pereira PR, de Almeida BA, de Andrade CP, Girotto-Soares A, Pavarini SP, Driemeier D, Soares JF, Sonne L (2021a) *Rangelia vitalii* molecular and histological quantification in tissues comparing crab-eating foxes (*Cerdocyon thous*) and domestic dogs. Ticks Tick Borne Dis 12:101731

de Lorenzo C, Boabaid FM, de Oliveira LGS, Bianchi MV, Felix ML, Armua-Fernandez MT, Soares JF, Venzal JM, Sonne L (2021b) *Rangelia vitalii* in free-living crab-eating foxes (*Cerdocyon thous*) in Uruguay. Ticks Tick Borne Dis 12:101765

de Sousa KC, Fernandes MP, Herrera HM, Benevenute JL, Santos FM, Rocha FL, Barreto WT, Macedo GC, Campos JB, Martins TF, de Andrade Pinto PC, Battesti DB, Piranda EM, Cancado PH, Machado RZ, Andre MR (2017a) Molecular detection of *Hepatozoon* spp. in domestic dogs and wild mammals in southern Pantanal, Brazil with implications in the transmission route. Vet Parasitol 237:37–46

de Sousa KCM, Calchi AC, Herrera HM, Dumler JS, Barros-Battesti DM, Machado RZ, Andre MR (2017b) Anaplasmataceae agents among wild mammals and ectoparasites in Brazil. Epidemiol Infect 145:3424–3437

de Sousa KCM, Herrera HM, Secato CT, Oliveira ADV, Santos FM, Rocha FL, Barreto WTG, Macedo GC, de Andrade Pinto PCE, Machado RZ, Costa MT, Andre MR (2017c) Occurrence and molecular characterization of hemoplasmas in domestic dogs and wild mammals in a Brazilian wetland. Acta Trop 171:172–181

de Sousa KCM, Fernandes MP, Herrera HM, Freschi CR, Machado RZ, Andre MR (2018) Diversity of piroplasmids among wild and domestic mammals and ectoparasites in Pantanal wetland, Brazil. Ticks Tick Borne Dis 9:245–253 de Sousa-Paula LC, Otranto D, Dantas-Torres F (2020) *Lutzomyia longipalpis* (Sand Fly). Trends Parasitol 36:796–797

de Souza VK, Dall'Agnol B, Souza UA, Webster A, Peters FB, Favarini MO, Mazim FD, da Rocha FL, Tirelli FP, Soares JF, Jardim MMD, Trigo TC, Reck J (2019) Detection of *Rangelia vitalii* (Piroplasmida: Babesiidae) in asymptomatic free-ranging wild canids from the Pampa biome, Brazil. Parasitol Res 118:1337–1342

Deem SL, Emmons LH (2005) Exposure of free-ranging maned wolves (*Chrysocyon brachyurus*) to infectious and parasitic disease agents in the Noel Kempff Mercado National Park, Bolivia. J Zoo Wildl Med 36:192–197

Deem SL, Davis R, Pacheco LF (2004) Serologic evidence of nonfatal rabies exposure in a free-ranging oncilla (*Leopardus tigrinus*) in Cotapata National Park, Bolivia. J Wildl Dis 40:811–815

Di Cataldo S, Hidalgo-Hermoso E, Sacristan I, Cevidanes A, Napolitano C, Hernandez CV, Esperon F, Moreira-Arce D, Cabello J, Muller A, Millan J (2020) Hemoplasmas are endemic and cause asymptomatic infection in the endangered Darwin's fox (*Lycalopex fulvipes*). Appl Environ Microbiol 86:101823

Di Cataldo S, Cevidanes A, Ulloa-Contreras C, Hidalgo-Hermoso E, Gargano V, Sacristan I, Sallaberry-Pincheira N, Penaloza-Madrid D, Gonzalez-Acuna D, Napolitano C, Vianna J, Acosta-Jamett G, Vicari D, Millan J (2021a) Mapping the distribution and risk factors of Anaplasmataceae in wild and domestic canines in Chile and their association with *Rhipicephalus sanguineus* species complex lineages. Ticks Tick Borne Dis 12:101752

Di Cataldo S, Cevidanes A, Ulloa-Contreras C, Sacristan I, Penaloza-Madrid D, Vianna J, Gonzalez-Acuna D, Sallaberry-Pincheira N, Cabello J, Napolitano C, Hidalgo-Hermoso E, Acosta-Jamett G, Millan J (2021b) Widespread infection with hemotropic Mycoplasmas in free-ranging dogs and wild foxes across six bioclimatic regions of Chile. Microorganisms 9:919

Dirzo R, Young HS, Galetti M, Ceballos G, Isaac NJB, Collen B (2014) Defaunation in the Anthropocene. Science 345:401–406

Dixon DM, Branda JA, Clark SH, Dumler JS, Horowitz HW, Perdue SS, Pritt BS, Sexton DJ, Storch GA, Walker DH (2021) Ehrlichiosis and anaplasmosis subcommittee report to the Tick-borne Disease Working Group. Ticks Tick Borne Dis 12:101823

Dobler G, Pfeffer M (2011) Fleas as parasites of the family Canidae. Parasite Vector 4:139

Dobson AP, Hudson PJ (1995) Microparasites: observed patterns in wild animal populations. In: Grenfell BT, Dobson AP (eds) Ecology of infectious diseases in natural populations. Cambridge University Press, Cambridge, pp 52–89

Drewe JA, O'Connor HM, Weber N, McDonald RA, Delahay RJ (2013) Patterns of direct and indirect contact between cattle and badgers naturally infected with tuberculosis. Epidemiol Infect 141:1467–1475

Dwyer RA, Witte C, Buss P, Goosen WJ, Miller M (2020) Epidemiology of tuberculosis in multi-host wildlife systems: implications for black (*Diceros bicornis*) and white (*Ceratotherium simum*) rhinoceros. Front Vet Sci 7:580476

Elbroch LM, Vickers TW, Quigley HB (2020) Plague, pumas and potential zoonotic exposure in the Greater Yellowstone Ecosystem. Environ Conserv 47:75–78

Elsmo EJ, Fenton H, Cleveland CA, Shock B, Cunningham M, Howerth EW, Yabsley MJ (2018) Necrotizing interstitial pneumonia and suppurative myocarditis associated with *Bartonella henselae* infection in three Florida pumas. J Vet Diagn Investig 30:728–732

Estevam LGTM, Fonseca AA, Silvestre BT, Hemetrio NS, Almeida LR, Oliveira MM, Silva SM, Ribeiro MFB, Silveira JAG (2020) Seven years of evaluation of ectoparasites and vector-borne pathogens among ring-tailed coatis in an urban park in southeastern Brazil. Vet Parasitol Reg St 21:100442

Farris ZJ, Gerber BD, Karpanty S, Murphy A, Wampole E, Ratelolahy F, Kelly MJ (2020) Exploring and interpreting spatiotemporal interactions between native and invasive carnivores across a gradient of rainforest degradation. Biol Invasions 22:2033–2047

Filoni C, Catao-Dias JL, Bay G, Durigon EL, Jorge RS, Lutz H, Hofmann-Lehmann R (2006) First evidence of feline herpesvirus, calicivirus, parvovirus, and *Ehrlichia* exposure in Brazilian free-ranging felids. J Wildl Dis 42:470–477

Filoni C, Catao-Dias JL, Cattori V, Willi B, Meli ML, Correa SH, Marques MC, Adania CH, Silva JC, Marvulo MF, Ferreira Neto JS, Durigon EL, de Carvalho VM, Coutinho SD, Lutz H, Hofmann-Lehmann R (2012) Surveillance using serological and molecular methods for the detection of infectious agents in captive Brazilian neotropic and exotic felids. J Vet Diagn Investig 24:166–173

Fiorello CV, Deem SL, Gompper ME, Dubovi EJ (2004) Seroprevalence of pathogens in domestic carnivores on the border of Madidi National Park, Bolivia. Anim Conserv 7:45–54

Fiorello CV, Noss AJ, Deem SL, Maffei L, Dubovi EJ (2007) Serosurvey of small carnivores in the Bolivian Chaco. J Wildl Dis 43:551–557

Flores-Ibarra M, Estrella-Valenzuela G (2004) Canine ecology and socioeconomic factors associated with dogs unvaccinated against rabies in a Mexican city across the US–Mexico border. Prev Vet Med 62:79–87

Fornazari F, Langoni H, Marson PM, Nobrega DB, Teixeira CR (2018) Leptospira reservoirs among wildlife in Brazil: beyond rodents. Acta Trop 178:205–212

Furtado MM, Gennari SM, Ikuta CY, Jacomo AT, de Morais ZM, Pena HF, Porfirio GE, Silveira L, Sollmann R, de Souza GO, Torres NM, Ferreira Neto JS (2015) Serosurvey of smooth *Brucella*, *Leptospira* spp. and *Toxoplasma gondii* in free-ranging jaguars (*Panthera onca*) and domestic animals from Brazil. PLoS One 10:e0143816

Furtado MM, Taniwaki SA, de Barros IN, Brandao PE, Catao-Dias JL, Cavalcanti S, Cullen L, Filoni C, Jacomo ATA, Jorge RSP, Silva NDS, Silveira L, Ferreira Neto JS (2017a) Molecular detection of viral agents in free-ranging and captive neotropical felids in Brazil. J Vet Diagn Investig 29:660–668

Furtado MM, Metzger B, de Almeida Jacomo AT, Labruna MB, Martins TF, O'Dwyer LH, Paduan KDS, Porfirio GEO, Silveira L, Sollmann R, Taniwaki SA, Torres NM, Neto JSF (2017b) *Hepatozoon spp.* infect free-ranging jaguars (*Panthera onca*) in Brazil. J Parasitol 103:243–250

Furtado MM, Taniwaki SA, Metzger B, Paduan KD, O'Dwyer HL, Jacomo ATD, Porfirio GEO, Silveira L, Sollmann R, Torres NM, Neto JSF (2017c) Is the free-ranging jaguar (*Panthera onca*) a reservoir for *Cytauxzoon* felis in Brazil? Ticks Tick Borne Dis 8:470–476

Garner MM, Lung NP, Citino S, Greiner EC, Harvey JW, Homer BL (1996) Fatal cytauxzoonosis in a captive-reared white tiger (*Panthera tigris*). Vet Pathol 33:82–86

Gilbertson MLJ, Onorato D, Cunningham M, VandeWoude S, Craft ME (2022) Paradoxes and synergies: optimizing management of a deadly virus in an endangered carnivore. J Appl Ecol 59:1548–1558

Gittleman JL (2001) Carnivore behavior, ecology, and evolution. Cornell University Press, p 620

Godfrey SS (2013) Networks and the ecology of parasite transmission: a framework for wildlife parasitology. Int J Parasitol Parasites Wildl 2:235–245

Gompper ME (2014a) The dog–human–wildlife interface: assessing the scope of the problem. In: Gompper ME (ed) Free-ranging dogs and wildlife conservation. Oxford University Press, New York, pp 9–54

Gompper ME (2014b) Free-ranging dogs and wildlife conservation. Oxford University Press, New York

Gonzalez-Acuna D, Ortega-Vasquez R, Rivera-Ramirez P, Cabello-Cabalin J (2003) A presumed case of canine distemper in a gray fox (*Pseudalopex griseus*) from central Chile. Z Jagdwiss 49:323–326

Gordon JC, Angrick EJ (1986) Canine parvovirus: environmental effects on infectivity. Am J Vet Res 47:1464–1467

Gottdenker NL, Streicker DG, Faust CL, Carroll CR (2014) Anthropogenic land use change and infectious diseases: a review of the evidence. EcoHealth 11:619–632

Greene CE (1998) Infectious diseases of the dog and cat, 2nd edn. W. B. Saunders, Philadelphia

Grenfell BT, Dobson AP (1995) Ecology of infectious diseases in natural populations. Cambridge University Press, Cambridge

Guegan JF, Ayouba A, Cappelle J, de Thoisy B (2020) Forests and emerging infectious diseases: unleashing the beast within. Environ Res Lett 15:083007

Hahn MB, Gurley ES, Epstein JH, Islam MS, Patz JA, Daszak P, Luby SP (2014) The role of landscape composition and configuration on *Pteropus giganteus* roosting ecology and nipah virus spillover risk in Bangladesh. Am J Trop Med Hyg 90:247–255

Hassell JM, Begon M, Ward MJ, Fevre EM (2017) Urbanization and disease emergence: dynamics at the wildlife-livestock-human interface. Trends Ecol Evol 32:55–67

Haydon DT, Cleaveland S, Taylor LH, Laurenson MK (2002) Identifying reservoirs of infection: a conceptual and practical challenge. Emerg Infect Dis 8:1468–1473

Haydon DT, Randall DA, Matthews L, Knobel DL, Tallents LA, Gravenor MB, Williams SD, Pollinger JP, Cleaveland S, Woolhouse MEJ, Sillero-Zubiri C, Marino J, Macdonald DW, Laurenson MK (2006) Low-coverage vaccination strategies for the conservation of endangered species. Nature 443:692–695

Hernandez FA, Manqui J, Mejias C, Acosta-Jamett G (2021) Domestic dogs and wild foxes interactions in a wildlife-domestic interface of north-central Chile: implications for multi-host pathogen transmission. Front Vet Sci 8:631788

Jones BA, Grace D, Kock R, Alonso S, Rushton J, Said MY, McKeever D, Mutua F, Young J, McDermott J, Pfeiffer DU (2013) Zoonosis emergence linked to agricultural intensification and environmental change. Proc Natl Acad Sci U S A 110:8399–8404

Jorge RS, Ferreira F, Ferreira Neto JS, Vasconcellos Sde A, Lima Ede S, Morais ZM, Souza GO (2011) Exposure of free-ranging wild carnivores, horses and domestic dogs to *Leptospira* spp in the northern Pantanal, Brazil. Mem Inst Oswaldo Cruz 106:441–444

Karanth KU, Srivathsa A, Vasudev D, Puri M, Parameshwaran R, Kumar NS (2017) Spatio-temporal interactions facilitate large carnivore sympatry across a resource gradient. Proc R Soc B-Biol Sci 284:20161860

Kilpatrick AM, Randolph SE (2012) Drivers, dynamics, and control of emerging vector-borne zoonotic diseases. Lancet 380:1946–1955

Knobel DL, Butler JR, Lembo T, Critchlow E, Gompper ME (2014) Dogs, disease, and wildlife. In: Gompper ME (ed) Free-ranging dogs and wildlife conservation. Oxford University Press, New York, pp 144–169

Lafferty KD, Gerber LR (2002) Good medicine for conservation biology: the intersection of epidemiology and conservation theory. Conserv Biol 16:593–604

Lange M, Kramer-Schadt S, Thulke HH (2016) Relevance of indirect transmission for wildlife disease surveillance. Front Vet Sci 3:110

Lembo T, Haydon DT, Velasco-Villa A, Rupprecht CE, Packer C, Brandao PE, Kuzmin IV, Fooks AR, Barrat J, Cleaveland S (2007) Molecular epidemiology identifies only a single rabies virus variant circulating in complex carnivore communities of the Serengeti. Proc R Soc B-Biol Sci 274:2123–2130

Lembo T, Hampson K, Haydon DT, Craft M, Dobson AP, Dushoff J, Ernest E, Hoare R, Kaare M, Mlengeya T, Mentzel C, Cleaveland S (2008) Exploring reservoir dynamics: a case study of rabies in the Serengeti ecosystem. J Appl Ecol 45:1246–1257

Levin ML, Killmaster LF, Zemtsova GE, Ritter JM, Langham G (2014) Clinical presentation, convalescence, and relapse of rocky mountain spotted fever in dogs experimentally infected via tick bite. PLoS One 9:e115105

Lopez G, Lopez-Parra M, Fernandez L, Martinez-Granados C, Martinez F, Meli ML, Gil-Sanchez JM, Viqueira N, Diaz-Portero MA, Cadenas R, Lutz H, Vargas A, Simon MA (2009) Management measures to control a feline leukemia virus outbreak in the endangered Iberian lynx. Anim Conserv 12:173–182

Lopez-Perez AM, Moreno K, Chaves A, Ibarra-Cerdena CN, Rubio A, Foley J, List R, Suzan G, Sarmiento RE (2019) Carnivore Protoparvovirus 1 at the Wild-Domestic Carnivore Interface in Northwestern Mexico. EcoHealth 16:502–511

Loss SR, Boughton B, Cady SM, Londe DW, McKinney C, O'Connell TJ, Riggs GJ, Robertson EP (2022) Review and synthesis of the global literature on domestic cat impacts on wildlife. J Anim Ecol 91:1361–1372

Lundgren DL, Ushijima RN, Sidwell RW (1963) Studies on infectious diseases in wild animals in Utah. V. Experimental rocky mountain spotted fever in the coyote, *Canis latrans* Lestes Merriam. Zoonoses Res 2:125–134

Luppi MM, Malta MCC, Silva TMA, Silva FL, Motta ROC, Miranda I, Ecco R, Santos RL (2008) Visceral leishmaniasis in captive wild canids in Brazil. Vet Parasitol 155:146–151

Martino PE, Samartino LE, Stanchi NO, Radman NE, Parrado EJ (2017) Serology and protein electrophoresis for evidence of exposure to 12 mink pathogens in free-ranging American mink (*Neovison vison*) in Argentina. Vet Q 37:207–211

McMichael AJ (2004) Environmental and social influences on emerging infectious diseases: past, present and future. Philos Trans R Soc Lond Ser B Biol Sci 359:1049–1058

Megid J, de Souza VAF, Teixeira CR, Cortez A, Amorin RL, Heinemman MB, Cagnini DQ, Richtzenhain LJ (2009) Canine distemper virus in a crab-eating fox (*Cerdocyon thous*) in Brazil: case report and phylogenetic analyses. J Wildl Dis 45:527–530

Meli ML, Cattori V, Martinez F, Lopez G, Vargas A, Palomares F, Lopez-Bao JV, Hofmann-Lehmann R, Lutz H (2010) Feline leukemia virus infection: a threat for the survival of the critically endangered Iberian lynx (*Lynx pardinus*). Vet Immunol Immunopathol 134:61–67

Metzger B, dos Santos Paduan K, Rubini AS, de Oliveira TG, Pereira C, O'Dwyer LH (2008) The first report of *Hepatozoon* sp. (Apicomplexa: Hepatozoidae) in neotropical felids from Brazil. Vet Parasitol 152:28–33

Michelazzo MDZ, de Oliveira TES, Viana NE, de Moraes W, Cubas ZS, Headley SA (2020) Immunohistochemical evidence of canine morbillivirus (canine distemper) infection in coatis (*Nasua nasua*) from Southern Brazil. Transbound Emerg Dis 67:178–184

Michelazzo MDZ, Martinelli TM, de Amorim VRG, Silva LE, Silva FHP, Xavier AAC, Cubas ZS, de Almeida RF, de Moraes W, Headley SA (2022) Canine distemper virus and canine adenovirus type-2 infections in neotropical otters (*Lontra longicaudis*) from Southern Brazil. Braz J Microbiol 53:369–375

Millan J, Travaini A, Zanet S, Lopez-Bao JV, Trisciuoglio A, Ferroglio E, Rodriguez A (2016) Detection of *Leishmania* DNA in wild foxes and associated ticks in Patagonia, Argentina, 2000 km south of its known distribution area. Parasite Vector 9:241

Millan J, Travaini A, Cevidanes A, Sacristan I, Rodriguez A (2019) Assessing the natural circulation of canine vector-borne pathogens in foxes, ticks and fleas in protected areas of Argentine Patagonia with negligible dog participation. Int J Parasitol Parasites Wildl 8:63–70

Miller S, Zieger U, Ganser C, Satterlee SA, Bankovich B, Amadi V, Hariharan H, Stone D, Wisely SM (2015) Influence of land use and climate on *Salmonella* carrier status in the small Indian mongoose (*Herpestes auropunctatus*) in Grenada, West Indies. J Wildl Dis 51:60–68

Mora M, Napolitano C, Ortega R, Poulin E, Pizarro-Lucero J (2015) Feline immunodeficiency virus and feline leukemia virus infection in free-ranging guignas (*Leopardus guigna*) and sympatric domestic cats in human perturbed landscapes on Chiloe Island, Chile. J Wildl Dis 51:199–208

Moseby KE, Peacock DE, Read JL (2015) Catastrophic cat predation: a call for predator profiling in wildlife protection programs. Biol Conserv 191:331–340

Müller R, Reuss F, Kendrovski V, Montag D (2019) Vector-borne diseases. In: Marselle M, Stadler J, Korn H, Irvine K, Bonn A (eds) Biodiversity and health in the face of climate change. Springer, Cham, pp 67–90

Murcia C (1995) Edge effects in fragmented forests—implications for conservation. Trends Ecol Evol 10:58–62

Murray KA, Daszak P (2013) Human ecology in pathogenic landscapes: two hypotheses on how land use change drives viral emergence. Curr Opin Virol 3:79–83

Nascimento COM, Silva MLCR, Kim PCP, Gomes AAB, Gomes ALV, Maia RCC, Almeida JC, Mota RA (2015) Occurrence of *Neospora caninum* and *Toxoplasma gondii* DNA in brain tissue from hoary foxes (*Pseudalopex vetulus*) in Brazil. Acta Trop 146:60–65

Nava AFD, Cullen L, Sana DA, Nardi MS, Ramos Filho J, Lima TF, Abreu KC, Ferreira F (2008) First evidence of canine distemper in Brazilian free-ranging felids. EcoHealth 5:513–518

Nietfeld JC, Pollock C (2002) Fatal cytauxzoonosis in a free-ranging bobcat (*Lynx rufus*). J Wildl Dis 38:607–610

Nikolin VM, Olarte-Castillo XA, Osterrieder N, Hofer H, Dubovi E, Mazzoni CJ, Brunner E, Goller KV, Fyumagwa RD, Moehlman PD, Thierer D, East ML (2017) Canine distemper virus in the Serengeti ecosystem: molecular adaptation to different carnivore species. Mol Ecol 26:2111–2130

Nishimura Y, Goto Y, Yoneda K, Endo Y, Mizuno T, Hamachi M, Maruyama H, Kinoshita H, Koga S, Komori M, Fushuku S, Ushinohama K, Akuzawa M, Watari T, Hasegawa A, Tsujimoto H (1999) Interspecies transmission of feline immunodeficiency virus from the domestic cat to the Tsushima cat (*Felis bengalensis euptilura*) in the wild. J Virol 73:7916–7921

Oakleaf JR, Kennedy CM, Baruch-Mordo S, West PC, Gerber JS, Jarvis L, Kiesecker J (2015) A world at risk: aggregating development trends to forecast global habitat conversion. PLoS One 10:e0138334

O'Brien SJ, Troyer JL, Brown MA, Johnson WE, Antunes A, Roelke ME, Pecon-Slattery J (2012) Emerging viruses in the Felidae: shifting paradigms. Viruses 4:236–257

O'Hanlon SJ, Rieux A, Farrer RA, Rosa GM, Waldman B, Bataille A, Kosch TA, Murray KA, Brankovics B, Fumagalli M, Martin MD, Wales N, Alvarado-Rybak M, Bates KA, Berger L, Boll S, Brookes L, Clare F, Courtois EA, Cunningham AA, Doherty-Bone TM, Ghosh P, Gower DJ, Hintz WE, Hoglund J, Jenkinson TS, Lin CF, Laurila A, Loyau A, Martel A, Meurling S, Miaud C, Minting P, Pasmans F, Schmeller DS, Schmidt BR, Shelton JMG, Skerratt LF, Smith F, Soto-Azat C, Spagnoletti M, Tessa G, Toledo LF, Valenzuela-Sanchez A, Verster R, Voros J, Webb RJ, Wierzbicki C, Wombwell E, Zamudio KR, Aanensen DM, James TY, Gilbert MTP, Weldon C, Bosch J, Balloux F, Garner TWJ, Fisher MC (2018) Recent Asian origin of chytrid fungi causing global amphibian declines. Science 360:621–627

Onuma SSM, Melo ALT, Kantek DLZ, Crawshaw PG, Morato RG, May JA, Pacheco TD, de Aguiar DM (2014) Exposure of free-living jaguars to *Toxoplasma gondii*, *Neospora caninum* and *Sarcocystis neurona* in the Brazilian Pantanal. Rev Bras Parasitol Vet 23:547–553

Onuma SS, Kantek DL, Crawshaw Junior PG, Morato RG, May-Junior JA, Morais ZM, Ferreira Neto JS, Aguiar DM (2015) Detection of *Leptospira* spp. and *Brucella abortus* antibodies in free-living jaguars (*Panthera onca*) in two protected areas of northern Pantanal, Brazil. Rev Inst Med Trop Sao Paulo 57:177–180

Orozco MM, Ceballos LA, Pino MD, Gurtler RE (2014) Local threats and potential infectious hazards to maned wolves (*Chrysocyon brachyurus*) in the southeastern Argentine Chaco. Mammalia 78:339–349

Ortega R, Mena J, Grecco S, Perez R, Panzera Y, Napolitano C, Zegpi NA, Sandoval A, Sandoval D, Gonzalez-Acuna D, Cofre S, Neira V, Castillo-Aliaga C (2021) Domestic dog origin of carnivore protoparvovirus infection in a rescued free-ranging guina (*Leopardus guigna*) in Chile. Transbound Emerg Dis 68:1062–1068

Parola P, Paddock CD, Socolovschi C, Labruna MB, Mediannikov O, Kernif T, Abdad MY, Stenos J, Bitam I, Fournier PE, Raoult D (2013) Update on tick-borne rickettsioses around the world: a geographic approach. Clin Microbiol Rev 26:657–702

Patz JA, Daszak P, Tabor GM, Aguirre AA, Pearl M, Epstein J, Wolfe ND, Kilpatrick AM, Foufopoulos J, Molyneux D, Bradley DJ (2004) Unhealthy landscapes: policy recommendations on land use change and infectious disease emergence. Environ Health Perspect 112:1092–1098

Paviolo A, De Angelo C, Ferraz KMPMB, Morato RG, Pardo JM, Srbek-Araujo AC, Beisiegel BD, Lima F, Sana D, da Silva MX, Velazquez MC, Cullen L, Crawshaw P, Jorge MLSP, Galetti PM, Di Bitetti MS, de Paula RC, Eizirik E, Aide TM, Cruz P, Perilli MLL, Souza ASMC, Quiroga V, Nakano E, Pinto FR, Fernandez S, Costa S, Moraes EA, Azevedo F (2016) A biodiversity hotspot losing its top predator: the challenge of jaguar conservation in the Atlantic Forest of South America. Sci Rep 6:37147

Pedersen AB, Jones KE, Nunn CL, Altizer S (2007) Infectious diseases and extinction risk in wild mammals. Conserv Biol 21:1269–1279

Peixoto PV, Soares CO, Scofield A, Santiago CD, Franca TN, Barros SS (2007) Fatal cytauxzoonosis in captive-reared lions in Brazil. Vet Parasitol 145:383–387

Pence DB, Ueckermann E (2002) Sarcoptic mange in wildlife. Rev Sci Tech Oie 21:385–398

Porfirio GEO, Santos FM, de Macedo GC, Barreto WTG, Campos JBV, Meyers AC, Andre MR, Perles L, de Oliveira CE, Xavier S, Andrade GB, Jansen AM, Herrera HM (2018) Maintenance of *Trypanosoma cruzi*, *T. evansi* and *Leishmania* spp. by domestic dogs and wild mammals in a rural settlement in Brazil-Bolivian border. International journal for parasitology. Parasites Wildl 7:398–404

Pysek P, Richardson DM (2010) Invasive species, environmental change and management, and health. Annu Rev Environ Resour 35:25–55

Quesnelle PE, Lindsay KE, Fahrig L (2014) Low reproductive rate predicts species sensitivity to habitat loss: a meta-analysis of wetland vertebrates. PLoS One 9:e90926

Ramirez-Pizarro F, Silva-de la Fuente C, Hernandez-Orellana C, Lopez J, Madrid V, Fernandez I, Martin N, Gonzalez-Acuna D, Sandoval D, Ortega R, Landaeta-Aqueveque C (2019) Zoonotic pathogens in the American mink in its southernmost distribution. Vector-Borne Zoonotic Dis 19:908–914.

Rendon-Franco E, Caso-Aguilar A, Jimenez-Sanchez NG, Hernandez-Jauregui DMB, Sandoval-Sanchez AL, Zepeda-Lopez HM (2012) Prevalence of anti-*Toxoplasma gondii* antibody in free-ranging ocelots (*Leopardus pardalis*) from Tamaulipas, Mexico. J Wildl Dis 48:829–831

Richini-Pereira VB, Marson PM, Hayasaka EY, Victoria C, da Silva RC, Langoni H (2014) Molecular detection of *Leishmania* spp. in road-killed wild mammals in the Central Western area of the State of Sao Paulo, Brazil. J Venom Anim Toxins Incl Trop Dis 20:27

Rodrigues TCS, Santos ALQ, Lima-Ribeiro AMC, Lemos FG, Azevedo FC, Arrais RC, Gomes DO, Tavares TCF (2015) Occurrence of antibodies against *Leptospira* spp. in free-ranging wild canids from the Brazilian savanna. Pesqui Vet Bras 35:734–740

Roelke-Parker ME, Munson L, Packer C, Kock R, Cleaveland S, Carpenter M, Obrien SJ, Pospischil A, HofmannLehmann R, Lutz H, Mwamengele GLM, Mgasa MN, Machange GA, Summers BA, Appel MJG (1996) A canine distemper virus epidemic in Serengeti lions (*Panthera leo*). Nature 379:441–445

Roque AL, Jansen AM (2014) Wild and synanthropic reservoirs of *Leishmania* species in the Americas. Int J Parasitol Parasites Wildl 3:251–262

Sacristan I, Acuna F, Aguilar E, Garcia S, Lopez MJ, Cabello J, Hidalgo-Hermoso E, Sanderson J, Terio KA, Barrs V, Beatty J, Johnson WE, Millan J, Poulin E, Napolitano C (2021a) Cross-species transmission of retroviruses among domestic and wild felids in human-occupied landscapes in Chile. Evol Appl 14:1070–1082

Sacristan I, Esperon F, Perez R, Acuna F, Aguilar E, Garcia S, Lopez MJ, Neves E, Cabello J, Hidalgo-Hermoso E, Terio KA, Millan J, Poulin E, Napolitano C (2021b) Epidemiology and molecular characterization of Carnivore protoparvovirus-1 infection in the wild felid *Leopardus guigna* in Chile. Transbound Emerg Dis 68:3335–3348

Sala OE, Chapin FS 3rd, Armesto JJ, Berlow E, Bloomfield J, Dirzo R, Huber-Sanwald E, Huenneke LF, Jackson RB, Kinzig A, Leemans R, Lodge DM, Mooney HA, Oesterheld M, Poff NL, Sykes MT, Walker BH, Walker M, Wall DH (2000) Global biodiversity scenarios for the year 2100. Science 287:1770–1774

Salkeld DJ, Stapp P (2006) Seroprevalence rates and transmission of plague (*Yersinia pestis*) in mammalian carnivores. Vector Borne Zoonotic Dis 6:231–239

Schnittger L, Ganzinelli S, Bhoora R, Omondi D, Nijhof AM, Florin-Christensen M (2022) The Piroplasmida Babesia, Cytauxzoon, and Theileria in farm and companion animals: species compilation, molecular phylogeny, and evolutionary insights. Parasitol Res 121:1207–1245

Sepulveda MA, Munoz-Zanzi C, Rosenfeld C, Jara R, Pelican KM, Hill D (2011) *Toxoplasma gondii* in feral American minks at the Maullin river, Chile. Vet Parasitol 175:60–65

Sepulveda MA, Singer RS, Silva-Rodriguez EA, Eguren A, Stowhas P, Pelican K (2014) Invasive American mink: linking pathogen risk between domestic and endangered carnivores. EcoHealth 11:409–419

Sepulveda-Garcia P, Raffo E, Medina-Vogel G, Munoz F, Munoz P, Alabi A, Navarrete-Talloni MJ, Goncalves LR, Califre de Mello VV, Machado RZ, Andre MR, Bittencourt P, Muller A (2021) Molecular survey of *Bartonella* spp. and hemoplasmas in American minks (*Neovison vison*). Transbound Emerg Dis 68:2094–2110

Shiokawa K, Llanes A, Hindoyan A, Cruz-Martinez L, Welcome S, Rajeev S (2019) Peridomestic small Indian mongoose: an invasive species posing as potential zoonotic risk for leptospirosis in the Caribbean. Acta Trop 190:166–170

Shpynov SN, Pozdnichenko NN, Gumenyuk AS, Skiba AA (2018) Genomosystematics of Rickettsiae. Infektsiya Immun 8:107–118

Sieg M, Sacristan I, Busch J, Terio KA, Cabello J, Hidalgo-Hermoso E, Millan J, Bottcher D, Heenemann K, Vahlenkamp TW, Napolitano C (2020) Identification of novel feline paramyxoviruses in guignas (*Leopardus guigna*) from Chile. Viruses-Basel 12:1397

Sillero-Zubiri C, King AA, Macdonald DW (1996) Rabies and mortality in Ethiopian wolves (*Canis simensis*). J Wildl Dis 32:80–86

Smith GC, Fooks AR (2006) Wildlife rabies control policy in Great Britain. Dev Biol (Basel) 125:113–118

Smith KF, Acevedo-Whitehouse K, Pedersen AB (2009) The role of infectious diseases in biological conservation. Anim Conserv 12:1–12

Soares JF, Dall'Agnol B, Costa FB, Krawczak FS, Comerlato AT, Rossato BC, Linck CM, Sigahi EK, Teixeira RH, Sonne L, Hagiwara MK, Gregori F, Vieira MI, Martins JR, Reck J, Labruna MB (2014) Natural infection of the wild canid, *Cerdocyon thous*, with the piroplasmid *Rangelia vitalii* in Brazil. Vet Parasitol 202:156–163

Souza TD, Turchetti AP, Fujiwara RT, Paixao TA, Santos RL (2014) Visceral leishmaniasis in zoo and wildlife. Vet Parasitol 200:233–241

Souza UA, Webster A, Dall'Agnol B, Peters FB, Favarini MO, Schott D, Zitelli LC, Mazim FD, Kasper CB, Ott R, Trigo TC, Reck J, Soares JF (2021) Ticks, mites, fleas, and vector-borne pathogens in free-ranging neotropical wild felids from southern Brazil. Ticks Tick Borne Dis 12:101706

Strauss A, White A, Boots M (2012) Invading with biological weapons: the importance of disease-mediated invasions. Funct Ecol 26:1249–1261

Stuen S, Granquist EG, Silaghi C (2013) *Anaplasma phagocytophilum*-a widespread multihost pathogen with highly adaptive strategies. Front Cell Infect Microbiol 3:31

Suzan G, Garcia-Pena GE, Castro-Arellano I, Rico O, Rubio AV, Tolsa MJ, Roche B, Hosseini PR, Rizzoli A, Murray KA, Zambrana-Torrelio C, Vittecoq M, Bailly X, Aguirre AA, Daszak P, Prieur-Richard AH, Mills JN, Guegan JF (2015) Metacommunity and phylogenetic structure determine wildlife and zoonotic infectious disease patterns in time and space. Ecol Evol 5:865–873

Swei A, Couper LI, Coffey LL, Kapan D, Bennett S (2020) Patterns, drivers, and challenges of vector-borne disease emergence. Vector Borne Zoonotic Dis 20:159–170

Thompson CS, Mangold AJ, Felix ML, Carvalho L, Armua-Fernandez MT, Venzal JM (2018) Molecular evidence of *Babesia* species in *Procyon cancrivorus* (Carnivora, Procyonidae) in Uruguay. Vet Parasitol Reg St 13:230–233

Tuomainen U, Candolin U (2011) Behavioral responses to human-induced environmental change. Biol Rev Camb Philos Soc 86:640–657

Uhart MM, Rago MV, Marull CA, Ferreyra HD, Pereira JA (2012) Exposure to selected pathogens in Geoffroy's cats and domestic carnivores from central Argentina. J Wildl Dis 48:899–909

Viana M, Mancy R, Biek R, Cleaveland S, Cross PC, Lloyd-Smith JO, Haydon DT (2014) Assembling evidence for identifying reservoirs of infection. Trends Ecol Evol 29:270–279

Viana M, Cleaveland S, Matthiopoulos J, Halliday J, Packer C, Craft ME, Hampson K, Czupryna A, Dobson AP, Dubovi EJ, Ernest E, Fyumagwa R, Hoare R, Hopcraft JG, Horton DL, Kaare MT, Kanellos T, Lankester F, Mentzel C, Mlengeya T, Mzimbiri I, Takahashi E, Willett B, Haydon DT, Lembo T (2015) Dynamics of a morbillivirus at the domestic-wildlife interface: canine distemper virus in domestic dogs and lions. Proc Natl Acad Sci U S A 112:1464–1469

Vitousek PM, DAntonio CM, Loope LL, Westbrooks R (1996) Biological invasions as global environmental change. Am Sci 84:468–478

Walker D, Fee SA, Hartley G, Learmount J, O'Hagan MJ, Meredith AL, de C Bronsvoort BM, Porphyre T, Sharp CP, Philbey AW (2016) Serological and molecular epidemiology of canine adenovirus type 1 in red foxes (*Vulpes vulpes*) in the United Kingdom. Sci Rep 6:36051

Williams ES, Thorne ET, Appel MJ, Belitsky DW (1988) Canine distemper in black-footed ferrets (*Mustela nigripes*) from Wyoming. J Wildl Dis 24:385–398

Woodroffe R (1997) The conservation implications of immobilizing, radio-collaring and vaccinating free-ranging wild dogs. In: Woodroffe R, Ginsberg JR, Macdonald D (eds) African wild dog: status survey and conservation action plan. IUCN/SSC Canid Specialist Group, World Conservation Union, Gland, pp 124–138

Zipkin EF, DiRenzo GV, Ray JM, Rossman S, Lips KR (2020) Tropical snake diversity collapses after widespread amphibian loss. Science 367:814–816

Chapter 11
Disease Ecology in Wild South American Camelids: Conservation Implications of a Long Cohabitation History with Exotic Ungulates

Pablo Moreno, M. Uhart, Maria Mercedes Cafrune, H. Ferreyra, F. Beltrán-Saavedra, M. V. Rago, G. Marcoppido, and P. Beldomenico

11.1 Introduction

South American camelids (SAC) comprise two wild species (WSAC), the guanaco (*Lama guanicoe* Müller 1776) and the vicuña (*Vicugna vicugna* Molina 1782), adapted to arid landscapes of the neotropical region (Fig. 11.1), and two domestic species (DSAC), llama (*Lama glama* Linnaeus 1758) and alpaca (*Lama pacos* Linneaus 1758), originating from the domestication from the guanaco and the vicuña, respectively, achieved by pre-Inca civilizations (Kadwell et al. 2001).

Guanacos have a wide distribution that extends from southern Chile and Argentina (where the greatest abundance of the species is found) to Perú in the north; in Bolivia and Paraguay there are small remnant populations (González and Acebes 2016; Baldi et al. 2016). Vicuñas are restricted to high-altitude environments distributed along the Andes mountain range, between 3000 and 5000 m above sea level in Peru, Bolivia, Chile, and Argentina (Acebes et al. 2018). Vicuñas are sedentary and have an activity pattern conditioned by their need to drink water daily

P. Moreno (✉)
Instituto Argentino de Investigaciones de Zonas Áridas (IADIZA), Consejo Nacional de Investigaciones Científicas y Técnicas, Mendoza, Argentina

Facultad de Ciencias Veterinarias y Ambientales, Universidad Maza, Mendoza, Argentina
e-mail: pmoreno@mendoza-conicet.gob.ar

M. Uhart
Karen C. Drayer Wildlife Health Center, School of Veterinary Medicine, University of California, Davis, CA, USA

M. M. Cafrune
Instituto de Investigación Animal del Chaco Semiárido, Área de Investigación en Salud Animal, Estación Experimental Agropecuaria Salta, Instituto Nacional de Tecnología Agropecuaria (INTA), Ministerio de Agricultura, Ganadería y Pesca (CIAP), Salta, Argentina

G. Acosta-Jamett, A. Chaves (eds.), *Ecology of Wildlife Diseases in the Neotropics*, https://doi.org/10.1007/978-3-031-50531-7_11

Fig. 11.1 Left: A female vicuña and 2 offspring at the Municipal Protected Area Pampa Tholar de las Vicuñas, in Villazon, Potosí, Bolivia (L. Fabián Beltrán ©Wildlife Conservation Society). Right: Group of guanacos feeding at La Payunia Provincial Reserve, Mendoza, Argentina (Pablo Moreno)

(Franklin 1982; Acebes et al. 2018). Guanacos are periodic drinkers and can survive without drinking water every day. Depending on the environmental conditions, guanaco populations can be sedentary or undergo seasonal migrations (Franklin 1982). Guanacos and vicuñas are gregarious species. The social organization of both species includes polygynous family groups (a male and females with young), groups of males, and solitary individuals; in the case of guanacos, there are also mixed groups (comprising individuals from the groups described above) (Franklin 1983). Male guanacos from family groups and territorial solitaries demarcate their territory by defecating in dung piles. For vicuñas, dung piles are communal and are used by

H. Ferreyra
Universidad Nacional de Villa María, Córdoba, Argentina

Administración de Parques Nacionales, Buenos Aires, Argentina

F. Beltrán-Saavedra
Wildlife Conservation Society, La Paz, Bolivia

Wildlife Conservation Society, Wildlife Health Program, New York, NY, USA

M. V. Rago
Instituto de Investigaciones en Biodiversidad y Medioambiente (INIBIOMA), Consejo Nacional de Investigaciones Cientícas y Técnicas, Universidad Nacional del Comahue, Neuquen, Argentina

G. Marcoppido
Instituto de Patobiologia Veterinaria IPVET UEDD INTA-CONICET-Instituto Nacional de Tecnología Agropecuaria, Castelar, Argentina

P. Beldomenico
Laboratorio de Ecología de Enfermedades, Instituto de Ciencias Veterinarias del Litoral (ICIVET LITORAL), Universidad Nacional del Litoral – Consejo Nacional de Investigaciones Científicas y Técnicas, Esperanza, Santa Fe, Argentina

Facultad de Ciencias Veterinarias, Universidad Nacional del Litoral, Esperanza, Santa Fe, Argentina individuals of both sexes from family groups (Franklin 1983). Both species groom themselves by performing dust baths in sites shared by several individuals (Franklin 1983).

The geographical distribution of WSAC has decreased since the colonization of America due to competition imposed by introduced domestic cattle and hunting pressure for their fine fibers. Currently, their distributions are fragmented into relatively isolated populations and concentrated in protected areas and private ranches (Baldi et al. 2016; Acebes et al. 2018). In the latest continental population estimate of guanacos (2016), a total population of 1,498,170 to 2,192,300 individuals was estimated, with a growing population trend, so the IUCN Red List classifies them as Least Concern globally (Baldi et al. 2016). However, from the analysis of the total distribution of guanacos, it seems that the populations of Perú, Bolivia, Paraguay, northern Chile, and northern Argentina are small and declining or slightly stable (Baldi et al. 2016). Meanwhile, the total population of vicuñas has been estimated at 473,297 to 527,691 individuals, with a growing population trend, and the last evaluation (2018) of the IUCN Red List categorized it as Least Concern (Acebes et al. 2018; Acebes and González 2021).

Both species of WSAC have been used by humans since ancient times (Wheeler 2012; Lichtenstein and Carmanchahi 2012). Their fiber is extremely fine and has a high market value, so some populations of both species are managed for live shearing in the wild. This sustainable management, used as a conservation strategy and for the benefit of indigenous and rural inhabitants, consists of the capture, shearing, and release of free-living individuals (Lichtenstein and Carmanchahi 2012). In addition, in some regions of Chile and Argentina, legislation allows the controlled harvest of guanacos (Valdebenito Díaz 2008; Plan de Manejo del Guanaco de la Provincia de Santa Cruz, Decreto Provincial 0032/15).

Leguía (1999), Wernery and Kaaden (2002), Aguirre and Cafrune (2007), and Fowler (2011) constitute the reference bibliography on health aspects and diseases in SAC, describing the etiology, morphology, and parasitic life cycles in these animals. It should be noted that the information available has been generated mainly in productive systems of DSAC (llamas and alpacas) through case reports of WSAC in captivity, and sometimes the epidemiological and pathophysiological knowledge of domestic ruminant diseases is extrapolated, assuming that pathogens behave similarly in SAC species.

WSAC harbor a variety of species-specific parasites; for example, the nematodes *Trichuris tenuis, Graphinema aucheniae, Spiculopteragia peruviana, Nematodirus lamae, Camelostrongylus mentulatus,* and *Lamanema chavezi*, as well as protozoa of the genera *Eimeria* and *Sarcocystis* have been identified (Leguía 1999). However, some pathogens of domestic livestock (cattle, sheep, and goats) can also infect neotropical camelids.

Knowledge about the distribution of pathogens, determining factors and infection dynamics, as well as their impact on WSAC populations, is very scarce (Ferreyra et al. 2022; Rago et al. 2022). Table 11.1 presents information on the literature on health studies carried out on populations of *V. vicugna* and *L. guanicoe* in the

Table 11.1 Summary of health studies in WSAC populations from Perú, Bolivia, Chile, and Argentina

Host population	Pathogens	Diagnostic techniques	Type of study
V(w), *n* = 101 Puno, Perú	*Toxoplasma gondii*	Indirect hemoagglutination test (IHA)	Serosurvey, descriptive, seroprevalence Pastor et al. (2003)
V(c), *n* = 200 Central and South Sierra, Perú	*Toxoplasma gondii*	Indirect fluorescent antibody test (IFAT) and western blot	Serosurvey, descriptive, seroprevalence, antibody titers Chávez-Velázquez et al. (2005)
V(w), *n* = 114 Puno, Perú	*Toxoplasma gondii, Neospora caninum*	Immunoblot, enzyme-linked immunosorbent assay (ELISA) and IFAT	Serosurvey, descriptive, seroprevalence Wolf et al. (2005)
V(w), *n* = 191 RNPGBD, Ayacucho, Perú	*Toxoplasma gondii*	IFAT	Serosurvey, analytical, seroprevalence Zuzunaga et al. (2006)
G(w), *n* = 132 (nine populations) Ayacucho, La Libertad, Ica, Arequipa, Tacna and Moquegua, Perú	*Graphinema aucheniae, Bunostomum sp., Ostertagia sp., Trichuris sp., Cooperia sp., Nematodirus sp., Mazamastrongylus peruvianus, Trichostrongylus sp., Eimeria lamae, E. alpacae, E. punoensis, E. macusaniensis*	Coproparasitology	Survey, descriptive, prevalence Castillo et al. (2008)
V(w), *n* = 120 Tacna, Perú	*Trichuris sp., Strongylus sp., Eimeria sp., Nematodirus sp., Capillaria sp., Amblyomma parvitarsum, Microthoracius praelongyceps*	Coproparasitology, ectoparasite identification	Survey, descriptive, prevalence Quispe García (2011)
V(w), *n* = 200 RNPGBD, Ayacucho, Perú	*Sarcoptes scabiei*	Clinical examination, mite identification	Survey, descriptive, prevalence, clinical presentation Gómez-Puerta et al. (2013)
V(w), *n* = 79 SAIS Túpac Amaru, Junin, Perú	*Neospora caninum, Toxoplasma gondii*	IFA and ELISA	Serosurvey, descriptive, seroprevalence Pinedo et al. (2014)
V(w), *n* = 81 V(c), *n* = 126 (nine populations) SABR, Arequipa, Perú	*Neospora caninum, Leptospira borgpetersenii* serovars Australis, Ballum, Bataviae, Djasiman, Hardjo, Javanica, Tarassovi, *Leptospira interrogans* serovars Autumnalis, Bratislava, Canicola, Copenhageni, Pomona, *Leptospira kirschneri* serovars Cynopteri, Grippotyphosa, *Leptospira noguchi serovars* Louisiana, Panama; *Leptospira weilii* serovar Sarmin, foot-and-mouth disease, bovine viral diarrhea virus, bovine herpesvirus type 1, bluetongue disease, *Brucella abortus, Brucella ovis, Brucella melitensis, Mycobacterium avium* subsp. *paratuberculosis*	ELISA: BVD, BHV-1, FMD, BTV, paratuberculosis, *Brucella abortus, B. ovis,* *B. melitensis, and Neospora caninum;* Microscopic agglutination test (MAT): *Leptospira* sp.; rose bengal test and Western blotting: *Brucella* spp. and *Neospora* sp.	Serosurvey, descriptive, seroprevalence Risco-Castillo et al. (2014)
V(w), *n* = 143 Yauli, Junin, Perú	*Fasciola hepatica*	Coproparasitology	Survey, descriptive, prevalence and intensity Samamé et al. (2016)
V(w), *n* = 25, 296 Ayacucho, Perú.	*Sarcoptes scabiei*	Clinical examination, mite identification	Survey, descriptive, prevalence, clinical presentation Bujaico Mauricio (2018)

(continued)

Table 11.1 (continued)

V(w), *n* = 208 Cajamarca, Perú	*Strongylida* suborder, *Nematodirus sp.*, *Trichuris sp.*, *Capillaria sp.*, *Moniezia sp.*, *Cooperia spp.*, *Trichostrongylus spp.*, *Ostertagia spp.*, *Oesophagostomum spp.*, *Haemonchus spp*, *Bunostomum spp.*	Coproparasitology, larval identification	Survey, descriptive, prevalence and intensity Curay Cabanillas (2018)
V(w), *n* = 733 (nine populations) Apurimac, Perú.	*Sarcoptes scabiei*	Clinical examination, mite identification	Survey, analytical, prevalence, clinical presentation Unzueta Lancho (2018)
V(w), *n* = 19, 066 RNPGBD, Ayacucho, Perú	*Sarcoptes scabiei*	Clinical examination, mite identification	Survey, descriptive, prevalence, clinical presentation Mayhua Mendoza (2021)
V(w), *n* = 777 V(c), *n* = 2049 (18 populations) Cusco, Perú	*Fasciola hepatica*, *Strongylida* suborder, *Nematodirus sp.*, *N. spathiger*, *Trichuris sp.*, *Eimeria spp.*, *Moniezia spp.*, *Sarcoptes scabiei.*	Coproparasitology, clinical examination, mite identification	Survey, descriptive, prevalence Angulo-Tisoc et al. (2021)
V(w), *n* = 36 ANMI Apolobamba, La Paz, Bolivia	*Eimeria punoensis*, *E. alpacae*, *Marshallagia sp.*, *Lamanema spp.*, *Nematodirus spp.*, *Strongylida* suborder, *Trichuris sp.*, *Capillaria sp.*, *Moniezia benedeni*, *Microthoracius mazzai*, *M. minor*, *Amblyomma parvitarsum*, *Sarcoptes scabiei var. aucheniae*. *antibodies against Brucella*, *antibodies against foot-and-mouth disease*	Coproparasitology, ectoparasite identification *Brucella*: buffered plate antigen test (BPA) and competitive ELISA (C-ELISA); foot -and-mouth disease antibodies: VIA agar gel immunodiffusion (IDGA-VIA)	Survey, serosurvey, analytical, prevalence and seroprevalence Beltrán-Saavedra et al. (2011)
V(w), *n* = 62 ANMI Apolobamba, La Paz, Bolivia	*Eimeria punoensis*, *E. alpacae*, *E. lamae*, *E. macusaniensis*, *Marshallagia spp.*, *Nematodirus spp.*, *Trichuris spp.*, *Capillaria spp.*, *L. chavezi*, *M. benedeni*, *M. expanza*, *Cooperia oncophora*, *Cooperia mcmasteri*, *Oesophagostomum columbianum*, *Ostertagia circumcincta*, *Trichostrongylus colubriformis*, *Trichostrongylus axei*, *Mazamastrongylus peruvianus*	Coproparasitology	Survey, descriptive, prevalence Condori (2012)
V(w), *n* = 98 (three populations) Potosí and Cochabamba, Bolivia	*E. punoensis*, *E. alpacae*, *E. peruviana*, *E. Lamae*, *Trichuris spp.*, *Marshallagia spp.*, *L. chavezi*, *Strongylida suborder*, *Capillaria sp.*, *Fasciola hepática*, *Moniezia benedeni*	Coproparasitology	Survey, analytical, prevalence and intensity Martela Mamani (2016)
V(w), *n* = 84 (seven populations) La Paz and Oruro, Bolivia	*Marshallagia spp.*, *Lamanema spp.* *Strongylida* suborder, *Nematodirus spp.*, *Trichuris spp.*, *Capillaria spp.*, *Moniezia benedeni*, *E. punoensis*, *E. alpacae*, *Microthoracius spp.*, *S. scabei var. aucheniae*, *A. parvitarsum.*	Coproparasitology, ectoparasite identification	Survey, analytical, prevalence and intensity Ruiz-Hurtado (2016)
V(w), *n* = 86 (feces); *n* = 92 (search of ectoparasites) (5 populations) ANMI Apolobamba, La Paz, Bolivia	*Eimeria punoensis*, *E. alpacae*, *E. lamae*, *E. macusaniensis*, *Trichuris spp.*, *Capillaria spp.*, *Strongyloides spp.*, *Nematodirus cf. battus*, *Nematodirus cf. spathiger*, *Lamanema sp.*, *Strongylida suborder*, *Marshallagia spp.*,	Coproparasitology, ectoparasite identification	Survey, descriptive, prevalence Beltrán-Saavedra and Mollericona (2019a)

(continued)

Table 11.1 (continued)

	Moniezia expansa, *M. benedeni*, *Sarcoptes scabiei*, *Microthoracius mazzai*, *M. praelongiceps*, *M. minor*, *Amblyomma parvitarsum*		
V(w), *n* = 68 (feces); *n* = 78 (search of ectoparasites) (three populations) ANMI Pampa Tholar de las Vicuñas, Villazón, Potosí, Bolivia	*Eimeria punoensis*, *E. alpacae*, *E. lamae*, *E. macusaniensis*, *Trichuris* spp., *Capillaria* spp., *Strongyloides* spp., *Nematodirus* spp., *Strongylida* suborder, *Marshallagia* spp., *Moniezia benedeni*, *Sarcoptes scabiei*, *Microthoracius mazzai*, *M. praelongiceps*	Coproparasitology, ectoparasite identification	Survey, descriptive, prevalence Beltrán-Saavedra & Mollericona (2019b)
V(w), *n* = 350 (feces); *n* = 331 (search of ectoparasites) (13 populations for endoparasite prevalence; 12 populations for ectoparasite prevalence) ANMI Apolobamba, La Paz, Bolivia	*Eimeria punoensis*, *E. alpacae*, *E. lamae*, *E. macusaniensis*, *Trichuris* spp., *Capillaria* spp., *Strongyloides* spp., *Nematodirus* cf. *battus*, *Nematodirus* cf. *spathiger*, *Lamanema* sp., *Strongylida* suborder, *Marshallagia* spp., *Moniezia expansa*, *M. benedeni*, *Sarcoptes scabiei*, *Microthoracius mazzai*, *M. praelongiceps*, *Amblyomma parvitarsum*	Coproparasitology, ectoparasit identification	Survey, descriptive, prevalence and intensity Beltrán-Saavedra & Mollericona (2020)
V(w), *n* = 139 (feces); *n* = 165 (search of ectoparasites) (ten populations) Colcha K, Potosí, Bolivia	*Eimeria punoensis*, *E. alpacae*, *E. lamae*, *E. macusaniensis*, *Trichuris* spp., *Capillaria* spp., *Nematodirus* cf. *battus*, *Marshallagia* spp., *Strongylida* suborder, *Sarcoptes scabiei*, *Microthoracius praelongiceps*, *Amblyomma parvitarsum*	Coproparasitology, ectoparasite identification	Survey, descriptive, prevalence Mollericona et al. (2020)
V(w), *n* = 21 (feces); *n* = 25 (search of ectoparasites) Wila Kollo, Oruro, Bolivia	*Eimeria punoensis*, *Trichuris* spp., *Nematodirus* cf. *spathiger*, *Marshallagia* spp., *Sarcoptes scabiei*, *Microthoracius praelongiceps*	Coproparasitology, ectoparasite identification	Survey, descriptive, prevalence Mollericona et al. (2021a)
V(w), *n* = 103 (two populations) Tomave, Potosí, Bolivia	*Eimeria punoensis*, *E. alpacae*, *E. lamae*, *E. macusaniensis*, *Trichuris* spp., *Capillaria* spp., *Strongyloides* spp., *Nematodirus* cf. *battus*, *Nematodirus* cf. *spathiger*, *Lamanema* sp., *Strongylida* suborder, *Sarcoptes scabiei*, *Microthoracius praelongiceps*, *Amblyomma parvitarsum*	Coproparasitological diagnostic, ectoparasite identification	Survey, descriptive, prevalence Mollericona et al. (2021b)
V(w), *n* = 376 (feces); *n* = 396 (search for ectoparasites) (14 populations) ANMI Apolobamba, La Paz, Bolivia	*Eimeria punoensis*, *E. alpacae*, *E. lamae*, *E. macusaniensis*, *Trichuris* spp., *Capillaria* spp., *Strongyloides* spp., *Nematodirus* cf. *battus*, *Nematodirus* cf. *spathiger*, *Lamanema* sp., *Strongylida* suborder, *Marshallagia* spp., *Moniezia expansa*, *M. benedeni*, *Sarcoptes scabiei*, *Microthoracius praelongiceps*, *M. minor*, *Amblyomma parvitarsum*	Coproparasitology, ectoparasite identification	Survey, descriptive, prevalence Mollericona (2022)
G(w) *n* = 253 Timaukel, Tierra del Fuego, Chile	*Echinococcus granulosus*	Macroscopic observation at necropsies	Survey, descriptive, prevalence Cunazza et al. (1981)
G(w), *n* = 371	*Sarcoptes scabiei*	Clinical examination, histopathology, mite	Survey, analytical, prevalence,

(continued)

Table 11.1 (continued)

Tierra del Fuego, Chile		identification	clinical presentation Alvarado (2004)
G(c), *n* = 15 Tierra del Fuego, Chile	*Nematodirus sp., Strongylida* suborder, *E. macusaniensis, Mycoplasma haemolamae*	Coproparasitology, blood smear analysis	Survey, descriptive prevalence Correa et al. (2011)
G(w) *n* = 501 Timaukel, Tierra del Fuego, Chile	*Mycobacterium avium* subsp. *paratuberculosis*	Fecal culture for real-time polymerase chain reaction (qPCR) and PCR typing	Survey, analytical, prevalence Salgado et al. (2009)
G(c) *n* = 89: Metropolitan, Magallanes and Antarctic regions, Chile; V(c) *n* = 106, Arica and Parinacota, Chile	Parainfluenza-3 virus	Hemagglutination inhibition test	Serosurvey, prevalence Cepeda et al. (2011)
G(w) *n* = 243 Patagonia, Chile	*Mycobacterium avium* subsp. *paratuberculosis*	Fecal culture for real -time polymerase chain reaction (qPCR)	Survey, descriptive, prevalence Collado et al. (2019)
G(w) *n* = 3715 Timaukel, Tierra del Fuego, Chile	*Echinococcus granulosus, Sarcocystis sp.*	Macroscopic observation at necropsies	Survey, descriptive prevalence Valdebenito (2008)
G(w) *n* = 322 (4 populations) Private ranches, PNPA and PNTP, Magallanes, Chile	*Mycobacterium avium* subsp. *paratuberculosis*	Fecal culture for real-time polymerase chain reaction (qPCR)	Longitudinal (seasonal), descriptive, prevalence Corti et al. (2022)
G(w), *n* = 3 Río Negro, Argentina	*Nematodirus spathiger, N. lanceolatus, N. filicolis, N. battus, Ostertagia ostertagi, Cooperia oncophora, C. mcmasteri, Trichostrongylus axei, T. vitrinus, Capillaria sp., Trichuris ovis, Skrjabinema ovis*	Adult helminth identification, coproparasitology	Survey, descriptive, intensity. (Larrieu et al. 1982)
G(w), *n* = 58 (4 populations) PM, Tierra del Fuego, Argentina.	*Eimeria spp., Nematodirus sp. Haemonchus sp., Ostertagia sp., Trichostrongylus sp, Cooperia sp., Marshallagia sp., Oesophagostomum sp., Chabertia sp.*	Coproparasitology	Survey, descriptive, prevalence (Navone & Merino 1989)
V(c), *n* = 187 AP, Jujuy, Argentina	*Fasciola hepatica*	Coproparasitology	Longitudinal (seasonal), descriptive, prevalence Cafrune et al. (1996)
G(w), *n* = 20 C2B, Chubut, Argentina	*Strongyloides sp., Nematodirus sp., Marshallagia sp., Trichostrongylus sp., Trichuris sp., Dictyocaulus sp.,* bluetongue virus, bovine viral diarrhea virus, bovine respiratory syncitial virus, equine herpesvirus 1, foot-and-mouth disease virus, infectious bovine rhinotracheitis virus, parainfluenza-3 virus, vesicular stomatitis virus, *Mycobacterium paratuberculosis, Leptospira interrogans, Brucella melitensis*	Coproparasitology, serum neutralization (SN): bovine viral diarrhea virus, bovine respiratory syncitial virus, equine herpesvirus 1, infectious bovine rhinotracheitis virus, vesicular stomatitis virus and parainfluenza-3 virus; C-ELISA and IDGA: bluetongue; IDGA: foot-and-mouth disease virus and *Mycobacterium paratuberculosis;* MAT: *Leptospira interrogans*	Survey, serosurvey, descriptive, prevalence, seroprevalence and intensity Karesh et al. (1998)
V(c), *n* = 69 AP, Jujuy, Argentina	*Trichuris tenuis*	Coproparasitology	Survey, descriptive, prevalence Cafrune et al. (1999)
G(c), *n* = 22	Rotavirus, coronavirus	Fecal and serum samples	Descriptive survey

(continued)

Table 11.1 (continued)

Río Negro, Argentina G(c), *n* = 30 Chubut, Argentina		analyzed by ELISA	Parreño et al. (2001)
G(c), *n* = 78 (three populations) Río Negro, Argentina. V (c), *n* = 73 (three populations) Catamarca and Salta, Argentina.	*Leptospira* serovars copenhageni (serogroup Icterohaemorrhagiae), castellonis (serogroup Ballum), canicola (serogroup Canicola)	Microscopic agglutination test (MAT)	Serosurvey, descriptive Llorente et al. (2002)
G(w), *n* = 12 C2B, Chubut, Argentina	*Nematodirus sp., Marshallagia sp., Eimeria spp., Trichuris tenuis, Moniezia expansa Sarcocystis sp., Dictyocaulus filaria*	Coproparasitology, adult helminth identification, histopathology	Case report, descriptive, pathology, prevalence and intensity Beldomenico et al. (2003)
V(w), *n* = 221 RLB, Catamarca, Argentina	*Trichuris sp., Capillaria sp., Eimeria spp., E. macusaniensis*	Coproparasitology	Survey, descriptive, prevalence and intensity. Cafrune et al. (2006)
G(w), *n* = 114 RPLP, Mendoza, Argentina.	*Eimeria spp., E. macusaniensis, Nematodirus sp., Trichuris sp., Capillaria sp., Cooperia sp.*	Coproparasitology	Longitudinal (seasonal), descriptive, prevalence Rago et al. (2008)
V(c), *n* = 161, G(w), *n* = 4 (three populations) Jujuy, Salta, Argentina	*Eimeria macusaniensis, E. ivitaensis*	Coproparasitology	Survey, descriptive, prevalence Cafrune et al. (2009b)
V(c), *n* = 171 G(w), *n* = 4 (four populations) Jujuy, Salta, Catamarca, Argentina	*Lamanema chavezi*	Coproparasitology	Survey, descriptive, prevalence and intensity. Cafrune et al. (2009a)
G(w), *n* = 140 La Payunia, Mendoza, Argentina.	*E. macusaniensis, E. ivitaensis, Nematodirus sp., Trichuris sp., Capillaria sp., Moniezia sp,* Trichostrongylidae.	Coproparasitology	Longitudinal (seasonal), descriptive, prevalence (Colombo et al. 2009)
G(w), *n* = 224 RPLLL, Mendoza, Argentina RPLP, Mendoza, Argentina.	*Fasciola hepatica*	Coproparasitology	Survey, descriptive, prevalence and intensity. Issia et al. (2009)
V(w) *n* = 128 Cieneguillas, Jujuy, Argentina	Rotavirus (RV), parainfluenza-3 virus (PI-3), bovine herpesvirus-1 (BHV-1), bovine viral diarrhea virus (BVDV-1)	ELISA: fecal samples for group A RV; SN: BHV-1 and BVDV-1; IHA: BPIV-3	Longitudinal (annual), descriptive, seroprevalence, antibody titers. Marcoppido et al. (2010)
G(c), *n* = 11 CICVyA, Buenos Aires, Argentina	Rotavirus (RV), parainfluenza-3 virus (PI-3), bovine herpesvirus-1 (BHV-1), bovine viral diarrhea virus (BVDV -1)	ELISA and RT-PCR: fecal samples for group A RV; virus neutralization (VN): BoHV-1 and BVDV-1; IHA: PI-3 virus; immunodiffusion (ID): BTV	Longitudinal (monthly), descriptive and antibody titers. Marcoppido et al. (2011)
G(sc,w) = 622 (14 populations) Chubut; Río Negro; Neuquén; Argentina	Coccidia oocysts, *Eimeria macusaniensis, Trichostrongylus spp., Cooperia spp., Nematodirus spathiger, N. oiriatianus, N. filicolis, N. abnormalis, Ostertagia trifurcata, O. ostertagi, Cooperia oncophora, Trichostrongylus*	Coproparasitology, ectoparasite identification	Survey, descriptive, prevalence and intensity. Olaechea et al. (2011)

(continued)

Table 11.1 (continued)

	colubriformis, Trichuris spp., Dictyocaulus spp., Moniezia spp., Sarcocystis spp, Microthoracius praelongiceps, M. minor, M. mazzai, Bovicola spp.		
V(w), *n* = 450 PBR, Jujuy, Argentina	*Microthoracius mazzai, M. praelongiceps, S. scabiei.*	Clinical examination, ectoparasite identification	Longitudinal (annual), descriptive, prevalence Arzamendia et al. (2012)
G(w) (three populations) RPLP, Mendoza, *n* = 234; Neuquén *n* = 68; Río Negro *n* = 54; Argentina	Bovine respiratory syncytial virus, infectious bovine rhinotracheitis, parainfueza-3 virus, rotavirus, bovine viral diarrhea, foot and mouth disease, bluetongue virus, enzootic bovine leukosis, *Leptospira spp., Brucella sp., and Mycobacterium paratuberculosis, Toxoplasma sp.*	ELISA: bovine respiratory syncytial virus and rotavirus; SN: infectious bovine rhinotracheitis; IHA: parainfueza-3 virus and *Toxoplasma* sp. and MAT: *Leptospira spp*	Longitudinal (annual), descriptive, seroprevalence Marull et al. (2012)
G(w), *n* = 154 RPLP, Mendoza, Argentina. G(w), *n* = 121 C2B, Chubut, Argentina.	*Eimeria spp., E. macusaniensis, E. ivitaensis, Nematodirus sp., Trichuris sp., Capillaria sp., Dictyocaulus filaria*	Coproparasitology	Longitudinal (seasonal), descriptive, prevalence Rago et al. (2012)
G(w), *n* = 675 RPLP, Mendoza, Argentina.	*Eimeria spp., E. macusaniensis, Nematodirus spp.*	Coproparasitology	Longitudinal (seasonal), analytical, abundance. Moreno et al. (2013)
V(c), *n* = 235 AP, Jujuy, Argentina	*Eimeria macusaniensis, E. ivitaensis, E. lamae, E. alpacae, E. punoensis*	Coproparasitology	Longitudinal (seasonal), analytical, prevalence and abundance. Cafrune et al. (2014)
G(w), *n* = 756 RPLP, Mendoza, Argentina.	*E. macusaniensis, E. ivitaensis, E. lamae, E. alpacae, E. punoensis, N. spathiger, N. helvetianus, Trichuris sp., Capillaria sp., M. benedeni, Strongyloides sp.*	Coproparasitology	Longitudinal (seasonal), analytical, prevalence and intensity. (Moreno, 2014)
V(w), *n* = 15 Antofagasta de la Sierra, Catamarca, Argentina. V(w), *n* = 17 Yaví, Jujuy, Argentina.	*Eimeria punoensis, E. macusaniensis, Lamanema chavezi, Trichuris sp., Capillaria sp., Nematodirus sp., Trichostrongylus sp., Chabertia sp., Teladorsagia sp., Haemonchus sp., Amblyomma parvitarsum*	Coproparasitology, larval identification, ectoparasite identification	Survey, descriptive, prevalence and intensity. Cafrune et al. (2015)
G(w), *n* = 756 RPLP, Mendoza, Argentina.	*E. macusaniensis, E. ivitaensis, E. lamae, E.alpacae, E. punoensis, N. spathiger, N. helvetianus, Trichuris sp., Capillaria sp., M. benedeni, Strongyloides sp.*	Coproparasitology	Longitudinal (seasonal), descriptive, parasite richness. Moreno et al. (2015)
V(c), *n* = 120 Rodeo, Jujuy, Argentina	*Sarcoptes scabiei.*	Clinical examination, histopathology	Survey, descriptive, prevalence, clinical presentation Aráoz et al. (2016)
V(w), *n* = 150 SC, Jujuy, Argentina	*Strongylida* suborder, *Nematodirus* spp.; coccidia oocytes and cestoda eggs.	Coproparasitology	Longitudinal (annual), descriptive, prevalence Marcoppido et al. (2016)
V(w), *n* = 21 AP, Jujuy, Argentina.	Subtypes of influenza A viruses: A (H1N1), A (H3N2) and A (H3N8)	IHA and ELISA	Serosurvey, descriptive, seroprevalence

(continued)

Table 11.1 (continued)

G(w), *n* = 89 RPLP, Mendoza, Argentina. G(c), *n* = 37 SCB, Río Negro, Argentina.			Barbieri et al. (2017)
V(w), *n* = 40 RLB, Catamarca, Argentina	*Eimeria lamae, E. alpacae, E. punoensis, Capillaria sp., Camelostrongylus sp., Haemonchus sp., Moniezia sp., Fasciola hepatica.*	Coproparasitology	Survey, descriptive, prevalence and Intensity. Cardozo (2019)
G(w), *n* = 33, RPSG, San Juan, Argentina. *n* = 22, RDC, San Juan, Argentina. *n* = 10, PPI, San Juan, Argentina.	*E. macusaniensis, E. ivitaensis, Eimeria spp., Nematodirus sp., Trichuris sp.*	Coproparasitology	Survey, descriptive, prevalence González-Rivas et al. (2019)
G(w), *n* = 55, RNVV, Mendoza, Argentina. *n* = 52, LD, Mendoza, Argentina.	*E. macusaniensis, E. ivitaensis, E. lamae, E. alpacae, E. punoensis, Nematodirus sp., Trichuris sp., Capillaria sp., M. benedeni, Lamanema chavezi, Fasciola hepatica.*	Coproparasitology	Longitudinal (seasonal), descriptive, prevalence, and mean intensity. Moreno et al. (2019)
G(w), *n* = 10 GüerAike, Santa cruz, Argentina	*Lamanema chavezi, Nematodirus spp., Capillaria spp., Trichuris spp.*	Coproparasitology, histopathology	Survey, descriptive Santana et al. (2020)
V(w), *n* = 602 G(w), *n* = 32 PNSG, San Juan, Argentina	*Sarcoptes scabiei*	Clinical examination, histopathology, and molecular diagnosis	Outbreak investigation, analytical, prevalence, clinical presentation Ferreyra et al. (2022)
V(w), *n* = 807 (12 populations) Jujuy, Argentina	*Sarcoptes scabiei*	Clinical examination, histopathology, mite morphology, and molecular diagnosis	Survey, descriptive, prevalence, clinical presentation. Sosa et al. (2022)

***Host population*: G = guanacos, V = vicuñas; (w) = wild, (c) = captive. *n* = sample size. Site, Country. *Pathogens*: include those researched in the original health studies (inclusion does not imply confirmation of presence or exposure when referring to viral and bacterial diseases). *Type of study*: pathogen survey, serosurvey, longitudinal sampling; descriptive/analytical; disease parameter assessed (prevalence, intensity, and abundance). References in gray literature are shaded gray.**

Type of study: Survey: cross-sectional sampling; Serosurvey: serologic cross-sectional study; Case report: description of disease event or die-off; Longitudinal: serial sampling at population level. Descriptive: descriptive statistical analysis; Analytical: statistical inference analysis; Outbreak investigation. Disease parameter assessed: Prevalence: number of hosts infected with the particular pathogen (or taxonomic group) divided by the number of hosts examined for that pathogen; Intensity: count of individuals/parasite stages of a particular parasite species in a single infected host; Abundance: the total number of individuals of a particular parasite species in a sample of a particular host species divided by the total number of hosts of that species examined (including both infected and uninfected hosts) (Bush et al. 1997).

ANMI Área Nacional de Manejo Integrado, *AP* Abra Pampa, *CICVyA* campo experimental del Centro de Investigaciones en Ciencias Veterinarias y Agronómicas, *PM* Península Mitre, *PNPA* Pali Aike National Park, *PNTP* Torres del Paine National Park, *PNSG* Parque Nacional San Guillermo, *PPI* Parque Provincial Ischigualasto, *RBP* Reserva de Biósfera Pozuelos, *RDC* Reserva de uso múltiples Don Carmelo, *RLB* Reserva Laguna Blanca, *RNPGBD* Reserva Nacional Pampa Galeras Bárbara D'Achille, *RNVV* Reserva Natural Villavicencio, *RPC2B* Reserva Provincial Cabo Dos Bahías, *RPLD* Reserva Provincial Laguna del Diamante, *RPLLL* Reserva Provincial Laguna de Llancanello, *RPLP* Reserva Provincial La Payunia, *RPSG* Reserva Provincial San Guillermo, *SABR* Salinas y Aguada Blanca National Reserve, *SAIS* Sociedad Agrícola de Interés Social, *SC* Santa Catalina, *SCB* San Carlos de Bariloche countries within their original distribution. This table summarizes the population features (host species, living conditions, locality, country, and sample size), methodological aspects (diagnostic techniques), pathogens investigated (genus and species), and characteristics of the studies carried out (type of study, sampling design, and disease descriptors).

The pathogens and disease descriptors included in the table are those reported in the original studies. Adequate interpretation requires considering potential biases in the sampling design, the limitations of the techniques used for the identification of pathogens and/or antibodies, and the sample size of each study.

From Table 11.1, it can be concluded that a significant number of WSAC population health studies are found in the gray literature. Most of them are reports that provide information on pathogens found in WSAC, with certain information on their frequency, but in general do not provide information on disease dynamics.

Many surveys with large sample sizes have been carried out in the context of sustainable live-shearing projects (ectoparasitism in vicuñas from Perú, Bolivia, and northern Argentina). During these procedures, the recurrent roundups for capture could produce disturbances in the social structure, and the confinement of groups of animals until shearing increases close contact, facilitating contagion and altering the natural dynamics of transmission. On the other hand, some health studies in WSAC have very low sample sizes, and few studies report the precision of quantitative descriptors (e.g., confidence intervals) (Bush et al. 1997). The interpretation of those studies must consider that they are descriptive reports of proportions of positive samples and that they may not be representative of the real prevalence of disease in the population. In addition, the information provided by seroprevalence studies is about previous exposure to specific pathogens but not about the timing of the infection, so it does not reflect the actual prevalence of the pathogen (Table 11.1).

Descriptive prevalence studies are the most frequent types, mainly of endoparasites and ectoparasites, which provide partial information on their regional distribution. Among those investigations, studies of gastrointestinal parasites predominate, some of which have explored potential associations between parasite presence, demographic features (sex, age of the host), and environmental characteristics (season and sampling site). Moreno (2014) analyzed associations between the abundance of endoparasites and extrinsic (season, weather conditions) and intrinsic determinants (sex, age, body condition, type, and size of social group) of guanacos from the most abundant population in the central region of Argentina. This study also investigated associations between parasite abundance and the concentration of fecal cortisol metabolites (as an indicator of stress) and with serum levels of natural antibodies (components of innate immunity) of guanacos, but neither stress levels nor natural antibody levels were associated with the parasite loads (Moreno 2014).

This chapter will next address the available knowledge on the most studied diseases at the population level in vicuñas and guanacos, their modulating factors, and the conservation impact.

11.2 Gastrointestinal Parasitism

11.2.1 Coccidiosis

It is common for WSAC to shed oocysts of the genus *Eimeria* (Apicomplexa: Eimeriidae) at low intensities without showing signs of disease, which is assumed to be subclinical infection (coccidiosis) with a low probability of representing a threat to the health of the host and its population (Cebra et al. 2007; Dubey 2019; Bowman 2020). *Eimeria* are obligate intracellular parasites whose life cycle includes asexual multiplication followed by two sexual reproduction phases in the intestinal cells of the host, where numerous infective sporozoites are produced and shed in the feces as oocysts (Melhorn 2008). The group of *Eimeria* species that affect SAC is specific. *Eimeria lamae, E. alpacae*, and *E. punoensis* have small ellipsoidal ovoid oocysts, which allow their identification by microscopic observations with detailed measurements. *Eimeria macusaniensis* oocysts are easily diagnosed because they are pear-shaped, large (106.6 × 80.5 μm), and dark brown with very thick walls. *Eimeria ivitaensis* oocysts are also easy to identify because they are large and ellipsoidal (88.8 × 51.8 μm) and brown and thick-walled (Cafrune et al. 2009b; Dubey 2018).

Its life cycle is direct, and transmission occurs by ingesting pasture contaminated with oocysts, which depend on temperature, environmental humidity, and available oxygen, determining the time necessary to acquire the ability to infect the host (Ballweber 2009). Dubey (2018) postulated that *E. macusaniensis* requires a prolonged time to sporulate due to the greater thickness of the walls of its oocysts, needing 9 days at 30 °C and 21 days at 18–25 °C; on the other hand, they do not sporulate at temperatures of 6 °C (Rohbeck 2006; Dubey 2019; Bowman 2020). The periods necessary for the sporulation of the remaining four species have not yet been described (Dubey 2018). *Eimeria macusaniensis* can maintain its infectivity for up to at least 7 years (Jarvinen 2008).

Eimeria lamae develops on the surface of the intestinal epithelium, unlike *E. macusaniensis* and *E. ivitaensis*, which parasitize the crypts of the jejunum and ileum (Guerrero et al. 1967; Palacios et al. 2006; Dubey 2018). *E. lamae* and *E. macusaniensis* are considered the most pathogenic species because the former destroys the epithelium, while the latter damages the crypts of the glands and inhibits cell regeneration (Leguía 1991; Guerrero et al. 1970; Palacios et al. 2006; Cebra et al. 2007). *Eimeria ivitaensis* has rarely been reported in association with clinical coccidiosis (Palacios et al. 2006). In WSAC, coinfection with multiple *Eimeria* species is common (Leguía 1991; Cafrune et al. 2014; Dubey 2019). Prepatent periods vary according to the species, being greater than 30 days for *E. macusaniensis*, 16–18 days for *E. alpacae*, and 10 days for *E. punoensis* and *E. lamae*; the length of the period between ingestion and shedding of *E. ivitaensis* oocysts is unknown (Dubey 2018, 2019).

Eimeria infections can be pathogenic depending on age, concurrent infections, environmental conditions, stress, and nutritional status (Dubey 2019).

Macroscopically, primary lesions of mucosal thickening, congestion, punctate plaques, and severe hemorrhagic enteritis are observed (Guerrero et al. 1967). These lesions are usually focal in the distal jejunum and ileum, although any region of the small intestine, cecum, and colon may be affected (Rosadio and Ameghino 1994; Palacios et al. 2006; Cebra et al. 2007; Johnson et al. 2009; Dubey 2018). Hyperplasia, nonsuppurative enteritis with fusion, and necrosis of villi are observed microscopically (Rosadio and Ameghino 1994; Johnson et al. 2009; Dubey 2018). At the metabolic level, the consequences are hypoproteinemia and hypoalbuminemia (Cebra et al. 2007).

In DSAC species, clinical coccidiosis has been reported to occur in very young animals and infrequently in adults (Ballweber 2009; Cebra et al. 2007; Dubey 2019). The severity of the disease is the result of the combined actions of the number of sporulated oocysts ingested, the nutritional status of the host, and climatic factors (e.g., temperature, humidity, oxygen pressure, and solar irradiation; Melhorn 2008). Milder clinical signs include increased fecal mucus and loose stools; diarrhea is an inconsistent finding, especially in adult camelids (Jarvinen 2008; Cebra et al. 2014). Clinical illness presents with lethargy, weakness, weight loss, anorexia, diarrhea, constipation, colic, and sudden death, often in association with stressful conditions (Schrey et al. 1991; Cebra et al. 2007; Chigerwe et al. 2007; Jarvinen 2008; Johnson et al. 2009; Rosadio et al. 2010; Dubey 2018).

Although for DSAC it has been reported that the most prevalent species among coccidia is *E. lamae*, studies of eimeriasis in WSAC found that *E. punoensis* and *E. alpacae* had the highest prevalence in guanacos from Perú and vicuñas from Bolivia and Argentina (Castillo et al. 2008; Beltrán-Saavedra et al. 2011; Cafrune et al. 2014; Dubey 2018). *Eimeria macusaniensis* is very frequently reported in WSAC populations, but its prevalence is generally intermediate and rarely exceeds 45% of parasitized animals (Rago et al. 2012; Moreno et al. 2019). Until now, *E. ivitaensis* has only been reported in the WSAC from Argentina, in populations of wild guanacos and in captive vicuñas, and its prevalence was always lower than 1% (Rago et al. 2012; Cafrune et al. 2014; Moreno et al. 2015). The only exceptions are those reported for populations of guanacos that inhabit central-western Argentina, where prevalence was between 3% and 10% (González-Rivas et al. 2019; Moreno et al. 2019).

Moreno (2014) studied the seasonal dynamics of eimeriasis in guanacos of La Payunia Provincial Reserve (RPLP, central-west Argentina) for 3 years and found that the five species of *Eimeria* had the highest prevalence during winter, a pattern consistent with other populations of guanacos in the Andean sector of the same region, such as Villavicencio (RNVV) and Laguna del Diamante (RPLD); however, in this last population, the prevalence of *Eimeria* spp. was higher during the summer (Moreno et al. 2019). Among the populations of WSAC studied, very marked seasonalities have been found for *E. macusaniensis*, with increases toward the season of highest prevalence (winter) that double or triple the proportion of parasitized animals, as was found in guanacos from RPLP, in adult vicuñas from northern Argentina in Abra Pampa (AP) and in guanacos of RPLD, respectively (Cafrune et al. 2014; Moreno 2014; Moreno et al. 2019). Cafrune et al. (2014) compared the dynamics of prevalence between autumn and spring in adult and juvenile captive vicuñas and found that *E. punoensis* had a higher prevalence during spring and *E. lamae* during autumn, and that *E. macusaniensis* and *E. alpacae* inverted the predominant season in both age groups (Cafrune et al. 2014). When analyzing the abundance of *E. lamae, E. alpacae*, and *E. punoensis* in the guanacos of the RPLP, important interannual differences were evidenced, explained by interactions in the prevailing weather conditions in the sampling season and in the immediately preceding season, suggesting that the scarcity of rainfall and high levels of relative humidity would facilitate the presentation of greater abundances in the following season (Moreno 2014). In turn, the minimum and maximum temperatures of the preceding season and of the sampling season proved to have an impact on the abundances of *Eimeria* spp. in guanacos from RPLP, suggesting that in temperate seasons (with elevated minimum temperatures and maximum temperatures that are not very high), there is a greater probability of high parasitic abundances (Moreno 2014).

In addition to environmental conditions, many other variables determine the intensity of *Eimeria* infection, including host age, number and infectivity of ingested oocysts, effects of immunity, nutritional status, and the presence of concurrent infections or presence of other stressors (Jarvinen 2008). Regarding the intrinsic variables of the host and their relationship with parasite loads, Marcoppido et al. (2016) found that juvenile vicuñas shed higher amounts of oocysts per gram of undifferentiated *Eimeria* in Santa Catalina (SC, northwestern Argentina). Age also played an important role in the abundance of *Eimeria* spp. of RPLP, where unlike what was reported for DSAC, strong evidence was found that animals younger than 1 year had lower abundance than adults (Moreno 2014). In contrast, the abundance of *E. macusaniensis* in lactating guanacos from the same population was higher than that in adults (Moreno 2014). Meanwhile, the prevalence of *E. alpacae* in vicuñas from Bolivia also showed a significant association with age (Ruiz Hurtado 2016). Likewise, AP juvenile vicuñas showed a significantly higher prevalence than adults for *E. macusaniensis*, while the abundance was significantly higher in juveniles for most coccidia species during autumn and spring, except for *E. ivitaensis* (Cafrune et al. 2014).

Host sex influenced the abundances of *Eimeria* spp. in guanacos from RPLP, where males had lower abundance than females ($p < 0.001$), but during winter and spring, males had significantly higher abundances than females (Moreno 2014). Meanwhile, in vicuñas from Bolivia, Beltrán-Saavedra et al. (2011) found that males had significantly higher prevalence than females.

Coccidiosis can also predispose individuals to microbial infections by *Clostridium perfringens, Escherichia coli*, coronavirus, rotavirus, *Giardia*, or *Cryptosporidium* (Dubey 2019). The existence of intraphylum and interphylum associations between gastrointestinal parasites in guanacos has also been observed. Synergistic associations were found between species of *Eimeria* and between *Eimeria* and *Nematodirus*, indicating that it is more likely that there are high-intensity infections by a species of *Eimeria* in individuals with high intensities of infection by other coccidia and/or by *Nematodirus* spp. (Moreno et al. 2013). The mechanism by which microparasites and helminths can interact synergistically involves the differentiated immune response triggered by each phylum. Protozoa stimulate responses induced by T-helper type 1 (Th1) and Th17 lymphocytes, while helminths induce Th2 responses, and the cytokines involved in each type of response downregulate the cytokines involved in the other type of response (Diaz and Allen 2007).

The immune response plays a relevant role in resistance to eimeriasis since it is generally self-limiting, although in highly contaminated environments, reinfection is constant, and there may be high parasitic loads (Jarvinen 2008; Fowler 2011). It has been reported that DSAC offspring reinfected with *E. macusaniensis* developed coccidiosis with a prolonged prepatent period, shorter patency duration, and less oocyst shedding (Jarvinen 1999). The immune response generated by the other *Eimeria* species in SAC is unknown (Dubey 2019).

11.2.2 Nematodiasis

Among the parasites that localize in the third stomach compartment of SAC, *Marshallagia*, a typical nematode of the abomasum of sheep and goats, has been identified in populations of vicuñas throughout their distribution in Bolivia (Beltrán-Saavedra et al. 2011; Condori et al. 2012; Martela Mamani 2016; Ruiz Hurtado 2016) and in populations of guanacos in southern Argentina (Larrieu et al. 1982; Navone and Merino 1989; Karesh et al. 1998; Beldomenico et al. 2003). Ruiz Hurtado (2016) found evidence that in vicuñas from La Paz and Oruro, the prevalence of *Marshallagia* spp. was significantly higher in the youngest animals. Unlike the pattern of elimination of eggs of the *Strongylida* suborder, which had the highest prevalence in vicuñas older than 3.5 years, when they are reaching sexual maturity (Arzamendia et al. 2018).

The direct cycle nematodes *Trichuris* and *Capillaria* have easily identifiable eggs and have been reported to parasitize several WSAC populations. Regarding the species of *Capillaria* that lives in the small intestine of the WSAC, some authors subscribe that the eggs found are identical to those diagnosed in cattle, assuming the same specific identity (Fowler 2011). However, paleoparasitological studies have confirmed their presence in South America since periods prior to the expansion of domestic cattle, identifying *Capillaria* and *Trichuris* eggs in mummified DSAC from Perú and at least three morphotypes of capillarid eggs in coprolites of camelids from southern Argentina (Leguía 1999; Fowler 2011; Taglioretti et al. 2015). When analyzing the prevalence of *Capillaria* sp. in managed vicuña populations in Bolivia in relation to age, Beltrán-Saavedra et al. (2011) showed that juveniles presented significantly higher prevalence than adult animals. Seasonality proved to be a determining factor for the abundance of *Capillaria* sp. in guanacos from RPLP, showing significantly higher levels during winter (Moreno 2014).

The genus *Nematodirus* mainly affects species of Cervidae and Bovidae from temperate and cold Holarctic environments, but it is one of the most diagnosed genera in sheep, cattle, and WSAC in South America (Anderson 2000; Suárez et al. 2007). Although *Nematodirus lamae* is a species of *Nematodirus* typical of the

SAC, *N. spathiger, N. lanceolatus, N. filicolis, N. battus, N. helvetianus, N. oiriatianus*, and *N. abnormalis* have also been identified parasitizing the small intestine of vicuñas from Perú and guanacos in sympatry with sheep and cattle in Argentina (Larrieu et al. 1982; Olaechea et al. 2011; Moreno et al. 2015; Angulo-Tisoc et al. 2021). The seasonal factor was also decisive for the dynamics of parasitism by *Nematodirus* in guanacos from the RPLP, with winter and spring abundances significantly higher than those of the rest of the seasons (Moreno 2014). A simultaneous study that analyzed the endoparasitic community in this population by means of larval cultures confirmed that the species present were *N. spathiger* and *N. helvetianus*, typical of sheep and cattle, without finding larvae of the SAC-specific species (Moreno et al. 2015). There were also significant differences in the abundance of *Nematodirus* between guanacos from sectors of the protected area with historically different land use (Moreno 2014). The influence of the site where the hosts live was also evidenced in vicuña populations in La Paz and Oruro, where significant differences were found between the prevalence of *Nematodirus* spp. at different sampling sites (Ruiz Hurtado 2016).

In the small intestine, the presence of *Lamanema chavezi* becomes important. This parasite, which has been reported to parasitize all species of SAC, is considered a specific nematode of this group. It is also one of the few SAC nematodes whose third- and fourth-stage parasitic larvae carry out enterohepatic migration, which gives it great pathogenicity. *Lamanema* sheds eggs with a typical shape that is easily differentiated from other species of nematodes in coprological diagnosis (Cafrune et al. 2009a; Santana et al. 2020).

In relation to large intestine nematodes, it has been suggested that *Trichuris tenuis* is the typical species of camelids. Its presence has been confirmed by identifying adult nematodes from vicuñas and guanacos from Salta and Chubut in Argentina (Cafrune et al. 1999; Beldomenico et al. 2003), although previous studies reported *Trichuris ovis* parasitizing DSAC from Perú and northern Chile (Guerrero 1960). However, most prevalence studies are based only on coprodiagnosis, and the morphometric characteristics of the eggs are not sufficient for specific identification (Cafrune et al. 1999; Moreno et al. 2015). Moreno (2014) showed that the abundance of *Trichuris* is affected by season in RPLP guanacos, where spring had the lowest abundance. An interaction with the age of the animals was also evidenced, since during that season, the offspring younger than 1 year had significantly higher abundances than guanacos of other age categories (Moreno 2014).

11.2.3 Cestodiasis

Guanacos and vicuñas have been reported as definitive hosts of the cestode species *Moniezia expansa* and *M. benedeni* (Bustinza et al. 1982; Beldomenico et al. 2003; Beltrán-Saavedra et al. 2011; Moreno et al. 2015; Martela Mamani 2016; Ruiz Hurtado 2016). The cycle of both cestodes requires the presence of oribatid mites as an intermediate host, which gives it a marked seasonality of infection in temperate climates (Fowler 2011). They are typical parasites of the small intestine of domestic cattle; *M. benedeni* more frequently parasitizes cattle and *M. expansa* sheep (Suárez et al. 2007). In the health study carried out in guanacos of RPLP, where *M. benedeni* was the only species of cestode identified, it was possible to show that their counts are influenced by several factors (Moreno 2014). In that protected area, *M. benedeni* was not found in all seasons, and when it was detected, the prevalence was below 10% and had a low mean intensity (2.85 HPG; Moreno 2014). Winter was the season with the highest abundance of *M. benedeni* eggs, while the lowest occurred during spring (Moreno 2014). The age of the host and traits of its social group also influenced shedding of *M. benedeni* eggs in the RPLP. Juvenile guanacos had the highest abundances, and an association was found between the size of the social group and parasitic abundances, being lower in guanacos from larger social groups (Moreno 2014).

In the study carried out by Moreno (2014), the parasitic abundances of the sheep and bovine helminth species *N. spathiger, N. helvetianus,* and *M. benedeni* were higher in the northwestern sector of the RPLP, although the cause could not be established. This sector includes an oil extraction site and cattle as the predominant livestock species, and in autumn-winter, livestock management drives an important influx of cattle and horses (Schroeder et al. 2013). In the northeast sector, on the other hand, where lower parasitic abundances were recorded, there was no oil-drilling impact, and the highest density of livestock was from goats and, to a lesser extent, cattle (Schroeder et al. 2013; Moreno 2014). The spatial variable also showed influences on population stress levels, with guanacos from the northwestern sector of the RPLP showing significantly higher concentrations of fecal cortisol metabolites than those found in the northeastern sector (Moreno 2014).

Boxes 11.1 and 11.2 summarize information on mass mortality events in WSAC populations where the contribution of parasitic diseases was identified. In both mortality events, infestations by livestock pathogens were documented, in addition to environmental and management particularities.

Box 11.1 Guanaco Die-off at Cabo Dos Bahías Provincial Reserve, Chubut, Argentina

- Cabo Dos Bahías is a 1188-ha provincial protected area in Argentine Patagonia, surrounded by one sheep ranch to the west, and the Patagonian Sea to the northeast, east, and southeast. Sheep from the neighboring ranch frequently trespass the protected area boundaries, interacting and competing with guanaco for habitat and food (Karesh et al. 1998; Baldi 1999).
- In 1995, Karesh et al. carried out a health assessment of 20 live guanacos from the population of circa 500 animals in the Reserve. In this study, they found no evidence of exposure to infectious diseases, including bluetongue virus, brucellosis, bovine respiratory syncitial virus, bovine viral diarrhea/mucosal disease, equine herpesvirus 1, infectious bovine rhinotracheitis,

(continued)

Box 11.1 (continued)

Mycobacterium paratuberculosis, foot-and-mouth disease, leptospirosis (17 serovars), parainfluenza-3, and vesicular stomatitis. Conversely, they reported fecal nematode ova belonging to *Strongyloides* sp. (5%), *Nematodirus* sp. (30%), *Marshallagia* sp. (10%), *Trichostrongylus* sp. (15%), *Trichuris* sp. (25%), and *Dictyocaulus* sp. larvae (5%). Blood parasites, ectoparasites, or signs of parasitic dermatitis were not found.

- Since the early winter of 1999, a severe drought affected the area of Cabo Dos Bahías and the availability of forage was severely reduced. In 2000, during the winter months, an extensive guanaco die-off occurred, and more than 300 guanacos died. Twelve guanacos were necropsied (five adult females, one adult male, five juvenile females, and one juvenile male), and the main findings reported by Beldomenico et al. (2003) were as follows:
  - *Eimeria macusaniensis* (in feces of 75% of necropsied animals, oocysts per gram (opg) <50), *Eimeria* spp. (83.3%, $I_{mean} = 3204$ opg, $I_{max} = 12{,}600$ opg), *Nematodirus* sp. (75% $I_{mean} = 58$ eggs per gram (epg), $I_{max} = 300$ epg), *Marshallagia* sp. (66.7%, $I_{mean} = 33$, $I_{max} = 200$), *T. tenuis* (41.7%, maximum parasites per host = 23), *M. expansa* (16.7%), and *Dictyocaulus filaria* (83.3%, mean parasites per host = 53, maximum parasites per host = 158).
  - Lesions resembling those caused by *Sarcocystis* sp. were found in tongue, heart, and skeletal muscles (66.7%).
  - The deaths were attributed to emaciation and starvation due to the long-term drought, competition with sheep, and extremely high guanaco densities. However, the livestock lungworm *D. filaria* was suggested to have a detectable effect on the health of the guanacos, according to the observed high burden of worms and lung congestion confirmed by histopathological examination.

Box 11.2 Description of an Outbreak of Mange in Vicuña and Guanaco Populations in a Protected Area, with Risk for Local Extinction Leading to Changes at the Landscape Scale

In Argentina, the two largest sympatric populations of vicuña and guanaco live in San Guillermo National Park (SGNP), created in 1999 in the province of San Juan, center-west of the country. The timeline of the event is as follows:

- 2009–2011: A provincial livestock incentive program was implemented, via which 156 llamas were delivered to farmers in areas adjacent to PNSG in San Juan. The llamas were brought from the north of the country, where DSAC are common livestock. Some introduced llamas were diagnosed with mange upon arrival in San Juan.

(continued)

Box 11.2 (continued)

- 2013: Population surveys in PNSG established that the densities were 11 vicuñas/km^2; 0.4 guanacos/km^2 (Ellis et al. 2017).
- 2014: The presence of mange was detected for the first time in vicuñas and guanacos of PNSG.
- 2014–2017: (*Outbreak study*; Ferreyra et al. 2020).
  - Vicuñas monitored by radio-collar were found with mange at an increasing rate of 0–100%.
  - 68% of vicuñas with mange were detected in some sectors of SGNP.
  - WSAC with mange predated by puma increased from 5% to 90%.
  - Densities of vicuñas decreased from 55% to 98%.
  - Density of guanacos showed a 95% reduction.
- 2017–2018: (*Study of the dynamics of the outbreak, its population effects in WSAC, and origin of the epidemic*; Ferreyra et al. 2022).
  - Vicuña density decreased by an additional 68%.
  - Guanaco density decreased by an additional 77%.
  - Among live animals, 24% of vicuñas and 33% of guanacos presented clinical signs of mange.
  - In carcasses, 94% of vicuñas and 85% of guanacos presented clinical signs of mange.
  - Macro- and microscopic diagnosis confirmed the hyperkeratotic clinical presentation, with the presence of abundant *S. scabiei* mites.
  - Genetic characterization (microsatellites) revealed that mites from guanacos and vicuñas were highly homologous with very little genetic variability, consistent with a common origin of the epidemic and recent occurrence.
- 2019:
  - No WSAC were observed in study transects, suggesting the local extinction of vicuñas and guanacos in the evaluated sectors.
- After the initial outbreak in SGNP, WSAC with mange were observed in areas adjacent to PNSG and in a neighboring protected area in La Rioja province.
- The authors who investigated the outbreak analyzed several factors that support the hypothesis that the outbreak originated from llamas introduced near SGNP (Ferreyra et al. 2022). A later ecological study postulated that the catastrophic reduction of WSAC in SGNP may have cascading effects on ecological communities, causing changes in the abundance and distribution of primary producers, predators (*Puma concolor*), scavengers (*Vultur gryphus*), and in the ecosystem in general (Monk et al. 2022).

11.2.4 Trematodiasis—Fascioliasis

The infestation by the trematode *Fasciola hepatica* has been described in its different forms (acute, subacute, and chronic) in DSAC raised in natural fields (Cafrune et al. 1996; Leguía 1999). On the other hand, the records of this parasitosis in WSAC are scarce and limited to the reports of Cafrune et al. (1996, 2004) in vicuñas, Olaechea and Abad (2005) and Issia et al. (2009) in guanacos. Parasitosis by *F. hepatica* in SAC can be subclinical as a probable consequence of low to moderate infestations (Cafrune et al. 1996). The occurrence of clinical fascioliasis is reported with increasing frequency in SAC (Leguía 1999; Cafrune et al. 2004; Olaechea and Abad 2005), sometimes concomitantly with other parasites that add their detrimental action, such as gastric nematodiasis (Cafrune et al. 2006). The minimum prepatent period in experimentally infested llamas was 56 days (Rickard and Foreyt 1992). The clinical form in adult vicuñas and guanacos occurred with a prevalence of less than 25% of infested individuals (Cafrune et al. 2004; Olaechea and Abad 2005).

11.3 Ectoparasitic Diseases

11.3.1 Sarcoptic Mange

Sarcoptic mange is a highly contagious skin disease caused by the presence and multiplication of the mite *Sarcoptes scabiei* (Leguía 1999; Fowler 2011). There are numerous reports of mange in SAC (Table 11.1). *Sarcoptes scabiei* var. *aucheniae* (Arthropoda: Sarcoptidae) has been described as a typical mite. However, varieties of *S. scabiei* have traditionally been named based on the host species where they were detected, despite the few morphological differences between mites found on different host species (Niedringhaus et al. 2019). There is currently no certainty about the specificity of *S. scabiei* collected from different mammalian species and whether there are multiple *Sarcoptes* species within or between host species (Arlian and Morgan 2017). *Sarcoptes scabiei* is a generalist parasite that can be transmitted between DSAC and WSAC and can cause epidemics with significant mortality in WSAC populations (Bornstein and de Verdier 2010; Arzamendia et al. 2012; Browne et al. 2022; Ferreyra et al. 2022).

Sarcoptic mange outbreaks in wildlife populations begin frequently with high prevalence and mortality, later becoming endemic with low prevalence and eventually disappearing (Ferreyra et al. 2022). There have been records of mange epidemics causing vast mortalities of SAC since the sixteenth century, and various studies suggest that the disease is endemic with low prevalence in some WSAC populations and is associated with sympatry with DSAC and/or live-shearing practices of wild populations (Dale and Venero 1977; Hofmann et al. 1983; Beltrán-Saavedra et al. 2011; Arzamendia et al. 2012; Gomez-Puerta et al. 2013; Ruiz Hurtado 2016;

Bujaico Mauricio 2018; Angulo-Tisoc et al. 2021; Mayhua Mendoza 2021; Ferreyra et al. 2022).

Although with heterogeneous prevalence, this disease has been reported in WSAC populations in Perú (Gomez-Puerta et al. 2013; Bujaico Mauricio 2018; Unzueta Lancho 2018; Angulo-Tisoc et al. 2021; Mayhua Mendoza 2021), Bolivia (Beltrán-Saavedra et al. 2011; Ruiz Hurtado 2016), Chile (Raedeke 1976; Alvarado et al. 2004; Vargas and Bonacic 2020), and Argentina (Arzamendia et al. 2012; Ferreyra et al. 2022; Sosa et al. 2022). In some scenarios, it is associated with sympatry with DSAC (Unzueta Lancho 2018; Mayhua Mendoza 2021). A recent increase in the distribution and prevalence of infection has been suggested in several of the WSAC range countries (Ferreyra et al. 2022). In addition, its high contagion rate and its consequences on a population scale generate growing concern for the conservation of guanacos and vicuñas, given the demonstrated capacity of the mite to cause unusual epidemics associated with strong population decline and risk of local extinctions (Ferreyra et al. 2020, 2022; Monk et al. 2022; Box 11.2). In Chile, guanacos and vicuñas were among the species most commonly recorded with abnormal alopecia, potentially caused by *S. scabiei*, in 9.62% of the protected areas, placing this disease as one of the main threats for some wildlife populations of that country (Montecino-Latorre et al. 2020).

The life cycle of *Sarcoptes* spp. is direct and occurs entirely on the skin of the host, but adult mites can survive in the environment on epithelial debris for up to 30 days (Fowler 2011). The adults copulate on the skin, and then the fertilized female advances, digging and depositing up to four eggs per day in her progress through the tunnels she digs in the lower layers of the epidermis (Leguía 1999; Fowler 2011). In alpacas from Perú, it was established that the complete cycle lasts between 20 and 25 days (Mayhua Mendoza 2021). By tunneling into the skin, *Sarcoptes* generates a hypersensitive dermatitis with mild to severe itching where small erythematous foci are generated with serous exudate that gives rise to cracked crusts (Bornstein and de Verdier 2010; Gomez-Puerta et al. 2013; Ferreyra et al. 2022). In camelids, the lesions begin in areas devoid of hair and predominate in their acute and chronic phases in the perineal area, between the toes, inner thighs, belly, foreskin, head, lips, and ears (Fowler 2011; Arzamendia et al. 2012; Alvarado et al. 2004; Gomez-Puerta et al. 2013; Unzueta Lancho 2018; Sosa et al. 2022). The lesions can progressively spread to other regions of the body, and the associated intense itching leads to the appearance of traumatic scratching lesions and secondary bacterial infections (Leguía 1999; Fowler 2011). The chronic form presents as hyperkeratosis with hard and dry scabs that come off with a large amount of fiber. Depending on the location of the lesions, difficulty walking or eating is observed, leading to a poor body condition, which can cause death and/or limit the ability to evade predators (Rojas 2004; Fowler 2011; Bornstein and de Verdier 2010; Ferreyra et al. 2022; Sosa et al. 2022).

A study of the San Guillermo National Park (SGNP) sarcoptic mange outbreak in WSAC developed a clinical characterization to infer the proportion of affected animals through visual estimates without capture or laboratory diagnosis, although it represents the minimum proportion of infected individuals in the population. In this way, Ferreyra et al. (2022) established three categories of increasing severity: early stage (obvious and severe and/or persistent scratching in animals in groups with other individuals in the advanced or severe stage), advanced stage (animals with difficulty walking and/or visible lesions on the limbs), and severe stage (individuals with alopecia and scabs spreading throughout the body).

The clinical presentation of sarcoptic mange in wildlife has in turn been classified by histopathological diagnosis, differentiating between the hyperkeratotic form, with high mite loads and a type I hypersensitivity response, with intense scratching and secondary skin lesions and infection, and the alopecic form, with low mite loads and loss of hair cover due to a type IV hypersensitivity response (Ferreyra et al. 2022). The clinicopathological presentation of sarcoptic mange outbreaks was compared by Aráoz et al. (2016) in llamas and vicuñas in captivity in northern Argentina, finding that in vicuña, the hyperkeratotic form occurred with an abundance of mites in severe clinical cases and with high lethality, while in llamas, the presentation was alopecic, with a low presence of mites and mild clinical cases, without mortality. These results coincided with the lesions found by Ferreyra et al. (2022) in the mange outbreak in vicuñas and guanacos in central-western Argentina, where the highly contagious hyperkeratotic form was recorded (Ferreyra et al. 2022). Similar results were found by Sosa et al. (2022) in a study of the occurrence of sarcoptic mange in free-ranging vicuñas of the high Andean plateau of Argentina, where animals showed focal to diffuse alopecic injuries, with hyperkeratosis, erythema, hyperpigmentation, exudation or bleeding, and the histopathological examination of skin biopsies showed extensive infiltration of the dermis with lymphocytes, neutrophils and eosinophils, hyperplasia at different stages, epidermis degeneration, and hyperkeratosis.

The main route of transmission is by direct contact, which in WSAC occurs in agonistic encounters between males, during copulation, or in interactions between mothers and offspring (Vilá 1992; Arzamendia et al. 2012; Ferreyra et al. 2022). The results of the mange outbreak investigation conducted by Ferreyra et al. (2022) suggest that, at least in that case, direct transmission allowed for the spread of the disease regardless of host density, disseminating mites even when the population size of WSAC decreased to near local extinction. In addition, the resistance of the mites in the environment and ethological aspects of WSAC, such as the use of wallows for dust baths and community dung piles, establish them as possible means of indirect contagion. Mite transmission could also occur through the shared use of bushes where animals scratch (Vilá 1992; Arzamendia et al. 2012; Browne et al. 2022; Ferreyra et al. 2022).

The temperature and environmental humidity of spring and summer favor the development of mites, which is when the most serious and extensive lesions are observed in DSAC, producing the greatest number of infections (Leguía 1999). In autumn and winter, the disease tends to be subclinical in DSAC since the parasites persist in body regions protected from sunlight, such as armpits, crotches, groins, and ears (Leguía 1999). Few studies have evaluated the seasonality of sarcoptic mange in WSAC populations, and their results are not consistent. In northern Chile, a higher prevalence was reported in wild vicuñas in spring-summer (CONAF 2014), while in the Parque Nacional San Guillermo (PNSG) mange outbreak study, there was no obvious seasonal trend (Ferreyra et al. 2022). Conversely, in populations of vicuñas in Perú, the highest prevalence of mange was found in the autumn-winter months, decreasing toward spring (Bujaico Mauricio 2018).

Animals infested for the first time develop partial immunity that is not fully protective and is manifested by reduced egg production and decreased mite population (Leguía 1999; Rojas 2004). Although current knowledge of mange in DSAC confirms that young animals are the most susceptible, in situations of stress, the disease can affect individuals of any age (Leguía 1999). In WSAC, no statistical association was found between the prevalence of disease and the age of the host. In vicuñas, a general trend is described in which offspring show the lowest prevalence, while adults present the highest prevalence (Hofmann et al. 1983; Unzueta Lancho 2018; Angulo-Tisoc et al. 2021). In the PNSG mange outbreak, evidence was found that offspring were more likely to present early-stage disease and less likely to present advanced stages compared to adults (Ferreyra et al. 2022). However, because offspring were seen suckling from seriously ill mothers, it is possible that lack of maternal care led to mortality in this age class before mange progressed or that there was detection bias due to difficulty in finding calf carcasses quickly eliminated by predators and scavengers (Ferreyra et al. 2022).

The sex of the animals has not been reported as an important determinant of mange in WSAC. In the DSAC literature, a trend toward males having a higher prevalence than females is reported, as well as in a study in adult vicuñas from Perú (Unzueta Lancho 2018). In a study of mange in guanacos from the southern tip of Chile, Alvarado et al. (2004) found negative correlations between the intensity of infection and parameters as weight, age, and sex and a positive correlation between intensity and the number of lesions. However, in opportunistic health studies that use postharvest necropsies of guanacos (southern Chile and Argentina), it should be considered that the regulations that authorize these procedures establish the criteria for selecting the animals to be slaughtered (mainly males), introducing sampling bias.

11.3.2 Ticks

The ixodid *Amblyomma parvitarsum* Neumann completes its cycle in three hosts, with the adult stages considered to prefer SAC as hosts, which are usually parasitized in the perianal and inguinal area (Estrada-Peña et al. 2005; Muñoz-Leal et al. 2014). The larvae and nymphal stages of this tick have been found parasitizing reptiles of the genus *Liolaemus* (Muñoz-Leal et al. 2016). The adult stages can be found in dung piles and on SAC in the rainy season and in reptiles of the genus *Liolaemus* in the transition between the rainy and dry seasons (Muñoz-Leal et al. 2014). The geographic distribution of *A. parvitarsum* includes the arid lands of the Andean high plateau region of Argentina, Bolivia, Chile, and Peru and the Argentin

Patagonia, where it has been found parasitizing SAC (Estrada-Peña et al. 2005; Muñoz-Leal et al. 2014).

Adults and nymphs of *A. parvitarsum* infected by the hemoparasite *Rickettsia* sp. have been detected in Chile, Argentina, and Perú (Ogrzewalska et al. 2016; Muñoz-Leal et al. 2018). The proportion of *A. parvitarsum* ticks infected by rickettsiae from Argentina and Chile was high (≥57.7%), while the overall positivity rate for ticks from Perú was 93.3% (Ogrzewalska et al. 2016; Muñoz-Leal et al. 2018). Until now, the infection of WSAC by *Rickettsia* has not been confirmed.

11.4 Infectious Diseases

Some investigations have confirmed current or previous exposure to various viruses, bacteria, and protozoa in WSAC. However, although estimates of the proportion of exposed animals in some populations have been published, it has not been possible to make inferences about the variables that determine the occurrence of diseases. A few studies have revealed evidence of the circulation of infectious agents in populations of vicuñas or guanacos. Karesh et al. (1998) and Risco-Castillo et al. (2014) found no evidence of exposure to viral agents in guanacos from Chubut (Argentina) and vicuñas from Arequipa (Perú), respectively, but Marull et al. (2012) detected antibodies to bovine respiratory syncytial virus, infectious bovine rhinotracheitis, parainfluenza-3 virus, and rotavirus in the serum of guanacos from Mendoza, Neuquén, and Río Negro in Argentina. Parreño et al. (2001) isolated a rotavirus (group A rotavirus) associated with a severe outbreak of diarrhea with high morbidity and mortality in captive guanaco calves in Argentine Patagonia. Those guanacos (captured from 1 day to 4 months of age, raised in captivity in small paddocks without their mothers, and fed with bovine milk substitute) presented acute diarrhea and associated symptoms and died within 2–6 days, confirming at necropsy the presence of *Escherichia coli* and *Salmonella* sp., with septicemia as the final cause of death. In addition, 95% (38/40) had very high rotavirus antibody titers, suggesting recent exposure (Parreño et al. 2001).

Marcoppido et al. (2010) studied evidence of exposure to pathogenic viruses of domestic cattle in wild vicuñas from Jujuy (northwestern Argentina), confirming that 100% (128/128) had been exposed to rotavirus, and reported lower prevalence for parainfluenza-3 virus (37%; 44/119), bovine herpesvirus-1 (0.78%; 1/128), and bovine viral diarrhea virus (0.78%; 1/128). Subsequently, Marcoppido et al. (2011) carried out a longitudinal study in captive guanaco offspring from Río Negro to evaluate the dynamics of antibody titers against viral diseases, also detecting evidence of antibodies against rotavirus (7/11) from the beginning of the study (possibly maternal) and subsequent seroconversion (4/11) with fecal elimination (1/11). At the beginning of the study, no antibodies against parainfluenza-3 virus were detected, but seropositive animals (11/11), although asymptomatic, were detected after 7 months, concluding that infections by this virus common in cattle could easily spread among young guanacos (Marcoppido et al. 2011). On the other hand,

Barbieri et al. (2017), using a hemagglutination inhibition assay, found sera positive for antibodies against H1-like influenza A virus in 36% (32/89) of guanacos sampled in the central-west and in 38% (8/21) of vicuñas sampled in northern Argentina, but none of the individuals manifested clinical signs of respiratory disease.

Regarding bacterial diseases in WSAC, Risco-Castillo et al. (2014) detected antibodies against *Leptospira* (serovars Pomona, Autumnalis, Copenhageni, and Bratislava) in 1.9% (4/207) of vicuñas studied in Perú, showing a higher risk of infection for females older than 2.5 years. In that survey, 1% were also seropositive for *Neospora caninum*, without evidence of clinical signs of disease. In Argentina, Llorente et al. (2002) detected seroprevalence of 3.8% (3/78) for *Leptospira* serovar copenhageni in guanacos from Río Negro; 16.4% (12/73), 28.7% (21/73), and 2.7% (2/73) for serovars copenhageni, castellonis, and pyrogenes, respectively, in vicuñas from Salta; while for vicuñas from Catamarca, seroprevalence of 6.6% (2/30) and 16.6% (5/30) were found for the copenhageni and castellonis serovars. Marull et al. (2012) detected antibodies against *Leptospira* spp. in wild populations of guanacos from Mendoza, Neuquén, and Rio Negro.

The causal agent of Johne's disease (paratuberculosis), *Mycobacterium avium* subsp. *paratuberculosis* (MAP), was confirmed in fecal culture of 4.2% (21/501) of guanacos analyzed in southern Chile (Salgado et al. 2009). The isolates were classified as bovine type and had a weak association with the age and sex of the individuals, and the sampled animals showed no signs of clinical disease (Salgado et al. 2009). The proportion of positive samples for MAP was associated with areas of sympatry between guanacos and sheep, suggesting that there may be horizontal transmission between these species of ungulates in Chilean Patagonia (Collado et al. 2019). However, another study on the dynamics of MAP infection in a sheep-guanaco interface in southern Chile showed that the intensity of pathogen shedding was higher in both species in summer, but shedding was higher in sheep than in guanaco across all seasons, supporting a higher competence of sheep as hosts for MAP (Corti et al. 2022).

11.5 Concluding Remarks

This chapter allows for the identification of aspects of the host (such as age, influenced by previous exposure and immunological maturity of the animals), ecological traits of the pathogens (host generalism, seasonality, and resistance to environmental conditions), and spatial factors (linked to land use and sympatry with domestic livestock) that are determining variables in the ecology of some of the parasitic diseases that affect WSAC populations.

Considering that for the distribution range of WSAC, it has been predicted that droughts will increase in the next two decades (they will be more frequent and severe), that extreme heat episodes could reach tolerance thresholds critical for health more frequently (IPCC 2022), and that knowledge about the ecology of diseases in WSAC is still very scarce (Rago et al. 2022), future research should consider potential associations between the epidemiological aspects, relevant environmental variables in host-parasite interaction, the impact of anthropogenic disturbance (oil extraction, mining, WSAC management, and use), and coexistence with domestic livestock.

Studies must be designed to elucidate disease dynamics, including systematic sampling that allows for the discrimination of associations between the different variables and applying diagnostic techniques with high sensitivity and specificity for the detection/identification of pathogens. It is also highly necessary to evaluate the impact of the identified pathogens (and their modulating factors) on the population dynamics of their camelid hosts. This information is critical for the identification of disease risk factors and gain knowledge about population resilience, enabling the prediction and prevention of the impact of diseases on WSAC populations, thus ensuring their long-term conservation.

References

Acebes P, González B (2021) *Vicugna vicugna* (Green Status assessment). The IUCN red list of threatened species 2021: e.T22956A2295620213. Accessed on 7 Mar 2022

Acebes P, Wheeler J, Baldo J, Tuppia P, Lichtenstein G, Hoces D, Franklin WL (2018) *Vicugna vicugna* (errata version published in 2019). The IUCN red list of threatened species 2018: e.T22956A145360542. https://doi.org/10.2305/IUCN.UK.2018-2.RLTS.T22956A145360542.en. Accessed on 16 Feb 2022

Aguirre DH, Cafrune MM (2007) Parasitosis de los camélidos sudamericanos. In: Suárez VH, Olaechea FV, Romero JR, Rossanigo CE (eds) Enfermedades Parasitarias de los Ovinos y otros Rumiantes Menores en el Cono Sur de América. INTA EEA Anguil Ediciones, pp 281–296

Alvarado L, Skewes O, Brevis C (2004) Estudio de sarna clínica en guanaco (*Lama guanicoe*) silvestre, en el sector centro-sur de Isla Tierra del Fuego, Chile. DVM Thesis, Facultad de Medicina Veterinaria, Universidad de Concepción, Chillán, Chile

Anderson RC (2000) Nematode parasites of vertebrates: their development and transmission. Cabi.

Angulo-Tisoc JM, Pacheco JI, Vélez V, García W, Castelo H, Gomez-Puerta LA (2021) Situación actual de la sarna e infecciones parasitarias en vicuñas (*Vicugna vicugna*) de la Región Cusco, Perú. Rev de Investig Vet del Perú 32(3): e20405

Aráoz V, Aguirre DH, Viñabal AE, Acuña AM, Abalos M, Micheloud J (2016) Descripción clínico-patológica en brotes de sarna sarcóptica en vicuñas (*Vicugna vicugna*) y llamas (*Lama glama*) de la provincia de Jujuy, INTA Argentina. X Reunión Argentina de Patología Veterinaria. Santa Fe

Arlian LG, Morgan MS (2017) A review of *Sarcoptes scabiei*: past, present and future. Parasit Vectors 10(1):1–22

Arzamendia Y, Neder LE, Marcoppido G, Ortiz F, Arce M, Lamas HE, Vilá BL (2012) Effect of the prevalence of ectoparasites in the behavioral patterns of wild vicuñas (*Vicugna vicugna*). J Camelid Sci 5:105–117

Arzamendia Y, Carbajo AE, Vilá B (2018) Social group dynamics and composition of managed wild vicuñas (*Vicugna vicugna vicugna*) in Jujuy, Argentina. J Ethol 36(2):125–134

Baldi RE (1999) The distribution and feeding strategy of guanacos in the Argentine Patagonia: a sheep-dependent scenario. University of London, University College London, United Kingdom

Baldi RB, Acebes P, Cuéllar E, Funes M, Hoces D, Puig S, Franklin WL (2016) *Lama guanicoe*. The IUCN red list of threatened species 2016: e.T11186A18540211. https://doi.org/10.2305/IUCN.UK.2016-1.RLTS.T11186A18540211.en. Accessed on 16 Feb 2022

Ballweber LR (2009) Ecto-and endoparasites of new world camelids. Veterinary Clinics of North America: Food Animal Practice, 25(2):295–310

Barbieri ES, Baumeister E, Romero S, Martínez Escribano JÁ, Puntel M, Parreño GV (2017) Natural infection by H1-like influenza A virus in South American camelids from Argentina: serological evidences. J Vet Healthc 1:21–38

Beldomenico PM, Uhart M, Bono MF, Marull C, Baldi R, Peralta JL (2003) Internal parasites of free-ranging guanacos from Patagonia. Vet Parasitol 118(1–2):71–77

Beltrán-Saavedra LF, Mollericona JL (2019a) Evaluación de la sarna, otros parásitos y de buenas prácticas durante el manejo en silvestría de vicuñas en las comunidades de Cañuhuma, Nube Pampa, Puyo Puyo, Plan Aeropuerto y Ucha Ucha, ANMIN Apolobamba, Bolivia, 2018. Informe técnico. Wildlife Conservation Society, La Paz, 24 p

Beltrán-Saavedra LF, Mollericona JL (2019b) Evaluación de la sarna, otros parásitos y de buenas prácticas durante el manejo en silvestría de vicuñas en comunidades del ANMI Pampa Tholar de las Vicuñas, Villazón, Potosí, Bolivia, 2018. Informe técnico. Wildlife Conservation Society, La Paz, 24 p

Beltrán-Saavedra LF, Mollericona JL (2020) Evaluación de la sarna y otros parásitos externos e internos durante capturas, esquilas y liberaciones de vicuñas en comunidades del ANMIN Apolobamba, La Paz – Bolivia, 2019. Informe técnico. Wildlife Conservation Society, La Paz, 15 p

Beltrán-Saavedra LF, Nallar-Gutiérrez R, Ayala G, Limachi JM, Gonzales-Rojas JL (2011) Health assessment of free-ranging vicuñas of the National Integrated Management Natural Area Apolobamba, Bolivia. Ecología en Bolivia 46(1):14–27

Bornstein S, de Verdier K (2010) Some important ectoparasites of Alpaca (*Vicugna pacos*) and Llama (*Lama glama*). J Camelid Sci 3:49–61

Bowman DD (2020) Georgis' parasitology for veterinarians E-Book. Elsevier Health Sciences

Browne E, Driessen MM, Cross PC, Escobar LE, Foley J, López-Olvera JR et al (2022) Sustaining transmission in different host species: the emblematic case of *Sarcoptes scabiei*. Bioscience 72(2):166–176

Bujaico Mauricio N (2018) Efecto de la prevalencia de la sarna (*Sarcoptes scabiei* var. *aucheniae*) en la producción y comercialización de la fibra de vicuña (*Vicugna vicugna*) en la comunidad campesina de Lucanas-Ayacucho. Tesis para optar al título de Ing. Zootecnista. Universidad Nacional de Huancavélica, Perú

Bush AO, Lafferty KD, Lotz JM, Shostak AW (1997) Parasitology meets ecology on its own terms: Margolis et al revisited. J Parasitol 83:575–583

Bustinza J, Sánchez C, Choque J (1982) Parásitos identificados en vicuñas de Pampa Galeras. In: Proceedings of the II Convención Nacional Multisectorial de Desarrollo de Camélidos Sudamericanos, Lima, Perú

Cafrune MM, Rebuffi GE, Gaido AB, Aguirre DH (1996) *Fasciola hepatica* in semi-captive vicuñas (*Vicugna vicugna*) in north west Argentina. Vet Rec 139(4):97

Cafrune MM, Aguirre DH, Rickard LG (1999) Recovery of *Trichuris tenuis* Chandler, 1930, from camelids (*Lama glama* and *Vicugna vicugna*) in Argentina. J Parasitol 85(5):961–962

Cafrune MM, Aguirre DH, Freytes I (2004) Fasciolosis en vicuñas (Vicugna vicugna) en semicautiverio de Molinos, Salta, Argentina, con notas de otros helmintos en este hospedador.

Cafrune M, Marín R, Salatin A, Pivotto R, Rigalt F, Vera R et al (2006) Coprología parasitaria en vicuñas (*Vicugna vicugna*) de la reserva de Laguna Blanca, Catamarca, Argentina. In: IV Congreso Mundial sobre Camélidos, Resúmenes y trabajos, vol 11

Cafrune MM, Marín RE, Rigalt FA, Romero SR, Aguirre DH (2009a) *Lamanema chavezi* (Nematoda: Molineidae): epidemiological data of the infection in South American camelids of Northwest Argentina. Vet Parasitol 166(3–4):321–325

Cafrune MM, Marín RE, Rigalt FA, Romero SR, Aguirre DH (2009b) Prevalence of *Eimeria macusaniensis* and *Eimeria ivitaensis* in South American camelids of Northwest Argentina. Vet Parasitol 162(3–4):338–341

Cafrune MM, Romero SR, Aguirre DH (2014) Prevalence and abundance of *Eimeria* spp. infection in captive vicuñas (*Vicugna vicugna*) from the Argentinean Andean Altiplano. Small Rumin Res 120(1):150–154

Cafrune MM, Viñabal AE, Romero SR, Rigalt FA, Aguirre DH (2015) Disímil infestación parasitaria en vicuñas (*Vicugna vicugna*) silvestres de dos áreas de la Puna Argentina. In: Congreso Latinoamericano de Especialistas en Pequeños Rumiantes y Camélidos Sudamericanos. 9. Congreso Argentino de Produccion Caprina. 2. Foro Nacional de Productores Caprinos, La Rioja, Argentina

Cardozo PA (2019) Caracterización de las especies parasitarias de ovinos, caprinos y camélidos sudamericanos en la Puna de Catamarca, Facultad de Ciencias Agrarias, Universidad Nacional de Mar del Plata. (Master dissertation)

Castillo H, Chávez A, Hoces D, Casas E, Rosadio R, Wheeler JC (2008) Contribución al estudio del parasitismo gastrointestinal en guanacos (*Lama guanicoe cacsilensis*). Rev de Investig Vet del Perú 19(2):168–175

Cebra CK, Valentine BA, Schlipf JW, Bildfell RJ, McKenzie E, Waitt LH, ... & Firshman AM (2007) Eimeria macusaniensis infection in 15 llamas and 34 alpacas. Journal of the American Veterinary Medical Association, 230(1), 94–100

Cebra C (2014) Disorders of the digestive system. Llama and Alpaca care, 477

Cepeda CP, Navarro C, Celedón MO (2011) Prospección serológica del virus parainfluenza 3 en camélidos sudamericanos en Chile. Arch Med Vet 43(2):177–179

Chávez-Velásquez A, Alvarez-Garcia G, Gómez-Bautista M, Casas-Astos E, Serrano-Martínez E, Ortega-Mora LM (2005) *Toxoplasma gondii* infection in adult llamas (*Lama glama*) and vicunas (*Vicugna vicugna*) in the Peruvian Andean region. Vet Parasitol 130(1–2):93–97

Chigerwe M, Middleton JR, Williams III F, Tyler JW, Kreeger JM (2007) Atypical coccidiosis in South American camelids. J Vet Diagn Invest, 19(1):122–125

Collado B, Salgado M, Corti P, Radic S, Moraga C (2019) Evidencia de interacción infectológica por *Mycobacterium avium* subsp. *paratuberculosis* entre guanacos y ovinos que co-pastorean la Patagonia. XXXII Jornadas Argentinas de Mastozoología, Puerto Madryn. Argentina

Colombo V, Rago V, Moreno P, Marull C, Perassi R, Macedo A, Gutierrez GL, Uhart M, Carmanchahi P, Blanco C, Beldomenico P (2009) Estacionalidad de endoparásitos en guanacos (*Lama guanicoe*) de La Reserva La Payunia, Mendoza. En V Congreso Argentino de Parasitología, Argentina. Asociación Parasitológica Argentina

CONAF (2014) Informe XXXI Reunión Ordinaria Comisión Técnico Administradora, Convenio de la Vicuña. Elaborado por Corporación Nacional Forestal (CONAF), región de Arica y Parinacota, Gobierno de Chile. Carlos Nassar San Martin (ed). 74 p. Spanish. Available from: HYPERLINK "http://intra.conaf.cl/" http://intra.conaf.cl/odules/contents/files/unit31/file/3bd8505b7bb11b0c5ad9c9f440a914b3.pdf

Condori W, Gutiérrez E, Mamani W, Guzmán J (2012) Determinación de parásitos gastrointestinales en vicuñas silvestres en dos comunidades del ANMIN Apolobamba La Paz, Bolivia. VI Congreso mundial de camélidos sudamericanos, Arica. Chile

Correa L, Zapata B, Soto-Gamboa M (2011) Gastrointestinal and blood parasite determination in the guanaco (*Lama guanicoe*) under semi-captivity conditions. Trop Anim Health Prod 44(1):11–15

Corti P, Collado B, Salgado M, Moraga CA, Radic-Schilling S, Tejeda C, Ruiz-Aravena M (2022) Dynamic of Mycobacterium avium subspecies paratuberculosis infection in a domestic–wildlife interface: domestic sheep and guanaco as reservoir community. Transbound Emerg Dis 69(4):e161–e174

Cunazza PC (1981) Extracción experimental de 100 guanacos en el sector Cameron, Tierra del Fuego. In: Claudio Venegas C, Claudio Cunazza P (eds) Actas de la Convención Internacional sobre Camélidos Sudamericanos. Inst. de la Patagonia y CONAF, Punta Arenas, Chile, pp 100–115

Curay Cabanillas JJ (2018) Helmintiasis en vicuñas (*Vicugna vicugna*) en el distrito de Contumazá, departamento de Contumazá-Cajamarca. Perú

Dale WE, Venero JL (1977) Insectos y ácaros ectoparásitos de la Vicuña en Pampa Galeras, Ayacucho. Rev Peru De Entolomogía 20(1):93–99

Díaz A, Allen JE (2007) Mapping immune response profiles: the emerging scenario from helminth immunology. Eur J Immunol, 37(12):3319–3326

Dubey JP (2018) A review of coccidiosis in South American camelids. Parasitol Res 117:1999–2013

Dubey JP (2019) Coccidiosis in South American Camelids. Coccidiosis in Livestock, Poultry, Companion Animals, and Humans, 153–158

Ellis V, Varela B, Fernández M, Chiaradia N, Kloster D, Mansilla A, Perrig P, Pritchard C, Middleton A, Sheriff M, Donadio E (2017) Evolución e impacto de un brote de sarna sarcóptica en las poblaciones de camélidos del Parque Nacional San Guillermo, San Juan. XXX Jornadas Argentinas de Mastozoología, Bahía Blanca. Argentina

Estrada-Peña A, Venzal JM, Mangold AJ, Cafrune MM, Guglielmone AA (2005) The Amblyomma maculatum Koch, 1844 (Acari: Ixodidae: Amblyomminae) tick group: diagnostic characters, description of the larva of A. parvitarsum Neumann, 1901, 16S rDNA sequences, distribution and hosts. Syst Parasitol, 60, 99–112

Ferreyra H, Donadío E, Uhart M (2020) Box 10.1. Un brote de sarna sarcóptica diezma las vicuñas del Parque Nacional San Guillermo, Argentina. In: González BA (ed) La Vicuña Austral. Facultad de Ciencias Forestales y de la Conservación de la Naturaleza, Corporación Nacional Forestal y Grupo Especialista en Camélidos Sudamericanos Silvestres, Santiago, Chile, pp i–ii

Ferreyra HDV, Rudd J, Foley J, Vanstreels RE, Martín AM, Donadio E, Uhart MM (2022) Sarcoptic mange outbreak decimates South American wild camelid populations in San Guillermo National Park, Argentina. PLoS One 17(1):e0256616

Fowler M (2011) Medicine and surgery of camelids. John Wiley & Sons, Hoboken

Franklin WL (1982) Biology, ecology, and relationship to man of the South American camelids. Mamm Biol South Am 6:457–489

Franklin WL (1983) Contrasting socioecologies of South America's wild camelids: the vicuña and the guanaco. In: Advances in the study of mammalian behavior, vol 7. The American Society of Mammalogists, pp 573–629

Gomez-Puerta LA, Olazabal J, Taylor CE, Cribillero NG, Lopez-Urbina MT, Gonzalez AE (2013) Sarcoptic mange in vicuna (*Vicugna vicugna*) population in Perú. Vet Rec 173(11):269

González BA, Acebes P (2016) Reevaluación del guanaco para la Lista Roja de la UICN: situación actual y recomendaciones a futuro. GECS News 6:15–21

González-Rivas CJ, Borghi CE, De Lamo DA (2019) Endoparásitos en guanaco (*Lama guanicoe*): Revisión de situación en Argentina y registros de la provincia de San Juan. Rev de Investig Vet del Perú 30(1):339–349

Guerrero C (1960) Helmintos en vicuñas (*Vicugna vicugna*). Rev Fac Med Vet, Lima 15:103–105

Guerrero C, Hernández J, Alva J (1967) Coccidiosis en alpacas. Rev Fac Med Vet UNMSM, Perú 21:59–68

Guerrero CA, Alva J, Bazalar H, Tabacchi L (1970) Infección experimental de alpacas con *Eimeria lamae*. Boletin Exptraordinario Instituto Veterinario de Investigaciones Tropicales y Altura 4:79–83

Hofmann R, Otte K, Ponce C, Ríos M (1983) El Manejo de la Vicuña Silvestre. Tomo II. GTZ, Eschoborn

IPCC (2022) In: Pörtner H-O, Roberts DC, Tignor M, Poloczanska ES, Mintenbeck K, Alegría A, Craig M, Langsdorf S, Löschke S, Möller V, Okem A, Rama B (eds) Climate change 2022: impacts, adaptation, and vulnerability. Contribution of Working Group II to the sixth assessment report of the Intergovernmental Panel on Climate Change. Cambridge University Press, Cambridge, New York, 3056 pp. https://doi.org/10.1017/9781009325844

Issia L, Pietrokovsky S, Sousa-Figueiredo J, Stothard JR, Wisnivesky-Colli C (2009) *Fasciola hepatica* infections in livestock flock, guanacos and coypus in two wildlife reserves in Argentina. Vet Parasitol 165(3–4):341–344

Jarvinen JA (1999) Prevalence of Eimeria macusaniensis (Apicomplexa: Eimeriidae) in midwestern Lama spp. The Journal of parasitology, 373–376

Jarvinen JA (2008) Infection of llamas with stored Eimeria macusaniensis oocysts obtained from guanaco and alpaca feces. J Parasitol, 94(4):969–972

Johnson AL, Stewart JE, Perkins GA (2009) Diagnosis and treatment of Eimeria macusaniensis in an adult alpaca with signs of colic. The Veterinary Journal, 179(3):465–467

Kadwell M, Fernandez M, Stanley HF, Baldi R, Wheeler JC, Rosadio R, Bruford MW (2001) Genetic analysis reveals the wild ancestors of the llama and the alpaca. Proc R Soc Lond Ser B Biol Sci 268(1485):2575–2584

Karesh WB, Uhart MM, Dierenfeld ES, Braselton WE, Torres A, House C et al (1998) Health evaluation of free-ranging guanaco (*Lama guanicoe*). J Zoo Wildl Med 29:134–141

Larrieu E, Bigatti R, Lukovich R, Eddi CS, Bonazzi EF, Gómez E et al (1982) Contribución al estudio del parasitismo gastrointestinal en guanacos (*Lama guanicoe*) y llamas (Lama glama). comunicación. Gaceta Veterinaria 44(374):958–960

Leguía G (1991) The epidemiology and economic impact of llama parasites. Parasitology Today, 7(2), 54–56

Leguía G (1999) Enfermedades Parasitarias de Camélidos Sudamericanos. In: Díaz Vargas F (ed) Primera edn. Editorial de Mar EIRL, Lima, 191 p

Lichtenstein G, Carmanchahi PD (2012) Guanaco management by pastoralists in the Southern Andes. Pastoralism: Res Policy Pract 2(1):1–16

Llorente P, Leoni L, Martínez Vivot M (2002) Leptospirosis en camélidos sudamericanos. Estudio de prevalencia serológica en distintas regiones de la Argentina. Arch Med Vet 34(1):59–68

Marcoppido G, Parreño V, Vilá B (2010) Antibodies to pathogenic livestock viruses in a wild vicuña (*Vicugna vicugna*) population in the Argentinean Andean Altiplano. J Wildl Dis 46(2):608–614

Marcoppido G, Olivera V, Bok K, Parreño V (2011) Study of the kinetics of antibodies titres against viral pathogens and detection of rotavirus and parainfluenza 3 infections in captive crias of guanacos (*Lama guanicoe*). Transbound Emerg Dis 58(1):37–43

Marcoppido GA, Schapiro JH, Morici G, Arzamendia Y, Vila BL (2016) Coproparasitological evaluation of nematodes and coccidia in a wild vicuña (*Vicugna vicugna*) population in the Argentinean Andean Altiplano

Martela Mamani W (2016) Identificacion de parasitos gastrointestinales en poblaciones de vicuña (*Vicugna Vicugna*) en tres regiones de Bolivia (dissertation)

Marull C, Rago MV, Uhart MM, Samartino L, Lomónaco M, Marcoppido G, Parreño V, Beldoménico P, Banco C, Paolichi F, Acuña F, Ferreyra H, Funes M, Carmanchahi P (2012) Estado sanitario de guanacos silvestres (*Lama guanicoe*) en Patagonia argentina. X Congreso Internacional de Manejo de Fauna Silvestre en la Amazonia y América Latina

Mayhua Mendoza P (2021) Estrategia para el control y tratamiento de la sarna (*Sarcoptes scabiei* var. *aucheniae*) en vicuñas (*Vicugna vicugna*) de la Reserva Nacional Pampa Galeras Barbara D'Achile-Ayacucho. (Magister dissertation)

Melhorn H (2008) Encyclopedia of parasitology. Springer Verlag, Berlin

Mollericona JL (2022) Evaluación de la salud en vicuñas (hematología, sarna y otros parásitos externos e internos) durante capturas, esquilas y liberaciones de vicuñas en comunidades del ANMIN Apolobamba, La Paz – Bolivia, 2021. Informe técnico. Wildlife Conservation Society, La Paz, 22 p

Mollericona JL, Mamani E, Maidana A (2020) Evaluación de parásitos en vicuñas (*Vicugna vicugna*) en la Asociación Regional de Comunidades Manejadoras de Vicuñas Colcha K, Potosí-Bolivia, 2019. Informe técnico. Wildlife Conservation Society, La Paz, 16 p

Mollericona JL, Coronel C, Torrez A (2021a) Evaluación de parásitos en vicuñas (*Vicugna vicugna*) de la comunidad Ayllu Collana Primero, ARCMV Wila Kollo Oruro-Bolivia, 2021. Informe técnico. Wildlife Conservation Society, La Paz, 12 p

Mollericona JL, Coronel C, Torrez A, Mamani J (2021b) Evaluación de parásitos en vicuñas (*Vicugna vicugna*) en la Asociación Regional de Comunidades Manejadoras de Vicuñas Tomave, Potosí-Bolivia, 2021. Informe técnico. Wildlife Conservation Society, La Paz, 14 p

Monk JD, Smith JA, Donadío E, Perrig PL, Crego RD, Fileni M et al (2022) Cascading effects of a disease outbreak in a remote protected area. Ecol Lett 25(5):1152–1163

Montecino-Latorre D, Napolitano C, Briceño C, Uhart MM (2020) Sarcoptic mange: an emerging threat to Chilean wild mammals? Perspect Ecol Conserv 18(4):267–276

Moreno PG (2014) Factores asociados a parasitismo gastrointestinal en guanacos silvestres (*Lama guanicoe*). Tesis de Doctorado. Facultad de Bioquímica y Ciencias Biológicas, Universidad Nacional del Litoral

Moreno PG, Eberhardt MAT, Lamattina D, Previtali MA, Beldomenico PM (2013) Intra-phylum and inter-phyla associations among gastrointestinal parasites in two wild mammal species. Parasitol Res 112(9):3295–3304

Moreno PG, Schroeder NM, Taraborelli P, Gregorio P, Carmanchahi PD, Beldomenico PM (2015) La comunidad de parásitos gastrointestinales de guanacos silvestres (*Lama guanicoe*) de la reserva provincial La Payunia, Mendoza, Argentina. Mastozool Neotrop 22(1):63–71

Moreno P, Laudecina A, Marinozzi A, Caballero S, Faccas C, Sosa J, Fernandez R, Vélez L, Albanese S, Linares C, Peña F (2019) Caracterización preliminar de las comunidades endoparasitarias de las poblaciones de guanacos (*Lama guanicoe*) de Áreas Protegidas de Mendoza. IV Jornadas de Conservación y Manejo de Áreas Naturales Protegidas, I Encuentro Binacional de Conservación de Humedales. Mendoza, Argentina

Muñoz-Leal S, González-Acuña D, Beltrán-Saavedra LF, Limachi JM, Guglielmone AA (2014) *Amblyomma parvitarsum* (Acari: Ixodidae): localities, hosts and host-parasite ecology. Exp Appl Acarol 62(1):91–104

Muñoz-Leal S, Tarragona EL, Martins TF, Martín CM, Burgos-Gallardo F, Nava S et al (2016) Liolaemus lizards (Squamata: Liolaemidae) as hosts for the nymph of Amblyomma parvitarsum (Acari: Ixodidae), with notes on Rickettsia infection. Exp Appl Acarol 70:253–259

Muñoz-Leal S, Martins TF, Luna LR, Rodriguez A, Labruna MB (2018) A new collection of *Amblyomma parvitarsum* (Acari: Ixodidae) in Peru, with description of a Gynandromorph and report of *Rickettsia* detection. J Med Entomol 55(2):464–467

Navone G, Merino M (1989) Contribución al conocimiento de la fauna endoparasitaria de *Lama guanicoe* Muller, 1776, de Península Mitre, Tierra del Fuego, Argentina. Bol Chil Parasitol 44:46–51

Niedringhaus KD, Brown JD, Sweeley KM, Yabsley MJ (2019) A review of sarcoptic mange in North American wildlife. Int J Parasitol Parasites Wildl 9:285–297

Ogrzewalska M, Nieri-Bastos FA, Marcili A, Nava S, González-Acuña D, Muñoz-Leal S et al (2016) A novel spotted fever group Rickettsia infecting *Amblyomma parvitarsum* (Acari: Ixodidae) in highlands of Argentina and Chile. Ticks Tick Borne Dis 7(3):439–442

Olaechea FV, Abad M (2005) An outbreak of fascioliasis in semicaptive guanacos (Lama guanicoe) in Patagonia (Argentina). First report. In XX International Conference of the World Association for the Advancement of Veterinary Parasitology. Christchurch, New Zealand.

Olaechea F, Larroza M, Raffo F (2011) Hallazgos parasitológicos en guanacos (*Lama guanicoe*) en el Laboratorio de Parasitología de la EEA INTA Bariloche (2001–2010). Rev Argent Prod Anim 31 Supl. 1:1–47

Palacios CA, Perales RA, Chavera AE, Lopez MT, Braga WU, Moro M (2006) *Eimeria macusaniensis* and *Eimeria ivitaensis* co-infection in fatal cases of diarrhoea in young alpacas (*Lama pacos*) in Perú. Vet Rec 158(10):344

Parreño V, Costantini V, Cheetham S, Blanco Viera J, Saif LJ, Fernández F et al (2001) First isolation of rotavirus associated with neonatal diarrhoea in guanacos (*Lama guanicoe*) in the Argentinean Patagonia region. J Veterinary Med Ser B 48(9):713–720

Pastor J, Chávez A, Casas E, Serrano E (2003) Seroprevalencia de *Toxoplasma gondii* en vicuñas de Puno. Rev de Investig Vet del Perú 14(1):79–82

Pinedo K, Chávez A, Rivera H, Pinedo R, Suárez F (2014) Frecuencia de *Toxoplasma gondii* y *Neospora caninum* en vicuñas (*Vicugna vicugna*) de la sierra central peruana mediante las técnicas de inmunofluorescencia indirecta y ELISA indirecta. Rev de Investig Vet del Perú 25(1):70–76

Quispe García HH (2011) Estudio de parásitos externos y gastrointestinales en vicuñas (*Vicugna vicugna mensalis*) en el anexo Mamuta de la provincia de Tarata en la Región de Tacna

Raedeke KJ (1976) El guanaco de Magallanes, Chile: distribución y biología. Corporación Nacional Forestal

Rago MV, Colombo VC, Moreno P, Beldomenico P, Marull C, Macedo A et al (2008) Convivencia forzada y parásitos en la Reserva Provincial La Payunia, Mendoza:¿desventajas para el guanaco. XXII Jornadas Argentinas de Mastozoología

Rago MV, Colombo VC, Marull C, Carmanchahi P, Moreno P, Macedo AM et al (2012) Comparación endoparasitaria entre poblaciones de guanacos silvestres (*Lama guanicoe*) en dos áreas protegidas de la Patagonia Argentina. In: X Congreso Internacional de Manejo de Fauna Silvestre en la Amazonía y Latinoamérica. 156 pp

Rago V, Marull C, Ferreyra H, Carmanchahi P, Uhart M (2022) Health Risks for Guanaco (Lama guanicoe) Conservation. In Guanacos and People in Patagonia: A Social-Ecological Approach to a Relationship of Conflicts and Opportunities (pp. 85–102). Cham: Springer International Publishing

Rohbeck S (2006) Parasitosen des Verdauungstrakts und der Atemwege bei Neuweltkameliden: Untersuchungen zu ihrer Epidemiologie und Bekämpfung in einer südhessischen Herde sowie zur Biologie von Eimeria macusaniensis

Rickard LG, Foreyt WJ (1992) Experimental fascioliasis in Lamas. J Helminthol Soc Wash 59(1):140–144

Risco-Castillo V, Wheeler JC, Rosadio R, García-Peña FJ, Arnaiz-Seco I, Hoces D et al (2014) Health impact evaluation of alternative management systems in vicuña (*Vicugna vicugna mensalis*) populations in Perú. Trop Anim Health Prod 46(4):641–646

Rojas M (2004) Nosoparasitosis de los rumiantes domésticos peruanos, 2nd edn. Martegraf, Lima

Rosadio RH, Ameghino EF (1994) Coccidial infections in neonatal Peruvian alpacas. The Veterinary record, 135(19):459–460

Rosadio R, Londoñe P, Pérez D, Castillo H, Véliz A, Llanco L et al (2010) *Eimeria macusaniensis* associated lesions in neonate alpacas dying from enterotoxemia. Vet Parasitol 168(1–2):116–120

Ruiz Hurtado CR (2016) Identificación y caracterización de la presencia de ectoparásitos y endoparásitos en vicunas (*Vicugna vicugna*) en comunidades de los departamentos de La Paz y Oruro (Magister dissertation)

Salgado M, Herthnek D, Bölske G, Leiva S, Kruze J (2009) First isolation of *Mycobacterium avium* subsp. *paratuberculosis* from wild guanacos (*Lama guanicoe*) on Tierra del Fuego Island. J Wildl Dis 45(2):295–301

Samamé LM, Chávez A, Pinedo R (2016) Fasciolosis en vicuñas (*Vicugna vicugna*) de la sierra central del Perú. Rev de Investig Vet del Perú 27(1):137–144

Santana JL, Martínez A, Soulés A, Milicevic F, Cafrune Wierna MM, Larroza MP, Robles CA (2020) Hepatitis parasitaria por Lamanema chavezi en guanacos (*Lama guanicoe*) faenados en la Provincia de Santa Cruz, Argentina. Sociedad de Medicina Veterinaria

Schrey CF, Abbott TA, Stewart VA, Marquardt WC (1991) Coccidia of the llama, Lama glama, in Colorado and Wyoming. Vet Parasitol 40(1–2):21–28

Schroeder NM, Ovejero R, Moreno PG, Gregorio P, Taraborelli P, Matteucci SD, Carmanchahi PD (2013) Including species interactions in resource selection of guanacos and livestock in N orthern P atagonia. J Zool 291(3):213–225

Sosa FE, Bertoni EA, Micheloud JF, Vallejo D, Medina MN, Olmos LH et al (2022) Occurrence of sarcoptic mange in free-ranging vicuñas (*Vicugna vicugna*) of the Andean high plateau region of Argentina. Parasitol Res 121:1587–1595

Suárez VH, Olaechea FV, Rossanigo C, Romero J (2007) Enfermedades parasitarias de los ovinos y otros rumiantes menores en el cono sur de América. Publicación técnica, 70

Taglioretti V, Fugassa MH, Sardella NH (2015) Parasitic diversity found in coprolites of camelids during the Holocene. Parasitol Res 114(7):2459–2464

Unzueta Lancho LA (2018) Sarna en vicuñas (*Vicugna vicugna*) en las provincias de Aymaraes y Andahuaylas de la región Apurímac. (MVZ dissertation)

Valdebenito Díaz M (2008) Estudio de prevalencia de hidatidosis en guanacos (*Lama guanicoe*) Tierra del Fuego, Chile

Vargas S, Bonacic C (2020) Nuevas Amenazas para la vicuña austral. In: González BA (ed) La Vicuña Austral. Facultad de Ciencias Forestales y de la Conservación de la Naturaleza, Corporación Nacional Forestal y Grupo Especialista en Camélidos Sudamericanos Silvestres, Santiago, Chile, pp i–ii

Vilá BL (1992) Vicuñas (*Vicugna vicugna*) agonistic behaviour during the reproductive season. In: Spitz F, Janeau G, Gonzalez G, Aulagnier S (eds) Ongulés/Ungulates, vol 91. SFEPM y IRGM, Toulosse, pp 475–482

Wernery U, Kaaden OR (2002) Infectious diseases in camelids. Georg Thieme Verlag, Stuttgart

Wheeler JC (2012) South American Camelids – Past, Present and Future. J Camelid Sci 5:1–24

Wolf D, Schares G, Cardenas O, Huanca W, Cordero A, Bärwald A et al (2005) Detection of specific antibodies to *Neospora caninum* and *Toxoplasma gondii* in naturally infected alpacas (*Lama pacos*), llamas (*Lama glama*) and vicuñas (*Lama vicugna*) from Perú and Germany. Vet Parasitol 130(1–2):81–87

Zuzunaga M, Chávez A, Li O, Evaristo R (2006) *Toxoplasma gondii* en vicuñas de la Reserva Nacional de Pampa Galeras. Rev de Investig Vet del Perú 17(2):173–177

Chapter 12
Infectious Diseases of Marine Mammals as Sentinels of Ecosystem Health in the Neotropics

Galaxia Cortés-Hinojosa and Mauricio Seguel

12.1 Introduction

Neotropical marine ecosystems harbor a substantial portion of global biodiversity and modulate planetary climate patterns (Sherman et al. 2017). These ecosystems also provide goods and services to nearly 80% of the population in Latin American and Caribbean countries and produce more than a third of the seafood on the planet (Gutierrez et al. 2017; Muñoz Sevilla and Le Bail 2017). Despite their importance, knowledge on the biological processes and health of these ecosystems is scarce and, in some areas, nonexistent, contrasting with the more studied and more protected neotropical terrestrial ecosystems. Part of this lack of knowledge is due to the size of these ecosystems and the long dispersion rates of most of their species, making it difficult to track and collect data from living organisms (Hazen et al. 2019). In this context, conspicuous, gregarious species with high site fidelity, such as marine mammals, are valuable allies to assess the health of marine ecosystems (Hazen et al. 2019). These charismatic species are also hosts of a wide range of pathogens that establish delicate equilibriums within the host (Bossart 2011). Many of these host-parasite relationships are sensitive to environmental pressures, which can facilitate parasite virulence and compromise marine mammal health while informing larger changes in the marine environment. Therefore, not only do marine mammals serve

G. Cortés-Hinojosa (✉)
Escuela de Medicina Veterinaria, Facultad de Agronomía e Ingeniería Forestal, Facultad de Ciencias Biológicas y Facultad de Medicina, Pontificia Universidad Católica de Chile, Santiago, Chile
e-mail: galaxia.cortes@uc.cl

M. Seguel
Department of Pathobiology, Ontario Veterinary College, University of Guelph, Guelph, ON, Canada

G. Acosta-Jamett, A. Chaves (eds.), *Ecology of Wildlife Diseases in the Neotropics*, https://doi.org/10.1007/978-3-031-50531-7_12

as sentinels, but their pathogens also provide valuable information on ecosystem status.

Marine mammals are key components of aquatic ecosystems throughout the neotropics, occupying high trophic levels in most cases. These trophic relationships make them sensitive to changes in lower trophic levels, a fundamental quality of good sentinel species. Sentinels are species that respond to ecosystem variability or change in a timely, measurable, and interpretable manner (Hazen et al. 2019). Additionally, sentinels provide information that would otherwise be difficult to access due to technical or spatiotemporal limitations. In the case of marine mammals, some species can be easily accessed and sampled on land, while others can be actively tracked and monitored at sea, providing valuable information on processes such as trophic ecology, demography, reproductive success, and health (Bossart 2011). These marine mammal population traits can inform changes in the marine ecosystem, such as prey abundance, marine productivity, and changes in species distribution. Without the use of sentinel species such as marine mammals, the acquisition of this information requires extensive assessment of multiple species over large temporal and spatial scales, which can be logistically challenging. Therefore, top predators such as marine mammals are important for the study of ocean ecosystems. In the context of predator-prey relationships that inform ecosystem conditions, marine mammal pathogens are also highlighted as excellent sentinels. For example, parasites feed on the tissues or food resources of the host, making them predators at a higher trophic level than their host. Therefore, their population traits can also inform the abundance and health of lower trophic levels, including marine mammals. In this chapter, we will discuss how marine mammals and the infectious organisms that exploit them can be valuable sentinels of marine ecosystem health. We will use a few well-studied systems in the neotropics as case studies to highlight that long-term and experimental approaches in these species become valuable tools to explore the mechanisms that underlie the connection between changes in marine ecosystems and the dynamics of infectious diseases in marine mammals.

12.2 Threats for Neotropical Marine Ecosystems and Marine Mammal Health

In the neotropics, at least ten large marine ecosystems (LMEs) are recognized based on geography, major hydrodynamic processes, and biodiversity (Muñoz Sevilla and Le Bail 2017). These ecosystems encompass a diverse arrangement of marine resources ranging from tropical coral reefs in the Caribbean Sea to cold water upwelling systems in the Humboldt Current and Patagonian shelf large marine ecosystems (Sherman et al. 2017). In most of these ecosystems, a diverse arrangement of marine mammal species occupies the top trophic levels. Neotropical LMEs harbor more than 60% global marine mammal diversity, including several endangered species, such as franciscana dolphin (*Pontoporia blainvillei*), West Indian manatee

(*Trichechus manatus*), marine otter (*Lontra felina*), blue whales (*Balaenoptera musculus*), and the most endangered marine mammal in the world, the vaquita (*Phocoena sinus*) (Jefferson et al. 2015). In all these cases, anthropogenic impacts limit the recovery and survival of species. However, these anthropogenic pressures also affect other components of marine ecosystems, such as primary productivity and biomass. In these cases, marine mammal abundance is a general indicator of the general health of their environment and the impact of human exploitation and pollution of oceans.

Threats to other global marine ecosystems are also major issues for neotropical LMEs. All these threats are directly or indirectly linked to anthropogenic pressures such as eutrophication, overfishing, bycatch, plastic and organic pollution, and climate change (Abelson et al. 2020). However, one of the most significant is global climate change, with increases in temperature and decreases in pH as the two key physicochemical factors associated with a cascade of events that threaten marine biodiversity (Henson et al. 2017; Lotze et al. 2019). Ocean acidification has been particularly detrimental in tropical LMEs. More acidic seawater affects the deposition of Ca^{+} in the skeleton of corals, the backbone of tropical marine ecosystems (Doney et al. 2020). Coral reefs serve as nurseries for fish and other vertebrate species; therefore, disease and loss of coral are associated with an overall decline in biomass and diversity (Doney et al. 2020). Similar to pH, temperature is a major determinant of nutrient circulation and/or influx in marine ecosystems, particularly for those that depend largely on upwelling for primary productivity (Lotze et al. 2019). In the neotropics, this is the case for the Humboldt Current LME and the California Current LME. In these areas, an increase in temperature is associated with a decrease in nutrient upwelling and a decline in primary productivity (Gutierrez et al. 2016; Espinosa-Carreón and Escobedo-Urías 2017). This leads to a decline in prey for top predators such as marine mammals, and in the case of species with limited dispersion capacity (central place foragers), this translates into a lower intake of high-quality, nutritious prey. These changes can escalate to lower reproductive success and decline of health in these species, as well as to changes in the dynamics of their parasites. Therefore, marine mammal population traits and the trends of their parasites are intimately linked to changes in these LMEs. However, the long-term consequences of climate change on marine mammal populations and their pathogens (viruses, bacteria, fungi, and parasites) are still unclear.

Marine pollution is a serious concern for biodiversity and health in neotropical LMEs. Although it is not fully understood how many of the organic and inorganic pollutants affect sentinel species, the evidence for some of the best studied substances is alarming. For instance, polychlorinated biphenyls and heavy metals have been associated with cancer, immunosuppression, and increased susceptibility to infectious diseases in marine mammals worldwide (Desforges et al. 2016). In the neotropics, research on the impact of these contaminants is accumulating, although concentrations of organic pollutants are usually lower compared to more industrialized regions of the world (Durante et al. 2016; Alava et al. 2020). In the case of heavy metals, some populations of neotropical cetaceans exhibit concentrations within the range known to cause immunotoxicity (Alava et al. 2020), raising concerns of potential health impacts on these species. Less clear is the role of microplastic pollution in the health of marine mammals, although evidence from invertebrates suggests that these particles can be incorporated into animal tissues, which could cause inflammation and other physiological changes. Additionally, microplastics can serve as sources or vectors for chemical pollutants such as bisphenol A, heavy metals, and polychlorinated biphenyls (PCBs) (Thushari and Senevirathna 2020), which have known endocrine and immune toxicity effects in mammalian cells (Oehlmann et al. 2009; Desforges et al. 2016). Microplastics can also adsorb and accumulate antibiotics, suggesting a potential impact on the mammalian microbiome (Fackelmann and Sommer 2019), a fundamental factor for immunity and health. For macroplastics, the impacts are usually direct and associated with mortality due to ingestion or entanglement (Jepsen and de Bruyn 2019; Kühn and van Franeker 2020). Macro- and microplastic pollution in LMEs is similar to those of other densely populated regions of the world (Perez-Venegas et al. 2021; Orona-Návar et al. 2022). Therefore, the impact of direct mortality and the potential health impacts of microplastics should be an important area of research in the neotropics as well as the study of policies to mitigate the release of plastics into the marine environment.

The impact of large-scale industrial fisheries is a serious concern for most global marine ecosystems. Fisheries can impact the health of marine mammals and other vertebrates through direct interactions during fishing activities or by overfishing key resources for top predators. The neotropics contains the largest single ecosystem producer of seafood on the planet, the Humboldt Current LME, which has faced historical overfishing, leading to the depletion of key resources for marine mammals (Gutierrez et al. 2016; Lima et al. 2020). Therefore, the decline or lack of recovery of certain species, such as the Peruvian fur seal and the South American sea lion, has been linked to overfishing (Cárdenas-Alayza et al. 2021). Similar concerns have been raised for other neotropical LMEs with strong fishing pressures, such as the Gulf of California LME and the Patagonian shelf LME, although there is still limited information on a direct link between the decline in commercial marine resources and marine mammal health (Arreguín-Sánchez et al. 2017; Muñoz Sevilla and Le Bail 2017). However, in these ecosystems, as in others around the world with large industrial fisheries, marine mammals suffer the consequences of direct interaction with fisheries through noise pollution or entanglement in fishing gear (Crespo et al. 1997; Arreguín-Sánchez et al. 2017).

12.3 Marine Mammal Infectious Diseases as Ecosystem Health Sentinels

Healthy ecosystems are usually highly productive and resilient and are formed by complex levels of organization and biodiversity (Rapport et al. 1998; Rapport and Maffi 2011). All these attributes are difficult to model and measure, but bioindicators or sentinel species contribute to assessing and evaluating these processes.

Sentinels inform the state of ecosystems through the adaptive and nonadaptive changes that ecosystem alteration forces on them. Therefore, depending on the sensitivity to change and the feasibility of collecting sufficient information, marine mammal species can be good or sufficient sentinels. According to the definitions by Hazen et al. (2019), good ecosystem sentinels are "conspicuous, easily accessible and provide ecosystem information across spatiotemporal scales" (Hazen et al. 2019). Sentinels should also reveal unobserved ecosystem components that are mechanistically linked to the sentinel species. Within this definition, several marine mammal species are highlighted as excellent sentinels. For instance, fur seals and sea lions (otariids) are large charismatic vertebrates that are relatively easy to access in coastal areas. Since these animals depend exclusively on marine resources, they are obligated to spend long periods in water and, given their foraging plasticity, occupy a wide range of marine ecozones from coastal to pelagic waters (Costa 2007). Therefore, the prey they consume and the conditions they experience in the ocean significantly affect their health and reproductive success (Costa 2007). Similarly, other marine mammal groups, such as cetaceans, also offer valuable information on marine ecosystems at even larger scales. For many of these species, especially large whales, their longevity and extreme migratory patterns can also inform historical or distant processes since they are among the few mammals in the world that travel from tropical to artic ecosystems (Roman et al. 2014). However, in contrast to otariids, direct sampling for health assessment in large cetaceans faces several logistical challenges. Fortunately, recent technological developments such as real-time tracking and refinement of laboratory techniques to use small environmental biological samples have contributed to advances in the understanding of cetacean health status in the wild.

Pathogens are one of the main concerns for marine mammal health, and there has been a recent surge in knowledge and understanding of pathogen-host dynamics among marine mammals in the neotropics. This portends a unique opportunity to use not only marine mammal population dynamics and life history traits as sentinels of ecosystem health but also their pathogens, given that they are also key components of marine ecosystems and could be at a higher trophic level than their host. Parasites can be considered predators since they consume host tissues or make use of resources consumed by the host. Therefore, when parasites infect a top predator and feed or replicate at the expense of host tissues, they occupy a top trophic level. This trophic position favors parasite impacts on many species within marine food webs. For instance, the oomycete that causes crayfish plague (*Aphanomyces astaci*) reduces crayfish (*Austropotamobius pallipes*) densities, thereby reducing predation pressure on amphipods, while the crayfish microsporidian parasite *Thelohania contejeani* reduces the ability of the crayfish to catch its amphipod prey (Hatcher et al. 2012). This highlights the important role of parasites in shaping species interactions. However, parasites not only impact ecosystems but are also very sensitive to changes in the ecosystem. For instance, a decline in Anisakis and strongyle nematode prevalence and burden in northern fur seals (*Callorhinus ursinus*) has been associated with a decline in the main prey item of fur seals, the Alaskan pollock, and an overall decline in fur seal density (Kuzmina et al. 2021). Therefore, similar to the

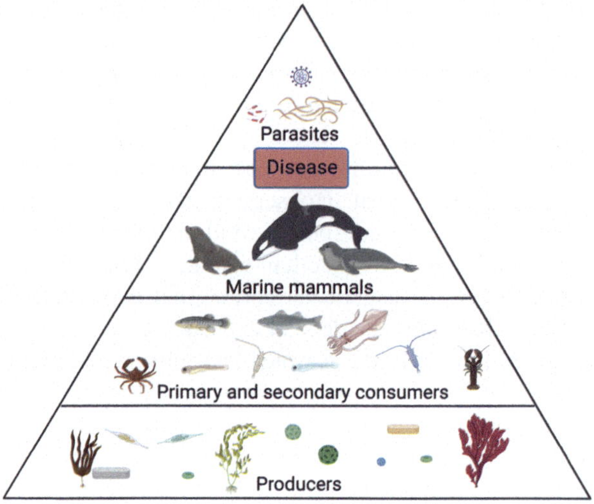

Fig. 12.1 Marine mammal parasites occupy the top trophic levels in marine food webs; therefore, they are highly sensitive to changes in lower trophic levels that also affect their host. The disease induced by these parasites is particularly suited as ecosystem sentinels since they arise from the interaction between the marine mammal host and the parasite. Different from parasite dynamics, which are harder to track, disease states are easier to assess and observe, even without direct handling of animals

case of other top predators, parasites act as sentinels of ecosystem conditions (Marcogliese 2005). However, although parasites are particularly suited to serve as sentinels based on trophic position and sensitivity to ecosystem change, they are small or microscopic lifeforms that require specialized tools to be detected and counted and to assess important ecological traits such as fitness. Fortunately, this disadvantage can be diminished if parasite dynamics and traits are assessed in terms of their impact on the host. Most parasites cause tissue damage, immune activation, disease, and/or mortality in a dose-dependent manner. These parasite effects on the host can be easier to detect and track compared to the assessment of parasite micro-dynamics. Therefore, monitoring infectious diseases in marine mammals paired with adequate detection and estimation of parasite burden can provide an interesting approximation of broader changes in the marine ecosystem (Fig. 12.1).

12.3.1 Marine Mammal Pathogens in the Neotropics

Marine mammal pathogens are a broad range of organisms ranging from the largest metazoan parasites among mammals (blue whale ascaris and cestodes) to small RNA viruses (e.g., morbillivirus). Compared to terrestrial mammals, less is known about the diversity and health impact of many parasite species, especially marine mammal species that are cryptic and difficult to study, such as pelagic or deep diving cetaceans. However, for pinnipeds and coastal cetaceans, there is a growing body of knowledge on these pathogens in the neotropics. Due to logistical reasons and economic and conservation concerns, these investigations have focused on a handful of species in the neotropics, including California sea lions, Galapagos sea lions, Guadalupe fur seals, South American sea lions and fur seals, bottlenose dolphins, manatees, Guiana dolphins, and Burmeister's porpoises (Table 12.1). Although many metazoan and microbial agents have been described in neotropical marine mammals, the evidence of disease associated with these pathogens is scarce. Therefore, a shorter list of pathogens within major taxa are considered an important health concern for marine mammals given their high morbidity and mortality. This also makes most of these organism species excellent sentinels for marine mammal and aquatic ecosystem health.

12.3.1.1 Viruses

Morbilliviruses are the most significant microbial pathogens of marine mammals worldwide. In the neotropics, however, there is still scarce knowledge on the impact of these viruses in free-ranging populations. The best documented case of the tremendous impact these viruses can have occurred in the Rio de Janeiro Bay in Brazil. In this location, hundreds of Guiana dolphins were stranded due to morbillivirus and other coinfections (Groch et al. 2018 and Cunha et al. 2021). Although morbilliviruses have been found in other cetacean species in the neotropics, their impact in these species is uncertain, as is the role of environmental stress in the presentation of outbreaks. Other viruses reported in neotropical marine mammals include poxviruses, herpesviruses, and papillomaviruses. In these groups, the health impact of poxviruses has been documented in more detail. Poxviruses cause skin lesions in cetaceans ranging from isolated tattoo-like lesions to widespread proliferative and ulcerative dermatitis. In association with other pathogens, these viruses have sometimes been indicated to cause poor health and could be contributory factors for stranding and death in cetaceans such as Burmeister's porpoises (*Phocoena spinipinnis*). They are commonly assumed to be a good estimator of cetacean population health since their lesions can be observed and graded through photographic screening, serving as a sentinel of population health. However, skin lesions can be caused by other pathogens, such as fungi, and careful observation and additional testing are recommended. Other viruses of medical importance in mammals, such as coronaviruses, circoviruses, anelloviruses, and adenoviruses, have been detected in neotropical marine mammals, although their role in clinical disease has yet to be confirmed.

Table 12.1 Infectious agents causing disease among marine mammals in the neotropics

Infectious agent group	Infectious agent	Host common name	Host species	Disease	Locations	References
Viruses	Cetacean morbillivirus	Guiana dolphin	*Sotalia guianensis*	Pneumonia, death	Rio de Janeiro, Brazil	Groch et al. (2018) and Cunha et al. (2021)
	Influenza virus	Cetacean, pinnipeds, and mustelids			Peru, Chile	Gamarra-Toledo et al. (2023), Leguia et al. (2023), and WAHIS (2023)
	Otariid herpesvirus 5	South American fur seal	*Arctocephalus australis*	Ulcerative dermatitis	Santa Catarina, Brazil	Sacristán et al. (2018b)
	Alpha herpesvirus (delphinid herpesvirus 10)	Guiana dolphin	*Sotalia guianensis*	Ulcerative and proliferative dermatitis	Brazil	Soto et al. (2012)
	Alpha herpesvirus (kogiid herpesvirus 2)	Dwarf sperm whale	*Kogia sima*	Ulcerative dermatitis	Brazil	Sacristan et al. (2019)
	Gamma herpesvirus (Iniid herpesvirus 1)	Bolivian river dolphin	*Inia boliviensis*	Proliferative dermatitis	Rio Guapore, Brazil	Exposto Novoselecki et al. (2021)
	Cetacean poxvirus	Burmeister's porpoise	*Phocoena spinipinnis*	Proliferative dermatitis	Peru, Chile, Brazil	Van Bressem et al. (1993) and Alvarado-Rybak et al. (2020)
	Papillomaviruses	Burmeister's porpoise	*Phocoena spinipinnis*	Genital papillomas		Van Bressem et al. (2007)

Bacteria	*Streptococcus*	Southern right whale	*Eubalaena australis*	Sepsis	Rio grande do Sul, Brazil	Bianchi et al. (2018)
		South American fur seal	*Arctocephalus australis*	Bronchopneumonia	Guafo Island, Chile	
	Brucella ceti	Several species		Encephalitis Abortion Discospondylitis	Peru, Brazil, Costa Rica	Guzmán-Verri et al. (2012), Hernández-Mora et al. (2017), Attademo et al. (2018), and Sánchez-Sarmiento et al. (2018, 2019)
	Mycoplasma spp.	Galapagos sea lions	*Zalophus wollebaeki*	Purulent rhinitis and bronchitis	Galapagos, Ecuador	Sarzosa et al. (2021)
Helminths	*Halocercus brasiliensis*	Guiana dolphin	*Sotalia guianensis*	Bronchopneumonia	Rio de Janeiro, Brazil	Demarque et al. (2020)
	Uncinaria sp.	South American fur seal	*Arctocephalus australis*		Chile	Seguel et al. (2017)
		South American sea lion	*Otaria byronia*		Argentina	Beron-Vera et al. (2004)
	Nasitrema sp.	Melon-headed whale	*Peponocephala electra*	Encephalitis, death	Trinidad and Tobago	Phillips and Suepaul (2017)

12.3.1.2 Bacteria

Several pathogenic bacteria infect marine mammals in the neotropics. However, *Brucella*, *Mycobacterium*, and *Leptospira* have concentrated most of the research effort given their zoonotic nature and the capacity of these bacteria to cause high morbidity and mortality among marine mammals. *Brucella ceti* is probably one of the most documented pathogenic bacteria of marine mammals in the neotropics. This bacterium causes encephalitis, abortion, discospondylitis, and sepsis in a wide range of cetacean species, and investigation and detection of these pathogens have been particularly active in Costa Rica (Hernández-Mora et al. 2017). *Mycobacterium pinnipedii* is endemic in Southern Hemisphere otariids, although it can infect cetaceans and other mammal species, including humans. This bacterium causes chronic granulomatous pneumonia that sometimes progresses to fatal pleuropneumonia with disseminated thoracic granulomas (De Amorim et al. 2014). Despite being recognized as a virulent mycobacterium, the population-level impact of this pathogen is not well understood. This is probably because animals develop a chronic disease that potentially affects older age categories, which can be harder to monitor at sea. Leptospirosis in marine mammals is usually associated with serovar Pomona (Colagross-Schouten et al. 2002). This bacterium causes subacute to chronic tubulointerstitial nephritis that can sometimes lead to renal failure and death. In the neotropics, the population-level impact of leptospirosis is not well understood compared to what has been documented in California sea lions in the California current ecosystem (Prager et al. 2020). In these populations, cyclic epidemics can significantly impact the recruitment of reproductive males into the population. In South America, *Leptospira interrogans* serovar Pomona causes lesions in South American sea lions (Sepulveda et al. 2015).

12.3.1.3 Fungi and Protozoa

The most recognized fungal pathogen in marine mammals, *Paracoccidioidomycosis ceti* (lobomycosis), is a dimorphic fungus associated with skin lesions and granulomatous and ulcerative dermatitis. At least three species of cetaceans have been identified, *Tursiops truncatus*, *Tursiops aduncus*, and *Sotalia guianenesis*, which include neotropical species. For many years, it was thought that this was the same agent as lobomycosis in humans and was considered a zoonotic agent. However, recent work has demonstrated that the cause of lobomycosis in cetaceans is a different fungus (Vilela et al. 2016). The presence of this skin pathology is associated with environmental stressors or dysfunction of the immune system (Reif et al. 2008, 2009).

Therefore, its monitoring is a clear example of a noninvasive assessment of the health of the dolphin population and ecosystem health. Reports of lobomycosis have occurred in Perú, Venezuela, Brazil, and Chile (Moreno et al. 2008; Van Bressem et al. 2008a, b; Bermudez et al. 2009).

In regard to Protozoa, the most relevant marine mammals belong to the family Sarcocystidae. For example, *Toxoplasma gondii* and *Sarcocystis neurona* are parasites of concern among marine mammals in North America, especially marine otters. *T. gondii* has been identified in the neotropics but not *S. neurona. T. gondii* can remain dormant within cysts or undergo replication in tissues and cause necrosis and inflammation, which can cause systemic disease and even death. In the neotropics, this type of systemic infection has been observed in Guiana and bottlenose dolphins, a Bryde's whale (*Balaenoptera brydei*), and South American fur seals in Brazil. Although antibodies and *T. gondii* cysts have been identified in other species, such as marine otters, their role as the agent of disease in these other neotropical species is yet unknown. Similarly, *Sarcocystis* sp. cysts are a common finding in marine mammal tissues; however, they are generally incidental.

12.3.1.4 Helminths

Helminths are metazoan parasites within the nematode, cestode, and trematode phyla. Helminths are one of the most documented parasites in marine mammals, although there are few studies describing the overall population- and individual-level health impact of these pathogens. Despite the large diversity of helminth species in marine mammals, a few species of lungworms (metastrongyles), tetrameres, hookworms, and respiratory sinus trematodes are highlighted as some of the most virulent members of this group. Metastrongyles such as *Halocercus brasiliensis* and *Halocercus* spp. cause bronchopneumonia and have been associated as contributors or causes of stranding in odontocete semipelagic and coastal cetaceans (Groch et al. 2020a, b). Additionally, in these cetacean species, pseudalid metastrongyles such as *Pseudalius inflexus* cause arteritis, although their role in mortality is not well established in the neotropics. In otariids, metastrongyles in the *Parafilaroides* genus cause bronchopneumonia and stranding, and in some neotropical pinnipeds, such as the Guadalupe fur seal, *Parafilaroides decorus* causes vasculitis and thrombosis (Seguel et al. 2018b). Tetrameres in the *Crassicauda* genus cause arteritis in several major blood vessels in several deep-diving cetacean species (*Ziphiidae whales*) worldwide. These vascular changes can lead to fatal aortic dissection in some cases, as has been recently found in Brazil (Jerdy et al. 2022). Among trematodes, *Nasitrema* species are highlighted as the most pathogenic. Nasitrema are respiratory sinuses parasites of cetaceans; however, they commonly migrate throughout tissues, including the brain, sometimes causing fatal encephalitis (Phillips and Suepaul 2017). The life cycle of most of these metazoan parasites is not fully elucidated, but epidemiological and trophic analyses suggest that most of them have indirect cycles with infective larvae present in marine mammal food items. Therefore, their overall prevalence and burden can be indicators of the diversity and abundance of certain prey items of marine mammals (Kuzmina et al. 2021). One exception of marine mammal parasites with an indirect life cycle is hookworms, which infect pups exclusively through colostrum (Seguel et al. 2018a). Hookworms are common parasites of terrestrial mammals; however, they also infect most otariid and some phocid species. All hookworms of marine mammals belong to the *Uncinaria* genus and cause hemorrhagic enteritis and anemia among pups. Hookworms are the main cause of pup mortality in several fur seal and sea lion rookeries around the globe, including important populations of fur seals and sea lions in the neotropics (Seguel and Gottdenker 2017). Other metazoan parasites of concern in marine mammals include anisakid gastric nematodes, hepatic and pancreatic flukes, and gastrointestinal acantocephalan. However, although these parasites can cause local inflammatory responses in infected tissues, their impact on individual or population health is uncertain.

12.4 Changes in the Marine Ecosystem and the Health of Marine Mammals in the Neotropics

Despite being exposed to a wide range of virulent pathogen species, marine mammals are among the most successful and long-living mammal groups on the planet. This probably reflects the fact that marine mammals, as their terrestrial counterparts, have evolved effective adaptations to establish an equilibrium with their pathogens and maximize fitness. These adaptations to infection can be generally categorized as resistance or tolerance, and both processes are mediated by the host immune system. Resistance involves an active immune response against pathogenic antigens that can boost effector mechanisms that decrease the burden of infection or eliminate pathogens from the host. A different strategy is tolerance, where the host response focuses on diminishing the tissue damage induced by the parasite, which is often achieved by allowing replication of the parasite. Resistance and tolerance immune mechanisms are context dependent, and their strength and dominance depend on the parasite species, infection burden, site of infection, and energy budget of the host. All these factors can be quite variable for marine mammals amid the complex dynamics of their infectious agents that come from land and the ocean and the seasonality and unpredictability of food resources in the marine environment. Therefore, even if the most evolutionarily stable defense strategy for some parasites is resistance, marine mammals are not always capable of initiating and sustaining these responses, which can lead to disbalanced host-parasite relationships and adverse health outcomes for the host.

Neotropical marine ecosystems are among the most productive on the planet; however, their primary productivity and overall biomass can be highly seasonal and unpredictable. For instance, the primary production in the Humboldt Current Large Marine Ecosystem and the California Current Ecosystem is dependent on the upwelling of cold, nutrient-rich waters (Gutierrez et al. 2016). Therefore, during periods of sea surface warming, upwelling is compromised, and primary productivity decreases. This leads to a decline in species at higher trophic levels, decreasing the total marine biomass. For marine mammals, this translates into dynamic changes in the diversity and availability of prey. However, marine mammal species can adapt to these changes with adequate foraging strategies to maximize the energy budget in these constrained environments (Costa 2007). These strategies involve the utilization of alternative prey items and migration to foraging areas with adequate food sources. However, for central place foragers, such as otariids, their reproductive strategies involve prolonged maternal attendance of pups on land. This means that, for these species, their reproductive success depends on adaptations to the resources found near their reproductive colonies. If these resources decline below a certain threshold, these marine mammals cannot compensate for the lower availability of resources and the energy budgets of adults, and therefore, their dependent pups are compromised. Sustained negative energy budgets can lead to improper immune resistance and tolerance mechanisms. For instance, California sea lions (*Zalophus californianus*), South American fur seals (*Arctocephalus australis*), and Galapagos sea lions (*Zalophus wollebaeki*) with lower body conditions and/or blood glucose have milder inflammatory responses (Brock et al. 2013; Vera-Massieu et al. 2015; Seguel et al. 2018c). Similarly, in years with lower availability of nutritious prey items for South American fur seals, T-cell reactivity and production of protective antibodies decrease, leading to a higher burden of hookworms and virulent bacteria such as streptococci (Seguel et al. 2018c).

12.4.1 Links Between Environmental Change and Marine Mammal Epidemics

Changes in climate and nutrient concentrations, as well as overfishing, can lead to decreases in marine productivity and the availability of prey for marine mammals. For marine mammals with limited spatiotemporal adaptability to these changes or when these changes are severe and widespread, the energy budget of individuals and populations can be decreased. This can lead to decreased immune function, increasing susceptibility to pathogens. Environmental stress from chemical or noise pollution and/or interaction with human activities such as fishing can also influence immune function and behavior, increasing susceptibility and exposure to pathogens (aggregation or larger groups). Epidemics will arise when susceptibility in a population reaches a determined threshold for a pathogen or when exposure of susceptible individuals is widespread. A combination of these two processes likely plays a role in most marine mammal epidemics (Fig. 12.2).

12.5 Selected Infectious Diseases

12.5.1 Dolphin Morbillivirus

Morbilliviruses are members of the Mononegavirales family, which includes negative single-stranded RNA viruses. This virus is one of the most relevant viruses in cetaceans and has caused massive mortality since the end of the 1980s, affecting cetaceans worldwide. For this taxon, the signs are generally detected in the postmortem examination and are related to respiratory symptoms due to secondary

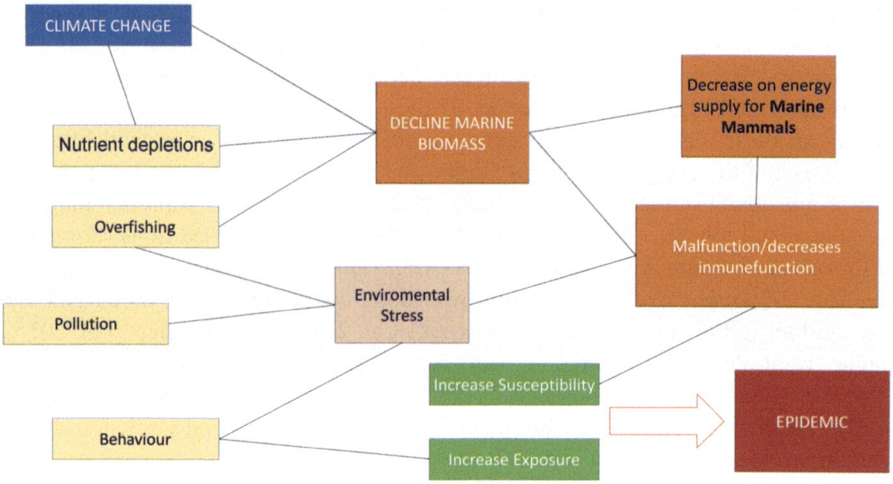

Fig. 12.2 Connection of environmental and host processes that can lead to epidemics in marine mammals

infections. The virus replicates mainly in lymphocytes, epithelial cells, and neurons. Viral replication in lymphoid tissue leads animals to depression of their immune system and the consequent invasion of tissues by other pathogens. The first report of morbillivirus in Brazil was from a single case in 2010 (Groch et al. 2014). Then, an unusual mortality event (UME) occurred between 2017 and 2018 on the coast of Brazil that affected 263 Guiana dolphins (*Sotalia guianensis*). In addition, between 2010 and 2017, eight stranded southern right whales (*Eubalaena australis*) were analyzed (Groch et al. 2020a; Groch et al. 2019). Cetacean morbillivirus was confirmed in these two species and a killer whale (*Orcinus orca*) (Groch et al. 2020b). Currently, five clades of cetacean morbillivirus have been recognized: porpoise morbillivirus (PMV), dolphin morbillivirus (DMV), pilot whale morbillivirus (PWMV), and Longman's beaked whale morbillivirus (BWMV). In the last decade, the fifth clade was associated with massive mortalities in Brazil and Australia (Kemper et al. 2016; Groch et al. 2020a). There are not other reports using molecular techniques of morbillivirus in South America's cetaceans. However, serological evidence of cetacean morbillivirus was reported in Peru in the late 1990s (Van Bressem et al. 1998).

12.5.2 *Influenza Virus*

Orthomyxoviruses are medium-sized (80–120 nm), segmented genome, pleomorphic, enveloped, negative-sense single-stranded viruses with intranuclear and intracytoplasmic replication. Since 2020, an epizootic of highly pathogenic avian influenza A (HPAI) (H5N1) clade 2.3.4.4b has been reported, initially affecting wild birds in Asia, Africa, and Europe. In late 2021, the virus was detected in Canada, and in 2022, cases were initially reported in wild birds in the USA. Then, cases of influenza affecting multiple mammals in the USA, including fox, skunks, and bears, and all cases related to avian-to-mammal transmission, most likely by consumption and/or direct contact with sick birds, were reported (Elsmo et al. 2023). In addition, two species of phocids were reported to be positive for H5N1 harbor (*Phoca vitulina*) and gray (*Halichoerus grypus*) seals in the North Atlantic. In this case, researchers believe that the exposure was because of a high accumulation of viral particles in the ecosystem, since no scavenger behavior has been reported for this species (Puryear et al. 2023). Phocids have been reported to be susceptible to several variants of influenza, and their mortality has been associated since the late 1970s (Lang et al. 1981; Webster et al. 1981; Fereidouni et al. 2016). Therefore, cases of influenza are relatively common.

On the virus migratory highways reaching South America by the end of 2022, cases of massive mortalities on seabirds were reported (Gamarra-Toledo et al. 2023; Leguia et al. 2023). Then, by the middle of the South American summer, the first report in otariids (*Otaria byrona*) and common dolphin (*Delphinus delphis*) from Perú occurred (Gamarra-Toledo et al. 2023, Leguia et al. 2023). Prior to 2023, only one peer review report on influenza virus in otariids was published, based solely on serological evidence author-informed exposure to influenza B and A in South American fur seals *(Arctophoca australis*) (Blanc et al. 2009). Currently, this virus has been confirmed in *Otaria byrona* from Peru and Chile. Furthermore, according to Chilean authorities, this virus has caused the death of South American sea lions, marine otters (*Lontra felina*), Burmeister's porpoise (Phocoena spinipinnis), and Chilean dolphin (Cephalorhynchus eutropia) (Table 12.2) (SERNAPESCA 2023; WAHIS 2023).

12.5.3 Poxvirus

This is large-size enveloped DNA virus with cytoplasmic replication and zoonotic potential. Cutaneous lesions caused by poxvirus in cetaceans are commonly referred to as "tattoo diseases" because they appear to form a tattoo, circle, or ring. These lesions are affected by multiple species of cetaceans. In general, these lesions are self-sufficient, although they can persist for an extended period. Recently, cases have been confirmed in Brazil (Sacristán et al. 2018a, c), and in Chile, Burmeister's porpoises have been reported with macroscopic and histological lesions compatible with a poxvirus infection but were not confirmed with molecular techniques (Alvarado-Rybak et al. 2020). In the poxvirus family, the genera *Orthopoxvirus* and *Parapoxvirus* are considered potentially zoonotic. Poxvirus in pinnipedes is called "sealpox" and has been reported in several pinnipeds. In California sea lions, the poxvirus that infects them belongs to the genus *Parapoxvirus* and consequently is zoonotic (Waltzek et al. 2012). A report of Peruvian South American Sea Lions with pox lesions occurred in the 1970s (Wilson and Poglayen-Neuwall 1971). In

Table 12.2 Stranded marine mammals and penguins in Chile during 2023

		Chilean states represented by their assigned number															Total by specie
Common name	Scientific name	1	2	3	4	5	6	7	8	9	10	11	12	14	15	16	
Ballena de aleta	*Balaenoptera physalus*								1								1
Ballena jorobada	*Megaptera novaeangliae*							1	2								3
Ballena picuda de Cuvier	*Ziphius cavirostris*						1										1
Ballena Sei	*Balaenoptera borealis*											3					3
Cachalote enano de cabeza corta	*Kogia breviceps*			1													1
Chungungo	*Lontra felina*	3		2	5	7			1						3		21
Delfín Chileno	*Cephalorhynchus eutropia*					2		8								3	13
Delfín común	*Delphinus delphis*							1									1
Delfin gris	*Grampus griseus*					1											1
Delfín naríz de botella	*Tursiops truncatus*					1									2		3
Elefante marino	*Mirounga leonina*			2	1		1		1					1	1		7
Foca leopardo	*Hydrurga leptonyx*				1												1
Lobo fino Austral	*Arctophoca australis*		1		2						1	1					5
Lobo fino de Juan Fernández	*Arctophoca philippii*					3	1	8								1	13
Lobo marino común	*Otariabyronia*	599	1038	528	324	82	207	79	306	15	47	1	5	15	1391	393	5030
Marsopa espinosa	*Phocoena spinipinnis*		1	1	2	1	1	1							9		16
Orca	*Orcinus orca*												1				1
Zifio de Arnoux	*Berardius arnuxii*												1				1
Grand total June 2023		603	1042	537	339	102	217	105	319	24	58	16	19	30	1421	413	5122

Data from SERNAPESCA (2023)

In clear orange, the marine mammals most affected are indicated. Not all animals tested positive for H5N1 addition, skin lesions of similar characteristics have been reported in sea lions in Chile, but no confirmation by molecular or histological techniques has been found in the literature at the time of the present review. Cases of poxvirus have been associated with animals in poor health and poxviral infection could be lethal (Wilson and Poglayen-Neuwall 1971).

12.5.4 Brucellosis

Brucella ceti and *Brucella pinnipedialis* affect marine mammals worldwide. *Brucella* is a gram-negative intracellular bacterium with zoonotic potential. *Brucella ceti* affects cetaceans, and *Brucella pinnipedialis* has been reported in multiple species of pinnipeds. Brucellosis in cetaceans has been associated with various clinical signs and pathological changes, such as abortions, placentitis, and endometritis, and with neurological changes, such as spinal spondylosis, meningoencephalitis, and meningitis. Its transmission is relatively small, and it is believed that it can be a contributing factor to pneumonia in *Stenella coeruleoalba* (González-Barrientos et al. 2010). In the neotropics, this pathogen has been reported in Peru, Brazil, and Costa Rica (Guzmán-Verri et al. 2012; Hernández-Mora et al. 2017; Attademo et al. 2018; Sánchez-Sarmiento et al. 2018, 2019). In Costa Rica, it is particularly relevant with multiple cases of stranded cetaceans with brucellosis. Furthermore, six species of cetaceans have been reported with *Brucella ceti*, and 70% of stranded striped dolphins (*Stenella coeruleoalba*) have brucellosis (Hernández-Mora et al. 2017). This makes this an endemic pathogen to these populations.

Brucellosis in pinnipeds has not been detected in the neotropics, and all cases are concentrated in the Northern Hemisphere. Interestingly, brucellosis has been detected in Antarctic pinnipeds, the Weddell seal, *Leptonychotes weddellii*, and the Antarctic fur seal, *Arctocephalus gazella* (Retamal et al. 2000). The lack of detection of brucellosis in some populations could be due to sample effort. However, the risk of connecting populations in Antarctica makes this a relevant pathogen for monitoring under climate change and possible changes in the distribution patterns of some species.

12.5.5 Tuberculosis

Tuberculosis is caused by acid-fast gram-positive bacilli belonging to the genus *Mycobacterium*. In cetaceans, *Mycobacterium* has been reported in multiple species; however, it does not belong to the *Mycobacterium tuberculosis* complex (Tryland et al. 2018). In cetaceans, most infections are related to environmental Mycobacterium (Clayton et al. 2012; Waltzek et al. 2012). There have been reports of abscesses, granulomatosis lesions, respiratory lesions, dermatitis, and panniculitis associated to Mycobacterium (Tryland et al. 2012). In the case of pinnipedes, tuberculosis is caused by *Mycobacterium pinnipedii*, which belongs to the *Mycobacterium tuberculosis* complex. This causes a chronic debilitating disease in pinnipeds and is considered endemic to otariids from South America (Forshaw and Phelps 1991; Bernardelli et al. 1996; Tryland et al. 2018). Its impact on the population remains unclear.

12.6 Remarks: The Need for Data

To generate more data on the effect of infectious diseases in wildlife populations, we need multidisciplinary teams. Some relevant data generated by biologists and ecologists are needed first to then be able to generate models of the effect of diseases on wildlife populations. In most species, there is no basic information on life history, so further analysis of the effect of infectious diseases is far from possible with traditional tools. For example, for dolphin morbillivirus, a virus that has caused mortalities across the globe, it has not been possible to obtain the basic reproductive number (R0) of the pathogen and to estimate its true impact at a population level. This type of knowledge is scarce in wildlife and it is even more difficult to generate in marine mammals that inhabit on large hard-to-explore ecosystems. However, estimation using the growth rate of cases has been carried out in other viruses to estimate R0 and could be considered for marine mammal outbreaks (Morris et al. 2015). During the 2013–2014 morbillivitus unusual mortality event in North America, a subsequent analysis generated an R0 using the partial date of the outbreak (Morris et al. 2015). However, the complexity of these diseases and their multispecies nature make it difficult to undertake reservoirs, and the population effect on the different species affects this virus (Jo et al. 2018). In the case of the outbreak reported in South America, this calculation has not been carried out, but this publication could be taken into consideration for the next outbreak or revisit the data obtained during the latest outbreaks.

12.7 Future for Ecosystem and Marine Mammal Population Health

Species related to the Humboldt upwelling large marine ecosystem are extremely resilient to environmental changes. They are evolutionarily adapted to overcome the effect of El Niño–Southern Oscillation (ENSO). There is evidence of the resilience of some species and the capacity for recovery. In the case of South American sea lions, they have been able to increase their population in most of their range, and their population is still growing until the most recent mortality events due to. However, scenarios of multiple impacts on the population could limit the capacity to recover for marine mammal populations. Furthermore, in addition to the environmental effect of El Niño, which reduces primary productivity, marine mammals sustain the effects of climate change and direct anthropogenic stressors such as overfishing and bycatch. Finally, the impact of pathogens or other potential threats could represent a key element that prevents the recovery of populations or even drives populations to decline. We need work in multidisciplinary teams to be able to model how pathogens under the scenario of climate change can affect ecosystem sentinels.

References

Abelson A, Reed DC, Edgar GJ, Smith CS, Kendrick GA, Orth RJ, Airoldi L et al (2020) Challenges for Restoration of Coastal Marine Ecosystems in the Anthropocene. Front Mar Sci 7:544105

Alava JJ, Calle P, Tirapé A, Biedenbach G, Cadena OA, Maruya K, Lao W et al (2020) Persistent Organic Pollutants and Mercury in Genetically Identified Inner Estuary Bottlenose Dolphin (Tursiops truncatus) Residents of the Guayaquil Gulf, Ecuador: Ecotoxicological Science in Support of Pollutant Management and Cetacean Conservation. Front Mar Sci 7:122

Alvarado-Rybak M, Toro F, Abarca P, Paredes E, Espanol-Jimenez S, Seguel M (2020) Pathological findings in cetaceans sporadically stranded along the Chilean coast. Front Mar Sci 7:684

Arreguín-Sánchez F, del Monte-Luna P, Zetina-Rejón MJ, Albáñez-Lucero MO (2017) The Gulf of California Large Marine Ecosystem: Fisheries and other natural resources. Environ Dev 22:71–77

Attademo FLN, Silva JCR, Luna FO, Ikeda J, Foppel EFC, Sousa GP, Bôaviagem-Freire AC, Soares RM, Faita T, Batinga MCA, Keid LB (2018) Retrospective survey for pathogens in stranded marine mammals in Northeastern Brazil: Brucella spp. infection in a Clymene dolphin (Stenella clymene). J Wildl Dis 54:151–155

Bermudez L, Van Bressem M-F, Reyes-Jaimes O, Sayegh AJ, Paniz-Mondolfi AE (2009) Lobomycosis in man and lobomycosis-like disease in bottlenose dolphin, Venezuela. Emerg Infect Dis 15:1301

Bernardelli A, Bastida R, Loureiro J, Michelis H, Romano MI, Cataldi A, Costa E (1996) Tuberculosis in sea lions and fur seals from the south-western Atlantic coast. Rev Sci Tech 15:985–1005

Beron-Vera B, Crespo EA, Raga JA, Pedraza SN (2004) Uncinaria hamiltoni (Nematoda: Ancylostomatidae) in South American sea lions, Otaria flavescens, from Northern Patagonia, Argentina. J Parasitol 90:860–863

Bianchi MV, Ehlers LP, Vargas TP, Lopes BC, Taunde PA, de Cecco BS, Henker LC, Vielmo A, Lorenzett MP, Riboldi CI, Lopes CE, de Amorim DB, Diaz-Delgado J, Snel GGM, Siqueira FM, Sonne L (2018) Omphalitis, urachocystitis and septicemia by Streptococcus dysgalactiae in a southern right whale calf Eubalaena australis, Brazil. Dis Aquat Org 131:227–232

Blanc A, Ruchansky D, Clara M, Achaval F, Le Bas A, Arbiza J (2009) Serologic evidence of influenza A and B viruses in South American fur seals (Arctocephalus australis). J Wildl Dis 45:519–521

Bossart GD (2011) Marine mammals as sentinel species for oceans and human health. Vet pathol 48(3):676–690. https://doi.org/10.1177/0300985810388525

Brock PM, Hall AJ, Goodman SJ, Cruz M, Acevedo-Whitehouse K (2013) Immune activity, body condition and human-associated environmental impacts in a wild marine mammal. PloS One 8(6):e67132

Cárdenas-Alayza S, Gutiérrez D, Tremblay Y (2021) Trends in sympatric otariid populations suggest resource limitations in the Peruvian Humboldt Current System. Mar Environ Res 169:105349

Clayton LA, Stamper MA, Whitaker BR, Hadfield CA, Simons B, Mankowski JL (2012) Mycobacterium abscessus pneumonia in an Atlantic bottlenose dolphin (Tursiops truncatus). J Zoo Wildl Med 43:961–965

Colagross-Schouten AM, Mazet JA, Gulland FM, Miller MA, Hietala S (2002) Diagnosis and seroprevalence of leptospirosis in California sea lions from coastal California. J Wildl Dis 38(1):7–17

Costa DP (2007) A conceptual model of the variation in parental attendance in response to environmental fluctuation: foraging energetics of lactating sea lions and fur seals. Aquatic Conserv: Mar Freshw Ecosyst 17:S44–S52

Crespo EA, Pedraza SN, Dans SL, Koen Alonso M, Reyes LM, García NA, Coscarella M et al (1997) Direct and Indirect Effects of the Highseas Fisheries on the Marine Mammal Populations in the Northern and Central Patagonian Coast. J Northw Atl Fish Sci 22:189–207

Cunha HA, Santos-Neto EB, Carvalho RR, Ikeda JMP, Groch KR, Díaz-Delgado J, Guari EB et al (2021) Epidemiological features of the first Unusual Mortality Event linked to cetacean morbillivirus in the South Atlantic (Brazil, 2017–2018). Mar Mam Sci 37:1375–1390

De Amorim DB, Casagrande RA, Alievi MM, Wouters F, De Oliveira LG, Driemeier D, ... Ferreira-Neto JS (2014) Mycobacterium pinnipedii in a stranded South American sea lion (Otaria byronia) in Brazil. J Wildl Dis 50(2):419–422

Demarque IDC, de Oliveira FCR, da Silveira LS, Barbosa LA, Ederli NB (2020) The lungworm, Halocercus brasiliensis (Nematoda: Pseudaliidae), from Guiana Dolphins Sotalia guianensis from Brazil with pathological findings. J Parasitol 106:254–260

Desforges J-PW, Sonne C, Levin M, Siebert U, De Guise S, Dietz R (2016) Immunotoxic effects of environmental pollutants in marine mammals. Environ Int 86:126–139

Doney SC, Busch DS, Cooley SR, Kroeker KJ (2020) The impacts of ocean acidification on marine ecosystems and reliant human communities. Annu Rev Environ Resour 45:83–112

Durante CA, Santos-Neto EB, Azevedo A, Crespo EA, Lailson-Brito J (2016) POPs in the South Latin America: Bioaccumulation of DDT, PCB, HCB, HCH and Mirex in blubber of common dolphin (Delphinus delphis) and Fraser's dolphin (Lagenodelphis hosei) from Argentina. Science of The Total Environment 572:352–360

Elsmo EJ, Wunschmann A, Beckmen KB, Broughton-Neiswanger LB, Buckles EL, Ellis J, Fitzgerald SD, Gerlach R, Hawkins S, Ip H (2023) Pathology of natural infection with highly pathogenic avian influenza virus (H5N1) clade 2.3. 4.4 b in wild terrestrial mammals in the United States in 2022. bioRxiv:2023.2003. 2010.532068

Espinosa-Carreón TL, Escobedo-Urías D (2017) South region of the Gulf of California large marine ecosystem upwelling, fluxes of CO2 and nutrients. Environ Dev 22:42–51

Exposto Novoselecki H, Catao-Dias JL, Ewbank AC, Navas-Suarez PE, Duarte-Benvenuto A, Lial HC, Costa Silva S, Sanchez-Sarmiento AM, Gravena W, da Silva VMF, Carvalho VL, Marmontel M, Bertozzi CP, Lanes Ribeiro V, Del Rio do Valle R, Marigo J, das Neves CG, Esperon F, Sacristan C (2021) Highly divergent herpesviruses in threatened river dolphins from Brazil. Sci Rep 11:24528

Fackelmann G, Sommer S (2019) Microplastics and the gut microbiome: How chronically exposed species may suffer from gut dysbiosis. Mar Pollut Bull 143:193–203

Fereidouni S, Munoz O, Von Dobschuetz S, De Nardi M (2016) Influenza virus infection of marine mammals. EcoHealth 13:161–170

Forshaw D, Phelps GR (1991) Tuberculosis in a captive colony of pinnipeds. J Wildl Dis 27:288–295

Gamarra-Toledo V, Plaza PI, Gutiérrez R, Inga-Diaz G, Saravia-Guevara P, Pereyra-Meza O, Coronado-Flores E, Calderón-Cerrón A, Quiroz-Jiménez G, Martinez P (2023) Mass mortality of marine mammals associated to highly pathogenic influenza virus (H5N1) in South America. bioRxiv:2023.2002. 2008.527769

González-Barrientos R, Morales JA, Hernández-Mora G, Barquero-Calvo E, Guzmán-Verri C, Chaves-Olarte E, Moreno E (2010) Pathology of striped dolphins (Stenella coeruleoalba) infected with Brucella ceti. J Comp Pathol 142:347–352

Groch KR, Colosio AC, Marcondes MCC, Zucca D, Diaz-Delgado J, Niemeyer C, Marigo J, Brandao PE, Fernandez A, Catao-Dias JL (2014) Novel cetacean morbillivirus in Guiana dolphin, Brazil. Emerg Infect Dis 20:511–513

Groch KR, Santos-Neto EB, Díaz-Delgado J, Ikeda JMP, Carvalho RR, Oliveira RB, Guari EB et al (2018) Guiana Dolphin Unusual Mortality Event and Link to Cetacean Morbillivirus, Brazil. Emerg Infect Dis 24:1349–1354

Groch KR, Groch KR, Kolesnikovas CKM, de Castilho PV, Moreira LMP, Barros CRMB, Morais CR, Renault-Braga EP, Sierra E, Fernandez A, Catao-Dias JL, Diaz-Delgado J (2019) Cetacean morbillivirus in Southern Right Whales, Brazil. Transbound Emerg Dis 66:606–610

Groch KR, Díaz-Delgado J, Santos-Neto EB, Ikeda JMP, Carvalho RR, Oliveira RB, Guari EB, Flach L, Sierra E, Godinho AI, Fernández A, Keid LB, Soares RM, Kanamura CT, Favero C, Ferreira-Machado E, Sacristán C, Porter BF, Bisi TL, Azevedo AF, Lailson-Brito J, Catão-Dias JL (2020a) The pathology of cetacean morbillivirus infection and comorbidities in Guiana dolphins during an unusual mortality event (Brazil, 2017–2018). Vet Pathol 57:845–857

Groch KR, Jerdy H, Marcondes MC, Barbosa LA, Ramos HG, Pavanelli L, Fornells LAM, Silva MB, Souza GS, Kanashiro MM, Bussad P, Silveira LS, Costa-Silva S, Wiener DJ, Travassos CE, Catão-Dias JL, Díaz-Delgado J (2020b) Cetacean morbillivirus infection in a killer whale (Orcinus orca) from Brazil. J Comp Pathol 181:26–32

Gutierrez D, Akester M, Naranjo L (2016) Productivity and sustainable management of the Humboldt current large marine ecosystem under climate change. Environmental Development 17:126–144

Gutiérrez TM, Castillo PJ, Naranjo BL, Akester MJ (2017) Current state of goods, services and governance of the Humboldt Current Large Marine Ecosystem in the context of climate change. Environ Dev 22:175–190. ISSN 2211-4645, https://doi.org/10.1016/j.envdev.2017.02.006. https://www.sciencedirect.com/science/article/pii/S2211464516302500

Guzmán-Verri C, González-Barrientos R, Hernández-Mora G, Morales JA, Baquero-Calvo E, Chaves-Olarte E, Moreno E (2012) Brucella ceti and brucellosis in cetaceans. Front Cell Infect Microbiol 2:3

Hatcher MJ, Dick JT, Dunn AM (2012) Diverse effects of parasites in ecosystems: linking interdependent processes. Front Ecol Environ 10:186–194

Hazen EL, Abrahms B, Brodie S, Carroll G, Jacox MG, Savoca MS, Bograd SJ (2019) Marine top predators as climate and ecosystem sentinels. Front Ecol Environ 17(10):565–574. https://doi.org/10.1002/fee.2125

Hernández-Mora G, Bonilla-Montoya R, Barrantes-Granados O, Esquivel-Suárez A, Montero-Caballero D, González-Barrientos R, Fallas-Monge Z, Palacios-Alfaro JD, Baldi M, Campos E, Chanto G, Barquero-Calvo E, Chacón-Díaz C, Chaves-Olarte E, Guzmán Verri C, Romero-Zúñiga JJ, Moreno E (2017) Brucellosis in mammals of Costa Rica: an epidemiological survey. PLoS One 12:e0182644

Henson SA, Beaulieu C, Ilyina T. John JG, Long M, Séférian R, Tjiputra J, et al (2017) Rapid emergence of climate change in environmental drivers of marine ecosystems. Nat Commun 8:14682

Jefferson TA, Webber MA, Pitman RL, Gorter U (2015) Marine mammals of the world: a comprehensive guide to their identification (Second edition.). Elsevier/AP, Academic Press is an imprint of Elsevier, London, UK; San Diego, CA, USA

Jerdy H, Werneck M, Barbosa L, Hauser-Davis RA, De-Oliveira-Nogueira CH, da Silveira LS (2022) First report on Phyllobothrium delphini infection and Crassicauda sp. parasitism resulting in osseous metaplasia in a Cuvier's beaked whale (Ziphius cavirostris) from the Brazilian region. Int J Parasitol: Parasites Wildl 17:60–64

Jepsen EM, de Bruyn PJN (2019) Pinniped entanglement in oceanic plastic pollution: A global review. Mar Pollut Bull 145:295–305

Jo, Wendy K, Jochen Kruppa, Andre Habierski, Marco van de Bildt, Sandro Mazzariol, Giovanni Di Guardo, Ursula Siebert et al (2018) "Evolutionary evidence for multi-host transmission of cetacean morbillivirus." *Emerging microbes & infections* 7(1):1–15

Kemper CM, Tomo I, Bingham J, Bastianello SS, Wang J, Gibbs SE, Woolford L, Dickason C, Kelly D (2016) Morbillivirus-associated unusual mortality event in South Australian bottlenose dolphins is largest reported for the Southern Hemisphere. R Soc Open Sci 3:160838

Kühn S, van Franeker JA (2020) Quantitative overview of marine debris ingested by marine megafauna. Mar Pollut Bull 151:110858

Kuzmina TA, Kuzmin Y, Dzeverin I, Lisitsyna OI, Spraker TR, Korol EM, Kuchta R (2021) Review of metazoan parasites of the northern fur seal (Callorhinus ursinus) and the analysis of the gastrointestinal helminth community of the population on St. Paul Island, Alaska. Parasitol Res 120:117–132

Lang G, Gagnon A, Geraci JR (1981) Isolation of an influenza A virus from seals. Arch Virol 68:189–195

Leguia M, Garcia-Glaessner A, Munoz-Saavedra B, Juarez D, Barrera P, Calvo-Mac C, Jara J, Silva W, Ploog K, Amaro L (2023) Highly pathogenic avian influenza A (H5N1) in marine mammals and seabirds in Peru. bioRxiv:2023.2003. 2003.531008

Lima M, Canales TM, Wiff R, Montero J (2020) The Interaction Between Stock Dynamics, Fishing and Climate Caused the Collapse of the Jack Mackerel Stock at Humboldt Current Ecosystem. Front Mar Sci 7:123

Lotze HK, Tittensor DP, Bryndum-Buchholz A, Eddy TD, Cheung WWL, Galbraith ED, Barange M et al (2019) Global ensemble projections reveal trophic amplification of ocean biomass declines with climate change. Proc Natl Acad Sci USA 116:12907–12912

Marcogliese DJ (2005) Parasites of the superorganism: Are they indicators of ecosystem health? Int J Parasitol 35:705–716

Moreno IB, Ott PH, Tavares M, Oliveira LR, Borba MR, Driemeier D, Nakashima SB, Heinzelmann LS, Siciliano S, Van Bressem M (2008) Mycotic dermatitis in common bottlenose dolphins (Tursiops truncatus) from southern Brazil, with a confirmed record of lobomycosis disease. IWC Scientific Committe

Morris SE, Zelner JL, Fauquier DA, Rowles TK, Rosel PE, Gulland F, Grenfell BT (2015) Partially observed epidemics in wildlife hosts: modelling an outbreak of dolphin morbillivirus in the northwestern Atlantic, June 2013–2014. J R Soc Interface 12(112):20150676

Muñoz Sevilla NP, Le Bail M (2017) Latin American and Caribbean regional perspective on Ecosystem Based Management (EBM) of Large Marine Ecosystems goods and services. Environ Dev 22:9–17. ISSN 2211-4645, https://doi.org/10.1016/j.envdev.2017.01.006. (https://www.sciencedirect.com/science/article/pii/S2211464516303037)

Oehlmann J, Schulte-Oehlmann U, Kloas W, Jagnytsch O, Lutz I, Kusk KO, Wollenberger L et al (2009) A critical analysis of the biological impacts of plasticizers on wildlife. Phil Trans R Soc B 364:2047–2062

Orona-Návar C, García-Morales R, Loge FJ, Mahlknecht J, Aguilar-Hernández I, Ornelas-Soto N (2022) Microplastics in Latin America and the Caribbean: A review on current status and perspectives. J Environ Manag 309:114698

Perez-Venegas DJ, Valenzuela-Sánchez A, Montalva F, Pavés H, Seguel M, Wilcox C, Galbán-Malagón C (2021) Towards understanding the effects of oceanic plastic pollution on population growth for a South American fur seal (Arctocephalus australis australis) colony in Chile. Environ Pollut 279:116881

Phillips ACN, Suepaul R (2017) Nasitrema Species: A Frequent Culprit in Melon-Headed Whale (Peponocephala electra) Strandings in Trinidad. Aquat Mamm 43:547–557

Prager KC, Buhnerkempe MG, Greig DJ, Orr AJ, Jensen ED, Gomez F, ... Lloyd-Smith JO (2020) Linking longitudinal and cross-sectional biomarker data to understand host-pathogen dynamics: Leptospira in California sea lions (Zalophus californianus) as a case study. PLoS Negl Trop Dis 14(6):e0008407

Puryear W, Sawatzki K, Hill N, Foss A, Stone JJ, Doughty L, Walk D, Gilbert K, Murray M, Cox E (2023) Highly pathogenic avian influenza A (H5N1) virus outbreak in New England Seals, United States. Emerg Infect Dis 29:786–791

Rapport DJ, Costanza R, McMichael AJ (1998) Assessing ecosystem health 13:6.

Rapport DJ, Maffi L (2011) Eco-cultural health, global health, and sustainability. Ecol Res 26:1039–1049

Reif JS, Fair PA, Adams J, Joseph B, Kilpatrick DS, Sanchez R, Goldstein JD, Townsend FI, McCulloch SD, Mazzoil M, Zolman ES, Hansen LJ, Bossart GD (2008) Evaluation and comparison of the health status of Atlantic bottlenose dolphins from the Indian River Lagoon, Florida, and Charleston, South Carolina. J Am Vet Med Assoc 233:299–307

Reif JS, Peden-Adams MM, Romano TA, Rice CD, Fair PA, Bossart GD (2009) Immune dysfunction in Atlantic bottlenose dolphins (Tursiops truncatus) with lobomycosis. Med Mycol 47:125–135

Retamal P, Blank O, Abalos P, Torres D (2000) Detection of anti-Brucella antibodies in pinnipeds from the Antarctic territory. Vet Rec 146:166–167

Roman J, Estes JA, Morissette L, Smith C, Costa D, McCarthy J, Nation JB et al (2014) Whales as marine ecosystem engineers. Front Ecol Environ 12:377–385

Sacristán C, Catão-Dias JL, Ewbank AC, Ferreira-Machado E, Neves E, Santos-Neto EB, Azevedo A, Laison-Brito J, De Castilho PV, Daura-Jorge FG, Simões-Lopes PC, Carballo M, García-Párraga D, Sánchez-Vizcaíno JM, Esperón F (2018a) Novel and highly sensitive SYBR® Green real-time pcr for poxvirus detection in odontocete cetaceans. J Virol Methods 259:45–49

Sacristán C, Esperón F, Ewbank AC, Costa-Silva S, Marigo J, Matushima ER, Kolesnikovas CKM, Catão-Dias JL (2018b) Identification of novel gammaherpesviruses in a South American fur seal (Arctocephalus australis) with ulcerative skin lesions. J Wildl Dis 54:592–596

Sacristán C, Esperón F, Marigo J, Ewbank AC, de Carvalho RR, Groch KR, de Castilho PV, Sánchez-Sarmiento AM, Costa-Silva S, Ferreira-Machado E, Gonzales-Viera OA, Daura-Jorge FG, Santos-Neto EB, Lailson-Brito J, de Freitas Azevedo A, Simões-Lopes PC, Neves CG, Catão-Dias JL (2018c) Molecular identification and microscopic characterization of poxvirus in a Guiana dolphin and a common bottlenose dolphin, Brazil. Dis Aquat Org 130:177–185

Sacristan C, Esperon F, Ewbank AC, Diaz-Delgado J, Ferreira-Machado E, Costa-Silva S, Sanchez-Sarmiento AM, Groch KR, Neves E, Dutra GHP, Gravena W, Da Silva VMF, Marcondes MCC, Colosio AC, Cremer MJ, Carvalho VL, Meirelles ACO, Marigo J, Catao-Dias JL (2019) Novel herpesviruses in riverine and marine cetaceans from South America. Acta Trop 190:220–227

Sánchez-Sarmiento AM, Carvalho VL, Meirelles ACO, Gravena W, Marigo J, Sacristán C, Costa-Silva S, Groch KR, Silva ND, Ferreira Neto JS, Catão-Dias JL (2018) Survey of Brucella spp. and Leptospira spp. antibodies in cetaceans and manatees of the Amazon basin and Atlantic Ocean, Brazil. Dis Aquat Organ 132:1–11

Sánchez-Sarmiento AM, Carvalho VL, Díaz-Delgado J, Ressio RA, Fernandes NCCA, Guerra JM, Sacristán C, Groch KR, Silvestre-Perez N, Ferreira-Machado E, Costa-Silva S, Navas-Suárez P, Meirelles ACO, Favero C, Marigo J, Bertozzi CP, Colosio AC, Marcondes MCC, Cremer MJ, Dos Santos Silva N, Ferreira Neto JS, Keid LB, Soares R, Sierra E, Fernández A, Catão-Dias JL (2019) Molecular, serological, pathological, immunohistochemical and microbiological investigation of Brucella spp. in marine mammals of Brazil reveals new cetacean hosts. Transbound Emerg Dis 66:1674–1692

Sarzosa MS, Duignan P, DeRango EJ, Field C, Rios C, Sanchez S, Espinoza E, Loyola A, Rueda D, Paez-Rosas D (2021) Occurrence of mycoplasmas in Galapagos sea lions (Zalophus wollebaeki) and their association with other respiratory pathogens. J Wildl Dis 57:623–627

Seguel M, Gottdenker N (2017) The diversity and impact of hookworm infections in wildlife. Int J Parasitol: Parasites Wildl 6:177–194

Seguel M, Munoz F, Navarrete MJ, Paredes E, Howerth E, Gottdenker N (2017) Hookworm infection in South American fur seal (Arctocephalus australis) pups: pathology and factors associated with host tissue damage and mortality. Vet Pathol 54:288–297

Seguel M, Muñoz F, Perez-Venegas D, Müller A, Paves H, Howerth E, Gottdenker N (2018a) The life history strategy of a fur seal hookworm in relation to pathogenicity and host health status. Int J Parasitol: Parasites Wildl 7:251–260

Seguel M, Nadler S, Field C, Duignan P (2018b) Vasculitis and Thrombosis due to the Sea Lion Lungworm, Parafilaroides decorus, in a Guadalupe Fur Seal (Arctocephalus philippii townsendi). J Wildl Dis 54:638–641

Seguel M, Montalva F, Perez-Venegas D, Gutierrez J, Paves HJ, Müller A, ... Gottdenker N (2018c) Immune-mediated hookworm clearance and survival of a marine mammal decrease with warmer ocean temperatures. Elife 7:e38432

Sepúlveda MA, Seguel M, Alvarado-Rybak M, Verdugo C, Muñoz-Zanzi C, Tamayo R (2015) Postmortem findings in four south American sea lions (Otaria byronia) from an urban colony in Valdivia, Chile. J Wildl Dis 51:279–282

SERNAPESCA (2023) Servicio Nacional de Pesca. Plataforma de influenza aviar. Downloaded 30 Apr 2023. http://www.sernapesca.cl/influenza-aviar

Sherman K, Sevilla NPM, Torres PÁ, Peterson B (2017) Sustainable development of latin american and the caribbean large marine ecosystems. Environmental Development, 22, 1–8. https://doi.org/10.1016/j.envdev.2017.04.001

Soto S, Gonzalez B, Willoughby K, Maley M, Olvera A, Kennedy S, Marco A, Domingo M (2012) Systemic herpesvirus and morbillivirus co-infection in a striped dolphin (Stenella coeruleoalba). J Comp Pathol 146:269–273

Thushari GGN, Senevirathna JDM (2020) Plastic pollution in the marine environment. Heliyon 6:e04709

Tryland M, Nymo I, Nielsen O, Nordoy E, Kovacs K, Krafft B, Thoresen S, Asbakk K, Osterrieder K, Roth S, Lydersen C, Godfroid J, Blix A (2012) Serum chemistry and antibodies against pathogens in antarctic fur seals, Weddell seals, crabeater seals, and Ross seals. J Wildl Dis 48:632–645

Tryland M, Larson AK, Nymo IH (2018) Bacterial infectious and diseases. In: Diereuf LA, Whitman KL (eds) CRC handbook of marine mammal medicine. CRC Press, Boca Raton

Van Bressem MF, Vanwaerebeek K, Reyes JC, Dekegel D, Pastoret PP (1993) Evidence of poxvirus in dusky dolphin (Lagenorhynchus-obscurus) and Burmeister porpoise (Phocoena-spinipinnis) from coastal Peru. J Wildl Dis 29:109–113

Van Bressem MF, Van Waerebeek K, Fleming M, Barrett T (1998) Serological evidence of morbillivirus infection in small cetaceans from the Southeast Pacific. Vet Microbiol 59:89–98

Van Bressem MF, Cassonnet P, Rector A, Desaintes C, Van Waerebeek K, Alfaro-Shigueto J, Van Ranst M, Orth G (2007) Genital warts in Burmeister's porpoises: characterization of Phocoena spinipinnis papillomavirus type 1 (PsPV-1) and evidence for a second, distantly related PsPV. J Gen Virol 88:1928–1933

Van Bressem M-F, Raga JA, di Guardo G, Jepson P, Duignan P, Siebert U, Barrett T, Santos M, Moreno I, Siciliano S (2008a) Emerging and recurring diseases in cetaceans worldwide and the role of environmental stressors. IWC Scientific Report

Van Bressem M-F, Van Waerebeek K, Flach L, Reyes JC, de Oliveira Santos M, Siciliano S, Echegaray M, Viddi F, Felix F, Crespo E (2008b) Skin diseases in cetaceans. International Whaling Commission

Vera-Massieu C, Brock PM, Godínez-Reyes C, Acevedo-Whitehouse K (2015) Activation of an inflammatory response is context-dependent during early development of the California sea lion. R Soc Open Sci 2(4):150108

Vilela R, Bossart GD, St Leger JA, Dalton LM, Reif JS, Schaefer AM, McCarthy PJ, Fair PA, Mendoza L (2016) Cutaneous granulomas in dolphins caused by novel uncultivated Paracoccidioides brasiliensis. Emerg Infect Dis 22:2063–2069

WAHIS (2023) World Organization for Animal Health, World animal health information system. https://wahis.woah.org/#/home

Waltzek TB, Cortés-Hinojosa G, Wellehan JF, Gray GC (2012) Marine mammal zoonoses: a review of disease manifestations. Zoonoses Public Health 59:521–535

Webster RG, Geraci J, Petursson G, Skirnisson K (1981) Conjunctivitis in human beings caused by influenza A virus of seals. N Engl J Med 304:911

Wilson TM, Poglayen-Neuwall I (1971) Pox in South American sea lions (Otaria-byronia). Can J Comp Med 35:174–177

Chapter 13
Tick-Borne Microorganisms in Neotropical Vertebrates

Sergio Bermúdez C., E. A. Zieman, E. Tarragona, T. F. Martins, A. A. Faccini-Martínez, R. Thomas, C. Guzmán-Cornejo, and S. Muñoz-Leal

13.1 Introduction

Ticks (Acari: Ixodida) are cosmopolitan ectoparasites of terrestrial and semi-aquatic vertebrates and harbor a plethora of microorganisms (Guglielmone et al. 2014; Guglielmone and Robbins 2018). As blood-feeding parasites, ticks are the primary vectors of pathogens for domestic and wild animals and the second most important vectors of disease to humans, surpassed only by mosquitoes (Nicholson et al. 2019). Ticks can also produce lesions of different degrees in the host, such as simple focal dermatitis to severe pathologies affecting different organs, paralysis, or death in some cases (Guglielmone et al. 2023). With few exceptions, the transmission of microorganisms (MOs) from a tick to a host occurs through salivary fluids inoculated during the meal (Nicholson et al. 2019). In the Neotropical Zoogeographical Region (NZR), ticks are known to transmit bacteria and protozoans (Barros-Battesti et al. 2006; Guglielmone et al. 2021). Nematodes, which are also transmitted by a few ticks (e.g., *Ornithodoros tartakovsky i* vector of *Acanthocheilonema vitae*), are unknown in neotropical ticks, probably because of lack of research. Ticks also transmit viruses to their hosts (Kazimírová et al. 2017), but similar to tick-borne

S. Bermúdez C.
Departamento de Investigación en Entomología Médica, Instituto Conmemorativo Gorgas de Estudios de la Salud, Panama City, Panama
e-mail: sbermudez@gorgas.gob.pa

E. A. Zieman
Department of Biological Sciences, Eastern Illinois University, Charleston, IL, USA
e-mail: eazieman@eiu.edu

E. Tarragona
Instituto de Investigación de la Cadena Láctea (IDICAL, INTA – CONICET), Instituto Nacional de Tecnología Agropecuaria, E.E.A. Rafaela, Rafaela, Santa Fe, Argentina
e-mail: tarragona.evelina@inta.gob.ar

G. Acosta-Jamett, A. Chaves (eds.), *Ecology of Wildlife Diseases in the Neotropics*, https://doi.org/10.1007/978-3-031-50531-7_13

nematodes, the evidence on viruses transmitted by ticks in the NZR is vague. Indeed, only two viruses have been isolated, Punta Salinas and Huacho strains from *Ornithodoros amblus* of Peru (Clifford et al. 1980), and recent studies have explored the viromes of *Antricola delacruzi* in Brazil (Blomström et al. 2019) and hard ticks (Ixodidae) in Colombia (Orozco et al. 2021). It is therefore not unexpected that the knowledge of tick-borne microorganisms in neotropical vertebrates also concentrates bacteria and protozoans rather than viruses or nematodes.

The study of tick-borne microorganisms in animal hosts can be performed by serological or genetic techniques or through the isolation of the agents. For instance, isolating bacterial agents facilitates the implementation of local strain antigens in serological assays. However, in the NZR, few organisms transmitted by native species of ticks have been isolated, with few exceptions of *Rickettsia* and *Borrelia* spp., which have been isolated in Brazil. Although isolates are available, few tick-borne agents have complete chromosomal genomes (Kneubehl et al. 2022) or have been successfully cultured (Hun et al. 2011; Pinter et al. 2008; Labruna et al. 2007, 2017; Pacheco et al. 2009; Bermúdez et al. 2021; Zaldívar et al. 2021). Detection of antibodies against *Rickettsia* spp. harbored by ticks has been performed in wild mammals; however, the results may yield cross-reactions and might be carefully interpreted (Miranda et al. 2011; Bermudez et al. 2017a, b; Londoño et al. 2017). The detection of antibodies against further agents is discussed below in the respective sections.

The bulk of data regarding tick-borne microorganisms in neotropical vertebrates comes through real-time and conventional polymerase chain reaction (PCR) and further sequencing of the amplicons. Using these approaches, we currently know that in the NZR, tick-transmitted microorganisms engage their cycles with mammals and birds. However, most of the data regarding tick-borne microorganisms are obtained by sequencing amplicons directly from tick-extracted genomic DNA and should be carefully interpreted. Indeed, the fact that ticks submitted to DNA extraction are often collected either on their hosts or free living precludes conclusions whether the detected microorganisms were associated with the tick or came from

T. F. Martins
Departamento de Medicina Veterinária Preventiva e Saúde animal, Faculdade de Medicina Veterinária e Zootecnia, Universidade de São Paulo, São Paulo, SP, Brazil

Instituto Pasteur, Secretaria de Estado da Saúde de São Paulo, São Paulo, SP, Brazil

A. A. Faccini-Martínez
Servicio de Infectología, Hospital Militar Central, Bogotá, Colombia

Servicios y Asesorías en Infectología – SAI, Bogotá, Colombia

R. Thomas · S. Muñoz-Leal (✉)
Departamento de Ciencia Animal, Facultad de Ciencias Veterinaria, Universidad de Concepción, Chillán, Chile
e-mail: sebamunoz@udec.cl

C. Guzmán-Cornejo
Laboratorio de Acarología, Departamento de Biología Comparada, Facultad de Ciencias, Universidad Nacional Autónoma de México, Ciudad de México, Mexico

the blood of their hosts. Moreover, to amplify and sequence amplicons does not prove that an agent was alive. Instead, it only gives evidence on the presence/absence of a determined microorganism in a sample.

This chapter aims to summarize and discuss the current knowledge on the microorganisms transmitted or potentially transmitted by ticks that have been detected or isolated in neotropical vertebrates with two exceptions. First, although *Coxiella burnetii* could be transmitted experimentally by *Amblyomma mixtum* in Panama (Rodaniche and Rodaniche 1949) and has been isolated from *Amblyomma tigrinum* in Argentina (Pacheco et al. 2013), ticks are not the primary source of infection for this cosmopolitan pathogen (Duron et al. 2015). Therefore, *C. burnetii* is not included in this chapter. Second, protozoans of the genus *Hepatozoon* are transmitted to vertebrates through the ingestion of infected ticks and other hematophagous invertebrates that harbor oocysts (Smith 1996). While in the NZR, DNA of *Hepatozoon* has been detected in wild vertebrates (i.e., reptiles, rodents, racoons, foxes), compelling evidence indicating that these protozoans are naturally transmitted by ticks from the region is lacking. Therefore, further studies should seek oocysts in the hemolymph of ticks but also mites, fleas, mosquitoes, and leeches to shed light on the vectors (Bennett et al. 1992; Smith 1996).

13.2 Hosts, Ticks, and Transmitted MO

Not all animals can harbor the agents that a given tick species can transmit during a meal. In fact, tick-borne MOs spread under natural conditions only if they infect competent or primary hosts. While the primary host is essential to maintain the transmission of a given MO and becomes chronically infected, secondary or incidental hosts are not essential for a MO's life cycle but contribute to its amplification (Barker and Reisen 2019). An amplifying vertebrate host (AVH) multiplies the MO and the number of infected vectors in an area and generally does not remain infected for long periods (Barker and Reisen 2019). The vector is defined as the "carrier" of a MO from one host to another (Barker and Reisen 2019). However, to be considered a vector, an arthropod must have intrinsic susceptibility to become infected, replicate, and transmit a given MO. That is to say, to become infected the vector must (1) acquire the MO by feeding on an AVH in which the MO circulates, (2) replicate the MO, and (3) transmit the MO to an uninfected susceptible host (Barker and Reisen 2019). This phenomenon is also called horizontal transmission. Vectors, especially ticks, also have an intrinsic amplifying mechanism called vertical transmission (VT). Two forms of VT occur: transstadial and transovarian. The first refers to the sequential passage of the MO acquired in one evolutionary tick stage (larva, nymph, or adult) to the next stage during molting. The second is related to the ability of the MO to colonize the ovarian tissue of female ticks and pass on to progeny during oogenesis (Barker and Reisen 2019).

Despite replicating inside ticks or susceptible vertebrates, tick-borne MOs are limited above all to the geographic range and seasonal biology of the vector. In the case of ticks, annual fluctuation is characterized by alternating periods of peak activity (host search, feeding, and post-feeding behavior) and inactivity (quiescence and diapause) (Sonenshine and Roe 2014). Hence, environmental variables such as temperature and saturation deficit rule tick biology and indirectly their harbored MO (Estrada peña et al. 2014). Tick-borne MOs discussed below propagate on vertical and horizontal pathways. The records in wild neotropical vertebrates are listed in Table 13.1.

13.2.1 Anaplasmataceae (Anaplasma, Ehrlichia, "Candidatus *Neoehrlichia*")

The Anaplasmataceae family belongs to the Alpha subclass of Proteobacteria and comprises host-associated bacteria that replicate intracellularly (Kersters et al. 2006). The family is composed of four genera (*Anaplasma*, *Ehrlichia*, *Neorickettsia*, and *Wolbachia*) (Yu and Walker 2006), with candidate genera being progressively described based on genetic characterizations (Kawahara et al. 2004; Eshoo et al. 2015). In particular, *Anaplasma*, *Ehrlichia*, and "*Candidatus* Neoehrlichia" are transmitted by ticks of the genera *Amblyomma*, *Dermacentor*, *Ixodes*, *Hyalomma*, or *Rhipicephalus* (Ixodidae) (Rar and Golovljova 2011). While under laboratory conditions transovarian transmission has been demonstrated for some species such as *Anaplasma platys* and *Anaplasma phagocytophilum* (Baldrige et al. 2009; Snellgrove et al. 2020), it is believed that most *Anaplasma* spp., *Ehrlichia* spp., and "*Ca.* Neoehrlichia" spp. do not infect tick progeny (Rar and Golovljova 2011). Instead, these three genera of intracellular bacteria invade mammalian blood or endothelial cells, causing a persistent infection that guarantees circulating bacteria when the vectors feed upon vertebrate hosts.

Although the majority of the evidence regarding the animals that these bacteria infect indicates that mammals and their associated ticks are the most important hosts globally, molecular detection of *Anaplasma* and *Ehrlichia* in bird organs or blood suggests that these bacteria could also be associated with avian hosts (Machado et al. 2012; Muñoz-Leal et al. 2019). Eight species are currently classified in the genus *Anaplasma* (Rar et al. 2021), six in *Ehrlichia* (Rar and Golovljova 2011; Aguiar et al. 2019), and three in "*Ca.* Neoehrlichia" (Rar and Golovljova 2011; Müller et al. 2018). However, the richness of the group seems to be far higher. In fact, multiple strains deserving further characterization have been described in ticks and vertebrates worldwide (Table 13.1). Particularly in the NZR, *Anaplasma* and *Ehrlichia* strains have been reported in domestic and wild animals (André et al. 2012; Widmer et al. 2011; Almeida et al. 2013; Soares et al. 2017a, b; de Oliveira et al. 2020; Lopes et al. 2018; Muñoz-Leal et al. 2019; Calchi et al. 2020; Orozco et al. 2020; Félix et al. 2021; Tarragona et al. 2022, 2023; Sebastian et al. 2022).

Genetic sequences of Anaplasmataceae obtained from scansorial and terrestrial xenarthrans diverge from known species and have been named "*Candidatus* Anaplasma brasiliensis" and "*Candidatus* Anaplasma amazonensis" (Calchi et al.

Table 13.1 Tick-borne microorganisms detected in wild animals from the Neotropical Zoogeographic Region

Microorganism	Method	Animal group	Host species	Country	References
Order Spirochetales					
Relapsing fever group					
Borrelia spp.	M	Mammalia/ Didelphimorphia	*Didelphis marsupialis*	Panama	Dunn and Clark (1933)
	M	Mammalia/Primates	*Saimiri oerstedii*	Panama	Dunn and Clark (1933)
	M	Mammalia/Cingulata	*Dasypus novemcinctus*	Panama	Dunn and Clark (1933)
	PCR	Mammalia/Rodentia	*Phyllotis xantophygus*	Chile	Thomas et al. (2020)
Borrelia sp. near *Borrelia venezuelensis*	PCR	Mammalia/Chiroptera	*Artibeus literatus*	Colombia	Mancilla-Agrono et al. (2022)
	PCR		*Platyrrhinus helleri*	Colombia	Mancilla-Agrono et al. (2022)
	PCR		*Mesophylla macconnelli*	Colombia	Mancilla-Agrono et al. (2022)
	PCR		*Rhynchonycteris naso*	Colombia	Mancilla-Agrono et al. (2022)
	PCR	Mammalia/Rodentia	*Microryzomys altissimus*	Colombia	Mancilla-Agrono et al. (2022)
Borrelia sp. near *Borrelia turicatae*	PCR	Mammalia/Rodentia	*Coendou rufescens*	Colombia	Mancilla-Agrono et al. (2022)
Bat group					
Borrelia sp.	M	Mammalia/Chiroptera	*Natalus tumidirostris*	Colombia	Marinkelle and Grose (1968)
	PCR		*Artibeus literatus*	Colombia	Mancilla-Agrono et al. (2022)
	PCR		*Carollia brevicauda*	Colombia	Mancilla-Agrono et al. (2022)
	PCR		*Sturnina erythropus*	Colombia	Mancilla-Agrono et al. (2022)
	PCR		*Phyllostomus discolor*	Colombia	Lopez et al. (2023)
	PCR		*Uroderma* sp.	Colombia	Lopez et al. (2023)
	PCR		*Glossophaga soricina*	Colombia	Lopez et al. (2023)
	PCR		*Carollia perspicillata*	Colombia	Muñoz-Leal et al. (2021) and Lopez et al. (2023)
	PCR		*Desmodus rotundus*	Brazil	Jorge et al. (2023)

(continued)

Table 13.1 (continued)

Microorganism	Method	Animal group	Host species	Country	References
Lyme group					
"*Candidatus* Borrelia paulista"	PCR	Mammalia/Rodentia	*Oligoryzomys* sp.	Brazil	Weck et al. (2022)
Borrelia sp.	PCR	Mammalia/Rodentia	*Rattus ratus*	Mexico	Solís-Hernández et al. (2016)
	PCR		*Mus musculus*	Mexico	Solís-Hernández et al. (2016)
	PCR		*Microtus mexicanus*	Mexico	Colunga-Salas et al. (2020a, b)
	PCR		*Neotoma mexicana*	Mexico	Colunga-Salas et al. (2020a, b)
	PCR		*Neotoma micropus*	Mexico	Colunga-Salas et al. (2020a, b)
	PCR		*Neotomodon alstoni*	Mexico	Colunga-Salas et al. (2020a, b)
	PCR		*Peromyscus leucopus*	Mexico	Colunga-Salas et al. (2020a, b)
	PCR		*Peromyscus maniculatus*	Mexico	Colunga-Salas et al. (2020a, b)
	PCR		*Oligoryzomys longicaudatus*	Chile	Thomas et al. (2020)
	PCR		*Thomasomys aureus*	Colombia	Mancilla-Agrono et al. (2022)
	ELISA/WB	Mammalia/Artiodactyla	*Odocoileus virginianus*	Mexico	Martinez et al. (1999)
Order Rickettsiales					
Anaplasma marginale	PCR	Mammalia/Artiodactyla	*Mazama gouazoubira*	Brazil	da Silveira et al. (2011)
	PCR		*Blastocerus dichotomus*	Argentina	Orozco et al. (2020)
Anaplasma bovis	PCR	Mammalia/Artiodactyla	*Mazama americana*	Trinidad	Sant et al. (2022)
	PCR			Brazil	Soares et al. (2017a, b)

Anaplasma phagocytophilum	PCR	Mammalia/Artiodactyla	*Mazama gouazoubira*	Brazil	da Silveira et al. (2011)
	PCR		*Pudu puda*	Chile	Santodomingo et al. (2023)
	PCR		*Odocoileus virginianus*	Mexico	Ojeda-Chi et al. (2019)
Anaplasma odocoilei	PCR	Mammalia/Artiodactyla	*Odocoileus virginianus*	Mexico	Ojeda-Chi et al. (2019)
	PCR		*Mazama temama*	Mexico	Ojeda-Chi et al. (2019)
	PCR		*Odocoileus virginianus*	Mexico	Ojeda-Chi et al. (2019)
	PCR		*Blastocerus dichotomus*	Argentina	Orozco et al. (2020)
Anaplasma sp. near *Anaplasma marginale*	PCR	Mammalia/Primates	*Sapajus apella*	Brazil	Candido et al. (2023)
Anaplasma sp. near *Anaplasma phagocytophilum*	PCR	Aves/Falconiformes	*Caracara plancus*	Brazil	Machado et al. (2012)
	PCR	Aves/Accipitriformes	*Coragyps atratus*	Brazil	Machado et al. (2012)
	PCR	Mammalia/Carnivora	*Leopardus tigrinus*	Brazil	André et al. (2012)
	PCR		*Speothos venaticus*	Brazil	André et al. (2012)
Anaplasma sp. near *Anaplasma platys*	PCR	Mammalia/Artiodactyla	*Blastocerus dichotomus*	Argentina	Orozco et al. (2020)
Anaplasma sp. near "*Candidatus* Anaplasma boleense"	PCR	Mammalia/Artiodactyla	*Blastocerus dichotomus*	Argentina	Orozco et al. (2020)

(continued)

Table 13.1 (continued)

Microorganism	Method	Animal group	Host species	Country	References
Anaplasma spp.	PCR	Mammalia/Artiodactyla	*Mazama gouazoubira*	Uruguay	Félix et al. (2020)
	PCR		*Mazama bororo*	Brazil	Félix et al. (2020)
	PCR		*Mazama americana*	Brazil	Soares et al. (2017a, b)
	PCR		*Dicotyles tajacu*	Brazil	Soares et al. (2017a, b)
	PCR		*Tayassu pecari*	Brazil	Soares et al. (2017a, b)
	PCR	Mammalia/Carnivora	*Procyon cancrivorus*	Argentina	Tarragona et al. (2023)
	PCR	Mammalia/Carnivora	*Nasua narica*	Brazil	Perles et al. (2023)
	IFA/PCR		*Cerdocyon thous*	Brazil	De Sousa et al. (2017)
	IFA/PCR		*Leopardus pardalis*	Brazil	De Sousa et al. (2017)
	IFA/PCR		*Nasua nasua*	Brazil	De Sousa et al. (2017)
	PCR	Mammalia/Pilosa	*Bradypus variegatus*	Brazil	Calchi et al. (2020)
	PCR		*Bradypus tridactylus*	Brazil	Soares et al. (2017a, b)
	PCR		*Choloepus didactylus*	Brazil	Calchi et al. (2020)
	PCR		*Tamandua tetradactyla*	Brazil	Calchi et al. (2020)
	PCR	Mammalia/Rodentia	*Hydrochoerus hydrochaeris*	Brazil	Vieira et al. (2022)
	PCR		*Hylaeamys megacephalus*	Brazil	Wolf et al. (2017)
	PCR		*Rattus rattus*	Brazil	Benevenute et al. (2017)
	PCR		*Akodon montensis*	Brazil	Benevenute et al. (2017)
	PCR		*Trichomys inermis*	Brazil	Benevenute et al. (2017)
	PCR		*Calomys expulsus*	Brazil	Benevenute et al. (2017)
	PCR		*Euryoryzomys russatus*	Brazil	Benevenute et al. (2017)
	PCR		*Akodon montesis*	Brazil	Benevenute et al. (2017)
	PCR		*Trichomys laurentius*	Brazil	Benevenute et al. (2017)
	PCR		*Cavia spixii*	Brazil	Benevenute et al. (2017)
	PCR		*Oligoryzomys nigripes*	Brazil	Benevenute et al. (2017)
	PCR		*Shiggurus cerqueierae*	Brazil	Benevenute et al. (2017)

"*Candidatus* Ehrlichia dumleri"	PCR	Mammalia/Carnivora	*Nasua narica*	Brazil	Perles et al. (2023)
"*Candidatus* Ehrlichia hydrochoerus"	PCR	Mammalia/Rodentia	*Hydrochoerus hydrochaeris*	Brazil	Vieira et al. (2022)
"*Candidatus* Ehrlichia pampeana"	PCR	Mammalia/Artiodactyla	*Mazama gouazoubira*	Uruguay	Félix et al. (2021)
Ehrlichia sp. near *Ehrlichia canis*	PCR	Mammalia/Artiodactyla	*Dicotyles tajacu*	Trinidad	Sant et al. (2022)
	PCR	Mammalia/Primates	*Callithrix* sp.	Brazil	Mafra et al. (2015)
	PCR		*Sapajus apella*	Brazil	Candido et al. (2023)
	PCR	Mammalia/ Didelphimorphia	*Didelphis* sp.	Brazil	Guimaraes et al. (2019)
	PCR	Mammalia/Carnivora	*Puma concolor*	Brazil	André et al. (2012)
	PCR		*Leopardus tigrinus*	Brazil	André et al. (2012)
	PCR		*Herpailurus yagouaroundi*	Brazil	André et al. (2012)
	PCR		*Leopardus pardalis*	Brazil	André et al. (2012)
	PCR		*Speothos venaticus*	Brazil	André et al. (2012)
	PCR		*Cerdocyon thous*	Brazil	André et al. (2012)
	PCR	Aves/Accipitriformes	*Coragyps atratus*	Brazil	Machado et al. (2012)
Ehrlichia sp. near *Ehrlichia chaffeensis*	PCR	Aves/Falconiformes	*Falco sparverius*	Brazil	Machado et al. (2012)
	PCR	Mammalia/Primates	*Mico melanurus*	Brazil	Candido et al. (2023)
	PCR	Mammalia/Artiodactyla	*Odocoileus virginianus*	Mexico	Ojeda-Chi et al. (2019)
	PCR		*Blastocerus dichotomus*	Argentina	Guillemi et al. (2018)
	PCR	Mammalia/Carnivora	*Leopardus tigrinus*	Brazil	André et al. (2012)
	PCR		*Leopardus pardalis*	Brazil	André et al. (2012)
	PCR		*Puma concolor*	Brazil	André et al. (2012)
	PCR		*Herpailurus yagouaroundi*	Brazil	André et al. (2012)
	PCR		*Cerdocyon thous*	Brazil	André et al. (2012)

(continued)

Table 13.1 (continued)

Microorganism	Method	Animal group	Host species	Country	References
Ehrlichia spp.	PCR	Aves/Sphenisciformes	*Spheniscus magellanicus*	Chile	Muñoz-Leal et al. (2019)
	PCR	Mammalia/Rodentia	*Trichomys laurentius*	Brazil	Benevenute et al. (2017)
	PCR		*Trichomys pahyurus*	Brazil	Benevenute et al. (2017)
	PCR		*Holochilus fosteri*	Brazil	Benevenute et al. (2017)
	PCR		*Clyomys laticeps*	Brazil	Benevenute et al. (2017)
	PCR		*Calomys cerqueiai*	Brazil	Benevenute et al. (2017)
	PCR		*Calomys expulsus*	Brazil	Benevenute et al. (2017)
	PCR		*Oecomys laticeps*	Brazil	Benevenute et al. (2017)
	PCR	Mammalia/Pilosa	*Myrmecophaga tetradactyla*	Brazil	Calchi et al. (2020)
	PCR		*Tamandua tetradactyla*	Brazil	Calchi et al. (2020)
	PCR		*Bradypus variegatus*	Brazil	Calchi et al. (2020)
			Bradypus tridactylus	Brazil	Soares et al. (2017a, b)
	PCR		*Dasypus novemcinctus*	Brazil	Calchi et al. (2020)
	PCR		*Choloepus* sp.	Brazil	Calchi et al. (2020)
	IFA/PCR	Mammalia/Carnivora	*Panthera onca*	Brazil	Widmer et al. (2011)
	PCR		*Cerdocyon thous*	Brazil	Almeida et al. (2013)
	PCR	Mammalia/Primates	*Mico melanurus*	Brazil	Candido et al. (2023)
	PCR	Mammalia/Didelphimorphia	*Gracilinanus agilis*	Brazil	Oliveira et al. (2020)
	PCR		*Didelphis albiventris*	Argentina	Tarragona et al. (2022)
			Didelphis albiventris	Brazil	Lopes et al. (2018)
	PCR	Mammalia/Artiodactyla	*Tayassu pecari*	Trinidad	Sant et al. (2022)
	PCR		*Dicotyles tajacu*	Brazil	Soares et al. (2017a, b)

Rickettsia amblyommatis	IFA/PCR	Aves/Columbiformes	*Leptotila rufaxilla*	Colombia	Cardona et al. (2022)
	IFA/PCR	Aves/Passeriformes	*Dendroica fuliginosa*	Colombia	Cardona et al. (2022)
	IFA/PCR		*Phacellodomus rufifrons*	Colombia	Cardona et al. (2022)
	IFA/PCR		*Tyrannus dominicensis*	Colombia	Cardona et al. (2022)
	IFA/PCR		*Turdus leucomelas*	Colombia	Cardona et al. (2022)
	IFA/PCR		*Euphonia laniirostris*	Colombia	Cardona et al. (2022)
	IFA/PCR		*Saltator orinocensis*	Colombia	Cardona et al. (2022)
Rickettsia bellii	IFA/PCR	Aves/Columbiformes	*Leptotila verreauxi*	Colombia	Cardona et al. (2022)
	IFA/PCR	Aves/Passeriformes	*Myiozetetes cayanensis*	Colombia	Cardona et al. (2022)
	IFA/PCR		*Tyrannus savana*	Colombia	Cardona et al. (2022)
Rickettsia felis	PCR	Mammalia/ Didelphimorphia	*Didelphis marsupialis*	Mexico	Sánchez-Montes et al. (2021)
	PCR		*Didelphis virginiana*	Mexico	Sánchez-Montes et al. (2021)
	PCR	Mammalia/Rodentia	*Ototylomys phyllotis*	Mexico	Sánchez-Montes et al. (2021)
	PCR		*Peromyscus yucatamicus*	Mexico	Sánchez-Montes et al. (2021)
	PCR		*Sigmodon toltecus*	Mexico	Sánchez-Montes et al. (2021)
	PCR		*Heteromys guaremi*	Mexico	Sánchez-Montes et al. (2021)
Rickettsia rhipicephali	IFA/PCR	Aves/Columbiformes	*Leptotila rufaxilla*	Colombia	Cardona et al. (2022)
	IFA/PCR	Aves/Passeriformes	*Pitangus sulphuratus*	Colombia	Cardona et al. (2022)
	IFA/PCR	Aves/Passeriformes	*Tyrannus melancholicus*	Colombia	Cardona et al. (2022)
	IFA/PCR	Aves/Passeriformes	*Euphonia laniirostris*	Colombia	Cardona et al. (2022)

(continued)

Table 13.1 (continued)

Microorganism	Method	Animal group	Host species	Country	References
Rickettsia parkeri	IFA/PCR	Aves/Columbiformes	*Colombina tapacolti*	Colombia	Cardona et al. (2022)
	IFA/PCR	Aves/Passeriformes	*Pitangus sulphuratus*	Colombia	Cardona et al. (2022)
	IFA/PCR		*Tyrannus dominicensis*	Colombia	Cardona et al. (2022)
	IFA/PCR		*Turdus leucomelas*	Colombia	Cardona et al. (2022)
	IFA/PCR		*Rhampocelus carbo*	Colombia	Cardona et al. (2022)
	IFA/PCR		*Thraupis palmarum*	Colombia	Cardona et al. (2022)
	PCR	Mammalia/Carnivora	*Cerdocyon thous*	Brazil	Dall'Agnol et al. (2017)
	PCR		*Lycalpex gymnocercus*	Brazil	Dall'Agnol et al. (2017)
Rickettsia rickettsii	IFA/PCR	Aves/Passeriformes	*Rhampocelus carbo*	Colombia	Cardona et al. (2022)
	IFA	Mammalia/Rodentia	*Hydrochoerus hydrochaeris*	Brazil	Yang et al. (2021)
	PCR		*Micrurus mexicanus*	Mexico	Sánchez-Montes et al. (2021)
	PCR		*Neotomodon alstoni*	Mexico	Sánchez-Montes et al. (2021)
	PCR		*Peromyscus beatae*	Mexico	Sánchez-Montes et al. (2021)
	PCR		*Peromyscus leucopus*	Mexico	Sánchez-Montes et al. (2021)
	PCR		*Peromyscus maniculatus*	Mexico	Sánchez-Montes et al. (2021)
	PCR		*Peromyscus megalops*	Mexico	Sánchez-Montes et al. (2021)
	PCR		*Rheithrodontomys fulvescens*	Mexico	Sánchez-Montes et al. (2021)
Rickettsia typhi	PCR	Mammalia/Rodentia	*Mus musculus*	Mexico	Sánchez-Montes et al. (2021)
	PCR		*Rattus rattus*	Mexico	Sánchez-Montes et al. (2021)

Rickettsia spp.	IFA	Mammalia/ Didelphimorphia	*Didelphis marsupialis*	Colombia	Londoño et al. (2017)
	IFA	Mammalia/Rodentia	*Hydrochoerus hydrochaeris*	Colombia	Miranda et al. (2011)
	IFA	Mammalia/ Didelphimorphia	*Didelphis marsupialis*	Panama	Bermúdez et al. (2017a, b)
	PCR		*Mus musculus*	Mexico	López-Pérez et al. (2022)
	PCR		*Rattus rattus*	Mexico	López-Pérez et al. (2022)
	PCR		*Peromyscus boylii*	Mexico	López-Pérez et al. (2022)
	PCR		*Peromyscus maniculatus*	Mexico	López-Pérez et al. (2022)
	PCR		*Dipodomys merrami*	Mexico	López-Pérez et al. (2022)
	I&C	Mammalia/Lagomorpha	*Sylvilagus* sp.	Costa Rica	Fuentes (1986)
	IFA	Mammalia/Carnivora	*Canis latrans*	Panama	Bermúdez et al. (2017a, b)
	PCR		*Canis latrans*	Mexico	López-Pérez et al. (2022)
	IFA/PCR	Aves/Passeriformes	*Pachyramphus polychopterus*	Colombia	Cardona et al. (2022)
"*Candidatus* Neoehrlichia chilensis"	PCR	Mammalia/Rodentia	*Abrothrix* spp.	Chile	Müller et al. (2018)
Order Piroplasmida					
Babesia sp. near *Babesia caballi*	IFA	Mammalia/Perissodactyla	*Tapirus pinchaque*	Ecuador	Castellanos (2013)
Babesia sp. near *Babesia duncani*	PCR	Mammalia/Artiodactyla	*Tayassu pecari*	Brazil	Soares et al. (2017a, b)
Babesia brasiliensis	M	Mammalia/ Didelphimorphia	*Didelphis marsupialis*	Brazil	Serra-Freire (1979)
	M		*Didelphis albiventris*	Brazil	Serra-Freire (1979)
	M		*Philander quica*	Brazil	Regendanz and Kikuth (1928)
Babesia sp. near *Babesia bovis*	PCR	Mammalia/Artiodactyla	*Mazama gouazoubira*	Brazil	da Silveira et al. (2011)
Babesia bigemina	PCR	Mammalia/Artiodactyla	*Blastocerus dichotomus*	Brazil	da Silveira et al. (2011)

(continued)

Table 13.1 (continued)

Microorganism	Method	Animal group	Host species	Country	References
Babesia sp.	PCR	Mammalia/Rodentia	*Cuniculus paca*	Brazil	Soares et al. (2017a, b)
	PCR		*Hydrochoerus hydrochaeris*	Brazil	Neves et al. (2022)
	PCR		*Thrichomys pachyurus*	Brazil	Wolf et al. (2017)
	PCR		*Phyllotis darwini*	Chile	Santodomingo et al. (2022a)
	PCR		*Abrothrix jelskii*	Chile	Santodomingo et al. (2022a)
	PCR	Mammalia/ Didelphimorphia	*Didelphis marsupialis*	Brazil	Soares et al. (2017a, b)
	PCR		*Didelphis marsupialis*	Brazil	Colle et al. (2019)
	PCR		*Monodelphis domestica*	Brazil	Wolf et al. (2017)
	M		*Didelphis albiventris*	Colombia	García (1945)
	PCR		*Didelphis albiventris*	Brazil	Gonçalves et al. (2021)
	PCR		*Didelphis aurita*	Brazil	Perles et al. (2023)
	M	Mammalia/Chiroptera	*Mormoops megalophylla*	Colombia	Marinkelle (1996)
	PCR	Mammalia/Artiodactyla	*Pudu puda*	Chile	Santodomingo et al. (2022b)
Cytauxzoon felis	PCR	Mammalia/Carnivora	*Leopardus pardalis*	Brazil	Soares et al. (2017a, b)
	PCR		*Puma concolor*	Brazil	de Paula et al. (2022)
	PCR		*Panthera onca*	Brazil	Guizelini et al. (2021)
Rangelia vitalii	PCR	Mammalia/Carnivora	*Cerdocyon thous*	Brazil	Fredo et al. (2015), Soares et al. (2014)
	PCR		*Chrysocyon brachyurus*	Brazil	Silveira et al. (2016a, b)
	PCR		*Lycalpex gymnocercus*	Brazil	Fredo et al. (2015)
Theileria terrestres	PCR	Mammalia/Perissodactyla	*Tapirus terrestris*	Brazil	Mongruel et al. (2022)
Theileria sp. near *Theileria capreoli*	PCR	Mammalia/Cingulata	*Dasyprocta* sp.	Brazil	Soares et al. (2017a, b)
Theileria sp. near *Theileria cervi*	PCR	Mammalia/Artiodactyla	*Mazama gouazoubira*	Brazil	da Silveira et al. (2011)
	PCR		*Mazama americana*	Brazil	Soares et al. (2017a, b)

Theileria cervi	PCR	Mammalia/Artiodactyla	*Blastocerus dichotomus*	Argentina	Sebastian et al. (2022)
Theileria sp. near *Theileria equi*	PCR	Mammalia/Cingulata	*Dasypus novemcinctus*	Brazil	Soares et al. (2017a, b)
Piroplasmida spp. near *Babesia* or *Theileria*	PCR	Mammalia/Rodentia	*Dasyprocta leporina*	Trinidad	Sant et al. (2022)
	PCR	Mammalia/Chiroptera	*Phyllostomus discolor*	Brazil	Ikeda et al. (2021)
	PCR		*Artibeus lituratus*	Brazil	Ikeda et al. (2021)
	PCR		*Artibeus planirostris*	Brazil	Ikeda et al. (2021)

[a]Abreviation: PCR, polymerase chain reaction; IFA, immunofluorescence assay; ELISA, enzyme-linked immunosorbent assay; WB, western blot; I&C, isolation and/or culture; M, microscopic observation of blood smears.

2020). Moreover, two yet-to-be isolated ehrlichiae, "*Candidatus* Ehrlichia hydrochoerus" and "*Candidatus* Ehrlichia dumleri", were proposed as novel taxa in Brazil based on multilocus genetic characterizations performed on capybara (*Hydrochoerus hydrochaeris*) and coati (*Nasua nasua*) blood (Vieira et al. 2022; Perles et al. 2022) Still, "*Candidatus* Ehrlichia pampeana" is a putatively new species detected in gray brocket deer (*Mazama gouazoubira*) in Uruguay (Felix et al. 2021). Captive and wild carnivores were also positive for Anaplasmataceae in Brazil (André 2018), indicating that strains closely related to *A. phagocytophilum* and *Ehrlichia chaffeensis* do infect or circulate among these animals (Table 13.1).

In the NZR and Paleartic Zoogeographical Regions, deer are important hosts of *A. phagocytophilum* and *E. chaffeensis*, two zoonotic agents transmitted chiefly by ticks of the genus *Ixodes* (Rar and Golovljova 2011; Rar et al. 2021). For instance, in North America (NZR), the white-tailed deer (*Odocoileus virginianus*) has been extensively studied because of its role as a reservoir of these two pathogens (Rar and Golovljova 2011; Rar et al. 2021). However, the white tailed deer has a wide distribution in the American Continent, extending towards Mexico and northern South America. Hence, it is not unexpected that genetic screenings aiming to detect DNA of *A. phagocytophilum* and *E. chaffeensis* also yielded positive results in Mexican populations of this deer species (Ojeda-Chi et al. 2019). However, further genetic characterizations are needed on these records since only short 16S rRNA gene sequences for both agents were obtained (Ojeda-Chi et al. 2019), which also the case for sequences of Anaplasmataceae (not available in GenBank) retrieved from marsh deer (*Blastocerus dichotomus*) in northern Argentina (Orozco et al. 2020) (Table 13.1). An extended genetic characterization would be appropriate to elucidate the diversity of strains that circulate in neotropical cervids (and other vertebrates). For example, studies regarding the genetic diversity of *A. phagocytophilum* have shown that the bacterium splits into four ecotypes scattered along the Northern Hemisphere (Rar et al. 2021). Interestingly, a fifth ecotype was recently characterized in pudus (*Pudu puda*), an associated tick species from Chile (Santodomingo et al. 2023), demonstrating that the bacterium occurs in similar ecological scenarios involving *Ixodes* spp. that parasitize deer and rodents in the Southern Hemisphere as well. Although in the NZR reports of *Anaplasma* including *A. phagocytophilum* and *Ehrlichia* strains do point rodents as carriers of the agents (Benevenute et al. 2017), the lack of extended genomic characterization and isolation of the bacteria precludes to understand the role of these mammals in the enzootic cycles of these bacteria.

While recent studies highlight that Anaplasmataceae agents that circulate in the Northern Hemisphere also thrive in similar ecological niches in South America (Müller et al. 2018; Santodomingo et al. 2023), others seem to challenge what is known regarding their common biological cycles. For instance, *Ehrlichia* spp. closely related to *Ehrlichia canis* have been recurrently detected in wild animals of the NZR, mainly in carnivores (André et al. 2012) but also in birds (Machado et al. 2012), primates (Mafra et al. 2015; Candido et al. 2023), marsupials (Guimaraes et al. 2019), and hogs (Sant et al. 2022). At least in carnivores, this fact has been confirmed by seroconversion against *E. canis* antigens (de Sousa et al. 2017).

Ehrlichia canis is a pathogen transmitted mainly by *Rhipicephalus sanguineus* sensu lato ticks that usually parasitize domestic dogs in tropical and subtropical regions of the globe (Rar and Golovljova 2011). Infection of wild canids with *E. canis* has been reported worldwide (André 2018). However, how *E. canis* can reach primates, marsupials, hogs, or even birds is still controversial. Nevertheless, the tick *R. sanguineus* sensu lato parasitizes synanthropic wild animals if available; therefore, it has been hypothesized that *E. canis* could infect uncommon hosts as a consequence of spillover from domestic dogs living nearby (André 2018). Although probable, this hypothesis needs further assessment.

Finally, evidence of the genus "*Ca.* Neoehrlichia" is incipient in neotropical mammals. In fact, the sole species corresponds to "*Candidatus* Neoehrlichia chilensis" characterized from organs of wild rodents in Chile (Müller et al. 2018). As "*Ca.* Neoehrlichia" bacteria are transmitted by *Ixodes* spp. (Rar and Golovljova 2011), a better understanding of the possible species existing in the NZR would be gained by targeted surveillance on the animals exploited by ticks of this genus.

13.2.2 Rickettsia

The genus *Rickettsia* comprises a group of obligate intracellular gram-negative rods that reside free inside the cytoplasm of eukaryotic host cells, with some species also reaching the nucleus (Fang et al. 2017). *Rickettsia* spp. harbored by ticks and other ectoparasites are usually classified into four phylogenetic groups: the spotted fever group (SFG), the typhus group (TG), the transitional group, and the ancestral group (Fang et al. 2017). The SFG is the largest one and includes several pathogenic species, such as *Rickettsia rickettsii*, *Rickettsia parkeri*, *Rickettsia africae*, *Rickettsia slovaca*, and other species of unknown pathogenicity. The TG is composed of two species, *Rickettsia prowazekii* and *Rickettsia typhi*, both human pathogens and transmitted by lice and fleas, respectively. The transitional group includes few species, some with clinical importance such as *Rickettsia akari* and *Rickettsia australis*. Finally, the ancestral group includes two nonpathogenic species, *Rickettsia bellii* and *Rickettsia canadensis* (Fang et al. 2017).

In the NZR, more than 20 species of *Rickettsia* have been reported in ticks and fleas, including formally described species, candidate species and strains, and the records of those detections are stated in Table 13.1 (Labruna et al. 2011a; Venzal and Nava 2011; Bermúdez and Troyo 2018; Charles et al. 2021). *Rickettsia rickettsii* is the most important infectious agent of the genus, which causes febrile symptoms that can be fatal in humans if not treated (Valbuena 2010; Hidalgo et al. 2013). In the NZR, species such as *R. bellii*, *R. rickettsii*, *R. parkeri*, *Rickettsia rhipicephali*, *Rickettsia amblyommatis*, and *Rickettsia felis* have been isolated (Labruna et al. 2004, 2007, 2017; Pinter and Labruna 2006; Szabó et al. 2013), and their antigens have been employed for the detection of anti-*Rickettsia* antibodies in birds, wild rodents, lagomorphs, marsupials, carnivores, and cervids (Fuentes 1986; Horta et al. 2009; Miranda et al. 2011; Rodriguez-Vivas et al. 2013; Londoño et al. 2017;

Bermúdez et al. 2017a, b; Ojeda-Chi et al. 2018; Serpa et al. 2021, Table 13.1). However, the evidence from serological tests gives little specificity for a determined rickettsial antigen (Hechemy et al. 1989). Therefore, many of these records only indicate exposure to a given *Rickettsia* spp. and do not always mean infection.

Most of the knowledge on the epidemiology of *R. rickettsii* in South America comes from studies developed in Brazil, where the bacterium is endemic. The recent spread and increase of capybara (*Hydrochoerus hydrochaeris*) populations in the State of São Paulo (Brazil) triggered substantial research to study the role of this large rodent as an AVH of *R. rickettsii*, which is transmitted to capybaras and humans by the vector tick *Amblyomma sculptum* (Luz et al. 2017). In the last decades, the abundance of sugarcane plantations to which capybaras have access for meals not only boosted populations of this rodent but also increased the abundance of *R. rickettsii*-infected *A. sculptum*, which populations currently overlapping urban areas in Brazil (Luz et al. 2017).

Similar to capybaras, opossums (*Didelphis aurita*) are also AVHs for the horizontal transmission of *R. rickettsii* (Horta et al. 2009), and the bacterium also infects *Amblyomma aureolatum*, in which it perpetuates transstadially and transovarially (Labruna et al. 2011b). Interestingly, larvae and nymphs of this tick are parasites of birds, and adults feed on carnivores (Luz et al. 2017; Barbieri et al. 2014). Although the transmission of *R. rickettsii* from infected *A. aureolatum* to birds or wild carnivores has not been demonstrated, domestic dogs were experimentally infected through *A. aureolatum* bites and served as amplifiers capable to infect naïve ticks (Binder et al. 2021). This fact suggests that wild carnivores that occur in areas where *R. rickettsii*-infected *A. aureolatum* thrive could also act as amplifier hosts. On the other hand, birds seem to be refractory to *Rickettsia* infection likely because of their elevated corporal temperature (Gillen 2014). Regarding rodents, the liver and spleen of several cricetid species were positive after screening for *Rickettsia* spp. using a highly sensitive real-time PCR protocol (Table 13.1); however, attempts to obtain further genetic data from the positive animals to identify the agents were unsuccessful, possibly because of the low amount of target DNA (Serpa et al. 2021). For instance, while sequences of *R. felis* were retrieved from DNA extracted from the spleen of a gracile opossum in Brazil (*Gracilinanus agilis*) (Serpa et al. 2021), the role of small mammals as amplifiers or reservoirs of *Rickettsia* spp. demands further research.

13.2.3 Borrelia

Tick-borne spirochetes in the genus *Borrelia* are globally distributed and thrive in enzootic cycles, switching infections between ticks and tick hosts (Margos et al. 2022). Currently, the genus is divided into three discrete phylogenetic groups: the Lyme disease group (*Borrelia burgdorferi* sensu lato, LDG), the relapsing fever group (RFG), and a group associated with echidna, reptiles, and birds (Binetruy et al. 2020; Margos et al. 2022; Lopez et al. 2023). Moreover, a novel monophyletic clade intercalating between the LDG and RFG species has been recently characterized upon blood and organs of neotropical bats (Muñoz-Leal et al. 2021; Lopez et al. 2023; Jorge et al. 2023). While in the majority of cases, LDG and RFG borreliae are transmitted by hard (Ixodidae) and soft bodied (Argasidae) ticks, respectively (Talagrand-Reboul et al. 2018), there are some exceptions. Indeed, few RFG spirochetes survive in hard ticks, which is the case of *Borrelia theileri* and *Borrelia miyamotoi,* transmitted by *Rhipicephalus microplus* and *Ixodes scapularis* respectively (Margos et al. 2022).

While the role of wild vertebrates in the biological cycle of LDG spirochetes has been widely studied in the endemic areas where Lyme disease occurs (i.e., Northern Hemisphere), less is known regarding wild animals as hosts of LDG spirochetes in the NZR. Although LDG spirochetes are yet to be isolated from vertebrate hosts in the NZG, molecular detections performed in rodents from Brazil, Chile, Colombia, and Mexico (Colunga-Salas et al. 2020a, b; Thomas et al. 2020; Mancilla-Agrono et al. 2022; Weck et al. 2022) suggest that as in the Northern Hemisphere, small mammals would be important hosts for these spirochetes. Interestingly, in the NZG most of the detected LDG *Borrelia* spp. correspond to previously uncharacterized taxa. A good example of this is "*Candidatus* Borrelia paulista," which was identified using a multilocus sequencing typing scheme from organs of an *Oligoryzomys* mouse in Brazil (Weck et al. 2022). Moreover, LDG *Borrelia* have been detected in cricetid rodents (Cricetidae) from Chile (Thomas et al. 2020) and Mexico (summarized in Colunga-Salas et al. 2020a, b). Notably, haplotypes of *Borrelia flaB* gene 99–100% identical with *B. burgdorferi* sensu stricto were described in native and exotic mice in Colombia (Mancilla-Agrono et al. 2022). However, the report needs further evaluation because sequences submitted to analyses were rather short (<274 pair bases), precluding a specific identification. Similarly, the molecular evidence pointing to a high prevalence (16.6–42.5%) of *Borrelia* DNA in organs from synanthropic rodent species (*Mus musculus*, *Rattus rattus*) in Colombia and Mexico (Solís-Hernández et al. 2016; Mancilla-Agrono et al. 2022) should be carefully interpreted, since other studies seeking DNA of these spirochetes in synanthropic rodents have shown null or low (<10%) prevalence (Hornok et al. 2015; Lau et al. 2020), and the detection of DNA does not imply that the rodents have a role in the *Borrelia* cycle.

In the Northern Hemisphere, *B. burgdorferi* sensu lato, the agent of Lyme disease, perpetuates in enzootic cycles involving *Ixodes* spp. as vectors that transmit spirochetes to rodents, which are common hosts for immature tick stages and competent hosts for spirochetes (Wolcott et al. 2021). Adult ticks involved in this enzootic cycle parasitize cervids, which in turn are refractory to *Borrelia* infection (Kurokawa et al. 2020). A similar epidemiological scenario occurs in southern South America, particularly in Chile, where *Borrelia chilensis*, a species of LDG, was isolated from *Ixodes stilesi* (Ivanova et al. 2014), a tick that parasitizes cricetid rodents and cervids such as pudus (*Pudu puda*) (Guglielmone et al. 2023). Interestingly, genetic screenings searching for *Borrelia* DNA in the blood and organs of pudus carrying *B. chilensis*-positive ticks were negative (Verdugo et al. 2017; Santodomingo et al. 2022), reinforcing the evidence that cervids are not reservoirs of LDG spirochetes. However, the animals could be exposed when infected ticks inoculate the spirochetes during the meal. The finding of LDG *Borrelia* antibodies in white-tailed deer (*Odocoileus virginianus*) in Mexico also supports such an exposure (Martínez et al. 1999).

Interest in studying RFG *Borrelia* in the NZR began in Panama and Colombia in the early twentieth century because of recurrent episodes of fever in workers of the Canal Zone and Emerald mines, respectively (Lopez et al. 2016; Faccini-Martínez and Botero-García 2016; Faccini-Martínez et al. 2022). In both cases, entering to dense tropical forests exposed people to novel agents transmitted by hematophagous parasites, including ticks. At the time, *Ornithodoros rudis* was identified in both foci as the vector or *Borrelia venezuelensis* (Muñoz-Leal et al. 2018). Although ticks biting humans were collected and spirochetes isolated, vertebrate tick hosts acting as reservoirs of *B. venezuelensis* in nature are still an unsolved question in the NZR, probably because of a lack of effort. Finding spirochetes in animals could be easier than expected since RFG *Borrelia* reach high concentrations in the blood of competent hosts, favoring their detection in blood smears observed with dark-field microscopy or stained with Giemsa. Indeed, early surveys seeking RFG spirochetes in Panama targeted the blood of primates, cingulates, and marsupials and detected the bacteria in squirrels, monkeys, opossums, and armadillos (Dunn and Clark 1933). However, at that time, the identity of the spirochetes was not determined by molecular tools. A gap of at least 80 years separates those studies from current findings in wild animals, which include the detection by PCR of a *Borrelia* sp. phylogenetically closely related with *B. venezuelensis* in organs and blood of bats and rodents from Colombia (Mancilla-Agrono et al. 2022). However, in that study, the generated sequences were short, precluding accuracy in the identification of the species.

Although the circulation of *B. venezuelensis* in bats needs further confirmation, recent studies unveiled a novel monophyletic group of *Borrelia* spp. in bats from Colombia and Brazil (Muñoz-Leal et al. 2021; Lopez et al. 2023; Jorge et al. 2023). In Colombia, hundreds of spirochetes were previously observed in blood smears of one of the positive species, Seba's short-tailed bat (*Carollia perspicillata*), suggesting that the spirochetes belonged to the RFG (Marinkelle and Grose 1968). However, phylogenetic studies now indicate that the group of spirochetes detected in neotropical bats compounds a sister group to the RFG species (Muñoz-Leal et al. 2021; Lopez et al. 2023; Jorge et al. 2023).

Species of RFG *Borrelia* that have been isolated from infected ticks collected in the wilderness correspond to *B. venezuelensis*, *Borrelia puertoricensis*, and "*Candidatus* Borrelia caatinga" (Muñoz-Leal et al. 2018; Bermúdez et al. 2021; Oliveira et al. 2023). The fact that ticks positive for *B. venezuelensis* (*O. rudis*) were collected in abandoned bird nests suggests that avian hosts could harbor the spirochete in Brazil. Nevertheless, bats and rodents could also occupy bird nests as shelters and become bitten by infected *O. rudis* (Muñoz-Leal et al. 2018), supporting that *B. venezuelensis* could infect bats and rodents as well. On the other hand, in Panama, agouties (*Dasyprocta punctata*) could be hosts for *B. puertoricensis* since ticks were collected from burrows of this rodent (Bermúdez et al. 2021). The third species, "*Ca*. B. caatinga", could be circulating in rodents (*Kerodon rupestris*), with which the infected *Ornithodoros* sp. associates in the Brazilian Caatinga (Oliveira et al. 2023).

Borrelia spp. in wild neotropical birds have not been documented. However, *Borrelia anserina* transmitted by *Argas miniatus*, a tick species with a distribution in the NZR (Muñoz-Leal et al. 2018), was isolated from domestic fowl infested by *A. miniatus* in Brazil (Marchoux and Salimbeni 1903; Ataliba et al. 2007). Therefore, localizing wild population of birds infested with *A. miniatus* should be taken into account to find possible bird species acting as reservoirs of *B. anserina* in the NZR.

*13.2.4 Piroplasmida (***Babesia, Cytauxzoon, Rangelia,** *and* **Theileria***)*

Piroplasmida (*phylum* Apicomplexa) is an order of hemoparasitic protists transmitted among vertebrates exclusively by ticks and is composed of four genera: *Babesia*, *Cytauxzoon*, *Rangelia*, and *Theileria* (França et al. 2014; Jalovecka et al. 2018). Piroplasmids have complex life cycles and alternate infections between ticks and vertebrates, and depending on the genus, they invade erythrocytes, leukocytes, or endothelial cells, provoking disease or not (Jalovecka et al. 2018). Merozoites, which multiplicate and invade different blood cells of their vertebrate hosts, are usually detected by microscopic observation of blood smears and constitute one of the asexual phases of parasites (Jalovecka et al. 2018). Sexual reproduction starts within the hosts' erythrocytes with the development of gametocytes that, after blood uptake, turn into gametes inside the tick gut (Jalovecka et al. 2018). Fertilization occurs, and the parasites' kinetics invade the tick salivary glands, undergo maturation, and finally shift into sporozoites, which are released into tick saliva and inoculated into the vertebrate host during the meal (Jalovecka et al. 2018). This general life cycle applies to *Babesia* and *Theileria* (Jalovecka et al. 2018) and presumably to *Cytauxzoon* and *Rangelia*. Piroplasmids have transstadial perpetuation and *Babesia*, *Theileria*, and *Rangelia* transovarial transmission in their vector ticks.

From a systematic point of view, the genus *Babesia* is paraphyletic and includes 11 groups (Ikeda et al. 2021; Santodomingo et al. 2022), with *Cytauxzoon* and *Theileria* intercalating between *Babesia* sensu *lato* and *Babesia* sensu stricto clades (Jalovecka et al. 2019), whereas the genus *Rangelia* is to date endemic from South America and composed of only one species (*Rangelia vitalii*) that infects canids and branches within the *Babesia* sensu stricto group (França et al. 2014; Soares et al. 2014). While cases of piroplasmosis in mammals caused by *Babesia*, *Theileria*, or *Cytauxzoon* have been reported in tropical, temperate, and cold regions globally and the parasites are transmitted by ticks of the genera *Ixodes*, *Dermacentor*, *Amblyomma*, or *Rhipicephalus* (Alvarado-Rybak et al. 2016; Sebastian et al. 2022), infection with *R. vitalii* has been reported only in wild carnivores from South America, and its vector corresponds to the hard tick *A. aureolatum* (Soares et al. 2014; Fredo et al. 2015; Silveira et al. 2016a, b). In addition to mammals, birds, including penguins, are also hosts of piroplasmids, specifically of the genus *Babesia* (Montero et al. 2016; Jalovecka et al. 2019). However, in the NZR, species of *Babesia* have been detected by PCR or observed in blood smears exclusively in wild (and domestic) mammals. Although their life cycles are still unsolved, only ticks of the Ixodidae family have been associated but not confirmed as putative vectors. A possible exception corresponds to *Ornithodoros guaporensis* (Argasidae) from Brazil, which was collected on a rodent and was positive for *Babesia* DNA (Wolf et al. 2017).

Babesia spp. are one of the most common blood parasites of mammals, and it is likely that every mammal that is parasitized by a tick could harbor a *Babesia* sp. (Jalovecka et al. 2019). In the NZR, rodents, marsupials, artiodactyls, perissodactyls, and bats have been positive for *Babesia* screenings, either molecularly or serologically using optical microscopy (Table 13.1). However, the following caveats should be considered. The detection of anti-*Babesia caballi* antibodies in mountain tapirs (*Tapirus pinchaque*) from Ecuador should be carefully interpreted (Castellanos 2013), since *B. caballi* is a common parasite of horses and cross-reactions with other (undescribed) *Babesia* spp. cannot be discarded. While the observation of *Babesia* in blood of opossums from Brazil and Colombia dates from decades ago (Regendanz and Kikuth 1928; Garcia 1945; Serra-Freire 1979), the alleged species, *Babesia brasiliensis*, currently lacks a genetic identity. Nevertheless, molecular detections indicate that *Babesia* spp. detected in opossums would be phylogenetically related to the "Western Babesia group" (Gonçalves et al. 2021) or placed elsewhere (Soares et al. 2017a, b; Colle et al. 2019). Recently, a group of *Babesia* spp. associated with South American marsupials was proposed based on a phylogenetic inference constructed with nearly complete sequences of the 18S rRNA gene (Perles et al. 2023). Although the clade is statistically highly supported, the authors named the group based on sequences retrieved from *Didelphis* spp. only. Therefore, the monophyly of the "South American marsupial group" should be reassessed in the future mainly because the region is particularly rich in marsupial species, and most of them have not been screened for *Babesia* spp. Finally, genetic screenings performed in hogs (*Tayassu pecari*) demonstrated the presence of a *Babesia* genotype 94% identical to *Babesia duncani* (Soares et al. 2017a, b). Because the sequenced gene (18S rRNA) is conserved, it is evident that this *Babesia* sp. likely corresponds to an undetermined taxon that merits further molecular work and morphological identification to be formally described.

Only two studies have targeted neotropical bats as hosts for piroplasmid species. Almost two decades ago, intraerythrocytic vacuoles compatible with *Babesia* merozoites were observed in the ghost-faced bat (*Mormoops megalophylla*) in Colombia (Marinkelle 1996). Moreover, undetermined piroplasmid species have been genetically characterized in bats from Brazil (Table 13.1), and phylogenetic analyses indicate the finding of putative novel species related to *Babesia* and *Theileria* genera (Ikeda et al. 2021). Because the two studies on bat-associated piroplasmids are impaired given that they lack either morphological or genetic data, future prospections in bats should combine both approaches to identify the agents.

In the Nearctic Zoogeographical Region, rodents and cervids are hosts of *Babesia microti* and *Babesia odocoilei*, two zoonotic species transmitted by *Ixodes* spp. (Yabesley and Shock 2013). Interestingly, recent studies performed in northern Chile demonstrated the presence of a *Babesia* sp. of the microti group in cricetid rodents (*Abrothrix jelskii*, *Phyllotis darwini*) (Santodomingo et al. 2022a), and a novel *Babesia* sp., preliminarily named *Babesia* sp. pudui, was genetically characterized from pudus' blood and associated ticks collected in the country's south (Santodomingo et al. 2022b). Remarkably, *Babesia* sp. pudui is phylogenetically closely related to *B. odocoilei*, which infects wild deer in the United States, a fact that supports the hypothesis that the ungulate-associated *Babesia* spp. form a monophyletic group (Hrazdilová et al. 2020). Deer often share the same landscape with cattle, and this situation has led to the spillover of exotic tick-borne pathogens because ticks that parasitize cattle (i.e., *Rhipicephalus microplus*) also feed on cervids (de Silveria et al. 2016a, b). In this regard, in the areas where novel tick-borne MOs have been detected on deer, herds of domestic ruminants that overlap geographically could be at risk risk.

Theileria spp. have been detected to a lesser extent in neotropical wild vertebrates, with only one species described through a multilocus analysis. Indeed, the description of *Theileria terrestris* from lowland tapirs (*Tapirus terrestris*) in Brazil showed that the sequencing larger sets of genes is useful to clearly separate species (Mongruel et al. 2022). Moreover, the analyses concluded that detections of a *Theileria* sp. near *Theileria equi* in lowland tapirs likely corresponded to this newly identified piroplasmid. In other neotropical mammals, such as armadillos and agouties, incomplete sequences (≈ 600 bp) of the 18S rRNA gene have matched *T. equi* and *Theileria capreoli* with 96–86% identity (Soares et al. 2017a, b). Therefore, current genetic evidence precludes a specific classification of the detected *Theileria* spp. Notably, 90% identity with the *Theileria* genus shown by short 18S rDNA sequences obtained from agouties of Trinidad implies the detection of a likely novel genus of Piroplasmida and not a *Theileria* sp., as stated (Sant et al. 2022). Nevertheless, short Piroplasmida 18S rDNA sequences 98–100% identical with *Theileria cervi* detected in red brocket (*Mazama americana*) and gray brocket (*Mazama gouazoubira*) in Brazil (de Silveira et al. 2016a, b; Soares et al. 2017a, b) support the circulation of that same *Theileria* species or a highly similar taxon. The circulation of *T. cervi* in South America was recently demonstrated through comparisons of nearly complete sequences of the 18S rRNA gene obtained from marsh deer (*Blastocerus dichotomus*) with symptoms of *Theileria* infection in Argentina (Sebastian et al. 2022).

Five species of *Cytauxzoon* have been described, all of which infect felids in the Nearctic (*C. felis*, *C. manul*, *C. europaeus*, *C. banethis*, and *C. ontrantorum*) (Panait et al. 2021), Palearctic (*C. felis*), and NZR (*C. felis*). *Cytauxzoon felis* is distributed in the midwestern and southeastern United States, where it is known to cause a highly fatal disease in domestic cats and other felids called cytauxzoonosis or bobcat fever (Reviewed by Wikander and Reif 2023). Cytauxzoonosis is an emerging veterinary disease in the United States, with cases increasing and expanding geographically (Miller and Davis 2013). Evidence indicates that domestic cats can be asymptomatically (subclinically) infected. Although these cats are healthy and have never been known to have cytauxzoonosis, the parasite is detectable on blood films and via polymerase chain reaction (Haber et al. 2007; Brown et al. 2008; Nagamori et al. 2016). Asymptomatically infected cats can transmit *C. felis* to tick vectors (Allen et al. 2019). Bobcats (*Lynx rufus*) are the reservoir or natural hosts of *C. felis* in the United States (Shock et al. 2011; Zieman et al. 2017). Bobcats appear to be healthy when infected with *C. felis*, with only rarely documented cases of cytauxzoonosis in bobcats (Nietfeld and pollock 2002). The vectors of *C. felis* in the United States are the lone star tick (*Amblyomma americanum*) (Reichard et al. 2010) and, less effectively, the American dog tick (*Dermacentor variabilis*) (Blouin et al. 1984). In the NZR, *C. felis* has been detected in domestic cats, native ocelots (*Leopardus pardalis*), puma (*Puma concolor*), and jaguars (*Panthera onca*) in Brazil (André et al. 2009; Soares et al. 2017a, b; Antunes et al. 2018; Paula et al. 2022). Additionally, two captive lions (*Panthera leo*) died of cytauxzoonosis in the first known documentation of *Cytauxzoon* spp. in the NZR (Peixoto et al. 2007). One known fatal case of cytauxzoonosis in a jaguar was described by Guizelini et al. (2021) in 2021 in Brazil. Maia et al. (2013) described mild disease in a domestic cat infected with *C. felis*; however, this cat was coinfected with *Mycoplasma haemominutum;* therefore, a singular cause of pathology cannot be determined. The vector of *C. felis* in the NZR is unknown, and neither known vectors from the United States are present. Further work is necessary to detect *C. felis* in other felid hosts and to determine what vector(s) is(are) responsible for transmission.

13.3 Final Remarks

Although ticks are the main vectors of MOs to wild vertebrates globally, there is little information on how parasites impact the health of their hosts in nature. Most of the knowledge on tick-borne MOs comes from mammals; tick-borne MOs hosted by birds have been poorly assessed, and little is known in reptiles or amphibians. While the detection of these agents has relied on the sequencing of specific genes from organs or blood samples, few of them have been isolated, a fact that prevents the design of specific serological assays or a neat sequencing of their complete genomes. This fact also hinders the understanding of ecoepidemiological pathways because detecting a MO by PCR only proves the occurrence of DNA and does not demonstrate active infection. With the advent of next-generation sequencing technologies, the genetic characterization of tick-borne microorganisms in neotropical vertebrates will generate larger datasets for genomic comparisons, and transcriptiomic data might shed light wether the agent was alive in the positive host. Moreover, the implementation of a metagenomic approach to identify blood meals could be a useful tool to unveil the trophic relationships of vector ticks and their host and elucidate which animals could be involved in a given pathogen's enzootic cycle. Finally, determining the number of naturally infected free-living vectors and hosts is only the first step in identifying a potential vector and host (primary, secondary, incidental, amplifying, or reservoir), demonstrating that the epidemiological universe of tick-borne MO in wild neotropical vertebrates is incipiently known.

References

Aguiar D, Pessoa Araújo Junior J, Nakazato L, Bard E, Aguilar-Bultet L, Vorimore F, Leonidovich Popov V, Moleta Colodel E, Cabezas-Cruz A (2019) Isolation and characterization of a novel pathogenic strain of Ehrlichia minasensis. Microorganisms 7(11):528. https://doi.org/10.3390/microorganisms7110528

Allen KE, Thomas JE, Wohltjen ML, Reichard MV (2019) Transmission of Cytauxzoon felis to domestic cats by Amblyomma americanum nymphs. Parasit Vectors 12:28

Almeida AP, Souza TD, Marcili A, Labruna MB (2013) Novel *Ehrlichia* and *Hepatozoon* agents infecting the crab-eating fox (*Cerdocyon thous*) in southeastern Brazil. J Med Entomol 50(3):640–646. https://doi.org/10.1603/me12272

Alvarado-Rybak M, Solano-Gallego L, Millán J (2016) A review of piroplasmid infections in wild carnivores worldwide: importance for domestic animal health and wildlife conservation. Parasit Vectors 9:538. https://doi.org/10.1186/s13071-016-1808-7

André MR, Dumler JS, Scorpio DG, Teixeira RH, Allegretti SM, Machado RZ (2012) Molecular detection of tickborne bacterial agents in Brazilian and exotic captive carnivores. Ticks Tick Borne Dis 3:247–253. https://doi.org/10.1016/j.ttbdis.2012.04.002

André MR (2018) Diversity of *Anaplasma* and *Ehrlichia/Neoehrlichia* agents in terrestrial wild carnivores worldwide: implications for human and domestic animal health and wildlife conservation. Front Vet Sci 5:293

André MR, Adania CH, Machado RZ, Allegretti SM, Felippe PAN, Silva KF, Nakaghi ACH, Dagnone AS (2009) Molecular detection of *Cytauxzoon* spp. in asymptomatic Brazilian wild captive felids. J Wildl Dis 45:234–237

Antunes TR, Silveira AW, Oliveira GG, Rezende AS, Azuaga LBS, Souza ML, Oliveira DR, Netto CRMC, Godoy KCS, Ramos CAN, Souza AI (2018) Infecção natural por *Cytauxzoon felis* em onça parda (*Puma concolor*) de vida livre proveniente da região sudoeste de Mato Grosso do Sul, Brasil. PUBVET 12(1):139

Ataliba AC, Resende JS, Yoshinari N, Labruna MB (2007) Isolation and molecular characterization of a Brazilian strain of *Borrelia anserina*, the agent of fowl spirpochaetosis. Res Vet Sci 83:145–149. https://doi.org/10.1016/j.rvsc.2006.11.014

Baldridge GD, Scoles GA, Burkhardt NY, Schloeder B, Kurtti TJ, Munderloh UG (2009) Transovarial transmission of *Francisella*-like endosymbionts and *Anaplasma phagocytophilum* variants in *Dermacentor albipictus* (Acari: Ixodidae). J Med Entomol 46(3):625–632. https://doi.org/10.1603/033.046.0330

Barker C, Reisen W (2019) Epidemiology of Vector-Borne Diseases. Medical and Veterinary Entomology (Third Edition), 33–49 pp. Academic Press. https://doi.org/10.1016/B978-0-12-814043-7.00004-2

Barros-Battesti DM, Arzua M, Bechara GH (2006) Carrapatos de Importância Médico-Veterinária da Região Neotropical: Um Guia Ilustrado para Identificação de Espécies. Vox/International Consortium on Ticks and Tick Borne Diseases (ICTTD-3)/Butantan, São Paulo, Brazil

Benevenute JL, Dumler JS, Ogrzewalska M, Roque ALR, Mello VVC, de Sousa KCM, Gonçalves LR, D'Andrea PS, de Sampaio Lemos ER, Machado RZ, André MR (2017) Assessment of a quantitative 5′ nuclease real-time polymerase chain reaction using groEL gene for *Ehrlichia* and *Anaplasma* species in rodents in Brazil. Ticks Tick Borne Dis 8(4):646–656. https://doi.org/10.1016/j.ttbdis.2017.04.011

Bennett G, Earlé RA, Penzhorn BL (1992) *Ornithodoros peringueyi* (Argasidae) and *Xenopsylla trispinis* (Siphonaptera), probable intermediate hosts of *Hepatozoon atticorae* of South African cliff swallow *Hirundo spilodera*. Can J Zool 70:188–190

Bermúdez CS, Troyo EA (2018) A review of the genus *Rickettsia* in Central America. Res Rep Trop Med 29(9):103–112

Bermúdez SE, Gottdenker N, Krishnvajhala A, Fox A, Wilder HK, González K, Smith D, López M, Perea M, Rigg C, Montilla S, Calzada JE, Saldaña A, Caballero CM, Lopez JE (2017b) Synanthropic mammals as potential hosts of tick-borne pathogens in Panama. PLoS One 14(12):e0226195

Bermúdez SE, Armstrong BA, Domínguez L, Krishnavajhala A, Kneubehl AR, Gunter SM, Replogle A, Petersen JM, Lopez JE (2021) Isolation and genetic characterization of a relapsing fever spirochete isolated from *Ornithodoros puertoricensis* collected in central Panama. PLoS Negl Trop Dis 15:e0009642

Binder LC, Ramírez-Hernández A, Serpa MCA, Moraes-Filho J, Pinter A, Scinachi CA, Labruna MB (2021) Domestic dogs as amplifying hosts of *Rickettsia rickettsii* for *Amblyomma aureolatum* ticks. Ticks Tick Borne Dis 12(6):101824. https://doi.org/10.1016/j.ttbdis.2021.101824

Binetruy F, Garnier S, Boulanger N, Talagrand-Reboul É, Loire E, Faivre B, Noël V, Buysse M, Duron O (2020) A novel *Borrelia* species, intermediate between Lyme disease and relapsing fever groups, in neotropical passerine-associated ticks. Sci Rep 10(1):10596. https://doi.org/10.1038/s41598-020-66828-7

Blomström AL, Luz HR, Öhlund P, Lukenge M, Brandão PE, Labruna MB, Berg M (2019) Novel viruses found in *Antricola* ticks collected in bat caves in the Western Amazonia of Brazil. Viruses 12(1):48

Blouin EF, Kocan AA, Glenn BL, Kocan KM, Hair JA (1984) Transmission of *Cytauxzoon felis* from bobcats *Felis rufus* to domestic cats by *Dermacentor variabilis*. J Wildl Dis 20:241–242

Brown HM, Latimer KS, Erikson LE, Cashwell ME, Britt JO, Peterson DS (2008) Detection of persistent *Cytauxzoon felis* infection by polymerase chain reaction in three asymptomatic domestic cats. J Vet Diagn Investig 20:485–488

Calchi AC, Vultão JG, Alves MH et al (2020) Ehrlichia spp. and Anaplasma spp. in Xenarthra mammals from Brazil, with evidence of novel 'Candidatus Anaplasma spp.' Sci Rep 10, 12615. https://doi.org/10.1038/s41598-020-69263-w

Cândido SL, de Assis Pereira N, de Oliveira Rosa Fonseca MJ, de Campos Pacheco R, Morgado TO, Colodel EM, Nakazato L, Dutra V, Vieira TSWJ, de Aguiar DM (2023) Molecular detection and genetic characterization of Ehrlichia canis and Ehrlichia sp. in neotropical primates from Brazil. Ticks Tick Borne Dis 14(4):102179. https://doi.org/10.1016/j.ttbdis.2023.102179

Cardona-Romero M, Martínez-Sánchez ET, Alvarez-Londoño J, Pérez-Cárdenas JE, Ossa-López PA, Castaño-Villa GJ, Binder LC, Faccini-Martínez ÁA, Rivera-Páez FA (2022) Seroprevalence and detection of Rickettsia spp. in wild birds of Arauca, Orinoquia region, Colombia. Vet Parasitol Reg Stud Rep 30:100720. https://doi.org/10.1016/j.vprsr.2022.100720

Castellanos A (2013) First report of positive serological response to the hemoparasite, *Babesia caballi*, in Mountain tapir. Tapir Cons 23(3):9

Charles RA, Bermúdez S, Banović P, Alvarez DO, Díaz-Sánchez AA, Corona-González B, Etter EMC, Rodríguez González I, Ghafar A, Jabbar A, Moutailler S, Cabezas-Cruz A (2021) Ticks and tick-borne diseases in Central America and the Caribbean: a one health perspective. Pathogens 10(2021):1273

Clifford C, Hoogstraal H, Radovsky F, Stiller D, Keirans J (1980) Ornithodoros (Alectorobius) amblus (Acarina: Ixodoidea: Argasidae): identity, marine bird and human hosts, virus infections, and distribution in Peru. J Parasitol 66(2):312–323

Colle AC, Mendonça RFB, Maia MO, Freitas LC, Witter R, Marcili A, Aguiar DM, Muñoz-Leal S, Labruna MB, Rossi RV, Pacheco RC (2019) Molecular survey of tick-borne pathogens in small mammals from Brazilian Amazonia. Braz J Vet Parasitol Jaboticabal 28(4):592–604. https://doi.org/10.1590/S1984-29612019086

Colunga-Salas P, Sánchez-Montes S, Volkow P, Ruíz-Remigio A, Becker I (2020a) Lyme disease and relapsing fever in Mexico: an overview of human and wildlife infections. PLoS One 15(9):e0238496. https://doi.org/10.1371/journal.pone.0238496. PMID: 32941463; PMCID: PMC7497999

Colunga-Salas P, Sánchez-Montes S, León-Paniagua L, Becker I (2020b) *Borrelia* in neotropical bats: detection of two new phylogenetic lineages. Ticks Tick Borne Dis 12(2):101642

da Silveira JA, Rabelo EM, Ribeiro MF (2011) Detection of *Theileria* and *Babesia* in brown brocket deer (Mazama gouazoubira) and marsh deer (Blastocerus dichotomus) in the State of Minas Gerais, Brazil. Vet Parasitol 177(1–2):61–66. https://doi.org/10.1016/j.vetpar.2010.10.044. Epub 2010 Oct 30. PMID: 21354704

Dall'Agnol B, Michel T, Weck B, Souza UA, Webster A, Leal BF, Klafke GM, Martins JR, Ott R, Venzal JM, Ferreira CAS, Reck J (2017) *Borrelia burgdorferi* sensu lato in *Ixodes longiscutatus* ticks from Brazilian Pampa. Ticks Tick Borne Dis 8:928–932

de Paula LGF, Weck BC, Neves LC, Paula WVF, Araújo LBM, Martins DB, Peres PCO, Labruna MB, Krawczak FS (2022) Natural infection and molecular detection of *Cytauxzoon felis* in a free-ranging *Puma concolor* in the state of Goiás, Brazil. Cienc Rural 52:10

de Sousa KCM, Calchi AC, Herrera HM, Dumler JS, Barros-Battesti DM, Machado RZ, André MR (2017) Anaplasmataceae agents among wild mammals and ectoparasites in Brazil. Epidemiol Infect 145(16):3424–3437. https://doi.org/10.1017/S095026881700245X

Dunn LH, Clark HC (1933) Notes on relapsing fever in Panama with special reference to animal hosts. Am J Trop Med Hyg 13:201–209

Duron O, Noël V, McCoy KD, Bonazzi M, Sidi-Boumedine K, Morel O, Vavre F, Zenner L, Jourdain E, Durand P, Arnathau C, Renaud F, Trape J, Biguezoton AS, Cremaschi J, Dietrich M, Léger E, Appelgren A, Dupraz M, Gómez-Díaz E, Diatta G, Dayo G, Adakal H, Zoungrana S, Vial L, Chevillon C (2015) The recent evolution of a maternally-inherited endosymbiont of ticks led to the emergence of the Q fever pathogen, *Coxiella burnetii*. PLoS Pathog 11(5):e1004892

Eshoo MW, Carolan HE, Massire C, Chou DM, Crowder CD, Rounds MA, Phillipson CA, Schutzer SE, Ecker DJ (2015) Survey of *Ixodes pacificus* ticks in California reveals a diversity of microorganisms and a novel and widespread Anaplasmataceae species. PLoS One 10(9):e0135828. https://doi.org/10.1371/journal.pone.0135828

Faccini-Martínez ÁA, Botero-García CA (2016) Regarding tick-borne relapsing fever in the Americas; some historical aspects of a forgotten disease in Colombia. Vet Sci 3(4):33

Faccini-Martínez ÁA, Silva-Ramos CR, Santodomingo AM, Ramírez-Hernández A, Costa FB, Labruna MB, Muñoz-Leal S (2022) Historical overview and update on relapsing fever group *Borrelia* in Latin America. Parasit Vectors 15:1–20

Fang R, Blanton LS, Walker DH (2017) Rickettsiae as emerging infectious agents. Clin Lab Med 37(2):383–400. https://doi.org/10.1016/j.cll.2017.01.009

Félix ML, Muñoz-Leal S, Carvalho LA, Queirolo D, Remesar S, Armúa-Fernández MT, Venzal JM (2020) Characterization of "*Candidatus* Ehrlichia Pampeana" in *Haemaphysalis juxtakochi* Ticks and Gray Brocket Deer (*Mazama gouazoubira*) from Uruguay. Microorganisms 9(10):2165

Félix ML, Muñoz-Leal S, Carvalho LA, Queirolo D, Alonso SR, Nava S, Armúa-Fernández MT, Venzal JM (2021) Molecular characterization of novel *Ehrlichia* genotypes in *Ixodes auritulus* from Uruguay. Curr Res Parasitol Vector Borne Dis 1:100022

França RT, Da Silva AS, Loretti AP, Mazzanti CM, Lopes ST (2014) Canine rangeliosis due to *Rangelia vitalii*: from first report in Brazil in 1910 to current day – a review. Ticks Tick Borne Dis 5(5):466–474. https://doi.org/10.1016/j.ttbdis.2014.04.005

Fredo G, Bianchi MV, De Andrade CP, De Souza SO, Leite-Filho RV, Bandinelli MB, Amorim DB, Driemeier D, Sonne L (2015) Natural infection of wild canids (*Cerdocyon thous* and *Lycalopex gymnocercus*) with the intraendothelial piroplasm *Rangelia vitalii* in Southern Brazil. J Wildl Dis 51(4):880–884. https://doi.org/10.7589/2014-12-283. Epub 2015 Aug 7. PMID: 26251988

Fuentes L (1986) Ecological study of Rocky Mountain spotted fever in Costa Rica. Am J Trop Med Hyg 35:192–196

García MR (1945) Un piroplasma del tipo "*Nuttallia equi*", parasito de "*Didelphis paraguayensis*" en Colombia. Rev Med Vet (Bogotá) 14:70–79

Gillen A (2014) Avaliação da competência de aves passeriformes *Turdus* sp como hospedeiro amplificador da bactéria *Rickettsia rickettsii* para carrapatos da espécie *Amblyomma aureolatum* em condições de laboratório. Coordenadoria de Controle de Doenças da Secretaria de Estado da Saúde, São Paulo. Dissertation. https://pesquisa.bvsalud.org/portal/resource/pt/lil-773062

Gonçalves LR, Paludo G, Bisol TB, Perles L, Oliveira LB, Oliveira CM, Silva TMV, Nantes WAG, Duarte MA, Santos FM, Porfírio GEO, Hirano LQL, Herrera HM, Barros-Battesti DM, Machado RZ, André MR (2021) Molecular detection of piroplasmids in synanthropic rodents, marsupials, and associated ticks from Brazil, with phylogenetic inference of a putative novel *Babesia* sp. from white-eared opossum (*Didelphis albiventris*). Parasitol Res 120:3537–3546. https://doi.org/10.1007/s00436-021-07284-8

Guglielmone AA, Robbins RG (2018) Hard ticks (Acari: Ixodida: Ixodidae) parasitizing humans. A global overview. Springer, Dordrecht/Heidelberg/New York/London

Guglielmone AA, Robbins RG, Apanaskevich DA, Petney TN, Estrada-Peña A, Horak IG (2014) The hard ticks of the world: (Acari: Ixodida: Ixodidae). Springer, Dordrecht/Heidelberg/New York/London

Guglielmone A, Nava S, Robbins R (2021) Neotropical hard ticks (Acari: Ixodida: Ixodidae). A critical analysis of their taxonomy, distribution, and host relationships. Springer, Cham

Guglielmone AA, Nava S, Robbins RG (2023) Geographic distribution of the hard ticks (Acari: Ixodida: Ixodidae) of the world by countries and territories. Zootaxa 5251(1):1–274

Guillemi EC, Orozco MM, Argibay HD, Farber MD (2018) Evidence of Ehrlichia chaffeensis in Argentina through molecular detection in marsh deer (Blastocerus dichotomus). Int J Parasitol Parasites Wildl 8:45–49. https://doi.org/10.1016/j.ijppaw.2018.12.004. PMID: 30619709; PMCID: PMC6312859

Guimarães A, Raimundo JM, Silva ATD, Carpintero FM, Pires JR, Benevenute JL, Machado RZ, André MR, Baldani CD (2019) Detection of a putative novel genotype of Ehrlichia sp. from opossums (Didelphis aurita) from Brazil. Rev Bras Parasitol Vet 28(1):140–144. https://doi.org/10.1590/S1984-296120180068. Epub 2018 Nov 8. PMID: 30427523

Guizelini CC, Nascimento CAR, Echeverria JT, Soares RL, Pimenta MM, de Deco-Souza T, Esteves FCB, Gomes DC (2021) Fatal infection caused by Cytauxzoon felis in a captive-reared jaguar (Panthera onca). Int J Parasitol Parasites Wildl 16:187–190. https://doi.org/10.1016/j.ijppaw.2021.10.002

Haber MD, Tucker MD, Marr HS, Levy JK, Burgess J, Lappin MR, Birkenheuer AJ (2007) The detection of *Cytauxzoon felis* in apparently healthy free-roaming cats in the USA. Vet Parasitol 146:316–320

Hechemy KE, Raoult D, Fox J, Han Y, Elliott LB, Rawlings J (1989) Cross-reaction of immune sera from patients with rickettsial diseases. J Med Microbiol 29(3):199–202

Hidalgo M, Faccini-Martínez AA, Valbuena G (2013) Rickettsiosis transmitidas por garrapatas en las Américas: avances clínicos y epidemiológicos, y retos en el diagnóstico. Biomedica 33(Sup1):161–178 https://revistabiomedica.org/index.php/biomedica/article/view/1466

Hornok S, Földvári G, Rigó K, Meli M, Gönczi E, Répási A, Farkas R, Papp I, Kontschán J, Hoffmann-Lehmann (2015) Synanthropic rodents and their ectoparasites as carriers of a novel haemoplasma and vector-borne, zoonotic pathogens indoors. Parasit Vectors 8:27. https://doi.org/10.1186/s13071-014-0630-3

Horta MC, Moraes-Filho J, Casagrande RA, Saito TB, Rosa SC, Ogrzewalska M, Canal RB, Pacheco RC, Martins TF, Matushima ER, Labruna MB (2009) Experimental infection of opossums *Didelphis aurita* by *Rickettsia rickettsii* and evaluation of the transmission of the infection to ticks *Alblyomma cajennense*. Vector Borne Zoonotic Dis 9(1):109–117

Hrazdilová K, Rybářovác M, Širokýa P, Votýpkad J, Zintlf A, Burgessg H, Steinbauerh V, Žákovčíki V, Modrý D (2020) Diversity of *Babesia* spp. in cervid ungulates based on the 18S rDNA and cytochrome c oxidase subunit I phylogenies. Infect Genet Evol 77:104060. https://doi.org/10.1016/j.meegid.2019.104060

Hun L, Troyo A, Taylor L, Barbieri AM, Labruna MB. (2011) First report of the isolation and molecular characterization of Rickettsia amblyommii and Rickettsia felis in Central America. Vector Borne Zoonotic Dis. 11(10):1395–7. https://doi.org/10.1089/vbz.2011.0641

Ikeda P, Menezes TR, Torres JM, Oliveira CE, Lourenço EC, Herrera HM, Machado RZ, André MR (2021) First molecular detection of piroplasmids in non-hematophagous bats from Brazil, with evidence of putative novel species. Parasitol Res 120:301–310. https://doi.org/10.1007/s00436-020-06985-w

Ivanova LB, Tomova A, González-Acuña D, Murúa R, Moreno CX, Hernández C, Cabello J, Cabello C, Daniels TJ, Godfrey HP, Cabello FC (2014) *Borrelia chilensis*, a new member of the *Borrelia burgdorferi* sensu lato complex that extends the range of this genospecies in the Southern Hemisphere. Environ Microbiol 16(4):1069–1080

Jalovecka M, Hajdusek O, Sojka D, Kopacek P, Malandrin L (2018) The complexity of piroplasms life cycles. Front Cell Infect Microbiol 23(8):248. https://doi.org/10.3389/fcimb.2018.00248

Jalovecka M, Sojka D, Ascencio M, Schnittger L (2019) *Babesia* life cycle – when phylogeny meets biology. Trends Parasitol 35:356–368. https://doi.org/10.1016/j.pt.2019.01.007

Jorge FR, Muñoz-Leal S, Oliveira GMB, Serpa MCA, Magalhães MML, Oliveira LMB, Moura FBP, Teixeira BM, Labruna MB (2023) Novel *Borrelia* genotypes from Brazil indicate a new group of *Borrelia* spp. associated with South American bats. J Med Entomol 60(1):213–217

Kawahara M, Rikihisa Y, Isogai E, Takahashi M, Misumi H, Suto C, Shibata S, Zhang C, Tsuji M (2004) Ultrastructure and phylogenetic analysis of 'Candidatus Neoehrlichia mikurensis' in the family Anaplasmataceae, isolated from wild rats and found in *Ixodes ovatus* ticks. Int J Syst Evol Microbiol 54:1837–1843. https://doi.org/10.1099/ijs.0.63260-0

Kazimírová M, Thangamani S, Bartíková P, Hermance M, Holíková V, Štibrániová I, Nuttall PA (2017) Tick-borne viruses and biological processes at the tick-host-virus interface. Front Cell Infect Microbiol 7:339. https://doi.org/10.3389/fcimb.2017.00339

Kersters, K., De Vos, P., Gillis, M., Swings, J., Vandamme, P., Stackebrandt, E., (2006) Introduction to the proteobacteria. In: Dworkin, M., Falkow, S., Rosenberg, E., Schleifer, K.-H., Stackebrandr, E. (Eds.), The Prokaryotes. A Handbook of the Biology of Bacteria. Vol 5: Proteobacteria: Alpha and Beta Subclasses, https://doi.org/10.1007/0-387-30741-9

Kneubehl AR, Krishnavajhala A, Muñoz-Leal S, Replogle AJ, Kingry LC, Bermúdez SE, Labruna MB, Lopez JE (2022) Comparative genomics of the Western Hemisphere soft tick-borne relapsing fever borreliae highlights extensive plasmid diversity. BMC Genomics 23(1):410

Kurokawa C, Lynn GE, Pedra JHF, Pal U, Narasimhan S, Fikrig E (2020) Interactions between *Borrelia burgdorferi* and ticks. Nat Rev Microbiol 18:587–600. https://doi.org/10.1038/s41579-020-0400-5

Labruna MB, Whitworth T, Bouyer DH, McBride J, Camargo LM, Camargo EP, Popov V, Walker DH (2004) *Rickettsia bellii* and *Rickettsia amblyommii* in *Amblyomma* ticks from the State of Rondônia, Western Amazon, Brazil. J Med Entomol 41(6):1073–1081

Labruna MB, Pacheco RC, Richtzenhain LJ, Szabó MP (2007) Isolation of *Rickettsia rhipicephali* and *Rickettsia bellii* from *Haemaphysalis juxtakochi* ticks in the state of São Paulo, Brazil. Appl Environ Microbiol 73(3):869–873. https://doi.org/10.1128/AEM.02249-06

Labruna MB, Mattar VS, Nava S, Bermudez S, Venzal JM, Dolz G, Abarca K, Romero L, Sousa R, Oteo J, Zavala-Castro J (2011a) Rickettsioses in Latin America, Caribbean, Spain and Portugal. Rev MVZ Córdoba 16(2):2435–2457

Labruna MB, Ogrzewalska M, Soares JF, Martins TF, Soares HS, Moraes-Filho J, Nieri-Bastos FA, Almeida AP, Pinter A (2011b) Experimental infection of *Amblyomma aureolatum* ticks with *Rickettsia rickettsii*. Emerg Infect Dis 17(5):829–834. https://doi.org/10.3201/eid1705.101524

Labruna MB, Krawczak FS, Gerardi M, Binder LC, Barbieri ARM, Paz GF, Rodrigues DS, Araújo RN, Bernardes ML, Leite RC (2017) Isolation of *Rickettsia rickettsii* from the tick *Amblyomma sculptum* from a Brazilian spotted fever-endemic area in the Pampulha Lake region, southeastern Brazil. Vet Parasitol Reg Stud Rep 8:82–85

Lau ACC, Qiu Y, Moustafa MAM, Nakao R, Shimozuru M, Onuma M, Mohd-Azlan J, Tsubota T (2020) Detection of *Borrelia* sensu lato and relapsing fever borrelia in feeding *Ixodes* ticks and rodents in Sarawak, Malaysia: new geographical records of *Borrelia yangtzensis* and *Borrelia miyamotoi*. Pathogens 9(10):846. https://doi.org/10.3390/pathogens9100846

Londoño AF, Acevedo-Gutiérrez LY, Marín D, Contreras V, Díaz FJ, Valbuena G, Labruna MB, Hidalgo M, Arboleda M, Mattar S, Solari S, Rodas JD (2017) Human prevalence of the spotted fever group (SFG) rickettsiae in endemic zones of Northwestern Colombia. Ticks Tick Borne Dis 8(4):477–482

Lopez JE, Krishnavahjala A, Garcia MN, Bermudez S (2016) Tick-borne relapsing fever spirochetes in the Americas. Vet Sci 3:E16

López Y, Muñoz-Leal S, Martínez C, Guzmán C, Calderón A, Martínez J, Galeano K, Muñoz M, Ramírez JD, Faccini-Martínez ÁA, Mattar S (2023) Molecular evidence of *Borrelia* spp. in bats from Córdoba Department, northwest Colombia. Parasit Vectors 16:5

López-Pérez AM, Chaves A, Sánchez-Montes S, Foley P, Uhart M, Barrón-Rodríguez J, Becker I, Suzán G, Foley J (2022) Diversity of rickettsiae in domestic, synanthropic, and sylvatic mammals and their ectoparasites in a spotted fever-epidemic region at the western US-Mexico border. Transbound Emerg Dis 69(2):609–622. https://doi.org/10.1111/tbed.14027

Lopes MG, Muñoz-Leal S, Lima JTR, Fournier GFSR, Acosta ICL, Martins TF, Ramirez DG, Gennari SM, Labruna MB (2018) Ticks, rickettsial and erlichial infection in small mammals from Atlantic forest remnants in northeastern Brazil. Ticks and Tick-borne Diseases 7, 380–385. https://doi.org/10.1016/j.ijppaw.2018.10.001

Luz HR, Faccini JLH, McIntosh D (2017) Molecular analyses reveal an abundant diversity of ticks and rickettsial agents associated with wild birds in two regions of primary Brazilian Atlantic Rainforest. Ticks Tick Borne Dis 8(4):657–665. https://doi.org/10.1016/j.ttbdis.2017.04.012

Machado RZ, André MR, Werther K, de Sousa E, Gavioli FA, Alves Junior JR. (2012) Migratory and carnivorous birds in Brazil: reservoirs for Anaplasma and Ehrlichia species? Vector Borne Zoonotic Dis. (8):705–708. https://doi.org/10.1089/vbz.2011.0803

Mafra C, Barcelos RM, Mantovani C, Carrizo J, Soares AC, Moreira HN, Maia NL, da Silva Fde F, e Silva VH, Boere V, e Silva Ide O (2015) Occurrence of Ehrlichia canis in free-living primates of the genus Callithrix. Rev Bras Parasitol Vet 24(1):78–81. https://doi.org/10.1590/S1984-29612015001. PMID: 25909257

Maia LM, Cerqueira Ade M, de Barros Macieira D, de Souza AM, Moreira NS, da Silva AV, Messick JB, Ferreira RF, Almosny NR (2013) Cytauxzoon felis and 'Candidatus Mycoplasma haemominutum' coinfection in a Brazilian domestic cat (Felis catus). Rev Bras Parasitol Vet 22:289–291

Mancilla-Agrono LY, Banguero-Micolta LF, Ossa-López PA, Ramírez-Chaves HE, Castaño-Villa GJ, Rivera-Páez FA (2022) Is Borrelia burgdorferi Sensu Stricto in South America? First molecular evidence of its presence in Colombia. Trop Med Infect Dis 7(12):428. https://doi.org/10.3390/tropicalmed7120428. PMID: 36548683; PMCID: PMC9788524

Margos G, Henningsson A, Markowicz M, Fingerle V (2022) Borrelia ecology and evolution: Ticks and hosts and the environment. Microorganisms 10:1513. https://doi.org/10.3390/microorganisms10081513

Marchoux E, Salimbeni A (1903) La Spirillose des Poules. Ann L'Institute Pasteur 17:569–580

Marinkelle CJ, Grose ES (1968) Species of Borrelia from a Colombian bat (Natalus tumidirostris). Nature 218:48. https://doi.org/10.1038/218487a0

Marinkelle CJ (1996) Babesia sp. in Colombian bats (Microchiroptera). J Wildl Dis 32(3):534–535. https://doi.org/10.7589/0090-3558-32.3.534

Martinez A, Salinas A, Martinez F, Cantu A, Miller DK (1999) Serosurvey for selected disease agents in white-tailed deer from Mexico. J Wildl Dis 35:799–803. https://doi.org/10.7589/0090-3558-35.4.799

Miller J, Davis CD (2013) Increasing frequency of feline cytauxzoonosis cases diagnosed in western Kentucky from 2001 to 2011. Vet Parasitol 198:205–208

Miranda J, Contreras V, Negrete Y, Labruna MB, Mattar S (2011) Vigilancia de la infección por Rickettsia sp en capibaras (Hydrochaeris hydrochaeris) un modelo potencial de alerta epidemiológica en zonas endémicas. Biomedica 31(2):216–221

Mongruel ACB, Medici EP, da Costa Canena A, Calchi AC, Perles L, Rodrigues BCB, Soares JF, Machado RZ, André MR (2022) *Theileria terrestris* nov. sp.: a novel Theileria in lowland tapirs

(*Tapirus terrestris*) from two different biomes in Brazil. Microorganisms 10(12):2319. https://doi.org/10.3390/microorganisms10122319. PMID: 36557572; PMCID: PMC9784709

Montero E, González LM, Chaparro A, Venzal J, Bertellotti M, Masero JA, Colominas-Ciuró R, Vidal V, Barbosa A (2016) First record of *Babesia* sp. in Antarctic penguins. Ticks Tick Borne Dis 7(3):498–501

Müller A, Monti G, Otth C, Sepúlveda P, Bittencourt P, Nachum-Biala Y, Gutiérrez R, Harrus S (2018) “Candidatus Neoehrlichia chilensis” sp. nov.: molecular detection and characterization of a novel Anaplasmataceae in wild rodents from Valdivia, southern Chile. Transbound Emerg Dis 65(2):357–362. https://doi.org/10.1111/tbed.12815. Epub 2018 Jan 24. PMID: 29363276

Muñoz-Leal S, Venzal JM, Nava S, Reyes M, Martins TF, Leite RC, Vilela VLR, Benatti HR, Ríos-Rosas D, Barros-Battesti DM, González-Acuña D, Labruna MB (2018) The geographic distribution of *Argas* (*Persicargas*) *miniatus* and *Argas* (*Persicargas*) *persicus* (Acari: Argasidae) in America, with morphological and molecular diagnoses from Brazil, Chile and Cuba. Ticks Tick Borne Dis 9(1):44–56. https://doi.org/10.1016/j.ttbdis.2017.10.009

Muñoz-Leal S, Lopes MG, Marcili A, Martins TF, González-Acuña D, Labruna MB (2019) *Anaplasmataceae*, *Borrelia* and *Hepatozoon* agents in ticks (Acari: Argasidae, Ixodidae) from Chile. Acta Trop 192:91–103

Muñoz-Leal S, Faccini-Martínez ÁA, Pérez-Torres J, Chala-Quintero SM, Herrera-Sepúlveda MT, Cuervo C, Labruna MB (2021) Novel *Borrelia* genotypes in bats from the Macaregua Cave, Colombia. Zoonoses Public Health 68:12–18. https://doi.org/10.1111/zph.12789

Nagamori Y, Slovak JE, Reichard MV (2016) Prevalence of Cytauxzoon felis infection in healthy free-roaming cats in north-central Oklahoma and central Iowa. JFMS Open Rep 2:2055116916655174

Neves LC, Sousa-Paula LC, Dias SA, da Silva BBF, Paula WVF, de Paula LGF, Pereira BG, Pádua GT, Borsanelli AC, Cardoso ERN, Krawczak FDS, Dantas-Torres F (2022) Detection of an undescribed *Babesia* sp. in capybaras and *Amblyomma* ticks in Central-Western Brazil. Animals 13(1):94. https://doi.org/10.3390/ani13010094

Nicholson WL, Sonenshine DE, Noden BH, Richard N, Brown RN (2019) Chapter 27 – ticks (Ixodida). In: Gary R, Mullen GR, Durden LA (eds) Medical and veterinary entomology, 3rd edn. Academic, pp 603–672., ISBN 9780128140437. https://doi.org/10.1016/B978-0-12-814043-7.00027-3

Nietfeld JC, Pollock C (2002) Fatal cytauxzoonosis in a free-ranging bobcat (*Lynx rufus*). J Wildl Dis 38:607–610

Ojeda-Chi MM, Rodriguez-Vivas RI, Esteve-Gasent MD, de León AP, Modarelli JJ, Villegas-Perez S (2018) Molecular detection of rickettsial tick-borne agents in white-tailed deer (*Odocoileus virginianus yucatanensis*), mazama deer (*Mazama temama*), and the ticks they host in Yucatan, Mexico. Ticks Tick Borne Dis 10(2):365–370

Ojeda-Chi MM, Rodriguez-Vivas RI, Esteve-Gasent MD, Pérez de León A, Modarelli JJ, Villegas-Perez S (2019) Molecular detection of rickettsial tick-borne agents in white-tailed deer (Odocoileus virginianus yucatanensis), mazama deer (Mazama temama), and the ticks they host in Yucatan, Mexico. Ticks Tick Borne Dis 10(2):365–370. https://doi.org/10.1016/j.ttbdis.2018.11.018. Epub 2018 Nov 27. PMID: 30503893

Oliveira GMB, da Silva IWG, da Cruz Ferreira Evaristo AM, de Azevedo Serpa MC, Silva Campos AN, Dutra V, Nakazato L, de Aguiar DM, Bahia Labruna M, Horta MC. (2020) Tick-borne pathogens in dogs, wild small mammals and their ectoparasites in the semi-arid Caatinga biome, northeastern Brazil. Ticks Tick Borne Dis (4):101409. https://doi.org/10.1016/j.ttbdis.2020.101409

Oliveira G, Muñoz-Leal S, Santodomingo A, Weck B, Faccini-Martínez Á, Horta M, Labruna M (2023) Novel Relapsing Fever Group Borrelia isolated from Ornithodoros ticks of the Brazilian Caatinga. Microorganisms 11:370. https://doi.org/10.3390/microorganisms11020370

Orozco MM, Argibay HD, Minatel L, Guillemi EC, Berra Y, Schapira A, Di Nucci D, Marcos A, Lois F, Falzone M, Farber MD (2020) A participatory surveillance of marsh deer (Blastocerus dichotomus) morbidity and mortality in Argentina: first results. BMC Vet Res 16(1):321. https://doi.org/10.1186/s12917-020-02533-x. PMID: 32873288; PMCID: PMC7465331

Orozco M, Gómez GF, Alzate JF, Isaza JP, Gutiérrez LA (2021) Virome analysis of three Ixodidae ticks species from Colombia: a potential strategy for discovering and surveying tick-borne viruses infection. Genet Evol 96:105103

Pacheco RC, Echaide IE, Alves RN, Beletti ME, Nava S, Labruna MB (2013) *Coxiella burnetii* in ticks, Argentina. Emerg Infect Dis 19:344–346

Pacheco R, Horta M, Pinter A, Moraes-Filho J, Martins T, Nardi M, Souza S, Souza C, Szabó M, Richtzenhain L, Labruna M (2009) Pesquisa de Rickettsia spp em carrapatos Amblyomma cajennense e Amblyomma dubitatum no Estado de São Paulo. Rev Soc Bras Med Trop 42(3):351–353. https://doi.org/10.1590/S0037-86822009000300023

Panait LC, Mihalca AD, Modry D, Jurankova J, Ionica AM, Deak G, Gherman CM, Heddergott M, Hodzic A, Veronesi F, Reichard M, Zieman EA, Nielsen CK, Jimenez-Ruiz FA, Hrazdilova K (2021) Three new species of Cytauxzoon in European wild felids. Vet Parasitol 290:109344

Paula L, Weck B, Neves L, Paula W, Araújo L, Martins, D et al (2022) Infecção natural e detecção molecular de Cytauxzoon felis em Puma concolor de vida livre no estado de Goiás, Brasil. Ciência Rural 52, e20210577. https://doi.org/10.1590/0103-8478cr20210577

Peixoto PV, Soares CO, Scofield A, Santiago CD, Franca TN, Barros SS (2007) Fatal cytauxzoonosis in captive-reared lions in Brazil. Vet Parasitol 145:383–387

Perles L, Herrera HM, Barreto WTG, de Macedo GC, Calchi AC, Machado RZ, André MR (2022) Multi-locus sequencing reveals putative novel Anaplasmataceae agents, 'Candidatus Ehrlichia dumleri' and Anaplasma sp., in ring-tailed coatis (Carnivora: Nasua nasua) from urban forested fragments at Midwestern Brazil. Microorganisms 10(12):2379. https://doi.org/10.3390/microorganisms10122379. PMID: 36557631; PMCID: PMC9784859

Perles L, Barreto WTG, Macedo GC, Calchi AC, Bezerra-Santos M, Mendoza-Roldan JA, Otranto D, Herrera HM, Barros-Battesti DM, Machado RZ, André MR (2023) Molecular detection of *Babesia* spp. and *Rickettsia* spp. in coatis (*Nasua nasua*) and associated ticks from midwestern Brazil. Parasitol Res 122(5):1151–1158. https://doi.org/10.1007/s00436-023-07815-5

Pinter A, Labruna MB (2006) Isolation of *Rickettsia rickettsii* and *Rickettsia bellii* in cell culture from the tick *Amblyomma aureolatum* in Brazil. Ann N Y Acad Sci 1078:523–529

Pinter A, Horta MC, Pacheco RC, Moraes-Filho J, Labruna MB (2008) Serosurvey of *Rickettsia* spp. in dogs and humans from an endemic area for Brazilian spotted fever in the State of São Paulo, Brazil. Cad Saúde Pública 24:247–252

Rar V, Golovljova I (2011) *Anaplasma*, *Ehrlichia*, and "Candidatus Neoehrlichia" bacteria: pathogenicity, biodiversity, and molecular genetic characteristics, a review. Infect Genet Evol 11(8):1842–1861. https://doi.org/10.1016/j.meegid.2011.09.019

Rar V, Tkachev S, Tikunova N (2021) Genetic diversity of *Anaplasma* bacteria: twenty years later. Infect Genet Evol 91:104833

Regendanz P, Kikuth W (1928) Sur un parasite du sang des «quica» (Metachirus quica), Nuttallia brasiliensis n. sp., et influence de la rate sur les infections latentes du sang. C. R. H. Seanc. Mem Soc Biol 98:1567–1569

Reichard MV, Edwards AC, Meinkoth JH, Snider TA, Meinkoth KR, Heinz RE, Little SE (2010) Confirmation of *Amblyomma americanum* (Acari: Ixodidae) as a vector for *Cytauxzoon felis* (Piroplasmorida: Theileriidae) to domestic cats. J Med Entomol 47:890–896

Rodaniche EC, Rodaniche A (1949) Studies on Q fever in Panama. Am J Hyg 49:67–75

Rodriguez-Vivas RI, Li AY, Ojeda-Chi MM, Trinidad-Martinez I, Rosado-Aguilar JA, Miller RJ, Pérez de León AA (2013) In vitro and in vivo evaluation of cypermethrin, amitraz, and piperonyl butoxide mixtures for the control of resistant *Rhipicephalus (Boophilus) microplus* (Acari: Ixodidae) in the Mexican tropics. Vet Parasitol 197(1–2, 18):288–296

Sánchez-Montes S, Colunga-Salas P, Lozano-Sardaneta YN, Zazueta-Islas HM, Ballados-González GG, Salceda-Sánchez B, Huerta-Jiménez H, Torres-Castro M, Panti-May JA, Peniche-Lara G, Muñoz-García CI, Rendón-Franco E, Ojeda-Chi MM, Rodríguez-Vivas RI, Zavala-Castro J, Dzul-Rosado K, Lugo-Caballero C, Alcántara-Rodríguez VE, Delgado-de la Mora J, Licona-Enríquez JD, Delgado-de la Mora D, López-Pérez AM, Álvarez-Hernández G, Tinoco-Gracia L, Rodríguez-Lomelí M, Ulloa-García A, Blum-Domínguez S, Tamay-Segovia P, Aguilar-Tipacamú G, Cruz-Romero A, Romero-Salas D, Martínez-Medina MA, Becker I

(2021) The genus *Rickettsia* in Mexico: current knowledge and perspectives. Ticks Tick Borne Dis 12(2):101633. https://doi.org/10.1016/j.ttbdis.2020.101633. Epub 2020 Dec 14. PMID: 33388556

Sant C, Seunarine D, Holder N, Maharaj K, Vaughan M, Harrus S, Gutierrez R, Nachum-Biala Y, Baneth G, Charles R, Pow-Brown P, Suepaul R, Georges K (2022) Tick-borne pathogens in neotropical animals in Trinidad, West Indies. Parasit Vectors 15(1):62. https://doi.org/10.1186/s13071-022-05184-z. PMID: 35183241; PMCID: PMC8858492

Santodomingo AM, Thomas RS, Quintero-Galvis JF, Echeverry-Berrio DM, la Fuente MCS, Moreno-Salas L, Muñoz-Leal S (2022a) Apicomplexans in small mammals from Chile, with the first report of the *Babesia microti* group in South American rodents. Parasitol Res 121(3):1009–1020. https://doi.org/10.1007/s00436-022-07452-4

Santodomingo A, Robbiano S, Thomas R, Parragué-Migone C, Cabello-Stom J, Vera-Otarola F, Valencia-Soto C, Moreira-Arce D, Moreno L, Hidalgo-Hermoso E, Muñoz-Leal S (2022b) A search for piroplasmids and spirochetes in threatened pudu (*Pudu puda*) and associated ticks from Southern Chile unveils a novel *Babesia* sp. and a variant of *Borrelia chilensis*. Transbound Emerg Dis 69(6):3737–3748. https://doi.org/10.1111/tbed.14743

Santodomingo A, Thomas R, Robbiano S, Uribe JE, Parragué-Migone C, Cabello-Stom J, Vera-Otarola F, Valencia-Soto C, Moreira-Arce D, Hidalgo-Hermoso E, Muñoz-Leal S (2023) Wild deer (*Pudu puda*) from Chile harbor a novel ecotype of *Anaplasma phagocytophilum*. Parasit Vectors 16(1):38. https://doi.org/10.1186/s13071-023-05657-9

Sebastian PS, Falzone MP, Lois MF, Sartori R, Zimmerman J, Tarragona EL, Nava S (2022) Phylogenetic position of *Theileria cervi* detected in *Blastocerus dichotomus* (Artiodactyla: Cervidae) with clinical symptoms from Argentina. Emerg Anim Species 5:100014. https://doi.org/10.1016/j.eas.2022.100014

Serpa MCA, Luz HR, Costa FB, Weck BC, Benatti HR, Martins TF, Correa LS, Ramirez DG, Rocha V, Dias TC, Correa LR, Brasil J, Brites-Neto J, Nievas AM, Suzin A, Monticelli PF, Moro MEG, Lopes B, Pacheco RC, Aguiar DM, Piovezan U, Szabó MPJ, Ferraz KMPMB, Percequillo AR, Labruna MB, Ramos VN (2021) Small mammals, ticks and rickettsiae in natural and human-modified landscapes: diversity and occurrence of Brazilian spotted fever in Brazil. Ticks Tick Borne Dis 12(6):101805. https://doi.org/10.1016/j.ttbdis.2021.101805

Serra-Freire NM (1979) *Babesia ernestoi* sp. n., in *Didelphis marsupialis* L., 1758, and *D. albiventris* Lund, 1841, in Brazil. Zentralbl Veterinarmed B 26:614–622. https://doi.org/10.1111/j.1439-0450.1979.tb00855.x

Shock BC, Murphy SM, Patton LL, Shock PM, Olfenbuttel C, Beringer J, Prange S, Grove DM, Peek M, Butfiloski JW, Hughes DW, Lockhart JM, Bevins SN, VandeWoude S, Crooks KR, Nettles VF, Brown HM, Peterson DS, Yabsley MJ (2011) Distribution and prevalence of *Cytauxzoon felis* in bobcats (*Lynx rufus*), the natural reservoir, and other wild felids in thirteen states. Vet Parasitol 175:325–330

Silveira JA, D'Elia ML, de Oliveira Avelar I, de Almeida LR, Dos Santos HA, de Magalhães Soares DF, Ribeiro MF, Dos Santos Lima W, Ecco R (2016a) Rangelia vitalii in a free-ranging maned wolf (*Chrysocyon brachyurus*) and co-infections. Int J Parasitol Parasites Wildl 5(3):280–285. https://doi.org/10.1016/j.ijppaw.2016.09.003

Silveira JAG, Oliveira CHS, Silvestre BT, Albernaz TT, Leite RC, Barbosa JD, Oliveira CMC, Ribeiro MFB (2016b) Molecular assays reveal the presence of *Theileria* spp. and *Babesia* spp. in Asian water buffaloes (*Bubalus bubalis*, Linnaeus, 1758) in the Amazon region of Brazil. Ticks Tick Borne Dis 7(5):1017–1023

Smith TG (1996) The genus *Hepatozoon* (Apicomplexa: Adeleina). J Parasitol 82:565–585

Snellgrove AN, Krapiunaya I, Ford SL, Stanley HM, Wickson AG, Hartzer KL, Levin ML (2020) Vector competence of *Rhipicephalus sanguineus* sensu stricto for *Anaplasma platys*. Ticks Tick Borne Dis 11(6):101517. https://doi.org/10.1016/j.ttbdis.2020.101517

Soares JF, Dall'Agnol B, Costa FB, Krawczak FS, Comerlato AT, Rossato BCD, Linck CM, Sigahi EKO, Teixeira RHF, Sonne L, Hagiwara MK, Gregori F, Vieira MIB, Martins JR, Reck J, Labruna MB (2014) Natural infection of the wild canid, *Cerdocyon thous*, with the piroplasmid *Rangelia vitalii* in Brazil. Vet Parasitol 202:156–163. https://doi.org/10.1016/j.vetpar.2014.02.058

Soares HS, Marcili A, Barbieri ARM, Minervino AHH, Malheiros AF, Gennari SM, Labruna MB (2017a) Novel *Anaplasma* and *Ehrlichia* organisms infecting the wildlife of two regions of the Brazilian Amazon. Acta Trop 174:82–87. https://doi.org/10.1016/j.actatropica.2017.07.006
Soares HS, Marcili A, Barbieri ARM, Minervino AHH, Moreira TR, Gennari SM, Labruna MB (2017b) Novel piroplasmid and *Hepatozoon* organisms infecting the wildlife of two regions of the Brazilian Amazon. Int J Parasitol Parasites Wildl 6(2):115–121
Solís-Hernández A, Rodríguez-Vivas RI, Esteve-Gassent MD, Villegas-Pérez SL (2016) Prevalencia de Borrelia burgdorferi sensu lato en roedores sinantrópicos de dos comunidades rurales de Yucatán, México. Biomedica 36(Sipl.1):109–117. https://doi.org/10.7705/biomedica.v36i3.3139
Sonenshine DE, Roe RM (2014) Overview: ticks, people, and animals. In: Sonenshine DE, Roe RM (eds) Biology of ticks. Oxford University Press, New York, pp 3–16
Szabó MP, Nieri-Bastos FA, Spolidorio MG, Martins TF, Barbieri AM, Labruna MB (2013) In vitro isolation from *Amblyomma ovale* (Acari: Ixodidae) and ecological aspects of the Atlantic rainforest Rickettsia, the causative agent of a novel spotted fever rickettsiosis in Brazil. Parasitology 140:719–728
Talagrand-Reboul E, Boyer PH, Bergström S, Vial L, Boulanger N (2018) Relapsing fevers: neglected tick-borne diseases. Front Cell Infect Microbiol 8:98
Tarragona EL, Flores FS, Lamattina D, Torrents J, Patrick S, Sebastian PS, Nava S (2022) Two novel *Ehrlichia* (Rickettsiales: Anaplasmataceae) strains detected in ticks (Ixodida, Ixodidae) and opossums (Didelphimorphia: Didelphidae) in Argentina. Ticks Tick Borne Dis 13(6):102043
Tarragona EL, Sebastian PS, Félix ML, Venzal JM (2023) Novel *Anaplasma* (Rickettsiales: Anaplasmataceae) strain and *Hepatozoon* sp. cf. *H. procyonis* (Apicomplexa, Hepatozoidae) detected in Procyon cancrivorus (Carnivora, Procyonidae) from Argentina, with note of tick-host association. Vet Res Commun. https://doi.org/10.1007/s11259-023-10099-w
Thomas R, Santodomingo AM, Muñoz-Leal S, Silva-de la Fuente MC, Llanos-Soto S, Moreno-Salas L, González-Acuña D (2020) Rodents as potential reservoirs for Borrelia spp. in northern Chile. Braz J Vet Parasitol 29(2):e000120. https://doi.org/10.1590/S1984-29612020029
Valbuena G (2010) Patogénesis de las rickettsiosis en las Américas. Rev MVZ 15(1):2004–2006
Venzal JM, Nava S (2011) El género Rickettsia como agente de zoonosis en el Cono Sur de Sudamérica. Rev Méd Urug 27(2):98–106
Verdugo C, Jiménez O, Hernández C, Álvarez P, Espinoza A, González-Acuña D (2017) Infection with *Borrelia chilensis* in *Ixodes stilesi* ticks collected from *Pudu puda* deer. Ticks Tick Borne Dis 8(5):733–740
Vieira TSWJ, Collere FCM, Ferrari LDR, Baggio RA, Lange RR, Ferrari MV, Duque JCM, Sanches GS, Pereira NA, Aguiar DM, Labruna MB, Vieira RFC (2022) Novel Anaplasmataceae agents *Candidatus* Ehrlichia hydrochoerus and *Anaplasma* spp. infecting capybaras, Brazil. Emerg Infect Dis 28(2):480–482
Weck BC, Serpa MCA, Labruna MB, Muñoz-Leal S (2022) A novel Genospecies of Borrelia burgdorferi Sensu Lato associated with cricetid rodents in Brazil. Microorganisms 10(2):204. https://doi.org/10.3390/microorganisms10020204
Widmer CE, Azevedo FC, Almeida AP, Ferreira F, Labruna MB (2011) Tick-borne bacteria in free-living jaguars (Panthera onca) in Pantanal, Brazil. Vector Borne Zoonotic Dis 11(8):1001–1005. https://doi.org/10.1089/vbz.2011.0619. Epub 2011 May 25. PMID: 21612532
Wikander YM, Reif KE (2023) Cytauxzoon felis: an overview. Pathogens 12:133
Wolcott KA, Margos G, Fingerle V, Becker NS (2021) Host association of *Borrelia burgdorferi* sensu lato: a review. Ticks Tick Borne Dis 12:101766
Wolf RF, Aragona M, Muñoz-Leal S, Pinto LB, Melo ALT, Braga IA, Costa JS, Martins TF, Marcili A, Pacheco RC, Labruna MB, Aguiar DM (2017) Novel *Babesia* and *Hepatozoon* agents infecting non-volant small mammals in the Brazilian Pantanal, with the first record of the ticks *Ornithodoros guaporensis* in Brazil. Ticks Tick-Borne Dis 7:449–456
Yabsley MJ, Shock BC (2013) Natural history of zoonotic *Babesia*: role of wildlife reservoirs. Int J Parasitol Parasites Wildl 2:18–31. https://doi.org/10.1016/j.ijppaw.2012.11.003

Yang SGNDS, Souza DDSE, Santiago ACDS, Silva RBS, Melo MA, Souza EAR, Labruna MB, Horta MC, Oliveira JB (2021) Rickettsial infection in free-ranging capybaras (Hydrochoerus hydrochaeris) and their ticks (Acari, Ixodidae) in the Caatinga and Atlantic forest biomes, Northeastern Brazil. Vet Parasitol Reg Stud Rep 26:100649. https://doi.org/10.1016/j.vprsr.2021.100649. Epub 2021 Sep 29. PMID: 34879960

Yu XJ, Walker DH (2006) The order Rickettsiales. In: Dworkin M, Falkow S, Rosenberg E, Schleifer KH, Stackebrandt E (eds) The prokaryotes. Springer, New York. https://doi.org/10.1007/0-387-30745-1_20

Zaldívar Y, Hernandez M, Domínguez L, Saenz L, Montilla S, de Antinori M, Krawczak FS, Bermúdez SE (2021) Isolation of *Rickettsia rickettsii* in Rocky Mountain spotted fever outbreak, Panama. Emerg Infect Dis 27:1245–1247

Zieman EA, Jimenez FA, Nielsen CK (2017) Concurrent examination of bobcats and ticks reveals high prevalence of *Cytauxzoon felis* in Southern Illinois. J Parasitol 103:343–348

Chapter 14
Internal Parasites and Their Impact on the Health and Conservation of Neotropical Vertebrates

Jaqueline Bianque de Oliveira, Gabriela Felix-Nascimento, Laís Kelly Amâncio Ribeiro Berenguer, Dênisson da Silva e Souza, Sofia Bernal-Valle, and Vanessa Campelo de Souza

14.1 Introduction

Approximately 30% of the global area is covered by forests, but these ecosystems and the biodiversity they host are both threatened by human actions associated with changes in land use and overexploitation of natural resources, impacting the balance of ecological processes (Daszak et al. 2000, 2001; Dunn and Hatcher 2015; OHHLEP et al. 2022). Habitat loss and fragmentation constitute the greatest threats to biological diversity since one of their effects is the isolation of small populations in forest remnants. This can result in greater exposure and susceptibility of wild populations to pathogens, which compromise their ecology, behavior, and reproduction, thus affecting their survival (Holmes 1996; Catenacci et al. 2016; Bueno et al. 2017).

Wild animals are hosts of a great diversity of pathogens, playing an important role in the reservoir system of these organisms, for which they act as maintainer or amplifier hosts (Daszak et al. 2000, 2001; Cunningham et al. 2003; Roque and Jansen 2014). A greater diversity of hosts can favor the maintenance of a pathogen, its high abundance (amplifying effect), or its reduction (dilution effect) (Gürtler and Cardenal 2015). The destruction of ecosystems and the expansion of areas occupied by human populations favor greater contact among wild fauna-domestic animals-humans, enabling the spillover of multi-host parasites. These, in turn,

J. B. de Oliveira (✉) · L. K. A. R. Berenguer · D. da Silva e Souza · V. C. de Souza
Universidade Federal Rural de Pernambuco (UFRPE), Recife, Pernambuco, Brazil
e-mail: jaqueline.boliveira@ufrpe.br

G. Felix-Nascimento
Universidade Federal do Vale do São Francisco (UNIVASF), Petrolina, Pernambuco, Brazil

S. Bernal-Valle
Programa de Pós-graduação em Ciência Animal (PPGCA), Universidade Estadual de Santa Cruz (UESC), Ilhéus, Bahia, Brazil

G. Acosta-Jamett, A. Chaves (eds.), *Ecology of Wildlife Diseases in the Neotropics*, https://doi.org/10.1007/978-3-031-50531-7_14

establish particular and dynamic interactions with each host species, which will act differently in the transmission cycle of these parasites on a spatiotemporal scale (Roque et al. 2008; Xavier et al. 2012; Aguirre 2017; Jansen et al. 2018; Oliveira et al. 2022).

Parasites constitute a significant portion of biodiversity in natural ecosystems, where they play an important role in the structure and organization of ecological communities, influencing competition between species, trophic interactions, and biological diversity (Holmes 1996; Marcogliese 2004; Hudson et al. 2006). The role of parasites in the structure and organization of ecological communities is due to differences in both susceptibility and host tolerance (Hudson et al. 2006). Although parasitism is a coevolutionary process that tends to be harmonious over time, this ecological relationship can be influenced by factors related to the parasite, the host, and the environment, negatively impacting the health and conservation of biodiversity and ecosystems (Holmes 1996; Jansen et al. 2015; Valenzuela-Sánchez et al. 2021). In some situations, human health can also be affected (Daszak et al. 2000, 2001; Cunningham et al. 2003; Cunningham 2005; Aguirre 2017). In this context, the approach to the impact of parasites on animal, human, and ecosystem health must be carried out under the prism of Conservation Medicine and One Health.

The anthropogenic dispersion of parasites, known as "pathogen pollution," is a threat to the conservation of biological diversity, both in wild and captive-breeding institutions and translocation-reintroduction programs (Daszak et al. 2000, 2001; Cunningham et al. 2003; Ramalho et al. 2009; Dunn and Hatcher 2015; Bienentreu and Lesbarrères 2020). In this chapter, we will present the ecoepidemiological aspects of some generalist enzootic parasites, which may pose a risk to the health and conservation of wildlife and ecosystems. For parasites with zoonotic potential, we discuss the risks to human health to demonstrate the complexity of control and prevention approaches.

14.2 Internal Parasites of Amphibians and Reptiles

Because they occupy an intermediate position in the food chain, sometimes preying on invertebrates or small vertebrates and sometimes serving as the prey of larger vertebrates, such as mammals or birds, vertebrates from the Amphibia and Reptilia classes host a wide diversity of parasites (Ávila and Silva 2010; Campião et al. 2014a). The decline in amphibian populations associated with diseases has been the main threat worldwide, but the true scale of the decline is still unknown, and the spatiotemporal patterns are still limited (Bienentreu and Lesbarrères 2020). Studies have revealed multifaceted and even conflicting patterns, which provide a complex picture that is often difficult to translate into conservation actions (Bienentreu and Lesbarrères 2020).

Aspects related to herpetofaunal health have been little studied, with most studies focusing on viral and fungal diseases, mainly for anurans, and relatively little attention has been given to helminths and protozoan parasites (Bienentreu and Lesbarrères 2020). The main factors that influence host-parasite dynamics in amphibians and reptiles are as follows: (i) host characteristics (species, body condition and reserves, developmental stage, behavior, habitat preferences, susceptibility, innate or adaptive immune response, and microbiome) and (ii) parasite characteristics (species, virulence, intensity of infection, infectivity, competitiveness, coinfection, order in which infection occurred by different helminth species, exposure time and natural history, coevolution, and infection). Additionally, biotic (host density, intra- and interspecific reservoirs, microhabitat characteristics, landscape characteristics, predation, resource availability, sources of stress, and species that make up the host community), abiotic and anthropogenic factors (temperature, humidity, seasonality, UV-B radiation, contaminants, chemical characteristics of water, and habitat fragmentation) are noteworthy (Bienentreu and Lesbarrères 2020; Herczeg et al. 2021; Valenzuela-Sánchez et al. 2021; Fonseca et al. 2023).

In populations of *Phrynops geoffroanus* in a river impacted by anthropogenic activities (e.g., domestic sewage and industrial, textile, and agricultural waste) in northeastern Brazil, it was demonstrated that the more impacted the river stretch, the lower the helminth richness (Fonseca et al. 2021a, 2023). In areas with a higher level of environmental alterations, a higher percentage of animals presented macro- and histological lesions related to parasitism by the nematodes *Spiroxys figueiredoi* and *Serpinema monospiculatus*, which, in addition to being more pathogenic, were more resistant to anthropic alterations (Fonseca et al 2021a, 2023). Trematodes were not found in animals from the most impacted areas, because they are more sensitive to anthropogenic disturbance, confirming their role as environmental bioindicators (Fonseca et al. 2021a).

Despite the environmental impacts, a great diversity of helminths has been recorded in *Crotalus durissus* snakes from urban areas in southeastern Brazil, with an emphasis on the heteroxene nematode *Ophidascaris arndti*, which showed high prevalence and was associated with multifocal granulomatous gastritis (Mello 2013). Other nematodes with high pathogenic potential have also been registered, such as *Kalicephalus* spp. and *Hexametra* spp. The mechanisms of pathogenicity and the impact of parasites on snake health are poorly understood. In this context, histopathology is highlighted as an important tool for understanding the pathogenicity of parasites at both the individual and population levels (Sena et al. 2018; Felix-Nascimento 2021; Felix-Nascimento et al. 2022; Fonseca et al. 2021a).

Parasitism by third-stage larvae (L3) of the nematode *Physaloptera* sp. determined inflammatory and necrotic lesions in anurans of the species *Leptodactylus macrosternum* in an area of conventional agriculture in the Caatinga biome in northeastern Brazil. Although this frog can act as a paratenic host for the parasite, the exposure of animals to pesticides and chemical fertilizers used in crops may be associated with the greater severity of the lesions found (Felix-Nascimento et al. 2022). In addition, animals from this and other studied areas (organic agriculture and an area of Caatinga *stricto* sensu) showed extremely high levels of essential minerals and heavy metals (Felix-Nascimento 2021). Despite this, although the study recorded histopathological alterations in the liver and cytogenotoxic alterations in the red blood cells of the animals from all three studied areas, these environments seem to provide the minimum conditions for maintaining the host-parasite relationship, considering that no marked differences were observed in the helminth fauna (Felix-Nascimento 2021).

Parasitism can indirectly affect physiological and behavioral processes in amphibians and reptiles, such as thermoregulation (Paranjpe et al. 2014), locomotion (Moretti et al. 2014, 2017), and vocalization (Madelaire et al. 2013). In specimens of *Hypsiboas prasinus* parasitized by helminths in an Atlantic Forest area in southeastern Brazil, males with lower parasite loads had higher vocalization rates. This is likely a result of the high energy cycles associated with maintaining high vocalization performance, which affects male reproductive efficiency (Madelaire et al. 2013). The negative impact of pulmonary infection by the nematode *Rhabdias fuelleborni* on locomotor performance and basal metabolic rate was recorded in *Rhinella icterica* in the Atlantic Forest in southeastern Brazil, which may be associated with reduced lung diffusion capacity and tidal volume, in addition to the indirect effect of reducing useful energy storage (Moretti et al. 2014). These data may represent a classic example of how parasites can modulate the distribution and abundance of hosts in an environment.

Changes in land use can increase morbidity and mortality due to the increase in parasites whose pathogenicity is density-dependent (McKenzie 2007). Accelerated changes in land use, associated with the erroneous idea that some ecosystems have low biodiversity, should be seen with concern, as should the low level of government incentives, mainly due to reduced investment in conservation strategies and in studying these environments (Portela et al. 2020). The diversity of pathogens and the abundance of host communities may increase or decrease due to habitat fragmentation of human origin (King et al. 2007; McKenzie 2007). In frogs of the families Hylidae and Leptodactylidae from southern Brazil, the prevalence, intensity, and abundance of helminths were higher in animals from areas with agricultural crops than in native pastures with cattle (Portela et al. 2020).

Currently, dilution and amplifying effects have received greater attention in studies of the parasitic ecology of herpetofauna. Although host density can potentially affect disease dynamics, only 3% of studies consider these effects (Bienentreu and Lesbarrères 2020). In lizards from the Atlantic Forest and naturally isolated forest enclaves, parasite richness was positively related to host specificity; that is, in communities that are rich in parasite species, they tended to parasitize one or a few host species (Teixeira et al. 2020). Furthermore, the parasites tend to be distributed in an aggregated way in a population, as observed in the chelonian *Mesoclemmys vanderhaegei* parasitized by hemogregarines in the Brazilian Cerrado (Goes et al. 2018).

The effects of community composition are important in translocation/introduction programs, as well as in the migration of amphibian species to previously unsuitable habitats induced by climate change (Bienentreu and Lesbarrères 2020). Changes in community diversity may be due to the introduction of new pathogens or changes in the dynamics of the pathogens present, arising from an increase or decrease in competent species (Bienentreu and Lesbarrères 2020). The consequences of introducing endoparasites into environments where they do not occur naturally are poorly studied (Ramalho et al. 2009). The generalist helminths of *Salvator merianae*, a lizard considered invasive on the island of Fernando de Noronha in Brazil, also began to parasitize the island's endemic reptile species, such as *Trachylepis atlantica* and *Amphisbaena ridley* (Ramalho et al. 2009). However, there are no reports of the pathogenicity of these parasites in this native species.

The host-parasite relationship under natural conditions in healthy environments tends toward equilibrium (Holmes 1996). However, in animals kept under human care, parasitism can be a threat, especially when the objective is the conservation of the host species. Stress generated by captivity, the reduction of space, and high density of animals in the enclosures can lead to an increase in parasite rates at pathogenic levels for these hosts, as observed in *Chelonoidis carbonarius* and *C. denticulatus,* parasitized by the nematode *Chapiniella variabilis* in Brazil (Freire et al. 2017, 2019). In these same species of chelonians kept under human care, infection by nematodes and trematodes was associated with macro- and histological changes (Freire et al. 2019). In *C. denticulatus* sold for meat consumption in Peru, reports have associated helminth infections with histological lesions (Julca et al. 2014).

Host species are the main filter for the structure of the parasite community, even when studying contrasting environments, given that the biological and biometric characteristics of animals can influence the ecological patterns of parasites (Campião and Dáttilo 2020; Felix-Nascimento 2021). In the anuran *Rhinella limensis* from the central coast of Peru, the prevalence and mean abundance of the nematode *Rhabdias fuelleborni* were associated with sex, with males showing higher rates of these ecological parameters (Chero et al. 2015). However, host body size is considered the best individual descriptor of host-parasite networks (Campião and Dáttilo 2020). In *Mesoclemmys vanderhaegei* from areas of the Brazilian Cerrado, a positive relationship between the prevalence of hemogregarines and the body size of the animals was demonstrated (Goes et al. 2018).

The helminth community is also influenced by foraging strategies (active or sit-and-wait) and the type of habit (arboreal, aquatic, terrestrial, or semiterrestrial) of frogs and lizards (Campião et al. 2015; Leung and Koprivnikar 2019). Infection by the nematodes *R. fuelleborni* and *Aplectana hylambatis*, for example, is more related to the foraging strategy of frogs than to the type of prey they capture (Chero et al. 2015). Anurans with terrestrial behavior, such as bufonids, generally have a low prevalence and medium intensity of digenetic trematodes due to the reduced ingestion of aquatic arthropods that are intermediate hosts. They also have greater exposure to monoxenous nematodes due to the ingestion of eggs and/or larvae or percutaneous penetration of infective larvae (Santos et al. 2013). On the other hand, arboreal anurans tend to be infected mainly by parasites that are transmitted via ingestion of the infective forms (Martins-Sobrinho et al. 2017; Sena et al. 2018), while leptodactylids (semiterrestrial) are infected by infective stages present both in the water (e.g., digenetics) and in soil (e.g., direct cycle nematodes) (Campião et al. 2015).

The aquatic habit of *Chelonia mydas* may also explain the predominance of trematode species in juvenile individuals of this turtle species off the Brazilian coast

(Werneck and Silva 2015). Juveniles of *C. mydas* have a lower species richness of parasites compared to adult specimens, which can be attributed to the shorter life span and, therefore, reduced opportunities to become infected (Werneck and Silva 2015). Diet can be a determining factor for the composition of the helminth parasite community, more specifically those whose infection occurs orally, either in anurans or lizards.

Worldwide, amphibians and reptiles can be hosts of helminths and protozoa with zoonotic potential, and in the neotropical region, the following parasites have been recorded: *Spirometra* spp., *S. erinaceieuropaei,* and *S. mansonoides* (Oda et al. 2016); *Trichinella zimbabwensis* and *T. papuae*; *Contracaecum* spp., *Anisakis* spp. and *Pseudoterranova* spp.; *Eustrongylides* spp., *Gnathostoma binucleatum, G. doloresi, G. hispidum, G. nipponicum, G. spingerum,* and *Angiostrongylus cantonensis* (Mendoza-Roldan et al. 2020). The transmission of these pathogens to humans can occur through ingestion of contaminated meat, direct contact with the feces of parasitized animals, and indirect contact due to contamination in the environment or through vectors (Mendoza-Roldan et al. 2020).

Sparganosis, caused by nematodes of the genus *Spirometra,* is considered an emerging disease in humans, and most cases have been detected in rural areas in South American countries (Oda et al. 2016). However, only a few of these records are related to reptiles and amphibians infected by these nematodes, and the pathogenic effects on these hosts are poorly studied, although larval migration can cause primary pathology in amphibians, in addition to predisposing them to the occurrence of other infections and diseases (Oda et al. 2016). In amphibians and squamata reptiles, there are different sites of infection (coelomic cavity, peritoneum, musculature, and different internal organs) (Mendoza-Roldan et al. 2020), and parasite rates may be higher in lizards (*Salvator merianae*) and snakes (*Erythrolamprus poecilogyrus sublineatus*, *E. semiaureus*, *Philodryas patagoniensis*, and *Xenodon dorbignyi*), which act as paratenic hosts due to predation by parasitized anurans (Oda et al. 2016).

In addition to the conservation of animal species and the spread of diseases among them, the study of potentially zoonotic parasites in amphibians and reptiles is also an aspect that needs to be further studied in the neotropical region, especially for those parasites that use these animals as intermediate or paratenic hosts. In studies of parasitic ecology, the occurrence of larval stages of unidentified endoparasites is common (Ávila and Silva 2010; Hamann et al. 2006, 2012; Campião et al. 2014b). Furthermore, the real role of amphibians and reptiles in the cycle of endoparasites with zoonotic potential is still uncertain or unknown (Mendoza-Roldan et al. 2020), and this may be related to the great difficulty in taxonomic identification of the larval stages of endoparasites, whether through morphology, molecular biology, or experimental infections for the study of adult stages.

On a local scale, the distribution of infectious agents is likely to be mediated by human activities, but the flow of water can facilitate the spread of pathogens in more complex landscapes. On a continental scale, amphibian pathogens are spread through commercial activities, such as bullfrog breeding, and this needs to be more rigorously prevented and monitored (Bienentreu and Lesbarrères 2020).

Future studies of disease ecology in amphibians and reptiles would need to include the characterization of the host community, life stage and species-specific effects, in addition to the identification of pathogens at the molecular level and an assessment of the environmental factors that potentially contribute to pathogen emergence (Bienentreu and Lesbarrères 2020). In general, what is observed among the studies published in the Neotropical region is an approach focused on ecological patterns or on the elaboration of species lists associated with a certain population or host community (e.g., Ávila and Silva 2010; Campião et al. 2014a; Campião and Dáttilo 2020; Portela et al. 2020). Understanding the parasite-host relationship, especially in free-ranging populations, requires a multidisciplinary approach and the cooperation of different professionals (taxonomists, veterinarians, biologists, and ecologists, among others).

14.3 Internal Parasites of Birds

Among parasitic diseases, trichomoniasis is one of the main threats to bird conservation, and it is considered an emerging disease on a global scale (Bunbury et al. 2007; Sansano-Maestre et al. 2009; Forzán et al. 2010; Amin et al. 2014). The flagellate protozoan *Trichomonas gallinae* is a parasite of the digestive tract, but it can also be found in the liver, spleen, lung, maxilla and mandible, orbital cavity, and ear canal (Joppert 2007; Ecco et al. 2012; Andery et al. 2013; Echenique et al. 2019). The severity of avian trichomoniasis is related to the susceptibility of the birds and the virulence of the strain (Joppert 2007; Forrester and Foster 2008; Forzán et al. 2010; Santos et al. 2011; Stimmelmayr et al. 2012; Merling de Chapa et al. 2021). Inflammation, caseification and necrosis are the tissue alterations responsible for the clinical manifestation, and death occurs because of difficulty swallowing food and water and/or respiratory failure (Joppert 2007; Forrester and Foster 2008; Ecco et al. 2012; Bruni et al. 2019).

Columbiform birds constitute the reservoir system of the protozoan and may be asymptomatic or present symptoms of varying severity (Forrester and Foster 2008; Rogers et al. 2016; Merling de Chapa et al. 2021). The dispersion of *T. gallinae* is mainly attributed to the species *Columba livia*, an invasive exotic species in several Neotropical countries (Sansano-Maestre et al. 2009; Castillo 2019; Merling de Chapa et al. 2021). There are data indicating that *C. livia* is present in urban areas of approximately 90% of cities worldwide (Aronson et al. 2014). For that reason, Castillo (2019) draws attention to the fact that due to the global emergence of avian trichomoniasis, the One Health approach must be applied to the control of *C. livia* populations, not only to ensure human health because of the zoonotic pathogens that these birds harbor but also because of the environmental impact and threat to the health of birds severely affected by pathogens such as *T. gallinae*.

The *T. gallinae* cycle is found mainly in trophic networks since transmission occurs through the consumption of parasitized prey. Although the viability of *T. gallinae* is reduced outside the host organism, oral transmission can occur through the consumption of food and water contaminated with the parasite's trophozoites and by fomites (Sansano-Maestre et al. 2009; Forzán et al. 2010; Amin et al. 2014; Rogers et al. 2016; Castillo 2019; Freitas et al. 2020; Merling de Chapa et al. 2021). Large populations of pigeons tend to be concentrated in grain storage and processing areas (Echenique et al. 2019). In wet grains, *T. gallinae* trophozoites can remain viable for up to 5 days, favoring the maintenance of the parasite cycle and the contamination of susceptible birds (Purple and Gerhold 2015). The food regurgitated by the adults to the young or the direct contact between beaks during feeding are situations that characterize vertical transmission (Joppert 2007; Sansano-Maestre et al. 2009; Forzán et al. 2010).

The diversity and geographic distribution of *T. gallinae* hosts characterize the conservation impact of avian trichomoniasis, especially in free-living populations of endangered species (Bunbury et al. 2007; Sansano-Maestre et al. 2009; Forzán et al. 2010; Amin et al. 2014; Bruni et al. 2019; Echenique et al. 2019). Trichomoniasis has been reported with increasing frequency in Falconiformes, Accipitriformes, Strigiformes, Psittaciformes, Passeriformes, Piciformes, and Galliformes (Joppert 2007; Bunbury et al. 2007; Forrester and Foster 2008; Sansano-Maestre et al. 2009; Forzán et al. 2010; Santos et al. 2011; Ecco et al. 2012; Andery et al. 2013; Amin et al. 2014; Bruni et al. 2019; Castillo 2019; Chavatte et al. 2019; Echenique et al. 2019; Freitas et al. 2020; Merling de Chapa et al. 2021). Around the world, trichomoniasis is seen as a threat to conservation, especially for birds of prey, a problem attributed to the expansion of the feeding habits of these birds that, due to the loss of their natural habitat, migrate to urban and peri-urban areas in search of shelter and food. In this context, columbids, which are more abundant in urban areas, become part of the diet of predators in greater frequency and quantity (Joppert 2007; Sansano-Maestre et al. 2009; Andery et al. 2013; Bruni et al. 2019; Echenique et al. 2019; Freitas et al. 2020; Merling de Chapa et al. 2021). Perhaps for this reason, morbidity and mortality from trichomoniasis in birds of prey are higher in urban areas than in rural areas (Andery et al. 2013). It is important to highlight that anthropization can affect the adaptive behavior of several groups of birds, thus contributing to a greater diversity of *T. gallinae* hosts (Bruni et al. 2019).

Little is known about the factors involved in the development of trichomoniasis in raptors (Merling de Chapa et al. 2021). Age, immune status, previous exposure, and virulence of the strains are determinants for the development of trichomoniasis in columbids (Joppert 2007; Merling de Chapa et al. 2021). A greater diversity of *T. gallinae* strains has been found in raptors, especially in those whose diet consists of parasitized columbids. In the Neotropics, mainly in Brazil, trichomoniasis has been recorded as the main infectious cause of mortality in birds of prey (Joppert 2007; Ecco et al. 2012; Andery et al. 2013; Bruni et al. 2019; Echenique et al. 2019; Freitas et al. 2020). Some of these birds come from areas where *T. gallinae* infection has been previously recorded in populations of *C. livia* (Andery et al. 2013). There are also records of infection in Passeriformes and Piciformes (Ecco et al. 2012; Freitas et al. 2020).

In some captive-breeding institutions and translocation-reintroduction programs, pigeons from population control programs are used to feed birds prey, which may represent a risk for the occurrence of the disease (Santos et al. 2011). Therefore, this is a practice that should be avoided. In addition, the diagnosis of *T. gallinae* infection in susceptible birds is essential in translocation/introduction programs to avoid risks to the health of released individuals, as well as the spread of the protozoan to native populations, compromising conservation efforts (Santos et al. 2011).

14.4 Internal Parasites of Mammals

14.4.1 Trypanosoma cruzi

The protozoan *Trypanosoma cruzi* is the etiologic agent of Chagas disease, a zoonotic neglected tropical disease (NTD) that affects approximately six to eight million people in 21 countries in the Americas, causing 14,000 deaths/year (WHO 2022a; Bergner et al. 2021; PAHO 2022). In nature, the protozoan is maintained by a system of mammalian reservoirs of several orders (Artiodactyla, Chiroptera, Primates, Carnivora, Rodentia, Cingulata, Pilosa, and Didelphimorphia) and transmitted by more than 150 species of triatomines (Hemiptera, Reduviidae) in the neotropical region (Roque et al. 2008; Cordovez and Guhl 2015; Jansen et al. 2018, 2020; Ihle-Soto et al. 2019; Lazo et al. 2019; Santos et al. 2019). This heterogeneity of vertebrate and invertebrate hosts, associated with habitat degradation and loss, results in differences in exposure and susceptibility to *T. cruzi* infection in the wild and peridomestic cycles (Orozco et al. 2013; Jansen et al. 2015; Lazo et al. 2019).

Because it is a heteroxene parasite with a great diversity of hosts, in which it infects almost all types of cells, this protozoan has been subjected to selective pressures that have resulted in large phenotypic and genetic variability (Roque et al. 2008; Costa et al. 2018; Jansen et al. 2020). Currently, seven subpopulations of *T. cruzi* are recognized, called discrete typing units (DTUs): TcI and TcII (most widely distributed), TcIII, TcIV, TcV, TcVI, and Tcbat (Lima et al. 2015; López-Cancino et al. 2015; Rendón et al. 2015; Brandão et al. 2019; Santos et al. 2019; Jansen et al. 2020). Of these, Tcbat has been recorded only in bats (Villena et al. 2018; Brandão et al. 2019; Lazo et al. 2019; Jansen et al. 2020). To date, there is no evidence that *T. cruzi* DTUs are associated with pathogenicity in humans or with biological and ecoepidemiological factors (Jansen et al. 2018, 2020; Brandão et al. 2019).

Despite the diversity of hosts, *T. cruzi* presents particular and dynamic patterns of interaction with some species of mammals, which result in differences in competence for maintenance (susceptibility, infection, and survival) and transmission (infectiousness and vector-host contact). Competence may be influenced by factors such as the occurrence of mixed DTU infections, coinfection with other pathogens, physical and nutritional status, and stress caused by unfavorable environmental conditions, factors that change over time and space (Orozco et al. 2013; Gürtler and

Cardenal 2015; Ihle-Soto et al. 2019; Jansen et al. 2018; Brandão et al. 2019; Santos et al. 2019).

The increased number of cases and outbreaks of oral transmission in humans in Latin America and the United States has drawn attention to new ecoepidemiological scenarios involving the sylvatic cycle through the participation of vectors and mammals that act as *T. cruzi* reservoirs (Rocha et al. 2013; Bustamante et al. 2014; Costales et al. 2015; Jansen et al. 2015, 2020; Bergner et al. 2021). As it is primarily an enzootic infection, the assessment of infection risks for humans must consider the triatomine fauna and the faunal composition of hosts, as well as the landscape and social conditions of each locality (Xavier et al. 2012; Bustamante et al. 2014; Ihle-Soto et al. 2019; Jansen et al. 2018, 2020).

The *T. cruzi* cycle in wildlife is embedded in trophic networks since transmission occurs mainly through the oral route through the predation of infected vectors or mammals. This is considered the oldest route of *T. cruzi* dispersion (Rocha et al. 2013; Jansen et al. 2018; Lazo et al. 2019; Brandão et al. 2019, 2020), whose infective potential seems to be favored by gastric juice (Roque et al. 2008). Despite the density of hair in some mammals acting as a barrier and hindering the cutaneous penetration of *T. cruzi*, the contamination of nests/shelters/refuges of animals with metacyclic forms of the protozoan, released in the feces of triatomines or in the secretion of the odoriferous glands of marsupials, is also likely to occur (Rocha et al. 2013; Jansen et al. 2015, 2018; Lazo et al. 2019). The participation of marsupials in the *T. cruzi* cycle is because these mammals are the oldest hosts and harbor all life stages of the protozoan cycle. Emphasis should also be given to some species of vampire bats in which the protozoan has already been detected in the salivary glands (Brandão et al. 2019; Lazo et al. 2019; Bergner et al. 2021). That is, marsupials and vampire bats can act as reservoirs and transmitters of the parasite to other mammals, including humans, and can transmit *T. cruzi* in the absence of triatomine vectors (Bustamante et al. 2014; Costales et al. 2015; Jansen et al. 2015, 2020; Villena et al. 2018; Bergner et al. 2021). The epidemiological importance of these mammals needs to be further studied.

In wildlife, different scenarios of enzootic infection can occur in the same forest fragment in different strata, depending on the faunal composition and ecological and behavioral characteristics of the hosts. This results in a lower or higher rate of infection and, consequently, the ability to act in the *T. cruzi* reservoir system (Jansen et al. 2015, 2018, 2020; Brandão et al. 2020). In an Atlantic Forest fragment in Brazil, where a *Leontopithecus rosalia* conservation program was carried out, TcII infection was detected only in this endangered primate species, while other sympatric mammals (marsupials and rodents) were infected by TcI. These data suggest that the transmission of *T. cruzi* can occur in distinct and independent cycles, regulated by as yet unknown factors (Lisboa et al. 2015; Jansen et al. 2018).

Despite the diversity of mammals that participate in the *T. cruzi* reservoir system, some species act as primary reservoirs, while others play secondary or dead-end roles, depending on the competence and rate of contact with the vector (Orozco et al. 2013; Jansen et al. 2015, 2018). According to Jansen et al. (2015), in wildlife, dispersal and maintenance of *T. cruzi* is performed by approximately 40% of infected mammals. In this context, Didelphimorphia (*Didelphis* spp. and *Philander* spp.), Carnivora (including coatis, skunks, ferrets, raccoons, and foxes), Primates (*Leontopithecus rosalia*, *L. chrysomelas* and *Sapajus* sp.), Cingulata (*Dasypus novemcinctus*), and Pilosa (*Tamandua tetradactyla*) seem to play a prominent role as primary reservoirs of the protozoan (Yeo et al. 2005; Mehrkens et al. 2013; Orozco et al. 2013; Jansen et al. 2017, 2018, 2020; Santos et al. 2019; Hodo et al. 2020; Ocanã-Mayorga et al. 2021).

Generalist species (in relation to habitat and diet), such as carnivores and marsupials, are considered bioaccumulators and dispersers of *T. cruzi* genotypes (Xavier et al. 2012; Rocha et al. 2013; Jansen et al. 2015; Brandão et al. 2019, 2020). Therefore, in urban and peri-urban forest fragments in endemic areas, systematic surveillance is necessary for the presence of synanthropic animals (such as marsupials, rodents, coatis, raccoons, and bats), which play an important role in the *T. cruzi* cycle (Xavier et al. 2012; Jansen et al. 2018; Trüeb et al. 2018; Yefi-Quinteros et al. 2018; Dronzino et al. 2019; Bergner et al. 2021; Hodo et al. 2020; Nantes et al. 2020).

In the Neotropical region, several species of wild and synanthropic rodents act as important reservoirs of *T. cruzi* (López-Cancino et al. 2015; Noya and González 2015; Yefi-Quinteros et al. 2018; Ihle-Soto et al. 2019; Jansen et al. 2020). However, in some contexts, wild rodents can exceptionally act as primary reservoirs of the protozoan, depending on the lifespan and extension of the home range, factors that influence a lower or higher exposure to infection (Jansen et al. 2015, 2018). Experimentally, some caviomorph rodent species are less resistant to infection and die. In nature, cardiac damage and the consequent oxygenation deficit could make rodents more susceptible to predation, favoring the transmission of *T. cruzi* to other mammals (Jansen et al. 2017).

The geographic distribution of *T. cruzi* subpopulations is not homogeneous. For this reason, captive-breeding and translocation/reintroduction programs can promote new infection scenarios (Bahia et al. 2017; Minuzzi-Souza et al. 2016; Jansen et al. 2020; Reis et al. 2020). Ecoepidemiological, animal health, and public health aspects related to *T. cruzi* infection in mammals kept in ex situ conservation institutions and biomedical research centers have been highlighted in several studies and include (i) the impact on health, especially of endangered species; (ii) interference in the results of studies with infected laboratory animals; (iii) the release of infected animals, which can result in the dispersion of the parasite in new areas, favoring infection during their stay in institutions or when they are released in environments where the parasite naturally circulates; and (iv) the transmission of the parasite to workers who deal directly with infected animals (Monteiro et al. 2006, 2010; Bahia et al. 2017; Hodo et al. 2018; Minuzzi-Souza et al. 2016; Reis et al. 2020).

In addition to being involved in the transmission cycle of *T. cruzi*, the health of some species may be affected by the parasite. In captivity, situations of stress and immunocompromise can boost *T. cruzi* infection and, consequently, negatively affect the health of animals, compromising conservation efforts, as well as the results of biomedical research with laboratory animals (Bahia et al. 2017). Despite this, there are still many gaps regarding the impact of *T. cruzi* infection on the health of host mammals (Bahia et al. 2017; Minuzzi-Souza et al. 2016; Bueno et al. 2017; Hodo et al. 2018, 2020; Santos et al. 2018; Reis et al. 2020).

Neotropical nonhuman primates (NHPs) of approximately 30 species of the Cebidae, Atelidae, Pitheciidae, and Aotidae families are natural hosts of *T. cruzi* (Bahia et al. 2017; Bueno et al. 2017; Minuzzi-Souza et al. 2016; Reis et al. 2020). In these mammals, although the infection is subclinical in most cases, changes in the heart and in serum biochemical markers can compromise survival, especially in endangered species that are released in translocation/reintroduction programs, as observed in specimens of *Leontopithecus rosalia* in remnants of Atlantic rainforest in Brazil (Monteiro et al. 2006, 2010).

In species of mammals of the orders Carnivora, Cetartiodactyla, Perissodactyla, Pilosa, and Primates kept in zoos in Brazil, natural infection by TcI was detected, including in threatened species of NHPs. Infection was also detected in triatomines of the species *Panstrongylus megistus*, found in several enclosures of parasitized and non-parasitized animals (Minuzzi-Souza et al. 2016; Reis et al. 2020). Some parasitized animals were born at the zoo and were offspring of non-parasitized mothers; this indicates that transmission may have occurred in the institution, which is surrounded by a forest fragment in which TcI infection had already been detected in *D. albiventris.* This may be the origin of the triatomines that colonized the enclosures and of the parasite. However, as most of the parasitized animals came from other institutions or from rescues carried out by environmental agencies, this is another possibility of introducing *T. cruzi* into the zoo. Therefore, it is extremely important to monitor the presence of pathogens systematically in animals kept in ex situ conservation and research institutions, as well as in wild populations around these institutions (Bahia et al. 2017; Minuzzi-Souza et al. 2016; Hodo et al. 2018; Reis et al. 2020).

Cardiac and other changes have been recorded in coatis (*Nasua nasua*) and marsupials (*D. marsupialis* and *D. virginiana*) naturally infected with *T. cruzi* in Brazil (Santos et al. 2018; Nantes et al. 2019) and Mexico (Villagrán et al. 2011; Carnevali et al. 2017). The alterations are suggestive of compromised health and conservation of these mammals, mainly in degraded ecosystems, as well as in urban and peri-urban areas. In the United States, cardiac lesions caused by TcI and TcIV infections have been reported in free-ranging coyotes and raccoons, and increased susceptibility has been demonstrated in coyotes (Hodo et al. 2020).

As it involves ecological aspects inherent to the human-animal-vector-ecosystem interface, Chagas disease is an example of a disease for which control must be approached in an integrative, multiprofessional, transdisciplinary, and intersectoral way in the context of One Health (Jansen et al. 2015; García et al. 2016; Dronzino et al. 2019; Rivera et al. 2020; Austen and Barbosa 2021). The disease should not be seen only as a "human parasitosis" due to the multiplicity of hosts of *T. cruzi* (Jansen et al. 2020).

14.4.2 Leishmania *spp.*

Leishmaniasis is a set of NTDs caused by several species of protozoa, including *Leishmania infantum* (syn. of *L. chagasi*), *L. braziliensis*, *L. amazonensis,* and *L. mexicana*, species that occur in the neotropical region. These trypanosomatids parasitize cells of the mononuclear phagocytic system of animals and humans. Visceral leishmaniasis (VL), caused by *L. infantum*, is the most severe disease and is endemic in 13 countries in the Americas. It is estimated that approximately 350 million people worldwide are at risk of infection, and more than one million new cases appear worldwide each year (WHO 2022b).

Enzootic and zoonotic cycles occur from the bite of female sandflies (Diptera, Psychodidae), which transmit *Leishmania* to a wide range of mammalian hosts. Although vector transmission is the most important, other transmission routes have already been recorded, such as vertical, sexual, and iatrogenic routes (Souza et al. 2014; Azami-Conesa et al. 2021). The relationship between host competence and sandfly food preference is an important factor in the spread of *Leishmania*, particularly in biodiverse environments (Vianna et al. 2016; Rivera et al. 2020). Different hosts have their own characteristics to attract vectors and transmit pathogens, and this involves factors that change on a space-time scale (Roque and Jansen 2014; Vianna et al. 2016; Brandão et al. 2019; Rivera et al. 2020).

In the Neotropical region, the potential reservoirs of *Leishmania*, responsible for the maintenance and transmission of the parasite in nature, are grouped in the orders Didelphimorphia, Carnivora, Pilosa, Cingulata, Rodentia, Primates, and Chiroptera (Courtenay et al. 2002; Dahroug et al. 2010; Quintal et al. 2011; Humberg et al. 2012; Lima et al. 2012; Shapiro et al. 2013; Roque and Jansen 2014; Cardoso et al. 2015; González et al. 2015; Oliveira et al. 2015; Pereira et al. 2017; Trüeb et al. 2018; Muñoz-García et al. 2019; Brandão et al. 2020; Moreno et al. 2020; Azami-Conesa et al. 2021). Although natural *Leishmania* infection has already been detected in a great diversity of Neotropical mammals, few studies have demonstrated the ecoepidemiological importance of these hosts in the maintenance and transmission of protozoa (Roque and Jansen 2014; Azami-Conesa et al. 2021). These studies are fundamental to controlling the expansion of VL, which unfortunately has emerged and re-emerged in developing countries in the Neotropical region (González et al. 2015; Moreno et al. 2020; Azami-Conesa et al. 2021). The occurrence of leishmaniasis has a significant dependence on wild hosts, so the surveillance of these species is important to detect the circulation of the parasite and thus anticipate cases in humans (Azami-Conesa et al. 2021; López et al. 2021).

The environmental degradation resulting from the expansion of urban centers and agricultural activities has contributed to the increase in the number of cases and outbreaks of VL, both in humans and in animals (Trüeb et al. 2018; Berenguer et al. 2021). Although the dog is considered the main reservoir of *L. infantum* in urban areas, the prevalence of infection in this animal is similar to or lower than that of synanthropic wild animals, such as marsupials, rodents, primates, and bats. This demonstrates the importance of research on the role of wild fauna in the maintenance and transmission of this parasite in different ecoepidemiological contexts (Quaresma et al. 2011; Quintal et al. 2011; Humberg et al. 2012; Mol et al. 2015; Pereira et al. 2017; Porfírio et al. 2018; Brandão et al. 2019).

Carnivores are widely studied because they are considered important reservoirs of *L. infantum* and because of the frequency of clinical cases, both in domestic and wild species (Souza et al. 2010; Roque and Jansen 2014; Souza et al. 2014; Porfírio et al. 2018; Azami-Conesa et al. 2021). Low infection rates have been recorded in wild canids, suggesting that the role of these animals in the parasite cycle may not be as important as that of other mammals (Courtenay et al. 2002; Trüeb et al. 2018; Brandão et al. 2020). However, by exploiting an extensive home range, the degradation of its natural habitat has favored a greater proximity of wild canids with dogs, favoring the sharing of generalist pathogens, such as *L. infantum* (Porfírio et al. 2018; Brandão et al. 2020).

Several factors are associated with the susceptibility of wild animals to infection by *L. infantum*, such as immune and nutritional status, age, reproductive stage, environmental and captivity stress, and infective dose of the parasite (Malta et al. 2010; Bueno et al. 2017; Azami-Conesa et al. 2021). Several species of carnivores, primates, marsupials, and rodents are considered the main hosts of *L. infantum*, some of which show susceptibility to infection in wildlife or under human care, although studies on the pathogenicity of this parasite in wild animals are still scarce (Malta et al. 2010; Souza et al. 2010, 2014; Lima et al. 2012; Reis et al. 2020; Azami-Conesa et al. 2021). Infected animals may not present symptoms, but when they do, the symptoms are usually similar to those presented by humans, dogs, and cats (Malta et al. 2010; Souza et al. 2010; Tenório et al. 2011; Humberg et al. 2012; Bueno et al. 2017; Azami-Conesa et al. 2021; Lima et al. 2021). Knowledge about the impact of infection on wild mammals is important regarding relevant aspects in the field of Conservation Medicine (Oliveira et al. 2019).

Detecting infection by *L. infantum* in wild animals under human care in ex situ conservation institutions in VL-endemic areas is an important measure of epidemiological monitoring, and it is also fundamental for knowledge about the impact of the disease on the health of animals, mainly in threatened species (Luppi et al. 2008; Malta et al. 2010; Souza et al. 2010; Bueno et al. 2017; Oliveira et al. 2019; Reis et al. 2020). In addition to posing a risk to other susceptible wild and synanthropic animals, infected animals in ex situ conservation and biomedical research institutions can also be a source of infection for professionals who deal with these animals and for visitors (Jusi et al. 2011). Therefore, despite the relevance of knowledge about the impact of VL on wildlife, this disease should not be seen only as a "human parasitic disease," and dealing with it in the multiprofessional, transdisciplinary, and multisectoral context of One Health is very important (Martín-Sánchez et al. 2020; Lopes et al. 2021).

There are few studies on the occurrence of leishmaniasis in Neotropical NHPs and on the role that these mammals play in the transmission of *Leishmania* (Oliveira et al. 2019; Azami-Conesa et al. 2021). Some Neotropical NHP species are used in experimental infections by *Leishmania* for studies with immunogens and immunopathology (Oliveira et al. 2019). Furthermore, natural infection by *Leishmania* has been detected both in wild populations and in zoos in endemic areas, the majority of which are asymptomatic animals (Malta et al. 2010; Souza et al. 2010, 2014; Lima et al. 2012; Bueno et al. 2017; Reis et al. 2020; Lopes et al. 2021). Although the innate resistance of neotropical primates to infection by *L. infantum* has already been recorded, situations of environmental stress and captivity can predispose these animals to the disease (Lopes et al. 2021). Although uncommon, death from *L. infantum* infection has already been recorded in *Callicebus nigrifrons*, an endangered species, in a zoo from a VL-endemic area in Brazil (Malta et al. 2010). The presence of mammals (wild, synanthropic, and domestic) and vectors in conservation institutions, as well as in their surroundings, can contribute to the maintenance of the transmission cycle, justifying the adoption of control and prevention measures (Malta et al. 2010; Souza et al. 2010; Minuzzi-Souza et al. 2016; Reis et al. 2020).

Infection by *L. infantum* has been reported in free-ranging canids and felids as well as in individuals kept in zoos and rescue and rehabilitation centers in VL-endemic areas in Brazil (Lima et al. 2009; Dahroug et al. 2010, 2011; Souza et al. 2010, 2014; Brandão et al. 2020; Reis et al. 2020; Lima et al. 2021). Domestic and wild felids are less susceptible to *Leishmania* infection (Azami-Conesa et al. 2021; Berenguer et al. 2021). Perhaps for this reason, wild felids of several species kept in two zoos in Brazil were asymptomatic (Dahroug et al. 2010; Reis et al. 2020). However, *L. infantum* infection associated with skin disease, eye lesions and onychogryphosis was recorded in *L. pardalis* at a wildlife rescue and rehabilitation center in Brazil (Lima et al. 2021).

Unlike felids, canids are more susceptible to infection by *L. infantum*, especially when kept under human care, as demonstrated in zoos and rescue and rehabilitation centers in Brazil, where canids present varied clinical symptoms (Luppi et al. 2008; Souza et al. 2010, 2014; Tenório et al. 2011; Lima et al. 2021). Age-related immunocompromise and captivity stress were considered decisive for the fatal outcome of infection in *Cerdocyon thous* and the endangered species *Chrysocyon brachyurus* and *Speothos venaticus* (Tenório et al. 2011; Souza et al. 2014). The presence of the parasite on the skin of *C. thous* reinforces the epidemiological importance of this canid in the transmission of *L. infantum* (Tenório et al. 2011; Souza et al. 2010, 2014; Brandão et al. 2020), while the participation of *C. brachyurus* and *S. venaticus* is still unknown (Souza et al. 2010; Mol et al. 2015; Brandão et al. 2020). In some Brazilian biomes, *C. thous*, *C. Brachyurus*, and *Lycalopex vetulus* are sympatric, which may favor the transmission of the protozoan (Brandão et al. 2020).

Although they are considered resistant and capable of eliminating the infection, some species of wild animals may be susceptible to infection without showing clinical signs but are capable of transmitting *Leishmania* to competent vectors (Mol et al. 2015; Oliveira et al. 2019). The canids *C. brachyurus* and *S. venaticus,* as well as Neotropical primates of the species *Leontopithecus rosalia* and *Sapajus apella,* which are naturally infected, are capable of transmitting *L. infantum* to the vector *Lutzomyia longipalpis*, although this results in low parasitic loads in sandflies (Mol et al. 2015; Oliveira et al. 2019). *Lutzomyia longipalpis* is the main vector of *L. Infantum* in the Neotropics, but other sandfly species, such as *L. cruzi*, *Migonemyia migonei*, and *Pintomyia fischeri* have been found in VL-endemic areas in which *L. longipalpis* is absent (Galvis-Ovallo et al. 2020). There are no studies on the infection of these potential vectors from naturally infected wild animals.

Animal trafficking, as well as the exchange of animals between conservation institutions, favors the spread of pathogens, even to non-endemic areas, which poses a risk to animal and human health. Therefore, the identification of pathogens is essential in programs for translocation/reintroduction, with special attention given to programs that involve threatened species (Souza et al. 2010, 2014; Bueno et al. 2017; Reis et al. 2020; Lima et al. 2021). The diagnosis and treatment of infected animals, in addition to the use of repellent collars, are recommended for the control of the disease and the prevention of transmission in conservation institutions.

14.4.3 Toxoplasma gondii

Toxoplasma gondii is an obligate intracellular protozoan that undergoes an entero-epithelial cycle in felids, its definitive hosts, and an extraintestinal cycle in practically all cell types of mammals and birds, domestic and wild, terrestrial and aquatic, which act as intermediate hosts (Dubey et al. 2020). The abundance of domestic and feral cats is the main factor involved in environmental contamination since these hosts are considered the main dispersers of *T. gondii* oocysts in terrestrial and aquatic ecosystems (VanWormer et al. 2014, 2016; Barros et al. 2018; Fonseca et al. 2021b). Studies on the ecoepidemiological importance of wild felids are still scarce (Silva et al. 2007; Cañón-Franco et al. 2013; Barros et al. 2018). Additionally, it is important to highlight the biotic potential of this protozoan, represented both by the large number of oocysts excreted in the primary infection of felids and by the physicochemical resistance of these oocysts for prolonged periods in the environment (Dubey et al. 2020).

In contrast to the high seroprevalence, *T. gondii* infection results in asymptomatic infection, and the clinical manifestation of toxoplasmosis is mainly associated with the immunocompromise of the infected hosts (Silva et al. 2007; Santos et al. 2014; Niehaus et al. 2020). However, the disease can manifest in immunocompetent individuals, and the severity of the clinical manifestation is associated with factors of the parasite (inoculum and infective stage, strain), of the host (immune response, ecological and feeding behavior) and of the environment (availability of the parasite in water, food, and soil) (Santos et al. 2017; Niehaus et al. 2020).

Currently, toxoplasmosis is considered an emerging disease in wildlife, constituting a threat to conservation, especially in Neotropical NHPs, as a cause of death in free-ranging animals or those that are part of in situ and ex situ conservation programs (Epiphanio et al. 2000, 2001, 2003; Espinosa-Avilés and Morales-Martínez 2007; Silva et al. 2007; Bernal et al. 2011; Casagrande et al. 2013; Catão-Dias et al. 2013; Santos et al. 2014, 2017; Assunção 2020). The recruitment and population growth rate of these mammals can be impacted by reproductive problems such as abortion and premature births (Catão-Dias et al. 2013; Epiphanio et al. 2003; Santos et al. 2014; Bueno et al. 2017).

The increasing population of domestic cats in urban and peri-urban areas and the presence of feral cats around and inside conservation areas are factors that may be related to the emergence of toxoplasmosis in wildlife (VanWormer et al. 2014, 2016; Bueno et al. 2017; Barros et al. 2018; Fonseca et al. 2021b). In animals used in biomedical research institutions, *T. gondii* infection can negatively influence the results of this research (Andrade et al. 2007). Toxoplasmosis has also had a significant impact on water and food security, being recorded in increasingly larger and more frequent outbreaks, such as the one that occurred in 2018 in humans in the city of Santa Maria, southern Brazil, which is considered the largest outbreak ever described in the world (Minuzzi et al. 2020). Due to its impact on human, animal, and environmental health, toxoplasmosis was one of the first parasitic diseases recognized as a model of prophylaxis from the perspective of One Health (Aguirre et al. 2019; Oliveira et al. 2022).

Systemic toxoplasmosis has been recorded as a cause of mortality in Neotropical NHPs, which have been shown to be more susceptible than Old World primates, possibly related to the former's arboreal behavior, which makes it impossible for animals to contact the infective forms of the parasite. This behavior probably hindered the development of a protective immune response, which is fundamental in the defense mechanism against intracellular protozoa such as *T. gondii* (Innes 1997; Epiphanio et al. 2000, 2001; Espinosa-Avilés and Morales-Martínez 2007; Bernal et al. 2011; Casagrande et al. 2013; Catão-Dias et al. 2013; Assunção 2020). In turn, habitat loss and maintenance under human care favor infection by *T. gondii*, making these primates even more susceptible (Espinosa-Avilés and Morales-Martínez 2007; Assunção 2020; Niehaus et al. 2020; Ehlers et al. 2022).

The few seroprevalence studies of infection in wild populations allow inferences to be drawn about the health risks to Neotropical primates, although the real impact of toxoplasmosis is still poorly understood (Bueno et al. 2017; Assunção 2020; Niehaus et al. 2020; Ehlers et al. 2022; Oliveira et al. 2022). In animals kept under human care, the seroprevalence of *T. gondii* is usually higher than that in free-living populations, perhaps due to the higher and longer contact rate due to the increased life expectancy of animals in captivity (Silva et al. 2007; Niehaus et al. 2020). Differences in the prevalence of anti-*T. gondii* may be related to susceptibility, contact rate or post-exposure time (Niehaus et al. 2020). Furthermore, serology may not be the most suitable tool to assess the presence of *T. gondii* in wild NHP populations, since in these animals, the disease usually has an acute lethal course, and the animals die before developing a humoral response (Niehaus et al. 2020; Paula et al. 2020; Oliveira et al. 2022).

In ex situ conservation institutions and research centers, the likely sources of infection for NHPs are contaminated water and food of animal and plant origin supplied to animals, proximity to wild felids and/or domestic and/or feral cats, ingestion of infected birds or rodents that invade the enclosures, and the dispersion of oocysts through the shoes/boots of professionals involved in handling the animals or by visitors (Epiphanio et al. 2000, 2001; Casagrande et al. 2013; Assunção 2020).

In wild populations of *Alouatta palliata* in Costa Rica, seroprevalence was associated with a higher rate of plant cover, which confers greater survival of infective *T. gondii* oocysts (Niehaus et al. 2020).

There are differences in susceptibility among Neotropical primate species, with high mortality in *Callithrix*, *Saguinus*, and *Leontopithecus* and moderate mortality in *Saimiri*, *Aotus*, *Ateles*, and *Alouatta* (Epiphanio et al. 2000, 2001; Pena et al. 2011; Casagrande et al. 2013; Catão-Dias et al. 2013; Niehaus et al. 2020; Ehlers et al. 2022; Oliveira et al. 2022). Mortality reports in neotropical NHPs in conservation institutions and/or research centers include asymptomatic cases, followed by sudden death, or nonspecific clinical signs (Epiphanio et al. 2000, 2001; Andrade et al. 2007; Bernal et al. 2011; Pena et al. 2011; Casagrande et al. 2013; Santos et al. 2014; Assunção 2020). Systemic toxoplasmosis varies from hyperacute to acute, rapidly evolving to death, and can affect animals individually or in outbreaks (Oliveira et al. 2022).

The few reports of toxoplasmosis as a cause of death in free-living primates refer to the species *Brachyteles arachnoides* (Santos et al. 2017), *Callithrix penicillata* (Assunção 2020), *Callithrix* spp. (Oliveira et al. 2022), and *Alouatta guariba* (Ehlers et al. 2022). All these reports are from animals in Brazil, a country where 25.18% of NHPs are on the Brazilian Red List as vulnerable, endangered, or critically endangered (Ehlers et al. 2022). In this respect, information on the causes of mortality in wild populations provides important supporting information for the implementation of surveillance measures for animal, human, and environmental health, which will also have an impact on the conservation of species maintained in ex situ conservation programs (Santos et al. 2017; Assunção 2020; Dubey et al. 2020; Ehlers et al. 2022; Oliveira et al. 2022).

14.4.4 Prosthenorchis elegans

Parasitism by gastrointestinal helminths is common in NHPs (Tavela et al. 2013; Martin-Solano et al. 2017; Ehlers et al. 2022). However, some highly pathogenic helminths, such as the acanthocephalan *Prosthenorchis elegans*, can be a threat to the health of insectivorous and omnivorous primates in the wild or under human care (Pérez García et al. 2007; Pissinatti et al. 2007; Wenz et al. 2010; Botero et al. 2011; Tavela et al. 2013; Catenacci et al. 2016; Martin-Solano et al. 2017; Oliveira et al. 2017). Animals kept under human care are more susceptible to *P. elegans* infection due to several factors related to the stress of captivity (Pissinatti et al. 2007; Botero et al. 2011; Tavela et al. 2013).

Prosthenorchis elegans is a acanthocephalan whose intermediate hosts are coprophagous beetles (*Lasioderma serricorne* and *Stegobium paniceum*) and cockroaches (*Blatella germanica*, *Blaberus fusca*, *Rhyparobia maderae*), insects that make up the diet of some primate species (Pérez García et al. 2007; Wenz et al. 2010). Anthropic landscapes and captive management can both favor the maintenance and intraspecific and interspecific transmission of this acanthocephalan

(Pissinatti et al. 2007; Wenz et al. 2010; Botero et al. 2011; Tavela et al. 2013; Martin-Solano et al. 2017; Ehlers et al. 2022). In natural environments, beetles are the main source of infection, although some cockroach species are present in the treetops of tropical forests (Wenz et al. 2010). Similar to coleoptera, cockroaches also have coprophagic behavior, which favors the participation of these insects in the transmission of the parasite.

Anthropization can change the population density of intermediate hosts, since in urbanized and peri-urbanized areas, vectors, especially cockroaches, are present in high densities, increasing the risk of infection for animals (Pissinatti et al. 2007; Wenz et al. 2010; Tavela et al. 2013; Catenacci et al. 2016). The availability of organic waste in and around conservation institutions provides a greater abundance of cockroaches, making these insects the main source of infection for animals kept under human care (Pissinatti et al. 2007; Botero et al. 2011; Martin-Solano et al. 2017).

Infection by *P. elegans* is among the main causes of morbidity and mortality in Neotropical NHPs under human care, which represents a threat to the conservation of endangered species that are part of translocation/reintroduction programs (Tavela et al. 2013). Ulcerative enteritis, which can result in intestinal perforation and peritonitis, is associated with attachment of the parasite's spinous proboscis to the mucosa of the ileum, cecum, and colon of hosts (Pérez García et al. 2007; Pissinatti et al. 2007; Alfaro et al. 2008; Sales et al. 2010; Wenz et al. 2010; Oliveira et al. 2017). The most frequent site of parasitism is the ileocecal valve, where high parasite loads can result in obstruction and intussusception (Pérez García et al. 2007; Alfaro et al. 2008; Catenacci et al. 2016). Sudden death in asymptomatic animals is not uncommon (Pérez García et al. 2007; Alfaro et al. 2008; Sales et al. 2010; Catenacci et al. 2016; Oliveira et al. 2017; Tavernard 2017).

In addition to the health risks, the ineffectiveness of antiparasitic drugs in controlling this acanthocephalan is highlighted, leaving the surgical removal of adult parasites as an alternative, a procedure that can result in death in the postoperative period (Pérez García et al. 2007; Tavernard 2017). Some antiparasitic drugs only cause a reduction in egg production, which to some extent reduces environmental contamination and, consequently, vector infection (Pérez García et al. 2007; Pissinatti et al. 2007; Martin-Solano et al. 2017). It is important to highlight the physical-chemical resistance of *P. elegans* eggs, which makes environmental control difficult, especially in ex situ conservation institutions.

In the Neotropical region, records of *P. elegans* and *Prosthenorchis* sp. in NHPs in the wild and under human care are abundant and concern both the parasites recovered at necropsies and the detection of eggs in parasitological examinations of stool (Parra et al. 2005; Tantaleán et al. 2005; Pérez García et al. 2007; Pissinatti et al. 2007; Alfaro et al. 2008; Sales et al. 2010; Wenz et al. 2010; Botero et al. 2011; Parr et al. 2013; Tavela et al. 2013; Catenacci et al. 2016; Corrêa et al. 2016; Martin-Solano et al. 2017; Oliveira et al. 2017; Tavernard 2017). Some of these reports refer to infection in endangered species, such as *Leontopithecus chrysomelas*, *L. chrysopygus*, *L. rosalia*, *Callithrix kuhlii*, *Saguinus niger*, *S. leucopus*, *Aotus nancymaae*, *Ateles paniscus*, *Cebuella pygmaea*, *Lagothrix lagotricha*, and *Saimiri oerstedii*

(Parra et al. 2005; Tantaleán et al. 2005; Pérez García et al. 2007; Pissinatti et al. 2007; Alfaro et al. 2008; Sales et al. 2010; Catenacci et al. 2016; Corrêa et al. 2016).

The risk of cross-transmission of generalist and pathogenic parasites, such as *P. elegans,* between species that have overlapping niches (share the same resources and shelters) is a cause for concern. This situation is worrying especially when it involves invasive species and threatened species, which may present differences in susceptibility to this acanthocephalan. (Sales et al. 2010; Parr et al. 2013; Tavela et al. 2013; Catenacci et al. 2016; Oliveira et al. 2017). Another risk situation occurs with animals under human care in institutions where sanitary management is deficient, favoring both the introduction of parasitized animals and the maintenance of transmission from the presence of intermediate hosts (Pérez García et al. 2007; Pissinatti et al. 2007; Botero et al. 2011). In this case, the diagnosis of *P. elegans* infection should be part of the clinical routine to avoid risks to the health of animals kept in these institutions, as well as to prevent the spread of this and other parasites to native populations in cases of animal release in translocation/reintroduction programs (Pissinatti et al. 2007; Sales et al. 2010; Parr et al. 2013; Tavela et al. 2013; Oliveira et al. 2017). Studies on the susceptibility of free-living populations are scarce, a gap that needs to be filled within the framework of Conservation Medicine.

14.5 Conclusions

Future studies of disease ecology in Neotropical vertebrates would need to include the characterization of the host community, life stage, and its species-specific effects, in addition to the assessment of environmental factors that potentially contribute to the emergence of the pathogen. Despite the increase in the number of studies in recent years, there are still significant gaps regarding the ecological and evolutionary relationships of parasites and the impact of parasitism on the health and conservation of Neotropical vertebrates, especially in the wild. Furthermore, as a threat to biodiversity conservation, biological invasions involving parasites should also be further explored in future studies. From the point of view of the One Health approach, animal, human, plant, and environmental health are connected and interdependent. Therefore, from this perspective, understanding the parasite-host-environment relationship requires a transdisciplinary approach and the cooperation of different professionals and sectors.

References

Aguirre AA (2017) Changing patterns of emerging zoonotic diseases in wildlife, domestic animals, and humans linked to biodiversity loss and globalization. ILAR J 58:315–318

Aguirre AA, Longcore T, Barbieri M et al (2019) The One Health approach to Toxoplasmosis: epidemiology, control, and prevention strategies. EcoHealth 16:378–390

Alfaro A, Morales J, Fallas S (2008) Ileitis y colitis piogranulomatosa en un mono ardilla (*Saimiri oerstedii*) asociada con *Prosthenorchis elegans*. Cienc Vet 26:81–86

Amin A, Bilic I, Liebhart D, Hess M (2014) Trichomonads in birds – a review. Parasitology 141:733–747

Andery DA, Ferreira Junior FC, Araújo AV et al (2013) Health assessment of raptors in triage in Belo Horizonte, MG, Brazil. Braz J Poult Sci 15:247–256

Andrade MCR, Coelho JMCO, Amendoeira MRR et al (2007) Toxoplasmosis in squirrel monkeys: histological and immunohistochemical analysis. Ciênc Rural 37:1724–1727

Aronson MFJ, La Sorte FA, Nilon CH et al (2014) A global analysis of the impacts of urbanization on bird and plant diversity reveals key anthropogenic drivers. Proc R Soc B 281:1–8

Assunção CF (2020) Toxoplasmose em primatas não humanos de vida livre e cativeiro no Distrito Federal, Goiás, Mato Grosso e Tocantins. Dissertação de Mestrado, Universidade de Brasília

Austen JM, Barbosa AD (2021) Diversity and epidemiology of bat Trypanosomes: a One Health perspective. Pathogens 10:1148. https://doi.org/10.3390/pathogens10091148

Azami-Conesa I, Gómez-Muñoz MT, Martínez-Díaz RA (2021) A systematic review (1990–2021) of wild animals infected with zoonotic *Leishmania*. Microorganisms 9:1101

Ávila RW, Silva RJ (2010) Checklist of helminths from lizards and amphisbaenians (Reptilia, Squamata) of South America. J Venom Anim Toxins Inclu Trop Dis 16:543–572. https://doi.org/10.1590/S1678-91992010000400005

Bahia M, Barros FNL, Magalhães-Matos PC et al (2017) *Trypanosoma cruzi* infection in captive Neotropical primates in the Brazilian Amazon. Am J Primatol 79:1–6

Barros M, Cabezo O, Dubey JP et al (2018) *Toxoplasma gondii* infection in wild mustelids and cats across an urban rural gradient. PLoS One 13:e0199085. https://doi.org/10.1371/journal.pone.0199085

Berenguer LKAR, Gomes CFCA, Nascimento JO et al (2021) *Leishmania infantum* infection in a domestic cat: a real threat or an occasional finding? Acta Parasitol 66:673–676

Bergner LM, Becker DJ, Tello C, Carrera JE, Streicker DG (2021) Detection of *Trypanosoma cruzi* in the saliva of diverse neotropical bats. Zoonoses Public Health 68:271–276

Bernal AM, Ramos IM, Garnica BS, Lezama JR (2011) Toxoplasmosis en una colonia de monos ardilla (*Saimiri sciureus*) en cautiverio en Cuernavaca, Morelos, México. Vet Méx 4:115–123

Bienentreu JF, Lesbarrères D (2020) Amphibian disease ecology: are we just scratching the surface? Herpetologica 76:153–166

Botero LC, Fernández A, Forero N, Rosas S, Soler-Tovar D (2011) Análisis retrospectivo de las enfermedades parasitarias del mono ardilla (*Saimiri sciureus*) en dos condiciones *ex situ* en el noroccidente de los Andes suramericanos. Rev Med Vet 22:85–93

Brandão EMV, Xavier SCC, Carvalhaes JG et al (2019) Trypanosomatids in small mammals of an agroecosystem in central Brazil: another piece in the puzzle of parasite transmission in an anthropogenic landscape. PathoGenetics 8:190. https://doi.org/10.3390/pathogens8040190

Brandão EMV, Xavier SCC, Rocha FL et al (2020) Wild and domestic canids and their interactions in the transmission cycles of *Trypanosoma cruzi* and *Leishmania* spp. in an area of the Brazilian Cerrado. Pathogens 9:818. https://doi.org/10.3390/pathogens9100818

Bruni MP, Echenique JVZ, Santos CC et al (2019) The raptor Chimango Caracara (*Milvago chimango*) (Aves: Falconiformes) – a new host for *Trichomonas gallinae* (protozoa: Trichomonadidae). Int J Parasitol Parasites Wildl 10:310–313

Bueno MG, Catão-Dias JL, Laroque PO et al (2017) Infectious diseases in free-ranging blonde capuchins, *Sapajus flavius*, in Brazil. Int J Primatol 38:1017–1031. https://doi.org/10.1007/s10764-017-9994-5

Bunbury N, Jones CG, Greewood AG, Bell DJ (2007) *Trichomonas gallinae* in Mauritian columbids: implications for an endangered endemic. J Wildl Dis 43:399–407

Bustamante DM, Urioste-Stone SM, Juárez JG, Pennington PM (2014) Ecological, social, and biological risk factors for continued *Trypanosoma cruzi* transmission by *Triatoma dimidiata* in Guatemala. PLoS One 9:e104599. https://doi.org/10.1371/journal.pone.0104599

Campião KM, Dáttilo W (2020) Biological drivers of individual-based anuran-parasite networks under contrasting environmental conditions. J Helminthol 94:1–7. https://doi.org/10.1017/S0022149X20000504

Campião KM, Silva RJ, Ferreira VL (2014a) Helminth parasite communities of allopatric populations of the frog Leptodactylus podicipinus from Pantanal, Brazil. J Helminthol 88:13–19. https://doi.org/10.1017/S0022149X12000557

Campião KM, Morais DH, Dias OT et al (2014b) Checklist of helminth parasites of Amphibians from South America. Zootaxa 3843:1–93. https://doi.org/10.11646/zootaxa.3843.1.1

Campião KM, Ribas A, Tavares LER (2015) Diversity and patterns of interaction of an anuran–parasite network in a neotropical wetland. Parasitology. https://doi.org/10.1017/S0031182015001262

Cañón-Franco WA, Araújo FAP, Gennari SM (2013) *Toxoplasma gondii* in small neotropical wild felids. Braz J Vet Res Anim Sci 50:50–67

Cardoso RM, Araújo NNSL, Romero GAS et al (2015) Expanding the knowledge about *Leishmania* species in wild mammals and dogs in Brazilian savannah. Parasit Vectors 8:171. https://doi.org/10.1186/s13071-015-0780-y

Carnevali V, Nogueda-Torres B, Villagrán-Herrera ME et al (2017) Prevalence of *Trypanosoma cruzi* and organ alterations in Virginia opossums (*Didelphis virginiana*) from Western Mexico – short communication. Acta Vet Hung 65:505–509

Casagrande RA, Silva TCE, Pescador CA et al (2013) Toxoplasmose em primatas neotropicais: Estudo retrospectivo de sete casos. Pesqui Vet Bras 33:94–98

Castillo CP (2019) *Columba livia* Gmelin, 1789 (Columbiformes: Columbidae): agent of change and threat to avian biodiversity in Huánuco district, Peru? An analysis based on the precautionary principle. Neotrop Helminthol 13:287–294

Catão-Dias JL, Epiphanio S, Kierulff MCM (2013) Neotropical primates and their susceptibility to *Toxoplasma gondii*: new insights for an old problem. In: Brinkworth JF, Pechenkina K (eds) Primates, pathogens, and evolution, developments in primatology: progress and prospects. Springer, New York, pp 253–289

Catenacci LS, Colosio AC, Oliveira LC et al (2016) Occurrence of *Prosthenorchis elegans* in free-living primates from the Atlantic forest of Southern Bahia, Brazil. J Wildl Dis 52:364–368

Chavatte JM, Giraud P, Esperet D, Place G et al (2019) An outbreak of trichomonosis in European greenfinches *Chloris chloris* and European goldfinches *Carduelis carduelis* wintering in Northern France. Parasite 26:1–12

Chero J, Cruces C, Iannacone J et al (2015) Parasitofauna of the Neotropical amphibian *Rhinella limensis* Werner, 1901 (Anura: Bufonidae) from Peruvian Central Coast. Neotrop Helminthol 9:87–102

Cordovez JM, Guhl F (2015) The impact of landscape transformation on the reinfestation rates of *Rhodnius prolixus* in the Orinoco Region, Colombia. Acta Trop 151:73–79

Corrêa P, Bueno C, Soares R, Vieira FM, Muniz-Pereira LC (2016) Checklist of helminth parasites of wild primates from Brazil. Rev Mex Biodivers 87:908–918

Costa AP, Ferreira JIGS, Silva RE et al (2018) *Trypanosoma cruzi* in Triatomines and wild mammals in the National Park of Serra das Confusões, Northeastern Brazil. Rev Soc Bras Med Trop 51:445–451

Costales JA, Jara-Palacios MA, Llewellyn MS et al (2015) *Trypanosoma cruzi* population dynamics in the Central Ecuadorian Coast. Acta Trop 151:88–93

Courtenay O, Quinnell RJ, Garcez LM, Dye C (2002) Low infectiousness of a wildlife host of *Leishmania Infantum*: the crab-eating fox is not important for transmission. Parasitology 125:407–414

Cunningham AA (2005) A walk on the wild side – emerging wildlife diseases. They increasingly threaten human and animal health. BMJ 331:1214–1215

Cunningham AA, Daszak P, Rodríguez JP (2003) Pathogen pollution: defining a parasitological threat to biodiversity conservation. J Parasitol 89(Suppl):S78–S83

Dahroug MAA, Almeida ABPF, Sousa VRF et al (2010) Leishmania (Leishmania) chagasi in captive wild felids in Brazil. Trans R Soc Trop Med Hyg 104:73–74
Dahroug MAA, Almeida ABPF, Sousa VRF et al (2011) The first case report of *Leishmania (Leishmania) chagasi* in *Panthera leo* in Brazil. Asian Pac J Trop Biomed 1:249–250
Daszak P, Cunningham AA, Hyatt AD (2000) Emerging infectious diseases of wildlife – threats to biodiversity and human health. Science 287:443–449
Daszak P, Cunningham AA, Hyatt AD (2001) Anthropogenic environmental change and the emergence of infectious diseases in wildlife. Acta Trop 78:103–116
Dronzino RN, Otomura FH, Gazarini J, Gomes ML, Toledo MJO (2019) *Trypanosoma* found in synanthropic mammals from urban forests of Paraná, Southern Brazil. Vector Borne Zoonotic Dis. https://doi.org/10.1089/vbz.2018.2433
Dubey JP, Murata FHA, Cerqueira-Cézar CK, Kwok OCH, Grigg ME (2020) Recent epidemiologic and clinical importance of *Toxoplasma gondii* infections in marine mammals: 2009–2020. Vet Parasitol 288:109296
Dunn AM, Hatcher MJ (2015) Parasites and biological invasions: parallels, interactions, and control. Trends Parasitol 31:189–199
Ecco R, Preis IS, Vilela DAR et al (2012) Molecular confirmation of *Trichomonas gallinae* and other parabasalids from Brazil using the 5.8S and ITS-1 rRNA regions. Vet Parasitol 190:36–42
Echenique JVZ, Soares MP, Bruni M et al (2019) Oral trichomoniasis in raptors in Southern Brazil. Pesqui Vet Bras 39:983–988
Ehlers LP et al (2022) Causes of death in neotropical primates in Rio Grande do Sul state, Southern Brazil. J Med Primatol 51:85–92
Epiphanio S, Guimarães MABV, Fedullo DL, Correa SHR, Catão-Dias JL (2000) Toxoplasmosis in golden-headed lion tamarins (*Leontopithecus chrysomelas*) and emperor marmosets (*Saguinus imperator*) in captivity. J Zoo Wildl Med 31:231–235
Epiphanio S, Sá LRM, Teixeira RHF, Catão-Dias JL (2001) Toxoplasmosis in a wild caught black lion tamarin (*Leontopithecus chrysopygus*). Vet Rec 17:627–628
Epiphanio S, Sinhorini IL, Catão-Dias JL (2003) Pathology of toxoplasmosis in captive New World primates. J Comp Pathol 129:196–204
Espinosa-Avilés D, Morales-Martínez MS (2007) Brote de toxoplasmosis en una colonia de monos ardilla (*Saimiri sciureus*) en cautiverio. Vet Méx 38:365–369
Felix-Nascimento G (2021) Parasitos e ecotoxicologia de *Leptodactylus macrosternum* Miranda-Ribeiro, 1926 (Anura: Leptodactylidae) em áreas com diferenças de uso do solo no domínio morfoclimático das Caatingas: aspectos taxonômicos, ecológicos e histopatológicos. Tese de Doutorado, Universidade Federal Rural de Pernambuco
Felix-Nascimento G, Vieira FM, Gomes ECA et al (2022) Morphological description of infective larval stage of *Physaloptera* (Spirurida: Physalopteridae), and histological lesions in the paratenic host *Leptodactylus macrosternum* (Anura: Leptodactylidae) in Caatinga biome, Brazil. Rev Mex Biodivers 93:e933772. https://doi.org/10.22201/ib.20078706e.2022.93.3772
Fonseca CF, Rojas NJJT, Vieira FM et al (2021a) Helminth fauna and histopathology associated with parasitic infections in *Phrynops geoffroanus* (Schweigger, 1812) (Testudines, Chelidae) in a Brazilian river subjected to anthropogenic activities. Vet Parasitol Reg Stud Rep 26:100639. https://doi.org/10.1016/j.vprsr.2021.100639
Fonseca FS, Mangini PR, Mello TJ et al (2021b) Feral cat population rises on Fernando de Noronha Archipelago: wildlife needs different cat control approaches and needs it now. Biodivers Bras 11:1–9
Fonseca CF, Silva IJS, Rodrigues MS, Silva BHS, Soares PC, Moura GJB, Oliveira JB (2023) Evaluation of metal exposure through the composition of essential and toxic micro-minerals in freshwater turtles (*Phrynops geoffroanus*) from a Brazilian river. Environ Sci Pollut Res. https://doi.org/10.1007/s11356-023-26127-w
Forrester DJ, Foster GW (2008) Trichomonosis. In: Atkinson CT, Thomas J, Hunter DB (eds) Parasitic diseases of wild birds. Wiley, New Jersey, pp 120–153

Forzán MJ, Vanderstichel R, Melekhovets YF, Mcburney S (2010) Trichomoniasis in finches from the Canadian Maritime provinces – an emerging disease. Can Vet J 51:391–396

Freire SM, Leal ARS, Knoff M et al (2017) *Chapiniella variabilis* (Nematoda) parasitizing *Chelonoidis carbonarius* and *C. denticulatus* (Testudinidae) in the state of Piauí. Rev Bras Parasitol Vet 26:359–365

Freire SM, Leal ARDS, Batista JF et al (2019) Evaluación histopatológica del intestino de jabutis (*Chelonoidis carbonarius* y *Chelonoidis denticulatus*) criados en cautiverio y parasitados por helmintos (*Chapiniella varibilis* y *Atractis thapari*). Rev Investig Vet del Perú 30:558–568. https://doi.org/10.15381/rivep.v30i2.14715

Freitas SSR, Martinello AF, Braz CHS et al (2020) Uso de nitazoxanida no tratamento de tricomoníase em aves: relatos de casos. Arch Vet Sci 25:79

Galvis-Ovallo F, Ueta AE, Marques GO et al (2020) Detection of *Pintomyia fischeri* (Diptera: Psychodidae) with *Leishmania infantum* (Trypanosomatida: Trypanosomatidae) promastigotes in a focus of visceral leishmaniasis in Brazil. J Med Entomol. https://doi.org/10.1093/jme/tjaa199

García MN, O'Day S, Fisher-Hoch S et al (2016) One Health interaction of Chagas Disease vectors, canid hosts, and human residents along the Texas–Mexico border. PLoS Negl Trop Dis 10:e0005074. https://doi.org/10.1371/journal.pntd.0005074

Goes VC, Brito ES, Valadao RM et al (2018) Haemogregarine (Apicomplexa: Adeleorina) infection in Vanderhaege's toad-headed turtle, *Mesoclemmys vanderhaegei* (Chelidae), from a Brazilian Neotropical savanna region. Folia Parasitol (Praha) 65:1–6. https://doi.org/10.14411/fp.2018.012

González K, Calzada JE, Saldaña A et al (2015) Survey of wild mammal hosts of cutaneous leishmaniasis parasites in Panamá and Costa Rica. Trop Med Health 43:75–78

Gürtler R, Cardenal MV (2015) Reservoir host competence and the role of domestic and commensal hosts in the transmission of *Trypanosoma cruzi*. Acta Trop 151:32–50

Hamann MI, Kehr AI, González CE (2006) Species affinity and infracommunity ordination of helminths of *Leptodactylus chaquensis* (Anura: Leptodactylidae) in two contrasting environments from northeastern Argentina. J Parasitol 92:1171–1179

Hamann MI, Kehr AI, González CE (2012) Biodiversity of trematodes associated with amphibians from a variety of habitats in Corrientes Province, Argentina. J Helminthol 87:286–300

Herczeg D, Ujszegi J, Kásler A et al (2021) Host–multiparasite interactions in amphibians: a review. Parasit Vectors 14:1–20. https://doi.org/10.1186/s13071-021-04796-1

Hodo CL, Wilkerson GK, Birkner EC, Gray SB, Hamer SA (2018) *Trypanosoma cruzi* transmission among captive nonhuman primates, wildlife, and vectors. EcoHealth 15:426–436

Hodo CL, Bañuelos RM, Edwards EE, Wozniak EJ, Hamer SA (2020) Pathology and discrete typing unit associations of *Trypanosoma cruzi* infection in coyotes (*Canis latrans*) and raccoons (*Procyon lotor*) of Texas, USA. J Wildl Dis 56:134–144

Holmes JC (1996) Parasites as threats to biodiversity in shrinking ecosystems. Biodivers Conserv 5:975–983

Hudson PJ, Dobson AP, Lafferty KD (2006) Is a healthy ecosystem one that is rich in parasites? Trends Ecol Evol 21:381–385

Humberg RM, Oshiro ET, Cruz MDSP et al (2012) *Leishmania chagasi* em gambás (*Didelphis albiventris*) em área urbana endêmica para leishmaniose visceral, Campo Grande, Mato Grosso do Sul, Brasil. Am J Trop Med Hyg 87:470–472

Ihle-Soto C, Costoya E, Correa JP et al (2019) Spatiotemporal characterization of *Trypanosoma cruzi* infection and discrete typing units infecting hosts and vectors from nondomestic foci of Chile. PLoS Negl Trop Dis 13:e0007170. https://doi.org/10.1371/journal.pntd.0007170

Innes EA (1997) Toxoplasmosis: comparative species susceptibility and host immune response. Comp Immunol Microbiol Infect Dis 20:131–138

Jansen AM, Xavier SCC, Roque ALR (2015) The multiple and complex and changeable scenarios of the *Trypanosoma cruzi* transmission cycle in the sylvatic environment. Acta Trop 151:1–15. https://doi.org/10.1016/j.actatropica.2015.07.018

Jansen AM, Xavier SCC, Roque ALR (2017) Ecological aspects of *Trypanosoma cruzi*: wild hosts and reservoirs. In: American trypanosomiasis Chagas disease. https://doi.org/10.1016/B978-0-12-801029-7.00011

Jansen AM, Xavier SCC, Roque ALR (2018) *Trypanosoma cruzi* transmission in the wild and its most important reservoir hosts in Brazil. Parasit Vectors 11:502. https://doi.org/10.1186/s13071-018-3067-2

Jansen AM, Xavier SCC, Roque ALR (2020) Landmarks of the knowledge and *Trypanosoma cruzi* biology in the wild environment. Front Cell Infect Microbiol 10:10. https://doi.org/10.3389/fcimb.2020.00010

Joppert AM (2007) Estudo retrospectivo das causas de morte de Falconiformes e Strigiformes de vida livre no município de São Paulo. Tese de Doutorado, Universidade de São Paulo

Julca R, Casas E, Chavera A et al (2014) Descripción anatomopatológica de lesiones por helmintos gastrointestinales en tortugas motelo (*Chelonoidis denticulata*). Rev Investig Vet del Perú 25:37–50

Jusi MMG, Starke-Buzetti, WA, Oliveira, TMFS, Tenório, MS, Sousa, LO, Machado, RZ (2011) Detecção sorológica e molecular de Leishmania spp. em animais selvagens do zoológico de Ilha Solteira, SP, Brasil. Revista Brasileira de Parasitologia Veterinária 20:219–222.

King KC, McLaughlin JD, Gendron AD et al (2007) Impacts of agriculture on the parasite communities of northern leopard frogs (*Rana pipiens*) in southern Quebec, Canada. Parasitology 134:2063–2080. https://doi.org/10.1017/S0031182007003277

Lazo JGP, Mayor P, Lescano AG (2019) Amazonian reservoir hosts of *Trypanosoma cruzi*. https://doi.org/10.5772/intechopen.86158

Leung TLF, Koprivnikar J (2019) Your infections are what you eat: how host ecology shapes the helminth parasite communities of lizards. J Anim Ecol 88:416–426. https://doi.org/10.1111/1365-2656.12934

Lima VM, Fattori KR, Michelin AF, Nogueira FS, Souza LO (2009) Evidence of *Leishmania* spp. antibodies and DNA in bush dogs (*Speothos venaticus*) in Brazil. J Zoo Wildl Med 40:91–94

Lima VMF, Santiago MEB, Sanches LC, Lima BD (2012) Molecular diagnosis of *Leishmania amazonensis* in a captive spider monkey in Bauru, São Paulo, Brazil. J Zoo Wildl Med 43:943–945

Lima L, Espinosa-Álvarez O, Ortiz PA et al (2015) Genetic diversity of *Trypanosoma cruzi* in bats, and multilocus phylogenetic and phylogeographical analyses supporting Tcbat as an independent DTU (discrete typing unit). Acta Trop 151:166–177

Lima VFS, Ramos RAN, Giannelli A et al (2021) Zoonotic parasites in wild animals such as carnivores and primates that are traded illegally in Brazil. Braz J Vet Med 43:e113720. https://doi.org/10.29374/2527-2179.bjvm113720

Lisboa CV, Monteiro RV, Martins AF, Xavier SC, Lima VS, Jansen AM (2015) Infection with Trypanosoma cruzi TcII and TcI in free-ranging populations of lion tamarins (*Leontopithecus* spp.): an 11-year follow-up. Mem Inst Oswaldo Cruz 110:394–402

Lopes KF, Delai RM, Zaniolo MM et al (2021) Urban capuchin monkeys *Sapajus nigritus* (Goldfuss, 1809) (Primates, Cebidae) as environmental bioindicators of Leishmaniasis. Transbound Emerg Dis:1–6. https://doi.org/10.1111/tbed.14247

López M, Erazo D, Hoyos J et al (2021) Measuring spatial co-occurrences of species potentially involved in *Leishmania* transmission cycles through a predictive and fieldwork approach. Sci Rep 11:6789. https://doi.org/10.1038/s41598-021-85763-9

López-Cancino SA, Tun-Ku E, Cruz-Felix HK et al (2015) Landscape ecology of *Trypanosoma cruzi* in the southern Yucatan Peninsula. Acta Trop 151:58–72

Luppi MM, Malta MCC, Silva TMA et al (2008) Visceral leishmaniasis in captive wild canids in Brazil. Vet Parasitol 155:146–151

Madelaire CB, Silva RJ, Gomes FR (2013) Calling behavior and parasite intensity in treefrogs, *Hypsiboas prasinus*. J Herpetol 47:450–455

Malta MCC, Tinoco HP, Xavier MN, Vieira ALS, Costa EA, Santos RL (2010) Naturally acquired visceral leishmaniasis in nonhuman primates in Brazil. Vet Parasitol 169:193–197

Marcogliese DJ (2004) Parasites: small players with crucial roles in the ecological theater. EcoHealth 1:151–164

Martín-Sánchez J, Rodríguez-Granger J, Morillas-Márquez F et al (2020) Leishmaniasis due to *Leishmania infantum*: integration of human, animal and environmental data through a One Health approach. Transbound Emerg Dis 00:1–12. https://doi.org/10.1111/tbed.13580

Martin-Solano S, Carrillo-Bilbao GA, Ramírez W et al (2017) Gastrointestinal parasites in captive and free-ranging *Cebus albifrons* in the Western Amazon, Ecuador. Int J Parasitol Parasites Wildl 6:209–218

Martins-Sobrinho PM, Silva WGO, Santos EG et al (2017) Helminths of some tree frogs of the families Hylidae and Phyllomedusidae in an Atlantic rainforest fragment, Brazil. J Nat Hist 51:1639–1648. https://doi.org/10.1080/00222933.2017.1337945

McKenzie VJ (2007) Human land use and patterns of parasitism in tropical amphibian hosts. Biol Conserv 137:102–116

Mehrkens LR, Shender L, Ayabsley MJ et al (2013) White-nosed coatis (*Nasua narica*) are a potential reservoir of *Trypanosoma cruzi* and other potentially zoonotic pathogens in Monteverde, Costa Rica. J Wildl Dis 49:1014–1018

Mello EM (2013) Endo e ectoparasitos de serpentes *Crotalus durissus* Linnaeus, 1758 (Viperidae) de algumas localidades de Minas Gerais. Dissertação de Mestrado, Universidade Federal de Minas Gerais

Mendoza-Roldan JA, Modry D, Otranto D (2020) Zoonotic parasites of reptiles: a crawling threat. Trends Parasitol 36:677–687

Merling de Chapa M, Auls S, Kenntner N, Krone O (2021) To get sick or not to get sick – *Trichomonas* infections in two *Accipiter* species from Germany. Parasitol Res 120:3555–3567

Minuzzi CE, Portella LP, Bräunig P et al (2020) Isolation and molecular characterization of *Toxoplasma gondii* from placental tissues of pregnant women who received toxoplasmosis treatment during an outbreak in southern Brazil. PLoS One 15:e0228442. https://doi.org/10.1371/journal.pone.0228442

Minuzzi-Souza TTC, Nitz N, Knox MB et al (2016) Vector-borne transmission of *Trypanosoma cruzi* among captive Neotropical primates in a Brazilian zoo. Parasit Vectors 9:39. https://doi.org/10.1186/s13071-016-1334-7

Mol JPS, Soave SA, Turchetti AP et al (2015) Transmissibility of *Leishmania infantum* from maned wolves (*Chrysocyon brachyurus*) and bush dogs (*Speothos venaticus*) to *Lutzomyia longipalpis*. Vet Parasitol 212:86–91

Monteiro RV, Baldez J, Dietz J, Baker A, Lisboa CV, Jansen AM (2006) Clinical, biochemical, and electrocardiographic aspects of *Trypanosoma cruzi* infection in free-ranging golden lion tamarins (*Leontopithecus rosalia*). J Med Primatol 35:48–55

Monteiro RV, Dietz JM, Jansen AM (2010) The impact of concomitant infections by *Trypanosoma cruzi* and intestinal helminths on the health of wild golden and golden-headed lion tamarins. Res Vet Sci 89:27–35

Moreno ES, Sabioni LA, Seixas MMMD, Souza Filho JAD, Marcelino AP, Shimabukuro PHF (2020) Evidence of a sylvatic enzootic cycle of *Leishmania infantum* in the State of Amapá, Brazil. Rev Soc Bras Med Trop 53:e20190169. https://doi.org/10.1590/0037-8682-0169-2019

Moretti EH, Madelaire CB, Silva RJ et al (2014) The relationships between parasite intensity, locomotor performance, and body condition in adult toads (*Rhinella icterica*) from the wild. J Herpetol 48:277–283

Moretti EH, Titon B, Madelaire CB et al (2017) Behavioral, physiological and morphological correlates of parasite intensity in the wild Cururu toad (*Rhinella icterica*). Int J Parasitol Parasites Wildl 6:146–154

Muñoz-García CL, Sánchez-Montes S, Villanueva-García C et al (2019) The role of sloths and anteaters as *Leishmania* spp. reservoirs: a review and a newly described natural infection of *Leishmania mexicana* in the northern anteater. Parasitol Res 118:1095–1101

Nantes WAG, Barreto WTG, Santos FM et al (2019) The influence of parasitism by *Trypanosoma cruzi* in the hematological parameters of the white ear opossum (*Didelphis albiventris*) from Campo Grande, Mato Grosso do Sul, Brazil. Int J Parasitol Parasites Wildl 9:16–20

Nantes WAG, Santos FM, Macedo GC (2020) Trypanosomatid species in *Didelphis albiventris* from urban forest fragments. Parasitol Res. https://doi.org/10.1007/s00436-020-06921-y

Niehaus C, Spínola M, Su C et al (2020) Environmental factors associated With *Toxoplasma gondii* Exposure in Neotropical Primates of Costa Rica. Front Vet Sci 7:583032. https://doi.org/10.3389/fvets.2020.583032

Noya BA, González ON (2015) An ecological overview on the factors that drives to *Trypanosoma cruzi* oral transmission. Acta Trop 151:94–102

Ocanã-Mayorga S, Bustillos JJ, Villacís AG et al (2021) Triatomine feeding profiles and *Trypanosoma cruzi* infection, implications in domestic and sylvatic transmission cycles in Ecuador. Pathogens 10:42. https://doi.org/10.3390/pathogens10010042

Oda FH, Borteiro C, Graça RJ et al (2016) Parasitism by larval tapeworms genus *Spirometra* in South American amphibians and reptiles: new records from Brazil and Uruguay, and a review of current knowledge in the region. Acta Trop 164:150–164

OHHLEP – One Health High-Level Expert Panel, Adisasmito WB, Almuhairi S et al (2022) One Health: a new definition for a sustainable and healthy future. PLoS Pathog 18:e1010537. https://doi.org/10.1371/journal.ppat.1010537

Oliveira FM, Costa LHC, Barros TL et al (2015) First detection of *Leishmania* spp. DNA in Brazilian bats captured strictly in urban areas. Acta Trop 150:176–181

Oliveira AR, Hiura E, Guião-Leite FL et al (2017) Pathological and parasitological characterization of *Prosthenorchis elegans* in a free-ranging marmoset *Callithrix geofroyi* from the Brazilian Atlantic Forest. Pesqui Vet Bras 37:1514–1518

Oliveira AR, Pinheiro GRG, Tinoco HPD et al (2019) Competence of nonhuman primates to transmit *Leishmania infantum* to the invertebrate vector *Lutzomyia longipalpis*. PLoS Negl Trop Dis 13:e0007313. https://doi.org/10.1371/journal.pntd.0007313

Oliveira AR, Ritter JM, Santos DO et al (2022) Pathology and epidemiology of fatal toxoplasmosis in free-ranging marmosets (*Callithrix* spp.) from the Brazilian Atlantic Forest. PLoS Negl Trop Dis 16:e0010782. https://doi.org/10.1371/journal.pntd.0010782

Orozco MM, Enriquez GF, Alvarado-Otegui JA et al (2013) New sylvatic hosts of *Trypanosoma cruzi* and their reservoir competence in the humid Chaco of Argentina: a longitudinal study. Am J Trop Med Hyg 88:872–882

Paranjpe DA, Medina D, Nielsen E et al (2014) Does thermal ecology influence the dynamics of side-blotched lizards and their microparasites? Integr Comp Biol 54:108–117

Parr NA, Fedigan LM, Kutz SJ (2013) A coprological survey of parasites in white-faced capuchins (*Cebus capucinus*) from Sector Santa Rosa, ACG, Costa Rica. Folia Primatol 84:102–114

Parra S, Patiño B, Stelle C (2005) Estudio coproparasitológico de tres especies de cébidos (*Aotus griseimembra*, *Cebus apella* y *Saguinus leucopus*) en cautiverio. Rev Colomb de Cienc Pecu 18:373

Paula NF, Dutra KS, Oliveira AR, Santos DO et al (2020) Host range and susceptibility to *Toxoplasma gondii* infection in captive neotropical and Old-world primates. J Med Primatol 49:202–210

Pena HFJ, Marvulo MFV, Horta MC et al (2011) Isolation and genetic characterization of *Toxoplasma gondii* from a red-handed howler monkey (*Alouatta belzebul*), a jaguarundi (*Puma yagouaroundi*), and a black-eared opossum (*Didelphis aurita*) from Brazil. Vet Parasitol 175:377–381

Pereira AAS, Ferreira EC, Lima ACVMR et al (2017) Detection of *Leishmania* spp. in silvatic mammals and isolation *of Leishmania* (*Viannia*) *braziliensis* from *Rattus rattus* in an endemic area for leishmaniasis in Minas Gerais State, Brazil. PLoS One 12:e0187704. https://doi.org/10.1371/journal.pone.0187704. eCollection 2017

Pérez García J, Ramírez DM, Hernández CA (2007) *Prosthenorchis* sp. en tities grises (*Saguinus leucopus*). Revisión de tema. Rev CES Med Vet y Zootec 2:51–57

Pissinatti L, Pissinatti A, Burity CHF, Mattos DG Jr, Tortelly R (2007) Ocorrência de Acanthocephala em *Leontopithecus* (Lesson, 1840), cativos: aspectos clínicopatológicos. Callitrichidae-Primates. Arq Bras Med Vet Zootec 59:1473–1477

Porfírio GEO, Santos FM, Macedo GC et al (2018) Maintenance of *Trypanosoma cruzi*, *T. evansi* and *Leishmania* spp. by domestic dogs and wild mammals in a rural settlement in Brazil-Bolivian border. Int J Parasitol Parasites Wildl 7:398–4040

Portela AAB, Santos TG, Anjos LA (2020) Changes in land use affect anuran helminths in the South Brazilian grasslands. J Helminthol 94:e206. https://doi.org/10.1017/S0022149X20000905

Purple KE, Gerhold RW (2015) Persistence of two Isolates of *Trichomonas gallinae* in simulated bird baths with and without organic material. Avian Dis 59:472–474

Pan American Health Organization. (2022). Menos de 10 % dos infectados com doença de Chagas recebem diagnóstico e tratamento oportunos https://www.paho.org/pt/noticias/13-4-2022-menos-10-dosinfectados-com-doenca-chagas-recebem-diagnostico-e-tratamento. Accessed on 13 April 2022.

Quaresma PF, Rêgo FD, Botelho HA et al (2011) Wild, synanthropic and domestic hosts of *Leishmania* in an endemic area of cutaneous leishmaniasis in Minas Gerais State, Brazil. Trans R Soc Trop Med Hyg 105:579–585

Quintal AP, Ribeiro ES, Rodrigues FP, Rocha FS, Floeter-Winter LM, Nunes CM (2011) *Leishmania* spp. in *Didelphis albiventris* and *Micoureus paraguayanus* (Didelphimorphia: Didelphidae) of Brazil. Vet Parasitol 176:112–119

Ramalho ACO, Silva RJ, Schwartz HO, Péres AK (2009) Helminths from an introduced species (*Tupinambis merianae*) and two endemic species (*Trachylepis atlantica* and *Amphisbaena ridleyi*) from Fernando de Noronha Archipelago, Brazil. J Parasitol 95:1026–1028

Reis FC, Minuzzi-Souza TCC, Neiva M et al (2020) Trypanosomatid infections in captive wild mammals and potential vectors at the Brasilia Zoo, Federal District, Brazil. Vet Med Sci 6:248–256

Rendón LM, Guhl F, Cordovez JM, Erazo D (2015) New scenarios of *Trypanosoma cruzi* transmission in the Orinoco region of Colombia. Mem Inst Oswaldo Cruz 110:283–288

Rivera RC, Bilal S, Michael E (2020) The relation between host competence and vector-feeding preference in a multihost model: Chagas and cutaneous leishmaniasis. Math Biosci Eng 17:5561–5583

Rocha FL, Roque ALR, Lima JS et al (2013) *Trypanosoma cruzi* infection in neotropical wild carnivores (Mammalia: Carnivora): at the top of the *T. cruzi* transmission chain. PLoS One 8:e67463. https://doi.org/10.1371/journal.pone.0067463

Rogers KH, Girard YA, Woods L, Johson CK (2016) Avian trichomonosis in spotted owls (*Strix occidentalis*): indication of opportunistic spillover from prey. Int J Parasitol Parasites Wildl 5:305–311

Roque ALR, Jansen AM (2014) Wild and synanthropic reservoirs of Leishmania species in the Americas. Int J Parasitol Parasites Wildl 3:251–262

Roque ALR, Xavier SCC, Rocha MG, Duarte ACM, D'Andrea PS, Jansen AM (2008) *Trypanosoma cruzi* transmission cycle among wild and domestic mammals in three areas of orally transmitted Chagas Disease outbreaks. Am J Trop Med Hyg 79:742–749

Sales IS, Ruiz-Miranda CR, Santos CP (2010) Helminths found in marmosets (*Callithrix penicillata* and *Callithrix jacchus*) introduced to the region of occurrence of golden lion tamarins (*Leontopithecus rosalia*) in Brazil. Vet Parasitol 171:123–129

Sansano-Maestre J, Garijo-Toledo MM, Gómez-Muñoz MT (2009) Prevalence and genotyping of *Trichomonas gallinae* in pigeons and birds of prey. Avian Pathol 38:201–207

Santos T, Oliveira J, Vaughan C, Santiago H (2011) Health of an *ex situ* population of raptors (Falconiformes and Strigiformes) in Mexico: diagnosis of internal parasites. Rev Biol Trop 59:1265–1274

Santos VGT, Amato SB, Borges-Martins M (2013) Community structure of helminth parasites of the "Cururu" toad, *Rhinella icterica* (Anura: Bufonidae) from Southern Brazil. Parasitol Res 112:1097–1103

Santos SV, Strefezzi RF, Pissinatti A et al (2014) Detection of Toxoplasma gondii in two southern wooly spider monkeys (*Brachyteles arachnoides* – Geoffroy, 1806) from the Rio de Janeiro primate center, Brazil. J Med Primatol 43:125–129

Santos SV, Pena HFJ, Gomes MT et al (2017) Fatal toxoplasmosis in a southern muriqui (*Brachyteles arachnoides*) from São Paulo state, Brazil: pathological, immunohistochemical, and molecular characterization. J Med Primatol 47:124–127

Santos FM, Macedo GC, Barreto WTG et al (2018) Outcomes of *Trypanosoma cruzi* and *Trypanosoma evansi* infections on health of Southern coati (*Nasua nasua*), crab-eating fox (*Cerdocyon thous*), and ocelot (*Leopardus pardalis*) in the Brazilian Pantanal. PLoS One 13:e0201357. https://doi.org/10.1371/journal.pone.0201357

Santos FM, Macedo GC, Barreto WTG et al (2019) The reservoir system for *Trypanosoma* (Kinetoplastida, Trypanosomatidae) species in large neotropical wetland. Acta Trop 199:105098. https://doi.org/10.1016/j.actatropica.2019.105098

Sena PA, Conceição BM, Silva PF et al (2018) Helminth communities of *Pithecopus nordestinus* (Anura: Phyllomedusidae) in forest remnants, Brazil. Herpetol Notes 11:565–572

Shapiro JT, Lima Junior MSC, Dorval ME, Oliveira FA, Matos MFC, Bordignon MO (2013) First record of *Leishmania braziliensis* presence detected in bats, Mato Grosso do Sul, southwest Brazil. Acta Trop 128:171–174

Silva JCR, Marvulo MFV, Dias RA et al (2007) Risk factors associated with seropositivity to Toxoplasma gondii in captive neotropical felids from Brazil. Prev Vet Med 78:286–295

Souza NP, Almeida ABPF, Freitas TPT et al (2010) *Leishmania (Leishmania) infantum chagasi* em canídeos silvestres mantidos em cativeiro, no Estado de Mato Grosso. Rev Soc Bras Med Trop 43:333–335

Souza TD, Turchetti AP, Fujiwara RT, Paixão TA, Santos RL (2014) Visceral leishmaniasis in zoo and wildlife. Veterinary 200:233–241

Stimmelmayr R, Stefani LM, Thrall MA et al (2012) Trichomonosis in free-ranging Eurasian collared doves (*Streptopelia decaocto*) and African collared dove hybrids (*Streptopelia risoria*) in the Caribbean and description of ITS-1 region genotypes. Avian Dis 56:441–445

Tantaleán M, Sánchez L, Gómez L, Huiza A (2005) Acantocéfalos del Perú. Rev Peru Biol 12:83–92

Tavela AO, Fuzessy LF, Silva VHD et al (2013) Helminths of wild hybrid marmosets (*Callithrix* sp.) living in an environment with high human activity. Rev Bras Parasitol Vet 22:391–397

Tavernard LC (2017) Relato de caso: *Prosthenorchis elegans* (parasita acantocéfalo) em *Saimiri* sp. (macaco-de-cheiro) em Centro de Primatologia Brasileiro. Monografia, Universidade Federal de Brasília

Teixeira AAM, Riul P, Brito SV et al (2020) Ecological release in lizard endoparasites from the Atlantic Forest, northeast of the Neotropical Region. Parasitology 147:491–500

Tenório MS, Souza LO, Paixão MS et al (2011) Visceral Leishmaniasis in a captive crab-eating fox *Cerdocyon thous*. J Zoo Wildl Med 42:608–616

Trüeb I, Portela RD, Franke CR et al (2018) *Trypanosoma cruzi* and *Leishmania* sp. infection in wildlife from urban rainforest fragments in northeast Brazil. J Wildl Dis 54:76–84

Valenzuela-Sánchez A, Wilber MQ, Canesa S et al (2021) Why disease ecology needs life-history theory: a host perspective. Ecol Lett 24:876–890

VanWormer E, Conrad PA, Miller MA, Melli AC, Carpenter TE, Mazet JAK (2014) Using molecular epidemiology to track *Toxoplasma gondii* from terrestrial carnivores to marine hosts: implications for public health and conservation. PLoS Negl Trop Dis 8:e2852

VanWormer E, Miller MA, Grigg ME, Rejmanek D, Carpenter TE, Mazet JAK (2016) Coastal development and precipitation drive pathogen flow from land to sea: evidence from a *Toxoplasma gondii* and felid host system. Sci Rep 6:29252

Vianna EN, Morais MHF, Almeida ASD, Sabroza PC, Reis IA, Dias ES, Carneiro M (2016) Abundance of *Lutzomyia longipalpis* in urban households as risk factor of transmission of visceral leishmaniasis. Mem Inst Oswaldo Cruz 111:302–310

Villagrán ME, Martínez-Ibarra JA, Diego JA (2011) Pathologial alterations and prevalence of *Trypanosoma cruzi* in opossums from western Mexico. Bol Malariol Salud Ambient 6:87–88

Villena FE, Gomez-Puerta LA, Jhonston EJ et al (2018) First report of *Trypanosoma cruzi* infection in salivary gland of bats from the Peruvian Amazon. Am J Trop Med Hyg 99:723–728

Wenz A, Heymann EW, Petney TN, Taraschewski HF (2010) The influence of human settlements on the parasite community in two species of Peruvian tamarin. Parasitology 137:675–684

Werneck MR, Silva RJ (2015) Helminth parasites of juvenile green turtles *Chelonia mydas* (Testudines: Cheloniidae) in Brazil. J Parasitol 101:713–716

WHO – World Health Organization (2022a) Chagas disease (American trypanosomiasis). http://www.who.int/en/news-room/fact-sheets/detail/chagas-disease-(american-trypanosomiasis). Accessed on 29 Aug 2022

WHO – World Health Organization (2022b) Leishmaniasis. https://www.who.int/leishmaniasis/en/. Accessed on 29 Aug 2022

Xavier SCC, Roque ALR, Lima VS et al (2012) Lower richness of small wild mammal species and Chagas Disease risk. PLoS Negl Trop Dis 6:e1647. https://doi.org/10.1371/journal.pntd.0001647

Yefi-Quinteros E, Muñoz-San Martín C, Bacigalupo A, Correa JP, Cattan PE (2018) *Trypanosoma cruzi* load in synanthropic rodents from rural areas in Chile. Parasit Vectors 11:171. https://doi.org/10.1186/s13071-018-2771-2

Yeo M, Acosta N, Llewellyn M (2005) Origins of Chagas disease: *Didelphis* species are natural hosts of *Trypanosoma cruzi* I and armadillo hosts of *Trypanosoma cruzi* II, including hybrids. Int J Parasitol 35:225–233

Correction to: Neotropical Primates and Humans: Risk of Bidirectional Parasite Transmission and Disease Sharing in Fragmented and Pristine Landscapes

Andrea Chaves, María del Carmen Villalobos-Segura, Juan Ricardo Sánchez Ayala, Óscar M. Chaves, Júlio César Bicca-Marques, Brenda Solórzano-García, Marco Antônio Barreto de Almeida, and Gustavo A. Gutiérrez-Espeleta

Correction to:
Chapter 9 in: G. Acosta-Jamett, A. Chaves (eds.), *Ecology of Wildlife Diseases in the Neotropics*, https://doi.org/10.1007/978-3-031-50531-7_9

The original version of this chapter was inadvertently published by placing incorrect affiliations to the authors. The affiliations have now been corrected and approved by the author.

The updated version of this chapter can be found at
https://doi.org/10.1007/978-3-031-50531-7_9

G. Acosta-Jamett, A. Chaves (eds.), *Ecology of Wildlife Diseases in the Neotropics*, https://doi.org/10.1007/978-3-031-50531-7_15

Index

G. Acosta-Jamett, A. Chaves (eds.), *Ecology of Wildlife Diseases in the Neotropics*, https://doi.org/10.1007/978-3-031-50531-7